Modeling and Simulation in Science, Technology and Engineering

Modeling and Simulation in Science, Technology and Engineering

Edited by Jaxon Wheeler

STATES
ACADEMIC PRESS
www.statesacademicpress.com

States Academic Press,
109 South 5th Street,
Brooklyn, NY 11249, USA

ISBN: 978-1-63989-359-1

Cataloging-in-Publication Data

Modeling and simulation in science, technology and engineering / edited by Jaxon Wheeler.
p. cm.
Includes bibliographical references and index.
ISBN 978-1-63989-359-1
1. Computer simulation. 2. Simulation methods. 3. Science--Computer simulation. 4. Science--Simulation methods.
5. Technology--Computer simulation. 6. Technology--Simulation methods. 7. Engineering--Computer simulation.
8. Engineering--Simulation methods. I. Wheeler, Jaxon.

QA76.9.C65 M63 2022
003.3--dc23

For information on all States Academic Press publications
visit our website at www.statesacademicpress.com

STATES
ACADEMIC PRESS

Contents

Preface VII

Chapter 1 **Soil Saturated Simulation in Embankment during Strong Earthquake by Effect of Elasticity Modulus** 1
Behrouz Gordan, Azlan Adnan and Mariyana A. K. Aida

Chapter 2 **Reliability Evaluation of Pavement Life-Cycle Assessment Model** 8
Jiale Huang, Fei Xiao and Yang Zhang

Chapter 3 **Numerical Study on Diffusion of Chloride and Induced Rebar Corrosion by Two-Dimensional Multiscale Approach** 16
Xi Tu, Jin Di, Cunjun Pang and Xiaoqing Xu

Chapter 4 **Synthesis of Spatially Correlated Earthquake Ground Motions based on Hilbert Transform** 36
Erlei Yao, Yu Miao and Guobo Wang

Chapter 5 **Modelling of Creep and Stress Relaxation Test of a Polypropylene Microfibre by using Fraction-Exponential Kernel** 45
Andrea Sorzia

Chapter 6 **Wind Load Reduction in Hollow Panel Arrayed Set** 52
Michalina Markousi, Dimitrios K. Fytanidis and Johannes V. Soulis

Chapter 7 **Numerical Simulation of Dam Break Flows using a Radial basis Function Meshless Method with Artificial Viscosity** 66
Elmiloud Chaabelasri

Chapter 8 **Traffic Multiresolution Modeling and Consistency Analysis of Urban Expressway based on Asynchronous Integration Strategy** 77
Liyan Zhang, Jian Ma, Bin Ran and Lixin Yan

Chapter 9 **A Note on Torsion of Nonlocal Composite Nanobeams** 96
Luciano Feo and Rosa Penna

Chapter 10 **Collapse Analysis of Transmission Tower Subjected to Earthquake Ground Motion** 101
Xiaohong Long, Wei Wang and Jian Fan

Chapter 11 **Interaction of Submerged Breakwater by a Solitary Wave using WC-SPH Method** 121
Afshin Mansouri and Babak Aminnejad

Chapter 12 **On Bending of Bernoulli-Euler Nanobeams for Nonlocal Composite Materials** 130
Luciano Feo and Rosa Penna

Chapter 13 **Initiation of Failure for Masonry Subject to In-Plane Loads through Micromechanics** **135**
V. P. Berardi

Chapter 14 **Scenario based Municipal Wastewater Estimation: Development and Application of a Dynamic Simulation Model** **141**
Yong Zhang, Tingsheng Zhao, Aijiao Zhou, Zhengzhu Zhang and Wen Liu

Chapter 15 **A Review on Radiation Damage in Concrete for Nuclear Facilities Facilities: From Experiments to Modeling** **154**
Beatrice Pomaro

Chapter 16 **Identification and Prioritization of "Black Spots" without using Accident Information** **164**
Mahmoudreza Keymanesh, Hasan Ziari, Samira Roudini and Ali Nasrollahtabar Ahangar

Chapter 17 **Modelling of Wind Pressure Coefficients on C-Shaped Building Models** **173**
Monalisa Mallick, Abinash Mohanta, Awadhesh Kumar and Vivek Raj

Chapter 18 **Development of Constitutive Model for Precast Prestressed Concrete Segmental Columns** **186**
M. Hafezolghorani Esfahani, F. Hejazi, R. Vaghei, E. Nikbakht and D. C. J. Tze

Chapter 19 **Numerical Modelling of Tailings Dam Thermal-Seepage Regime Considering Phase Transitions** **197**
Aniskin Nikolay Alekseevich and Antonov Anton Sergeevich

Chapter 20 **System Identification based Proxy Model of a Reservoir under Water Injection** **207**
Berihun M. Negash, Lemma D. Tufa, M. Ramasamy and Mariyamni Bt. Awang

Chapter 21 **Numerical Investigation on Vortex Shedding from a Hydrofoil with a Beveled Trailing Edge** **216**
Seung-Jae Lee, Jun-Hyeok Lee and Jung-Chun Suh

Chapter 22 **Consistent Weighted Average Flux of Well-Balanced TVD-RK Discontinuous Galerkin Method for Shallow Water Flows** **229**
Thida Pongsanguansin, Montri Maleewong and Khamron Mekchay

Permissions

List of Contributors

Index

Preface

A model is a representation of a system made of concepts, mathematical formulas and calculations to predict and understand the data. A model can be physical or conceptual. Simulation is the implementation of model to study the behavior and performance of a system. It allows systems to optimize and predict. Modeling and simulation have broad spectrum of applications in other areas. Some of these are system engineering, training and education support, and analysis support. Scientific models are used to explain and predict behavior of real objects, for instance, predicting model for weather forecasting, drug design, urban planning, describing dual nature of wave, Bohr atomic model, DNA double helix model, projecting health outcomes of diseases, etc. Simulation is majorly used in analyzing and experimentation such as in performance tuning or optimizing, safety engineering, testing, training, education and video games. This book elucidates new techniques and their applications in a multidisciplinary manner. It studies, analyzes and upholds the pillars of modeling and simulation and its utmost significance in modern times. The extensive content of this book provides the readers with a thorough understanding of the subject.

The researches compiled throughout the book are authentic and of high quality, combining several disciplines and from very diverse regions from around the world. Drawing on the contributions of many researchers from diverse countries, the book's objective is to provide the readers with the latest achievements in the area of research. This book will surely be a source of knowledge to all interested and researching the field.

In the end, I would like to express my deep sense of gratitude to all the authors for meeting the set deadlines in completing and submitting their research chapters. I would also like to thank the publisher for the support offered to us throughout the course of the book. Finally, I extend my sincere thanks to my family for being a constant source of inspiration and encouragement.

Editor

1

Soil Saturated Simulation in Embankment during Strong Earthquake by Effect of Elasticity Modulus

Behrouz Gordan, Azlan Adnan, and Mariyana A. K. Aida

Faculty of Civil Engineering, Universiti Teknologi Malaysia, 81310 Skudai, Johor, Malaysia

Correspondence should be addressed to Behrouz Gordan; bh.gordan@yahoo.com

Academic Editor: Hongyi Li

The dynamic analysis process was started after failure in some embankments during an earthquake. In this context, maximum displacement was reported at the crest based on interaction between structure and reservoir. This paper investigates the dynamic behavior of short embankment on soft soil. For this purpose, numerical analysis was carried out using ANSYS13 program based on finite-element method. Simulated models were vibrated by strong earthquake, so the peak ground acceleration (PGA) and duration were 0.65 g and 5.02 seconds, respectively. The comparison results were discussed in key points of plane strain analysis based on modulus ratio between saturated embankment and foundation. As concluded, the modulus ratio between 0.53 and 0.66 led to having a minimum value of horizontal displacement, relative displacement in vertical direction, and shear stress. Consequently, the shear stress was increased while the modulus ratio was decreased. Finally, to avoid more rigidity in the embankment on the soft soil, optimum modulus ratio was recommended at 0.66 in order to reduce the probabilistic of body cracks at the crest with respect to homogeneous behavior during an earthquake.

1. Introduction

Dynamic analysis is one of the main aspects for embankment designing. As a lesson learned, some huge damages such as different types of body cracks were recorded during an earthquake. It seems that a specific attention to complete research in this area is one of the major concerns for geotechnical designers. According to the literature, depth review in this category demonstrated that since the beginning of 1920 until 1960, pseudostatic method was well known. With full respect for this technique, this method was very simple and indicated the weak performance in order to cover some factors such as nature of the slope-forming and foundation material. Based on deformation characteristics, the sliding block method was completely presented by Newmark in 1965 [1]. Among other methods, shear beam method was presented by Mononobe [2]. As stated by Gazetas [3], an improved inhomogeneous shear beam model was investigated and it was concluded that the shear modulus was variable value in earth dams or rock fill dams and increased with 2/3 power of depth from the crest. To estimate the dynamic response of an embankment, the finite-element method was successfully presented by Clough et al. [4] using plan-strain analysis (2D). In this study, materials for modeling were consisted of linearly elastic, homogeneous, and isotropic. Later, this study was developed by other researchers using finite-element method and finite-difference method to cover some factors such as nonlinear, inelastic, nonuniform, and anisotropic behavior of materials under dynamic loads. As stated by Zeghal et al. [5], a local-global finite-element method was proposed to determine the nonlinear behavior of earth dams. Ming et al. [6] investigated fully coupled analysis of failure and remediation of lower San Fernando dam. The significant reasons for dam failure were reported in this study. In terms of earthquake engineering research, quick development in the computer programs directly led to the comprehensive analysis. For instance, several programs [7–11] were widely used for dynamic analysis of embankment. As presented by Namdar et al. [12], the assessment of embankment behavior was investigated using shaking table and finite-element method. The good agreement between analytical and physical modeling was obtained in this study. According to study from Wang et al. [13],

TABLE 1: A dimension of models.

$H = 30$ m	$K = 15.00$ m	$X = 108.92$ m	$R = 36.37$ m	$H1 = 21.00$ m	$L = 9.00$ m	$Z = 15.00$ m	$W = 5.00$ m	$\theta = 30$

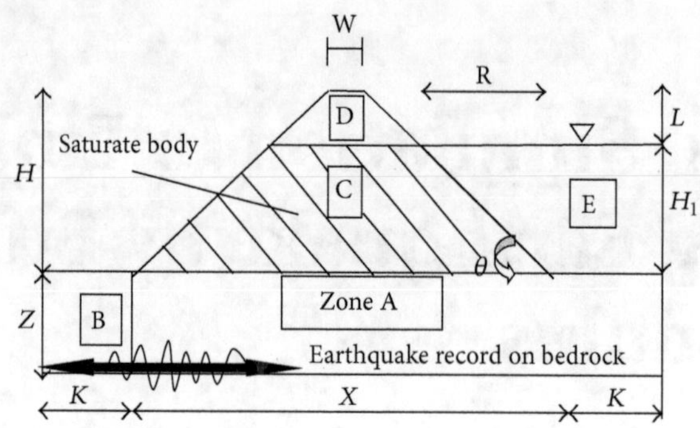

FIGURE 1: A modeling dimension according to Table 1 for plan strain (2D).

the geotextile-reinforced soil in embankment body was studied using centrifuge modeling, and they discussed embankment behavior during an earthquake. According to Zhu et al. [14] and Brinkgreve et al. [15], a two-dimensional seismic stability for a levee embankment was possible using PLAXIS program based on finite-element method. After all, as mentioned, the maximum displacement during earthquake was at the crest, as regards the soil amplification and interaction between dam and reservoir. In terms of dynamic behavior with respect to flexibility, the main role was featured by elasticity modulus. This paper tried to evaluate the effect of material properties based on parametric study in the saturate embankment on dynamic performance in order to find the best modulus ratio between saturated zone and loose sand foundation.

2. Modeling Process

This process includes some phases such as introducing program, element, boundary conditions, parametric dimension, material properties, meshing, key points, and earthquake record which are described below.

2.1. Introduce ANSYS Program and Element. This program is one of the comprehensive software's under finite-element method, and 100000 coed line are available, as related to the computer-aided engineering (CAE). In fact, this program is very famous with strong ability to analyze in different engineering purposes. In the present study, "solid42" element for embankment and foundation was used. In addition, "fluid79" was applied for water element according to help menu. Both elements were recommended for plan strain (2D). In fact, the lateral strain in the simulated models assumed zero.

2.2. Boundary Conditions. In terms of boundary condition for modeling, earthquake record in the horizontal direction of the base of model was used with respect to displacement-time. For this case, NAGAN record with acceleration-time converted to displacement time using SISMOSOFT3 program. Vertical displacement was assumed zero in the bedrock, and earthquake duration was 5.02 seconds with substep (0.02 second). Both lines of vertical boundary were restricted by zero value in vertical direction, and 0.01 meters for horizontal displacement using trial-error procedure in order to create transient elasticity property for wave during vibration.

2.3. Modeling Parameters. Table 1 presents the dimension of simulated models in the parametric study. As seen, the height of embankment is short, and slope is moderate in order to satisfy slope stability. In addition, the width of the crest is five meters with respect to compaction and the minimum road width.

Figure 1 illustrates the parametric dimension of the simulated models. As shown, the bedrock was coupled by earthquake record. Moreover, based on water level in reservoir, two parts such as saturated zone (C) and unsaturated zone (D) were separated in models. In addition, water tank located in the right side of model, as marked (E).

2.4. Material Properties. Three materials were used for modeling such as water, foundation, and embankment. The structural body was the clay soil with homogeneous characteristics. Foundation material was the soft loose sand saturated. Help menu recommends for modeling the incompressible water; water properties were also presented in Table 2. This table presents the material property in different zone using three properties of the dam body. It is worth noting that the solid material was simulated using hardening isotropic bilinear. In addition, the friction coefficient was used in models in order to establish interface effect between different materials. The amplitude of material properties in different zone of the model for the present parametric study was applied [16]. The nonlinear behavior of soil was utilized using some factors such as density, elasticity modulus, Poisson's ratio, yield stress, and tangent modulus. Table 2 presents the material properties in different zones (A, B, C, D, and E). As can be seen in this table, the huge elasticity modulus was applied for incompressible behavior of water as mentioned earlier.

TABLE 2: Material properties.

Zone	Density (Kg/m^3)	Elasticity modulus (Kg/m^2)	Poisson ratio	Yield stress (Kg/m^2)	Tangent modulus	Friction coefficient
A-foundation (loose-sand saturate)	800	3E6	0.25	1.20E4	0.01	0.30
B-foundation (loose-sand saturate)	800	3E6	0.25	1.20E4	0.01	0.001
C1-embankment (clay saturation)	900	2E6	0.45	8.00E3	0.01	0.001
C2-embankment (clay saturation)	900	16E5	0.45	3.20E3	0.01	0.001
C3-embankment (reinforcement)	900	8E5	0.45	6.40E3	0.01	0.001
D-embankment (unsaturated)	1900	4E5	0.30	1.60E3	0.01	—
E-water	1000	1E12	0.01	—	—	—

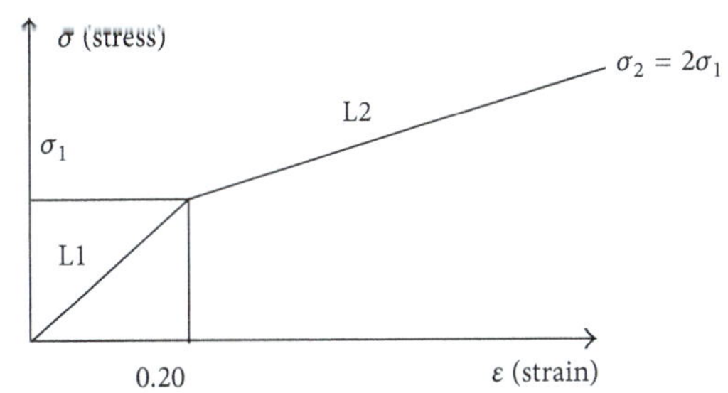

FIGURE 2: Nonlinear property of soil in bilinear method.

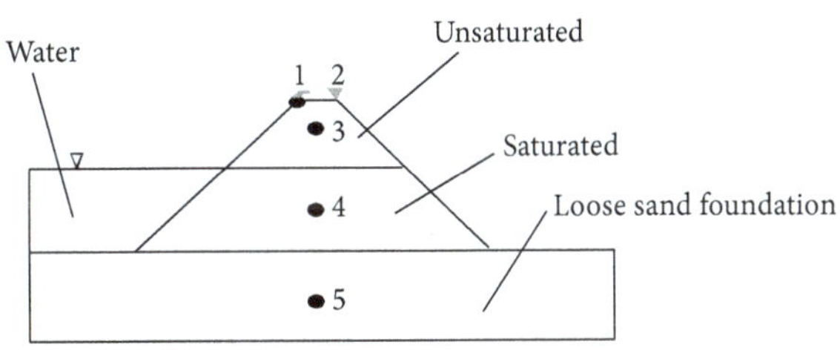

FIGURE 3: Key points (1–5) in models.

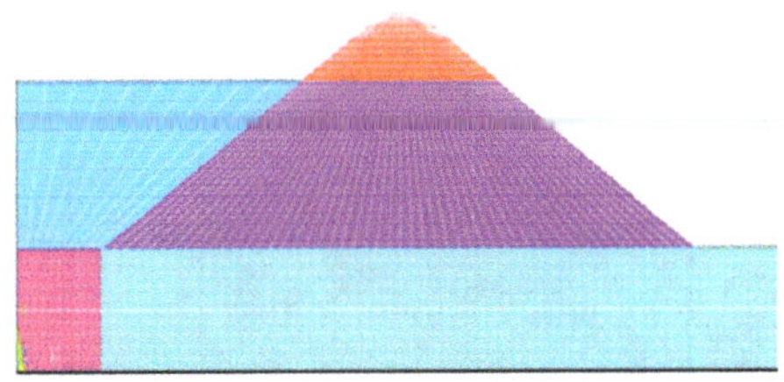

FIGURE 4: Regular meshing method.

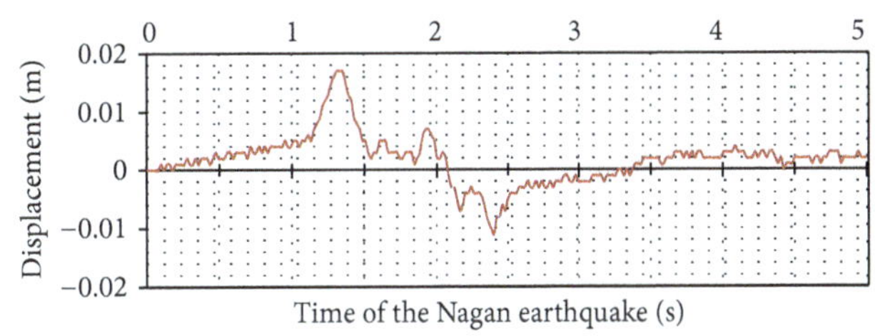

FIGURE 5: Input data from Nagan earthquake record according to displacement in the vertical axis (m) and time in horizontal axis (seconds).

Figure 2 shows the nonlinear behavior of soil based on two stages. It can be seen that line L1 for elastic and line L2 to evaluate elastic-plastic and plastic condition, respectively. Line L1 is related to linear performance with respect to elasticity modulus, elastic stress according to 0.2% strain, and Poisson's ratio. Line L2 is related to the nonlinear behavior (elastic-plastic and plastic) with tangent modulus; both lines show bilinear isotropic hardening. It is noted that ultimate stress is double in comparison to elastic stress. In this figure, horizontal and vertical axes are, respectively, strain and stress.

2.5. Meshing and Key Points. According to Figure 3, five key points 1–5 have been selected to evaluate existing data. Both of them are at the crest, and other points are in the middle of zones in D, C, and A. Figure 4 shows the regular meshing in models. This mesh is used in order to interface between both different lines with same nodes.

2.6. Earthquake Record. In terms of earthquake vibration, all models were carried out based on time history analysis using Nagan earthquake. For this case, PGA (peak ground acceleration) and duration are, respectively, 0.65 g and 5.02 seconds. In terms of input data for ANSYS, this record was converted to displacement-time. For converting, SISMOSOFT3 program from Berkeley University was used. Figure 5 shows displacement-time for Nagan earthquake after converting.

As shown, the maximum and minimum displacements are 16.5 mm and 11 mm. In this figure, horizontal axis shows the earthquake duration (substep equal to 0.02 seconds). Vertical axis also shows the distribution of displacement based on meter measurement unit.

3. Results and Analysis

Numerical analyses were carried out based on scope of study and research methodology as explained earlier. After analysis

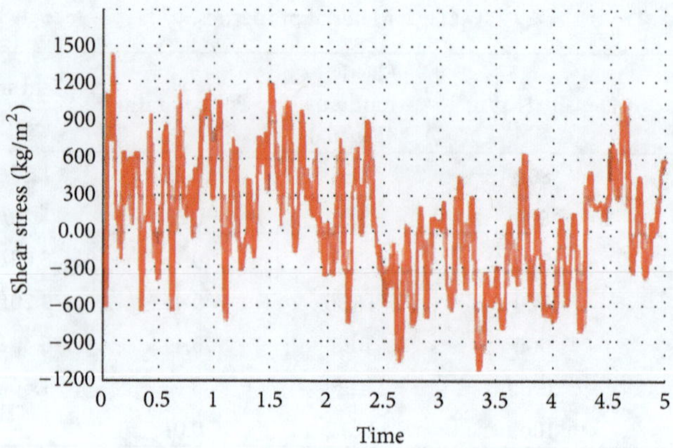

FIGURE 6: Shear stress in point 4 from model 1; vertical axis is value of shear stress, and horizontal axis is time of earthquake.

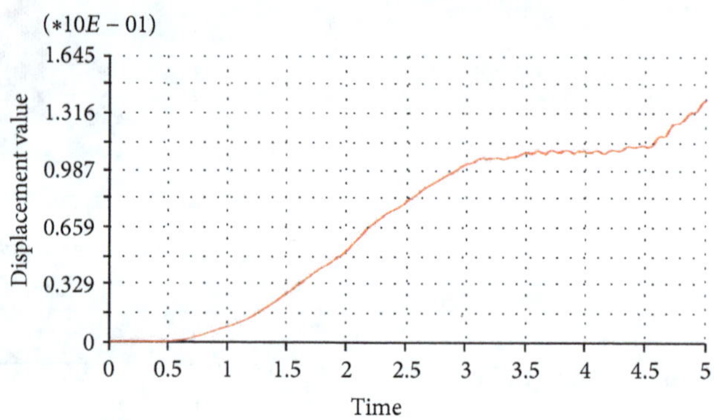

FIGURE 7: Horizontal displacement of point 1 in model 1; vertical axis is value of displacement (meter), and horizontal axis is time of earthquake (seconds).

in order to discuss, some results such as displacement distribution in both directions and shear stress were significantly compared. First of all, Figures 6 and 7 show the nonlinear behavior of material under dynamic load for shear stress and displacement. Figure 8 shows the distribution of shear stress for different models for the same key points. As observed, these factors were changed in each time with positive or negative position. It means that the transient analysis was obviously performed in models. Figures 6 and 7 illustrated the shear stress of point 4 in model 1 and horizontal displacement of point 1 in model 1, respectively.

However, this trend was obtained for structural behavior during an earthquake, as expected. This role was repeated in all key points as introduced earlier. This ability was one of the best engineering earthquake purposes using Ansys program. Due to the results classification, one modulus ratio (λ) was defined according to the following:

$$\lambda = \frac{\text{Elasticity modulus in embankment}}{\text{Elasticity modulus in foundation}}. \quad (1)$$

This ratio was the simple relationship between elasticity modulus in the saturated embankment and soft soil for foundation. Figure 8 shows the distribution of horizontal displacement in models. As seen, displacement in the first model was less than in second model, with respect to increasing elasticity modulus. This value was dramatically increased in the third model in comparison to other models. In fact, displacement distribution depends on the modulus ratio, and the maximum ratio led to the homogeneous behavior during the strong earthquake. Therefore, horizontal displacement can be decreased while system was limited to the isotropic behavior. The maximum displacement was exposed at key point 4, as located at the middle of saturated embankment. In this case, the influence of water interaction according to the static pressure in the 1/3 of the height was very important. A raise of the modulus ratio led to the increase of horizontal displacement in the middle of foundation zone. However, the minimum horizontal displacement was featured in the key point 5 for all models. Hence, the suitable ratio was obtained in the first model to control horizontal dynamic displacement of the short saturated embankment. Based on the displacement distribution in the vertical direction, the relative displacement for both edges of the crest can play main role for controlling body cracks. Figure 9 shows the vertical displacement in the models. As compared, the maximum and minimum relative displacement were, respectively, found in the second model and first model. As seen, this factor was reduced while the modulus ratio was increased. The reduction of this ratio led to the increased flexibility and settlement. It is worth noting that maximum horizontal displacement occurred at the crest for

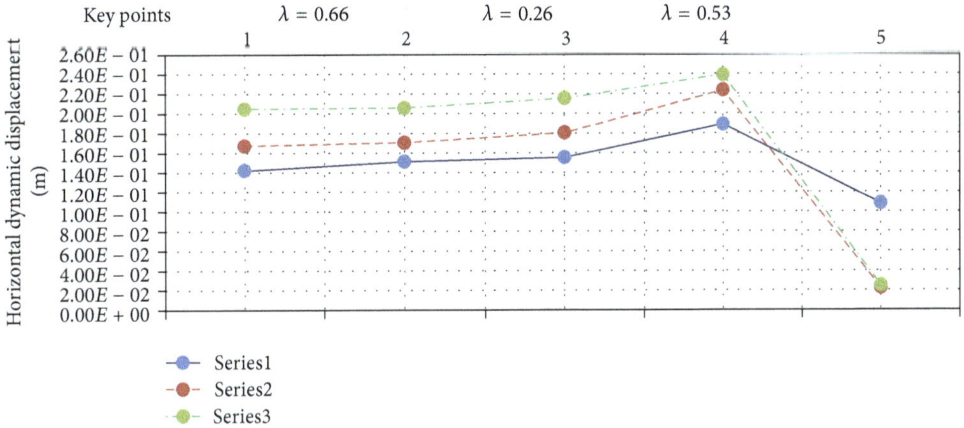

FIGURE 8: Horizontal displacement at the end of earthquake; horizontal and vertical axes are key points and dynamic displacement (meter).

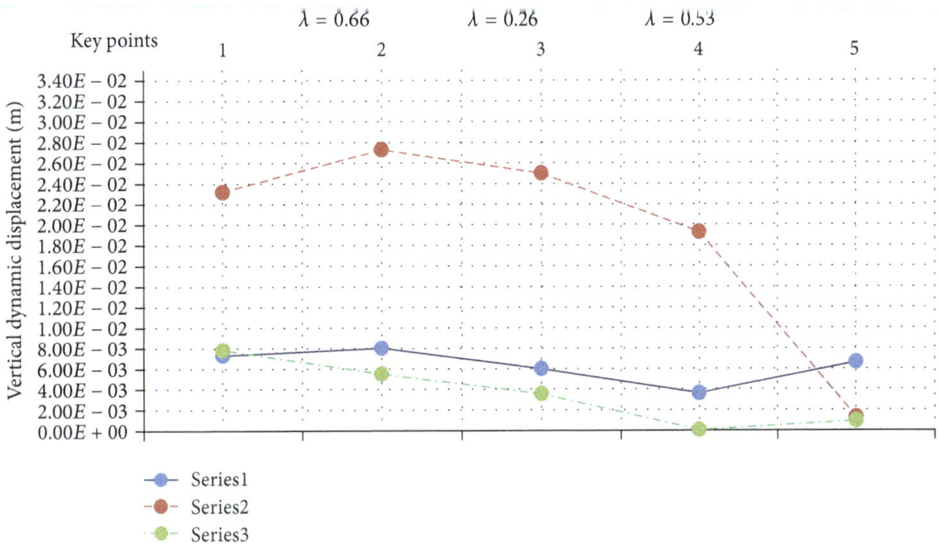

FIGURE 9: Vertical dynamic displacement at the end of earthquake for models, horizontal and vertical axis show key points and vertical displacement (meter), respectively.

all models, and there was good agreement between numerical result and case studies reported in the literature.

Figure 10 illustrates that the body cracks occurred at the crest in terms of relative displacement during vibration [17]. Consequently, the first model represented the appropriate aspects to control vertical dynamic displacement of the short saturated embankment. In addition, the prediction of cracks and damage requires nonlinear dynamic analyses [17]. After the above discussion, the shear stress distribution in models was shown in Figure 11.

The maximum shear stress was placed in the middle of the foundation as expected. Shear stress increased, while modulus ratio decreased. It was revealed that stress on the foundation was four times in the first model in comparison to other models. In brief, the structural behavior indicated the suitable modulus ratio equal to 0.66. As mentioned earlier, the first model also was the best position in order to minimize relative displacement. Finally, modulus ratio that equals 0.66 indicated the flexible behavior when the foundation was placed on the loose soft soil.

4. Conclusion

In this study, the short saturated embankment was evaluated during the strong earthquake while it was coupled by foundation with loose sand. The modulus ratio between saturated embankment and foundation showed the main role to choose the best performance. As a result, based on comparison of the displacement in both direction and shear stress on some key points of the simulated model using ANSYS program, the suitable modulus ratio was 0.66. Finally, the good agreement was revealed between results based on numerical analysis and case studies according to literature. More attempt to find this ratio for tall embankment is the best suggestion to investigate at next.

FIGURE 10: Cracking and inelastic deformations of the crest in an embankment dam caused by the 2001 Buhl earthquake in India.

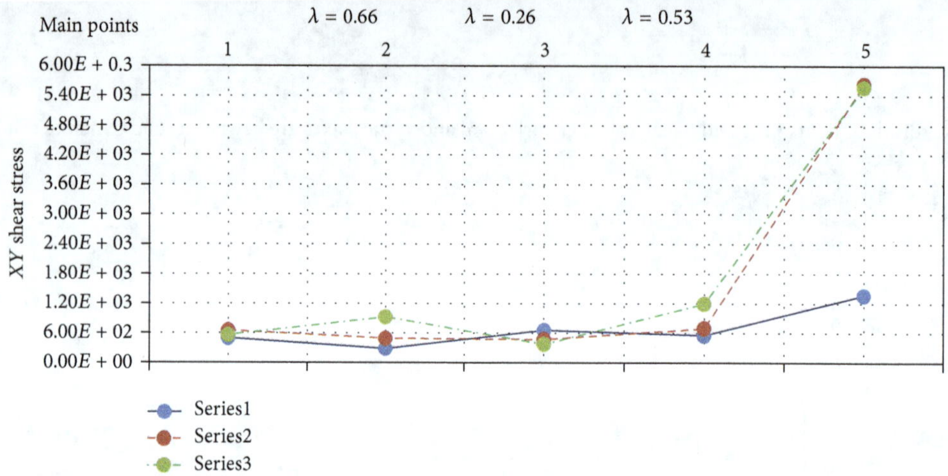

FIGURE 11: *XY* Shear stress in the end of earthquake for models; horizontal axis is main points, and vertical axis is shear stress (Kg/m^2).

Conflict of Interests

The authors declare that there is no conflict of interests regarding the publication of this paper.

Acknowledgment

This study is made possible by the support of the International Doctorate Fellowship of Universiti Teknologi Malaysia, and it is very much appreciated.

References

[1] N. M. Newmark, "Effects of earthquakes on dams and embankments," *Geotechnique*, vol. 15, no. 2, pp. 139–160, 1965.

[2] N. Mononobe, A. Takata, and M. Matumura, "Seismic stability of the earth dam," in *Proceedings of the 2nd Congress on Large Dams (CLD '36)*, Washington, DC, USA, 1936.

[3] G. Gazetas, "A new dynamic model for earth dams evaluated through case histories," *Soils and Foundations*, vol. 21, no. 1, pp. 67–78, 1981.

[4] R. W. Clough and A. K. Chopra, *Earthquake Stress Analysis in Earth Dams*, Department of Civil Engineering, University of California, 1965.

[5] M. Zeghal and A. M. Abdel-Ghaffar, "Local-global finite element analysis of the seismic response of earth dams," *Computers and Structures*, vol. 42, no. 4, pp. 569–579, 1992.

[6] H. Y. Ming and X. S. Li, "Fully coupled analysis of failure and remediation of Lower San Fernando Dam," *Journal of Geotechnical and Geoenvironmental Engineering*, vol. 129, no. 4, pp. 336–349, 2003.

[7] J. L. Borges, "Microstructural weathering of sedimentary rocks by freeze-thaw cycles: experimental study of state and transfer parameters," *Comptes Rendus Geoscience*, vol. 31, no. 8, pp. 665–676, 2004.

[8] A. Yildiz, "Numerical analyses of embankments on PVD improved soft clays," *Advances in Engineering Software*, vol. 40, no. 10, pp. 1047–1055, 2009.

[9] G. A. Ordóñez, *SHAKE2000: A Computer Program for the 1D Analysis of Geotechnical Earthquake Engineering Problems*, User's Manual, Geomotions, LLC, Lacey, Wash, USA, 2011.

[10] R. Noorzad and M. Omidvar, "Seismic displacement analysis of embankment dams with reinforced cohesive shell," *Soil Dynamics and Earthquake Engineering*, vol. 30, no. 11, pp. 1149–1157, 2010.

[11] Z. F. Xia, G. Ye, J. Wang, B. Ye, and F. Zhang, "Fully coupled numerical analysis of repeated shake-consolidation process of earth embankment on liquefiable foundation," *Soil Dynamics and Earthquake Engineering*, vol. 30, no. 11, pp. 1309–1318, 2010.

[12] A. Namdar and A. K. Pelko, "Seismic evaluation of embankment by shaking table test and finite element method," *The Pacific Journal of Science and Technology*, vol. 11, no. 2, pp. 699–707, 2010.

[13] L. Wang, G. Zhang, and J.-M. Zhang, "Centrifuge model tests of geotextile-reinforced soil embankments during an earthquake," *Geotextiles and Geomembranes*, vol. 29, no. 3, pp. 222–232, 2011.

[14] Y. Zhu, K. Lee, and G. H. Collison, "A 2D seismic stability and deformation analysis," in *Proceeding of the Geo-Frontiers*, pp. 1121–1135, Austin, Tex, USA, January 2005.

[15] R. B. J. Brinkgreve, R. Al-Khoury, K. J. Bakker et al., *PLAXIS 2D-Version 8 Full Manual*, Delft, The Netherlands, 2002.

[16] M. D. Braja, *Advanced Soil Mechanics*, CRC Press, 2013.

[17] M. Wieland, "Analysis aspects of dams subjected to strong ground shaking," *International Water Power & Dam Construction*, 2008.

2

Reliability Evaluation of Pavement Life-Cycle Assessment Model

Jiale Huang[1] **Fei Xiao**[2] **and Yang Zhang**[2]

[1]*School of Civil Engineering and Mechanics, Huazhong University of Science and Technology, Wuhan 430074, China*
[2]*Jiangsu Research Institute of Building Science Co. Ltd, Nanjing 210008, China*

Correspondence should be addressed to Fei Xiao; xiaofeihust@163.com

Academic Editor: Ricardo Perera

Inventory reliability of the life-cycle assessment (LCA) model highly depends on the data quality and normally exhibits significant uncertainty. A rigorous statistical methodology was established to capture and quantify the inherent uncertainties linked to the results of the LCA model. Two sources of uncertainty, data quality and model, were identified. The former was captured by converting the deterministic value to probability density function using beta distribution according to the evaluation matrix of data quality; the latter was assessed by prescribing variation interval through defining uncertainty factor. The functional equivalent pavement structures were designed, and the corresponding energy consumption and CO_2 emission were calculated by the LCA model. A 10% variation was observed for the LCA results and within 30-year analysis span, at the 95% confidence level, and environmental burdens of cement pavement are higher than those of asphalt pavements while the comparison between the two asphalt pavements is not significant statistically. Therefore, the established statistical methodology is capable of capturing the uncertainty of the LCA model and quantifying the reliability the LCA results.

1. Introduction

At present, energy conservation and emission reduction in transportation engineering have become the most important aim in China development planning. However, it is just a macroscopic one. Therefore, a data base about the CO_2 emission and energy consumption is needed to accurately assess the effect of technologies for energy conservation and emission reduction. Besides, developing an evaluation template on the bituminous pavement construction and the benefit of environmental protection of maintenance technology is also necessary [1]. Conducting studies on the evaluation system of technologies for energy conservation and emission reduction can provide theoretical foundation and technical support for evaluating the effect of technologies for energy conservation and emission reduction. Therefore, it is extremely necessary to carry out the relevant studies.

The rapid development of road construction brings about serious environmental issues, such as energy consumption, CO_2 emission, and land occupancy [2]. Thus, quantifying and reducing the influence of roadway on environment in its lifetime becomes gradually important and attracts wide attention [3, 4]. As a tool for assessing environmental implication, the life assessment (LCA) has been widely used in roadway engineering and becomes the main tool for analysis of influence of roadway on environment [5–7]. However, some weaknesses exist in the theory of LCA, particularly the uncertainness of data quality and model [8]. The accuracy of LCA for roadway or else objects depends on data and model parameters with high quality. Due to the absence of data and their reliability, the data utilized in LCA are not necessarily the expected one. Therefore, even though optimal data can be acquired or certain existing data can be combined, a trade-off has to be adopted. For example, the authors have collected numerous information and found that the consumption intensity for cement production is 4.6–7.3 MJ/kg and that for bitumen production is 0.7–6.0 MJ/kg. The great difference is shocking and resulted from the differences between system boundary, manufacturing technique, and manufacturing process in the local area. Besides, for bitumen material, the environmental influence depends on the distribution of different chemical products, such as petrol, diesel oil, and plastics.

The difference of distribution style results in significantly different results [9–11]. In the process of LCA establishment, the assumption and simplification are involved, which also affect the reliability of assessment results of environmental influence.

Therefore, based on a practical case, a LCA model was developed and the influence of roadway on the environment was carried out. The uncertainness of assessment results of the LCA model is captured from the perspective of data quality and model assumption to establish a rigorous methodology for reliability assessment of results of the LCA model.

2. The Establishment of LCA Model

Roadway is a complex structure and susceptible to several external factors, such as climate, transportation intensity, and the production and transportation of material. Thus, the authors took advantage of the existing data model and established a LCA model which includes the following module:

(1) Material production module: consideration on the process involved in the material production
(2) Transportation module: description about the environmental influence during the material transportation
(3) Construction maintenance module: consideration on the oil consumption of facilities and its environmental influence
(4) Postprocessing module: different postprocessing scenes can be used according to the system boundary and the setup of consumption

The function unit of the LCA model is set in an analysis cycle of 30 years and the service performance is applicable to a roadway with single lane and a length of 1000 m.

3. Methodology for Reliability Evaluation

The application of LCA depends on large amounts of data input which defined the characters of the model such as specification for road design, input and output stream of material energy, and environmental impact. The results of LCA are highly dependent on the reliability of input parameters with a certain degree of uncertainty inevitably. This investigation establishes corresponding methodologies for quantitative assessment from data uncertainty and model uncertainty.

3.1. Data Uncertainty. The factors of data uncertainty are divided into experiential evaluation, measurement error, and nonrepresentativeness which includes data obsoleting and excessive technical differences, even data missing. Ideally, the solution to data uncertainty is establishing corresponding probability density function by collected data and putting it into LCA. However, the measurement data, samples, and experience that we need are usually rare. In this case, the descriptive index was employed to assess the quality of data, which forms the data quality evaluation matrix as shown in Table 1.

The data quality index (DQI) can be obtained by the established data quality system matrix. The DQI can be converted to probability density function as input parameters of uncertainty analysis, specifically shown as follows:

$$\begin{aligned} f(x;\alpha,\beta,a,b) = {} & [1/(b-a)] \times \{\Gamma(\alpha+\beta)/\Gamma(\alpha)\times\Gamma(\beta)\} \\ & \times [(x-a)/(b-a)]^{\alpha-1} \\ & \times [(b-x)/(b-a)]^{\beta-1} \quad (a \le x \le b), \end{aligned} \tag{1}$$

where α and β are the shape parameters of distributions and a and b are the interval end points.

The reasons why choosing beta function are shown as follows:

(1) The shape and probability of the distribution established by shape parameters are unknown, and endpoint parameters limit the range of prefabrication
(2) The distribution shape of the beta function is various due to the shape and endpoint parameters.

3.2. Model Uncertainty. The inherent uncertainty of the model associated with the decision affects the analysis results of the LCA. The model refers to the concept of generalization and has a wide range of extensions. It can be either a specific regression model, for example, prediction of asphalt mixing energy consumption, or input parameter assumptions, such as the amount of asphalt used in grading design and the organization of construction platoons. For the variability of LCA model input values, such as the ratio of oil and stone and transportation distance, it can be described by defining "uncertainty factor (UF)" where the uniform distribution and lognormal distribution are common distribution forms.

For the uniform distribution, the minimum and maximum values of the input parameters M_x are defined using the following equation:

$$\begin{cases} M_{x,\min} = \dfrac{M_{x,\mathrm{mod}}}{\mathrm{UF}_x}, \\ M_{x,\max} = \mathrm{UF}_x \times M_{x,\mathrm{mod}}, \end{cases} \tag{2}$$

where $M_{x,\min}$, $M_{x,\max}$, and $M_{x,\mathrm{mod}}$ are the minimum, maximum, and modus, respectively.

For the lognormal distribution, the fluctuation range of the input parameters M and UF are defined in the following equation:

$$p\left(\frac{M_{x,\mathrm{mod}}}{\mathrm{UF}_x} < M_{x,\mathrm{mod}} < \mathrm{UF}_x \times M_{x,\mathrm{mod}}\right) = 0.95, \tag{3}$$

where the parameters are defined in Equation (2).

4. Case Analysis

This case analyzed the energy consumption and carbon emissions of highways cement pavement structure and asphalt pavement structure during their life cycle under the

Table 1: Data quality evaluation matrix.

Quality grade	Judgement standard						
	Selection standard	Technology correlation	Location correlation	Data sources	Acquiring methods	Time	Representation of data
5	Transparent, verified, and generally applicable	Representative process and representative technology	Represent value of local area	Verified data from independent source	Actual measurement	<3 years	Long time and large sample
4	Transparent, verified, and no generally applicable	Representative process and similar technology	Average value of local area	Verified data from relevant source	Measurement-based calculation	<6 years	Long time and small sample
3	Transparent, no verified, and no generally applicable	Representative process and different technology	Other area with similar production process	Unverified data from independent source	Hypothesis-based calculation	<10 years	Short time and large sample
2	Selective transparent and no selection standard	Similar process and similar technology	Other area with slightly similar production process	Unverified data from irrelevant source	Expert estimation	<15 years	Short time and small sample
1	Unknown selection standard	Similar process and different technology	Unknown location	Unverified data from relevant source	Calculation with no guaranteed quality	>15 years	Cannot be obtained

same traffic volume and weather conditions and demonstrated the reliability of the calculation results.

4.1. Waterway Asphalt Pavement Structure Design. For highways with asphalt pavement structures, the design period is generally 15 years. For cement highway structures, the design period is generally 30 years. In order to make the environmental impact analysis consistent, the functional unit is set to provide an acceptable road service performance for a single-lane one-kilometer-long road surface in a 30-year analysis cycle. The design index of the road structure tends to and does not exceed the design tolerance value. For asphalt pavement structural design, there are two situations:

(1) The design life of asphalt pavement structure is set to be 30 years to achieve a certain degree of long-life asphalt pavement design.

(2) The asphalt pavement design period is 15 years. Assuming that, in the 15th year, the basement structure of the asphalt pavement was not damaged, only the asphalt surface structure was resurfaced by milling.

In order to ensure the functional equivalence of the cement and asphalt pavement structures, the pavement structure needs to be carefully designed so that the control index of the pavement structure approaches its limit at the end of the design life cycle, ensuring that the pavement structure has neither "excessive design" nor "insufficient design". Under this guiding principle, the equivalent design parameters of cement and asphalt pavement structures are shown in Table 2.

Although the asphalt pavement designed for 30 years of life will not suffer structural damage, it is still necessary to periodically perform conservation actions to ensure the functionality of the surface structure. At present, China has not established a strict decision-making system for pavement maintenance. Therefore, the conservation manual in California is referred to maintain the three types of pavement structure [13]. The maintenance plan is as follows:

(1) For an asphalt pavement structure with a design life of 30 years, a 2.5 cm thick overlaying in thin layer is laid at the end of the 20th year, and a 4 cm-thick pavement structure is milling-resurfaced at the end of the 24th year

(2) For an asphalt pavement structure with a design life of 15 years, a 6 cm-thick pavement structure is milling-resurfaced at the end of the 15th year, a 4 cm-thick overlaying in thin layer is laid at the end of the 20th year, and a 4 cm-thick pavement structure is milling-resurfaced at the end of the 25th year

(3) For a cement pavement structure, the pavement structure is replaced with the cement board at the end of the 15th year and the end of the 25th year, respectively

4.2. Calculation Based on LCA Model. The environmental impact on each module has been computed based on the LCA model constructed before. The calculation factors include energy consumption and CO_2 emissions, using unit mass standard coal (kgce) and kg-CO_2 as their unit, respectively.

4.2.1. Material Producing Module. Environmental impact of unit productivity on road materials has been determined through document retrieval, theoretical calculations, and on-site inspections, and the result is shown in Table 3.

According to the pavement structure design maintenance parameters and Table 3, the environmental impact on material producing module can be obtained.

TABLE 2: Converting matrix of probability density function [12].

DQI		5.0	4.5	4.0	3.5	3.0	2.5	2.0	1.5	1.0
Parameters of beta distribution	Shape function	(5,5)	(4,4)	(3,3)	(2,2)	(1,1)	(1,1)	(1,1)	(1,1)	(1,1)
	Interval endpoint	10	15	20	25	30	35	40	45	50

TABLE 3: Design parameters of pavement structure.

Pavement structure	Upper layer (cm)	Middle layer (cm)	Lower layer (cm)	Base (cm)	Subbase (cm)	Designed deflection (0.01 mm)	Calculated deflection (0.01 mm)
Asphalt (30 years of design)	6	8	12	40	25	14.0	13.5
Asphalt (15 years of design)	4	6	8	30	25	17.2	16.8
Cement	Surface course (cm)	Base course (cm)	Fatigue stress of surface course (MPa)		Maximum multiple stress of surface course (MPa)	Flexural tensile strength of surface course (MPa)	
	30	20	4.89		2.94	5.0	

4.2.2. *Transportation and Construction Module.* As for the calculation of the environmental impact on transportation and construction modules, we choose JTG/T B06-02-2007 and JTG/T B06-03-2007 as the fundamental documents. The main steps are as follows: first, determining the number of per-unit-production of mechanical shifts according to the construction process flow stipulated in the budget quota; second, determining the total energy consumption by the energy consumption parameters of the unit machinery and equipment specified by the mechanical station cost quota; the last but not the least, calculating per-unit energy consumption.

4.2.3. *Postprocessing Module.* The asphalt pavement and cement pavement reach the end of design life after the 30-year life cycle. Thus, for the asphalt pavement, the pavement layer needs to be milled and the RAP needs to be transported to mixing building, and for the cement pavement, the pavement layer needs to be smashed and the RCM needs to be transported to the accumulation point, where the transportation distance is assumed to be 25 km.

4.2.4. *Calculation Results of LCA Model.* Calculation of the environmental impact of the cement asphalt pavement structure under life cycle with the LCA model is done, and the results are shown in Table 4.

It can be seen that the stage of producing pavement material consumed the most energy and released the most CO_2 in all structures with any pavement. By comparing the structure with the asphalt pavement (15-year or 30-year design) and the cement pavement, the structure with the cement pavement consumes less energy and emits less CO_2, which means the cement pavement structure has a greater probability of reducing the environmental impact during the life cycle.

In this paper, two life-cycle asphalt pavement structures were designed. The asphalt roads with a design life of 15 years have less impact on environment than those with a design life of 30 years, but the difference is not so significant. Also, such conclusion is closely dependent on the parameter assumptions in the LCA modeling process and needs further confirmation.

4.3. *Reliability Analysis for LCA Model.* There remains some uncertainty during modeling the LAC model due to data quality defects and model parameters assumptions, which are needed to be evaluated.

4.3.1. *Data Uncertainty Analysis.* With the data quality assessment methodology established in Section 3.1, the quality of road material environmental data was evaluated. Taking bituminous materials as an example, the energy consumption of matrix asphalt production is 170.34 kgce.

The calculation for energy consumption of base-asphalt production uses the calculation model of Eurobitume 2011 and the data of China Statistical Yearbook as reference. Therefore, the evaluation scores for the indicators according to Table 1 are {3.0, 5.0, 4.0, 4.0, 3.0, 5.0, 3.0}, and thus its DQI is 3.85. By using Table 5 as a reference, the parameter of beta function can be de defined. Its shape function α, β is (3,3), and the interval endpoint is (−20%, +20%). The probability density function of the production of matrix asphalt materials is shown in Figure 2.

According to the probability density function established in Figure 2, the energy consumption value and reliability of the matrix asphalt materials can be evaluated. Analogously, a probability density function curve can be established for the production of carbon emissions from matrix asphalt materials and other road material, providing a data source for the Monte Carlo analysis.

4.3.2. *Model Uncertainty Analysis.* During modeling, parameters like transportation distance and asphalt mixture

Table 4: Energy consumption and CO_2 emission of pavement materials in China.

Material	Energy consumption (kgce/t)	CO_2 emissions (kg/t)	Data source
Asphalt	170.34	287.51	《China Energy Statistical Yearbook 2013》 [14]
Modified asphalt (3.5% SBS)	311.61	485.54	《China Energy Statistical Yearbook 2013》and Eurobitume [11] calculation
Emulsified asphalt (60% DPU)	118.55	205.98	《China Energy Statistical Yearbook 2013》and Eurobitume [11] calculation
Stone	3.71	9.02	Investigation
Comment	136	377.06	《China Energy Statistical Yearbook 2013》
Industrial water	0.0857	0.24	《Comprehensive Energy Consumption Calculation Principles 2008》 [15]
Steel	674	1868.67	《China Energy Statistical Yearbook 2013》
Asphalt heating	0.04	0.09	Investigation
Aggregate heating	10.9	24.2	Investigation
Asphalt mixture mixing	1.62	3.99	Investigation
Cement concrete mixing	4.09	8.93	Investigation

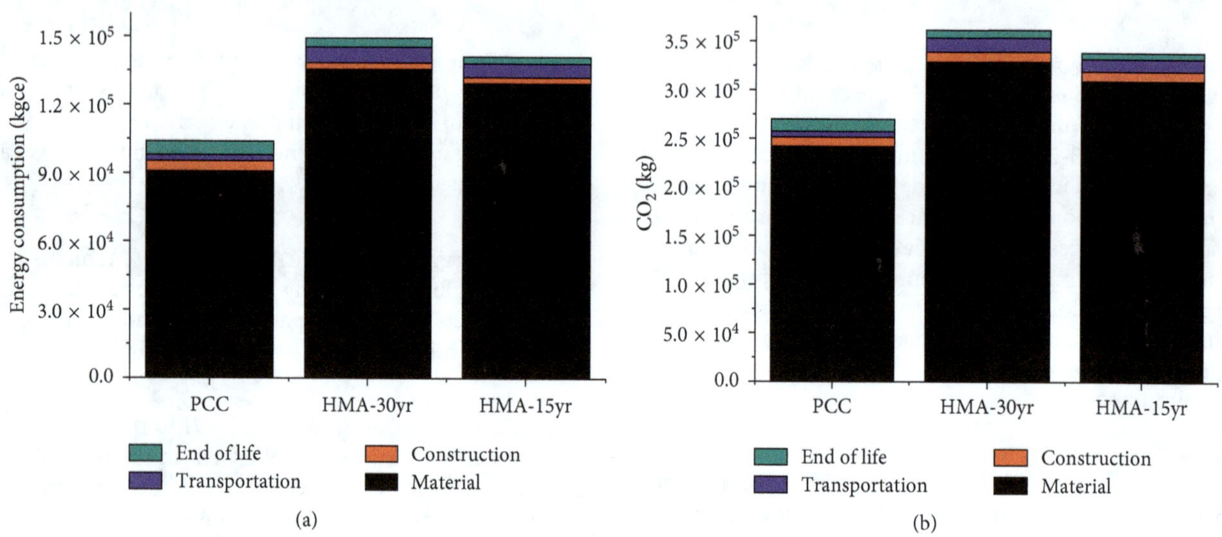

Figure 1: (a) Energy consumption and (b) CO_2 emission of the three pavement structures.

Table 5: Life-cycle inventories of the three pavement structures.

Cement			Asphalt (15 years of design)			Asphalt (30 years of design)		
Modules	Energy consumption (kgce)	CO_2 (kg)	Modules	Energy consumption (kgce)	CO_2 (kg)	Modules	Energy consumption (kgce)	CO_2 (kg)
Material production	90773	241737	Material production	129326	309219	Material production	135508	328646
Construction	4432	9416	Construction	2716	9887	Construction	2914	10373
Transport	2758	5859	Transport	6024	12799	Transport	6797	14443
Postprocessing	5958	12658	Postprocessing	2931	6228	Postprocessing	3908	8303

With the data in Table 5, the analysis chart can be drawn, shown in Figure 1.

mixing energy consumption are introduced, and the influence of those parameters on the LCA calculation results is needed to be examined. Those parameters and their fluctuation range are specified in Tables 6–7.

4.3.3. Results of LAC Model Uncertainty Analysis. The Monte Carlo analysis was performed based on the parameters inputted in Sections 4.3.1 and 4.3.2. After 50 times of iteration, the results are shown in Figure 3.

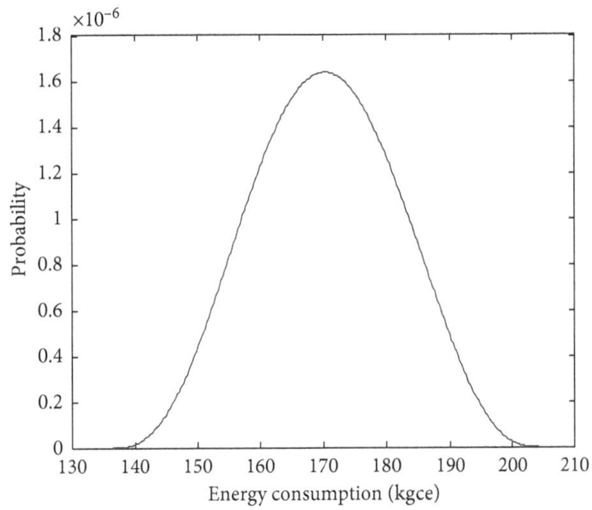

FIGURE 2: Probability density function of base asphalt production energy consumption.

TABLE 6: Parameter uncertainties of the asphalt pavement LCA model.

Stage	Parameters	Uncertainty[1]	Distribution form
Material production	Asphalt production	—	Probability density function
	Aggregate production	UF = 1.62	Lognormal distribution
	Mixing	UF = 1.47	Lognormal distribution
	SBS amount[2]	UF = 1.09	Uniform distribution
	Asphalt amount	UF = 1.09	Uniform distribution
	Asphalt aggregate heating	UF = 1.09	Uniform distribution
Construction stage	Shift	UF = 1.09	Uniform distribution
Transportation stage	Transport distance	UF = 1.41	Uniform distribution
	Carbon emission rate of transportation energy	UF = 1.14	Uniform distribution
Postprocessing stage	Shift	UF = 1.09	Uniform distribution
	Transport distance	UF = 1.41	Uniform distribution

Note. 1: uncertainty parameters determined by engineering investigation (the subscripts are similar); 2: the asphalt amount adjusted with SBS amount.

TABLE 7: Parameter uncertainties of the cement pavement LCA model.

Stage	Parameters	Uncertainty[1]	Distribution form
Material production	Cement production	—	Probability density function
	Aggregate production	UF = 1.62	Lognormal distribution
	Mixing	UF = 1.47	Lognormal distribution
	Cement amount	UF = 1.09	Uniform distribution
	Steel amount	UF = 1.09	Uniform distribution

Note. 1: uncertainty parameters in the remaining stages are consistent with those in Table 2.

In Figure 3, the left Figure 3(a) shows the average energy consumption and its standard deviation for different scenarios, and Figure 3(b) shows the probability density function for each scenario. From Figure 3(a), the energy consumption of the PCC scheme is smaller than the HMA scheme generally. It can be found that the coefficient of variation of the PCC program, 30-year, and 15-year design of the HMA program are 7.4%, 9.0%, and 9.1%, respectively, when it is used as the indicator of energy fluctuation range. The volatility of the PCC program is slightly less than HMA program. Since the maximum volatility of all programs is about 10%, which is quite limited, the energy consumption indicators of different programs have a good stability.

Define R as the ratio of the energy consumption of design 1 to the energy consumption of design 2, and then P (R<1) means the probability that R is less than 1. According to Figure 3(b), 50000 Monte Carlo analyses were performed, and the results are shown in Figure 4.

It can be seen from Figure 4 that the probability that the energy consumption of the cement pavement structure is less than the 15-year design asphalt pavement structure is 97%, and the probability that the energy

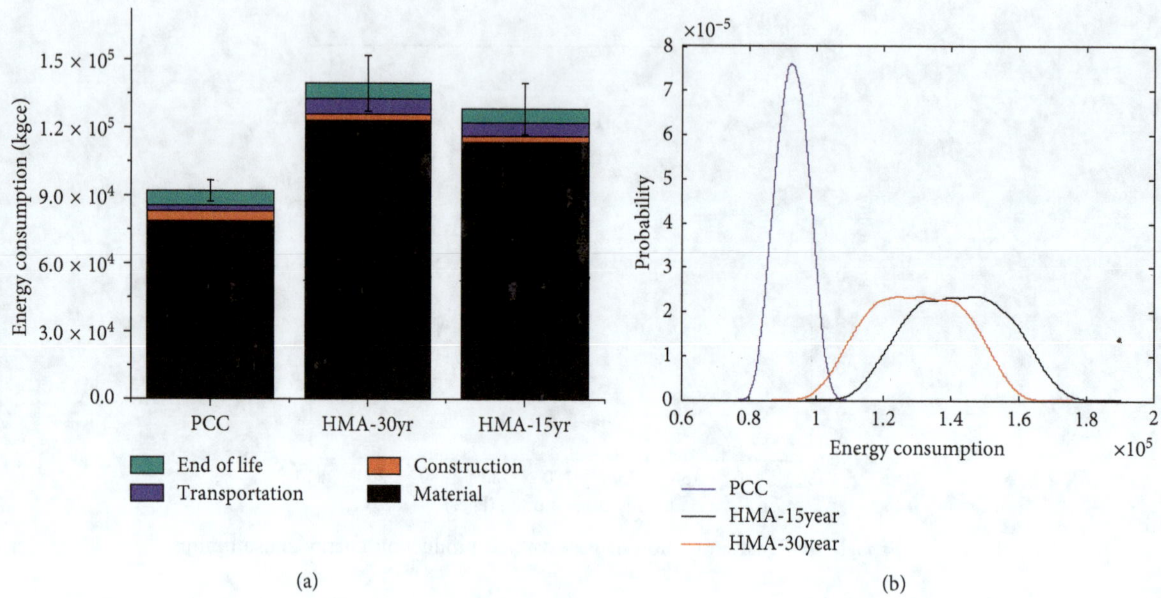

FIGURE 3: Energy consumption comparisons of the three pavement structures.

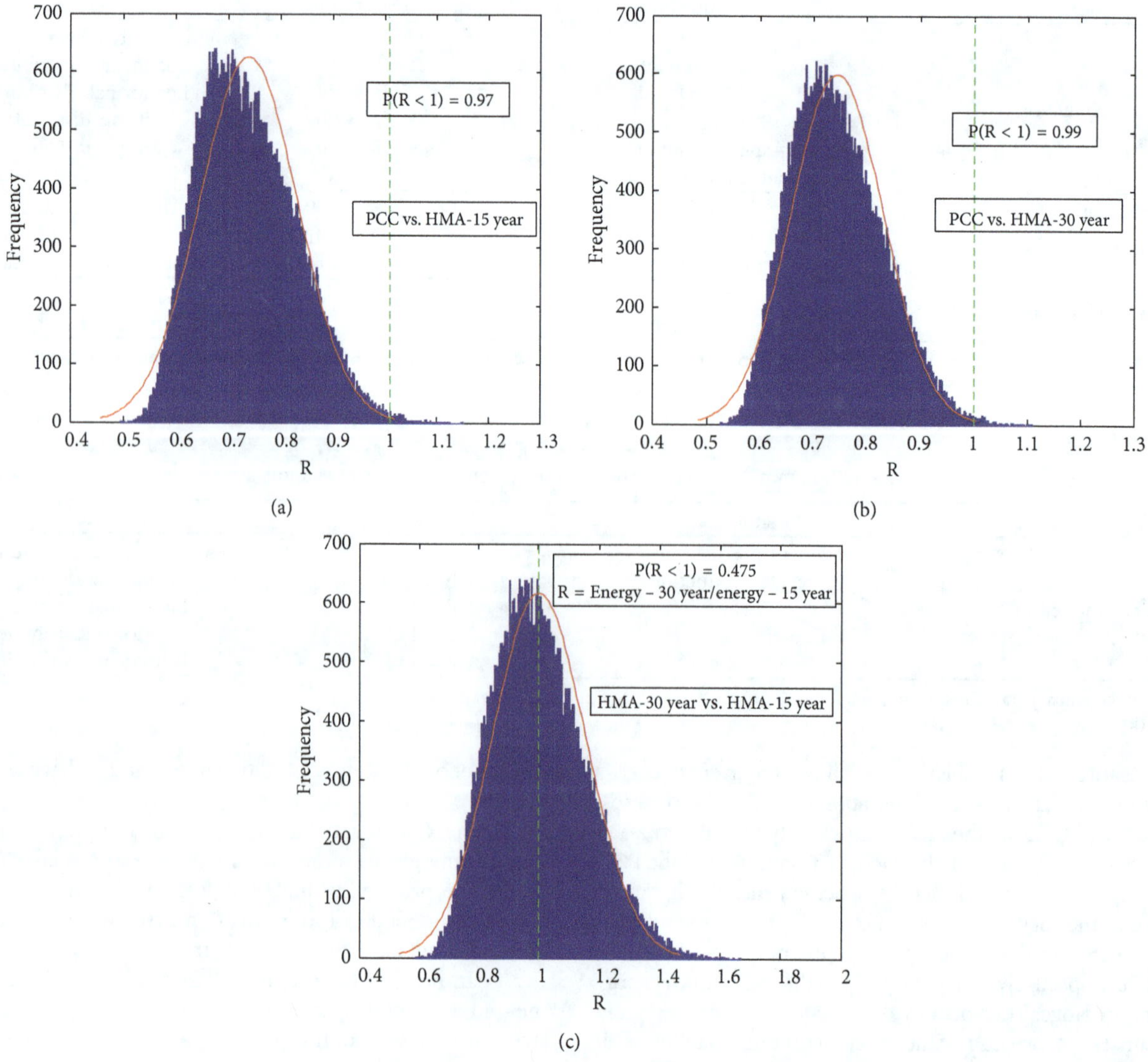

FIGURE 4: Energy consumption ratio probability distribution of the three pavement structures.

consumption of the cement pavement structure is less than the 30-year design asphalt pavement structure is 99%. Under the conditions set in this study, the cement pavement structure consumes less energy during the life cycle at the 95% confidence level. Comparing these two asphalt pavement designs, the probability that the energy consumption of the 30-year design asphalt pavement structure is less than that of the 15-year design asphalt pavement structure is 47.5%. It is difficult to have sufficient reason to detail the difference in environmental impact between the two structural designs. Similar conclusions can be drawn for carbon emissions indicators and will not be repeated here.

5. Conclusion

(1) A road LCA model with complete system boundary has been constructed. This model can not only describe the environmental impact of the road during service, but also can be used to assess the environmental impact of road materials in a measurable way.

(2) A rigorous method for evaluating the reliability of LCA model calculation results is constructed, which can be used to evaluate the reliability of LCA model settlement results under data quality and model assumptions.

(3) Under the established research background, the environmental impact of the cement pavement structure is less than the asphalt pavement structure at the 95% confidence level. The average energy consumption and carbon emission of the cement pavement is 21.6%–28.2% lower than the asphalt pavement during the life cycle.

(4) The reliability analysis shows that the results of the LCA model are stable and the fluctuation range is about 10%.

(5) Under the research background setup, there is no obvious environmental difference between the 15-year design life asphalt pavement and the 30-year design asphalt pavement at the 95% confidence level.

Conflicts of Interest

The authors declare that they have no conflicts of interest.

References

[1] N. J. Santero, *Pavements and the environment: a life-cycle assessment approach*, Ph.D. thesis, University of California, Berkeley, CA, USA, 2009.

[2] B. Yu, *Environmental implications of pavement: a life cycle view*, Ph.D. thesis, University of South Florida, Tampa, FL, USA, 2013.

[3] H. Zhang, G. A. Keoleian, M. D. Lepech, and A. Kendall, "Life-cycle optimization of pavement overlay systems," *Journal of Infrastructure Systems*, vol. 16, no. 4, pp. 310–322, 2010.

[4] Y. Huang, R. Bird, and O. Heidrich, "Development of a life cycle assessment tool for construction and maintenance of asphalt pavements," *Journal of Cleaner Production*, vol. 17, no. 2, pp. 283–296, 2009.

[5] B. Yu and Q. Lu, "Life cycle assessment of pavement: methodology and case study," *Transportation Research Part D: Transport and Environment*, vol. 17, no. 5, pp. 380–388, 2012.

[6] J. Santos, A. Ferreira, and G. Flintsch, "A life cycle assessment model for pavement management: methodology and computational framework," *International Journal of Pavement Engineering*, vol. 16, no. 3, pp. 268–286, 2015.

[7] L. Cheng, G. J. Yan, D. Z. Chen et al., "Quantitative investigation on energy conservation and emission reduction related to warm mix asphalt (WMA)," *Chinese Journal of Environmental Engineering*, vol. 4, no. 9, pp. 2151–2155, 2010.

[8] N. J. Santero, E. Masanet, and A. Horvath, "Life-cycle assessment of pavements part II: filling the research gaps," *Resources, Conservation and Recycling*, vol. 55, no. 9-10, pp. 810–881, 2011.

[9] A. L. Nicholson, E. A. Olivetti, J. R. Gregory et al., "End-of-life LCA allocation methods: open loop recycling impacts on robustness of material selection decisions," in *Proceedings of IEEE International Symposium on Sustainable Systems and Technology*, Phoenix, AZ, USA, December 2009.

[10] Y. Huang, A. Spray, and T. Parry, "Sensitivity analysis of methodological choices in road pavement LCA," *International Journal of Life Cycle Assessment*, vol. 18, no. 1, pp. 93–101, 2013.

[11] Eurobitume, *Life Cycle Inventory: Bitumen*, The European Bitumen Association, Brussels, Belgium, 2nd edition, 2012.

[12] K. G. Canter, D. J. Kennedy, D. C. Montgomery et al., "Screening stochastic life cycle assessment inventory models," *International Journal of Life Cycle Assessment*, vol. 7, no. 1, pp. 18–26, 2002.

[13] University of California Pavement Research Center (UCPRC), *Life-Cycle Cost Analysis Procedures Manual, Prepared for the California Department of Transportation Division of Design and the Pavement Standards Team*, University of California Pavement Research Center (UCPRC), Davis, CA, USA, 2007.

[14] Energy Statistics Division of National Bureau of Statistics, *China Energy Statistical Yearbook 2013*, China Statistics Press, Beijing, China, 2013.

[15] National Development and Reform Commission Resource Conservation and Environmental Protection Department, GB/T2589-2008, *General Principles for Calculation of the Comprehensive Energy Consumption*, China Standards Press, Beijing, China, 2008.

3

Numerical Study on Diffusion of Chloride and Induced Rebar Corrosion by Two-Dimensional Multiscale Approach

Xi Tu,[1,2] Jin Di,[1,2] Cunjun Pang,[1,2] and Xiaoqing Xu[1,2]

[1]*Key Laboratory of New Technology for Construction of Cities in Mountain Area (Chongqing University), Ministry of Education, Chongqing, China*
[2]*College of Civil Engineering, Chongqing University, Chongqing, China*

Correspondence should be addressed to Xi Tu; tuxi@cqu.edu.cn

Guest Editor: Qing-feng Liu

Modeling approach for mesoscopic model of concrete depicting mass transportation and physicochemical reaction is important since there is growing demand for accuracy and computational efficiency of numerical simulation. Mesoscopic numerical simulation considering binder, aggregate, and interfacial transition zone (ITZ) generally produces huge number of DOFs, which is inapplicable for full structure. In this paper, a two-dimensional multiscale approach describing three-phase structure of concrete was discussed numerically. An effective approach generating random aggregate in polygon based on checking centroid distance and intersection of line segment was introduced. Moreover, ITZ elements were built by parallel expanding the edge of aggregates on inner side. By combining mesoscopic model including full-graded aggregate and macroscopic model, cases related to diffusivity and width of ITZ, volume fraction, and grade of aggregate were studied regarding the consideration of multiscale compensation. Result clearly showed that larger analysis model in multiscale model expanded the diffusion space of chloride ion and decreased chloride content in front of rebar. Finally, this paper addressed some noteworthy conclusions about the chloride distribution and rebar corrosion regarding the configuration of rebar diameter, concrete cover, and exposure period.

1. Introduction

Corrosion of rebar induced by chloride ion could significantly deteriorate the serviceability of concrete structures [1]. The protective film of rebar was depassivated by chloride ion penetrated from surface of concrete and thus corrosion of rebar initiates once the chloride content on surface of rebar reaches a certain thresholding value, defined as critical chloride content. Afterwards, corrosion of rebar keeps propagating and induces further cracking and spalling due to expansion of rust. Therefore, it is important to accurately assess the diffusion process of chloride ion within concrete, which is meaningful for further evaluation of durability of reinforced concrete structures.

In the view of mesoscopic numerical simulation, concrete was recognized as the heterogeneous composite of three phases, including cement paste, aggregate, and interfacial transition zone (ITZ) [2]. ITZ was identified as a fine cement paste zone enclosing aggregate and rebar, providing higher water-to-cement ratio, higher porosity, and lower cement content compared with normal bulk cement paste regions. Thus ITZ was generally considered as an individual phase [3–5]. Though the porosity of ITZ decreases with increase of distance from aggregate surface [6], ITZ was simplified as a homogenous thin layer in most studies. Significant influence on the overall behavior of concrete attributes to ITZ due to its fraction of the total paste volume, such as diffusivity and strength [4, 6–8].

In the view of mesoscopic numerical simulation, due to large amount of nodes and elements, an optimized approach was that three-phase composite of concrete was simplified into two-phase composite including cement paste and homogenized equivalent aggregate [4, 7, 9–13]. Diffusivity of ITZ was transferred by means of deriving the analytical solution of general effective chloride diffusivity, which means modeling of ITZ is unnecessary.

Generally, modeling and analysis for cement-based material were categorized in microscopic scale, mesoscopic scale, and macroscopic scale [14, 15]. Recently a number of theory research and modeling efforts have been devoted to studying multiscale modeling. Multiscale modeling was usually defined as the integrated numerical process of transferring mechanical and chemical response between lower scale and higher scale [16, 17]. A significant advantage of multiscale modeling is optimizing computing loading and enhancing the precision. According to available literature, researchers generally focused on the transferring process of material response between scales.

As the major purpose of multiscale modeling to reduce calculation loading, heterogeneous model was always modeled for transferring damage and mass transportation of critical regions from the macroscale to the mesoscale which provided coupling of subdomains [18]. Balance of accuracy and efficiency is the major consideration of multiscale modeling for numerical simulation. By means of macroscopic model substituting for a part of mesoscopic model, multiscale modeling is able to provide relatively smaller size of mesoscopic model and generate the same filling rate of aggregate and less amount of aggregate, which is no doubt beneficial for raising success rate of meshing [19].

For mesoscopic structure of concrete, both of size and shape of aggregates are generated as given design. Regarding two-dimensional space, general process was divided into two steps, including modeling of geometric model and meshing. Angular aggregate was generated based on inscribed convex polygon within elongated ellipse [20]. Wang et al. proposed a procedure for generating random aggregate structures for angular aggregates by means of Monte Carlo sampling, which was compatible for concave polygon [21]. An inevitable algorithm was intersection check of convex polygons, which was achieved by detecting space independence of two polygons [22, 23]. In terms of meshing, mortar plus smaller aggregates embedding coarse aggregate was discretized into finite elements by free meshing [20, 24] or uniform background grid [25–27].

Another major factor for durability of reinforced concrete structures is corrosion of steel rebar. Researchers devoted efforts on laboratorial research by means of natural experiments [28–32] and accelerated experiments [33–38] about the process of chloride diffusion and corrosion. On the other hand, some types of simplified approximate law for corrosion were introduced. Biondini introduced a reasonable linear damage model for corrosion of rebar under aggressive agent, which evaluates corrosion rate according to the real-time chloride content surrounding rebar [39–42]. Despite the complex principle of corrosion, based on the natural damage mode of corrosion, this linear damage model is able to provide acceptable description of damage process.

This paper discussed two important aspects regarding multiscale modeling for numerical simulation of chloride ion diffusing within concrete. Firstly, a comprehensive modeling approach for concrete considering both multiscale modeling including interfacial transition layer (ITL) and three-phase mesoscale structure including ITZ. Secondly, by means of the multiscale numerical simulation tool, the influence of ITZ on diffusion of chloride ion within concrete in mesoscale was studied in detail, where ITZ was modeled as an individual phase in FE model for chloride diffusion in concrete. Besides, the corrosion of rebar in straight edge of concrete induced by diffusion of chloride ion was also calculated to study the time evolution of corrosion process.

2. Multiscale Modeling and Discussion

2.1. Multiscale Modeling Theory. In mesoscopic numerical simulation for chloride diffusion, usually 100~200 mm size of specimen in square or rectangle was considered. Within this typical size of analysis space, enough number of aggregates with various sizes were included for in-depth study. Within limited analysis space, chloride ion would able to penetrate and saturate the whole space within the assessing period, indicating the whole space was inadequate for further simulation. However, it would be impossible to increasing the size of analysis space due to two reasons. First, several times of original size of analysis space will generate much more nodes and elements, especially for three-dimensional problem. Second, to control computing loading, changing the grading curve and ignoring fine aggregate will cause inaccuracy of numerical simulation.

In this paper, a scheme of multiscale modeling was introduced in terms of balancing amount of nodes and elements and calculation accuracy (Figure 1). In the view of multiscale modeling, based on original mesoscopic model including multiphase components which was defined as core part, a macroscopic compensation model, which was defined as compensation part, was modeled. The aim of introducing compensation part was creating addition space to absorb the chloride within core part and adjusting the distribution of chloride content within core part, which means redistribution of chloride in larger space.

Due to different definition of nodal chloride content and diffusion coefficient in mesoscopic model and macroscopic model, the two types of models cannot be directly combined and analyzed simultaneously. In this paper, interfacial transition layer (ITL) was introduced to connect both models at the interface which consists of a group of nodes and elements, shown in Figure 1. ITL was designed as a banded shape transition zone allowing chloride ion diffusing from core part into compensation part. Within each time step, core part and compensation part were analyzed sequentially and the chloride content at interface of the two parts was transferred based on the function of ITL.

It is worth noting that the most important precondition is the mechanism of ITL for transporting the chloride ion, which generates equivalent diffusion and distribution of chloride within the model on the same scale. ITL could be composed by a row of common nodes occupied by both mesoscopic model and macroscopic model, as well as an amount of elements with certain thickness stuck by both models.

The process of transferring chloride ion within ITL in one-dimension was illustrated in Figure 2. For initial state, analysis space of three phases including core part, compensation part, and ITL was modeled, respectively. Boundary condition was set at surface of concrete, which was the top of core part shown in Figure 2.

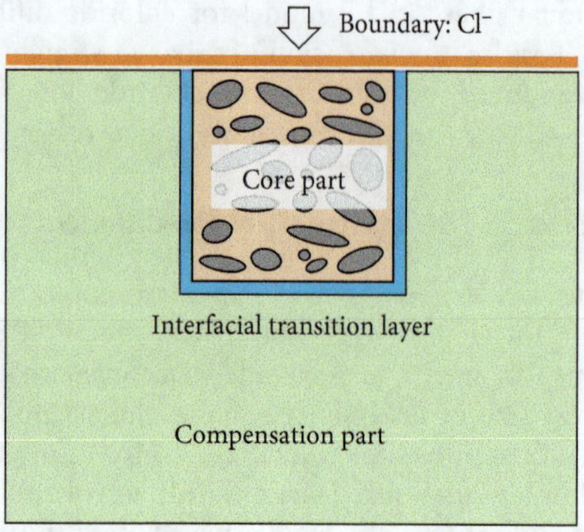

FIGURE 1: Two-dimensional multiscale model with core part and compensation part for simulating chloride diffusion within concrete.

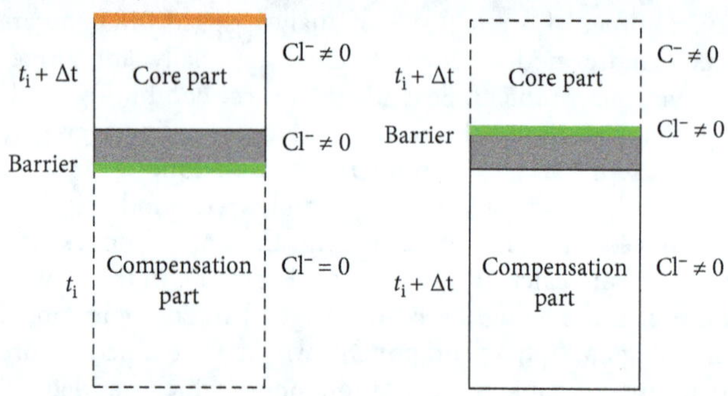

FIGURE 2: The two stages of chloride diffusion at the interfacial transition layer connecting macroscopic and mesoscopic model. (a) First stage. (b) Second stage.

This mixed model behaved in sequential multiscale process. For the mixed case that core part is in mesoscale and compensation part in macroscale, equivalent chloride content in ITL will be calculated and transferred. Definitions of chloride content in both parts were different. Chloride content in macroscopic model was measured in concrete, and the value in mesoscopic model was measured in cement. Therefore, a necessary procedure was included that the value of chloride content in ITL should be calculated for each stage. Simulation of chloride diffusion within mixed model was divided into two stages of diffusion and two steps of conversion.

(1) The first stage of diffusion: core part and ITL were connected as common interface while ITL and compensation part were disconnected. With one time step, chloride ion penetrated into core part from boundary condition. During current stage, diffusion of chloride ion restrained within core part and ITL. The chloride content of core part and ITL rose up from previous content while compensation part kept unchanged.

Find out all of the elements with the centroid located within the elements from compensation model.

Calculate the weighted mean content of all these elements in core model. This content is the represented content for the element in compensation model.

(2) Conversion from mesoscale to macroscale within ITL: the simulated result in region in core part covered by ITL was converted from mesoscale to macroscale within this step. For the instance of center bottom of concrete boundary, compensation part was modeled as simply one-dimensional model. Within core part, elements with their centroid included in interfacial transfer layer were identified (the region enclosed by red dashed line in Figure 3). Weighted mean value of chloride content of these involved elements was calculated for the nodal value of ITL in macroscale.

(3) The second stage of diffusion: a new barrier was inserted at common border of core part and ITL which was equal to cut down the connection of these parts. The existing barrier at common border of compensation part and ITL was removed. During this stage, chloride redistributes within the region including ITL and compensation part.

(4) Conversion from macroscale to mesoscale within ITL: as the inverse process of the 2nd step, the nodal content of mesoscopic core part was interpolated from macroscopic compensation part in the form of regular rectangle element. As was illustrated in Figure 4, the outer rectangle represents the uniform quadrilateral model in compensation part and the target node denotes the node included within core part. Four nodes of the quadrilateral were numbered in counterclockwise.

2.2. Analytical Study for Finite Space with Exposed and Time-Dependent Boundary. In following section, this paper discussed about the feasibility of multiscale model in analytical approach. Due to the sectional design of rebar in concrete member, some research studies on chloride diffusion were transformed within two-dimensional space equivalently. For the case on center of bottom of concrete section, concrete and outer atmosphere were divided into two infinite half-space into which the boundary plane divides the three-dimensional space. Regarding numerical simulation, for the same configuration of boundary condition and diffusivity of concrete, all points in concrete at the same depth from the boundary behave in the same characteristics of diffusion. In simplification, the chloride diffusion in three-dimensional and two-dimensional space was equivalent to one-dimensional one. Therefore, the problem was simplified to one-dimensional macroscopic model, and the analytical solution for finite space with both exposed and sealing boundary was discussed. Usually diffusion of chloride within solution was expressed by linear Fick's second law:

$$\frac{\partial C}{\partial t} = D \cdot \nabla^2 C, \quad (1)$$

where D is the diffusion coefficient of concrete. C denotes the chloride content. Considering the boundary condition of $C(0,t) = C_s$ and $C(\infty,t) = C_0$. As was widely known, according to Laplace transform, the analytical solution of above partially derivative equation was

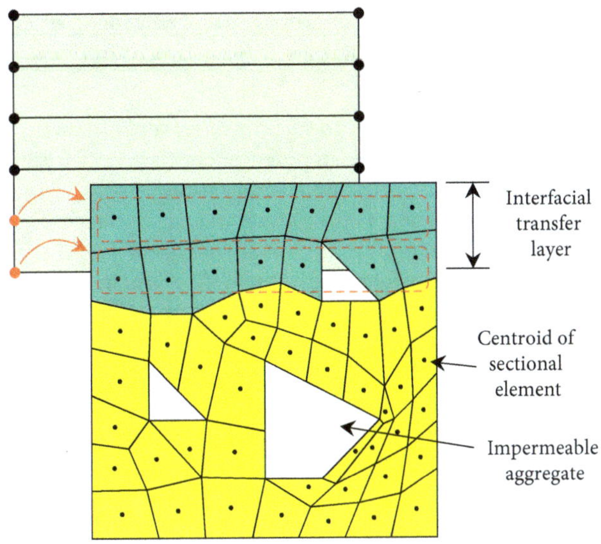

FIGURE 3: First step of conversion: calculation of nodal chloride content of compensation part from involved elements in core part (red dashed: coverage of nodes in compensation part; blue: involved elements in core part; yellow: uninvolved elements).

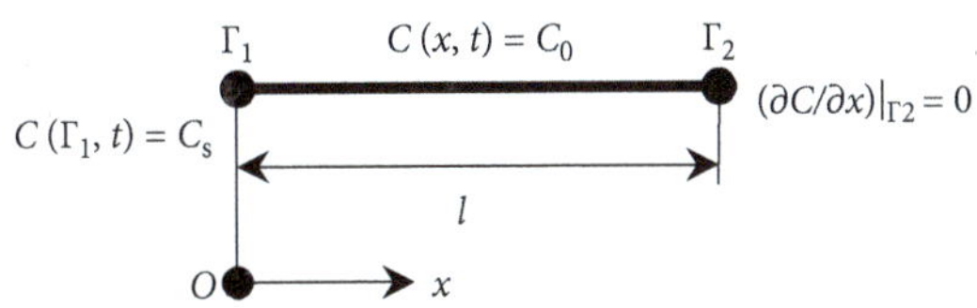

FIGURE 4: Analysis model for analytical deduction.

$$C(x,t) = C_0 + (C_s - C_0)\left[1 - \mathrm{erf}\left(\frac{x}{2\sqrt{Dt}}\right)\right], \tag{2}$$

where erf is Gauss error function. The limitation of the above solution is the premise of infinite space for boundary condition, which is suitable for concrete members with large thickness compared with thickness of concrete cover, but inaccurate for typical beams and slabs. For finite space, three-dimensional analytical solution by means of separation of variables was discussed and deduced by Li et al. [43]. For one-dimensional space, assuming the total length of diffusion space is l. x axis is the longitudinal direction of the one-dimensional model. Initial condition for diffusion space, $C(x,0) = C_0$ and boundary condition, exposed boundary $C(\Gamma_1, t) = C_s$, and sealing boundary $\partial C/\partial x|_{\Gamma_2} = 0$ were all considered. The schematic illustration is shown in Figure 4.

The simplified function of chloride content including the variables of time and depth in one-dimensional space according to Li's solution was expressed as follows:

$$C = C_s + (C_0 - C_s)\sum_{n=1}^{\infty}\frac{4}{(2n-1)\pi}\sin\left(\frac{2n-1}{2l}\pi x\right)e^{-D(((2n-1)/2l)\pi)^2 t}. \tag{3}$$

If zero chloride content for initial condition, $C_0 = 0$, was considered, equation (3) was simplified as given below:

$$C = C_s\left[1 - \sum_{n=1}^{\infty}\frac{4}{(2n-1)\pi}\sin\left(\frac{2n-1}{2l}\pi x\right)e^{-D(((2n-1)/2l)\pi)^2 t}\right]. \tag{4}$$

The aim of following section was to deduce the expression of analytical solution of chloride content with both exposed boundary and time-dependent boundary according to Li's deduction process and prove that the development in core model with predefined time-dependent boundary condition is same as the part in full model. For the consideration of compensation for chloride diffusion, a part of the full model was extracted and called as *Core Model*. The schematic illustration of core model was shown in Figure 5. The length of core model was l_1 while assuming the length of full model was l_2 $(l_2 > l_1)$.

Being different with the precondition of Li's solution [43], the boundary condition Γ_2 of core model was modified into the function of time-dependent boundary Γ_2' strictly defined by equation (4) set at l_1, expressed in equation (5). The other parameters and definition were the same as before. For simplification, zero chloride content for initial condition, $C_0 = 0$, was considered.

$$C(t)|_{\Gamma_2} = C_s\left[1 - \sum_{n=1}^{\infty}\frac{4}{(2n-1)\pi}\sin\left(\frac{2n-1}{2l_2}\pi l_1\right)e^{-D((2n-1)/2l_2\pi)^2 t}\right]. \tag{5}$$

Considering a temporary variable V,

$$V = C - C_s. \tag{6}$$

Apart from initial condition and boundary condition, the time-dependent boundary Γ_2' was expressed as

$$V|_{x=l_1} = -C_s\sum_{n=1}^{\infty}\frac{4}{(2n-1)\pi}\sin\left(\frac{2n-1}{2l_2}\pi l_1\right)e^{-D((2n-1)/2l_2\pi)^2 t}. \tag{7}$$

By means of separation of variables, V was expressed as

$$V(x,t) = X(x)\cdot T(t). \tag{8}$$

By introducing boundary condition of exposed surface Γ_1, the general solution of X was simplified as

$$X(x) = B\sin(\alpha x). \tag{9}$$

For time-dependent boundary Γ_2', by combining equations (7) and (9), we obtained the expression of α:

$$\alpha_n = \frac{2n-1}{2l_2}\pi, \quad (n = 1, 2, \ldots). \tag{10}$$

Thus, particular solution of X was

$$X_n(x) = k_n\sin\left(\frac{2n-1}{2l_2}\pi x\right). \tag{11}$$

The particular solution of T was solved as

$$T_n = -d_n e^{-D\alpha_n^2 t}. \tag{12}$$

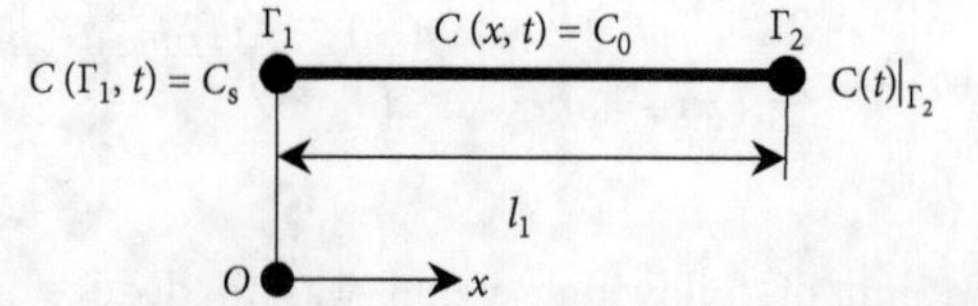

FIGURE 5: Illustration of core model for diffusion with compensation.

Thus, general solution of V was

$$V_n = e_n \sin\left(\frac{2n-1}{2l_2}\pi x\right)e^{-D\alpha_n^2 t}. \tag{13}$$

where e_n was the nth order of unknown expression and $e_n = k_n \cdot d_n$. Considering initial condition ($t = 0$), we obtained the following from equation (13):

$$\sum_{n=1}^{\infty} e_n \sin\left(\frac{2n-1}{2l_2}\pi x\right) = -C_s. \tag{14}$$

Due to orthogonality of the following function series,

$$\int_0^l \sin(\alpha_m x)\sin(\alpha_n x)dx = \begin{cases} 0, & m \neq n, \\ l/2, & m = n, \end{cases} \tag{15}$$

Combining equations (14) and (15), we obtained:

$$e_n = -\frac{4C_s}{(2n-1)\pi}. \tag{16}$$

By combining equations (6), (13), and (16), finally we obtained:

$$C = C_s\left[1 - \sum_{n=1}^{\infty}\frac{4}{(2n-1)\pi}\sin\left(\frac{2n-1}{2l_2}\pi x\right)e^{-D\left(\left((2n-1)/2l_2\pi\right)\right)^2 t}\right],$$

$$(x \in [0,\ l_1]). \tag{17}$$

Therefore, even considering the time-dependent boundary, the development of diffusion within core model was exactly the same as the original model, which proved the feasibility of compensation for chloride diffusion.

3. Mesoscopic Modeling Approach

3.1. Generation of Two-Dimensional Random Aggregate. In two-dimensional space, the major consideration of generating mesoscopic model is requirement of both grading of aggregates and its volume fraction in concrete. Generally, the mesoscopic structure of aggregate is randomly generated by dispersing particles in given shape and size in a limited square or rectangle space. This process was fulfilled by two essential steps, including constructing qualified aggregate and dispersing the aggregate. An applicable approach introduced in this paper was depicted as follows.

(1) Construction of aggregate in given shape and size regarding grading of concrete: the convex polygon of an applicable aggregate is connected along the points on the curve of a randomly generated ellipse, the area of which meets the requirement of design particle size. Then, the preliminary inscribed polygon is enlarged to match the same area of designed ellipse.

Two additional points to guarantee reasonable shape of aggregate should be noted. In practical concrete engineering, elongated aggregate (with dimension ratio over 2 : 1) should be avoided considering its poor structural performance. Thus, in aggregate generation program, all vertices of polygon are well distributed on ellipse in terms of the included angle defined by the connecting line of vertices of polygon and its centroid (Figures 6(a) and 6(b)). Another measurement is controlling the area occupancy of polygon within the design ellipse (Figures 6(c) and 6(d)). Detailed values for minimum included angle and minimum area occupancy are designed by domestic specification for aggregates.

(2) Dispersing aggregate by judgement of overlapping of aggregates: within two-dimensional space, for one newly dispersed aggregate, checking its involvement within given rectangle analysis region and overlapping with all existing aggregates are required. The first condition can be met by checking whether the maximum and minimum coordinates of polygon are covered by the rectangle. The detailed judgement for overlapping of convex polygon was described as follows. Terms quoted are shown in Figure 7.

  (i) For each pair of aggregates, judge whether sum of outer radius ($R_0 + R_0'$) is larger than distance of centroids (d_{agg}). Here, outer radius is defined as the maximum radical distance from any vertex to centroid. If no ($(R_0 + R_0') \le d_{agg}$), there is no overlapping of aggregate and the newly dispersed aggregate is accepted. If yes ($(R_0 + R_0') \ge d_{agg}$), possible overlapping cannot be avoided and continue to next step for further check.
  (ii) Judge whether sum of inner radius ($R_i + R_i'$) is smaller than distance of centroids (d_{agg}). Here, inner radius is defined as the minimum radical distance from any side to centroid. If no ($(R_i + R_i') > d_{agg}$), solid overlapping appears and a redispersion of new aggregate is required. If equal ($(R_i + R_i') = d_{agg}$), a rare case representing parallel sides of aggregates is inappropriate which is also unaccepted. If yes ($(R_i + R_i') < d_{agg}$), intersected sides cannot be avoided continue to next step for further check.
  (iii) Judge whether the vertices of polygon representing aggregate are inclusion of the existing aggregate (Figure 8). If yes, overlapping is proved and a redispersion of new aggregate is required. If no, a complicated case of overlapping that intersection polygons without any vertex located within each polygon still cannot be assured and continue to next step.
  (iv) The last step is judging whether there are any shared nodes of background meshing grid for both polygons (Figure 9). If yes, overlapping is proved and a redispersion of new aggregate is

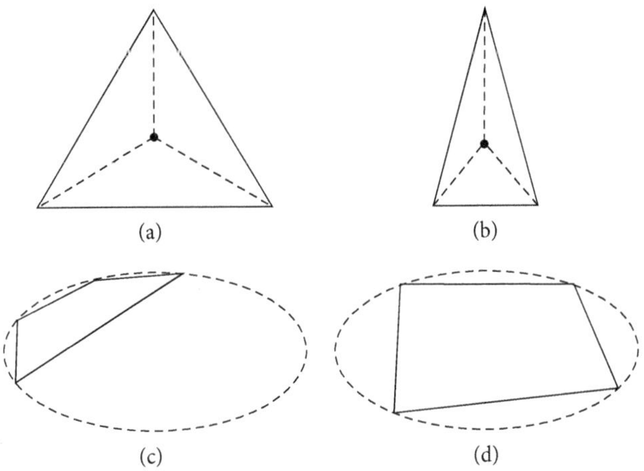

Figure 6: Generating polygon of aggregate based on ellipse. (a) Acceptable aggregate. (b) Elongated aggregate. (c) Low area proportion. (d) Fair area proportion.

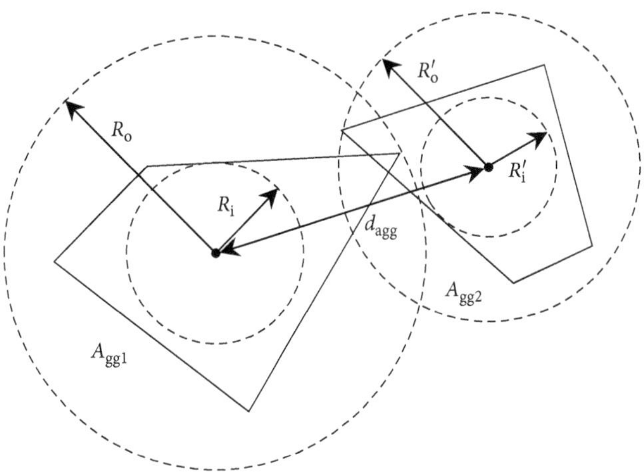

Figure 7: Inner radius, outer radius, and distance of centroids of two intersected aggregates.

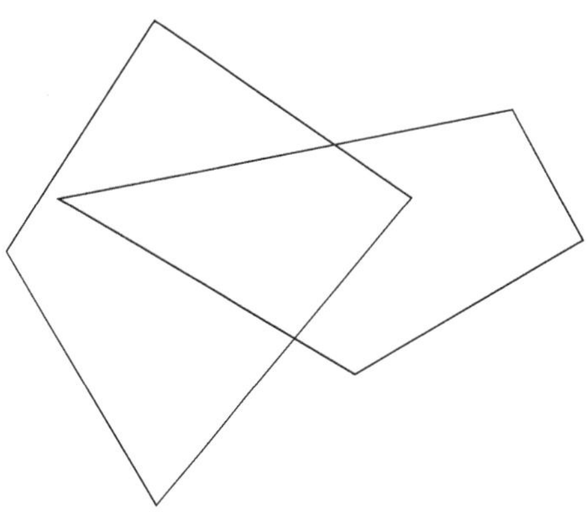

Figure 8: Inclusion of vertices owned by dispersed aggregate in existing aggregate.

required. If no, the newly dispersed aggregate is accepted.

The above process was illustrated in Figure 10. Two typical models with elongated aggregate are shown in Figure 11.

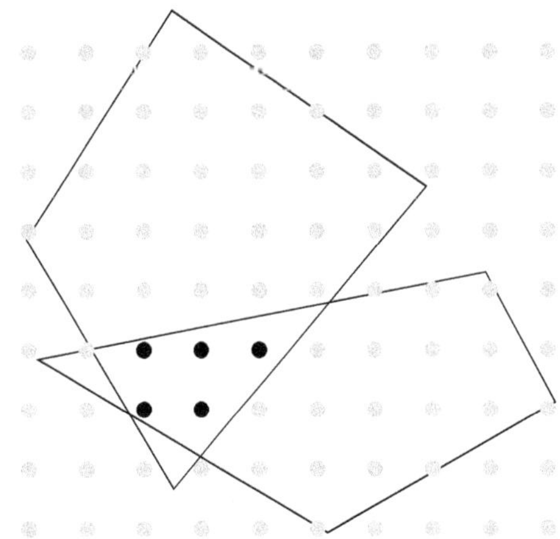

Figure 9: Shared nodes of background meshing grid for both polygons.

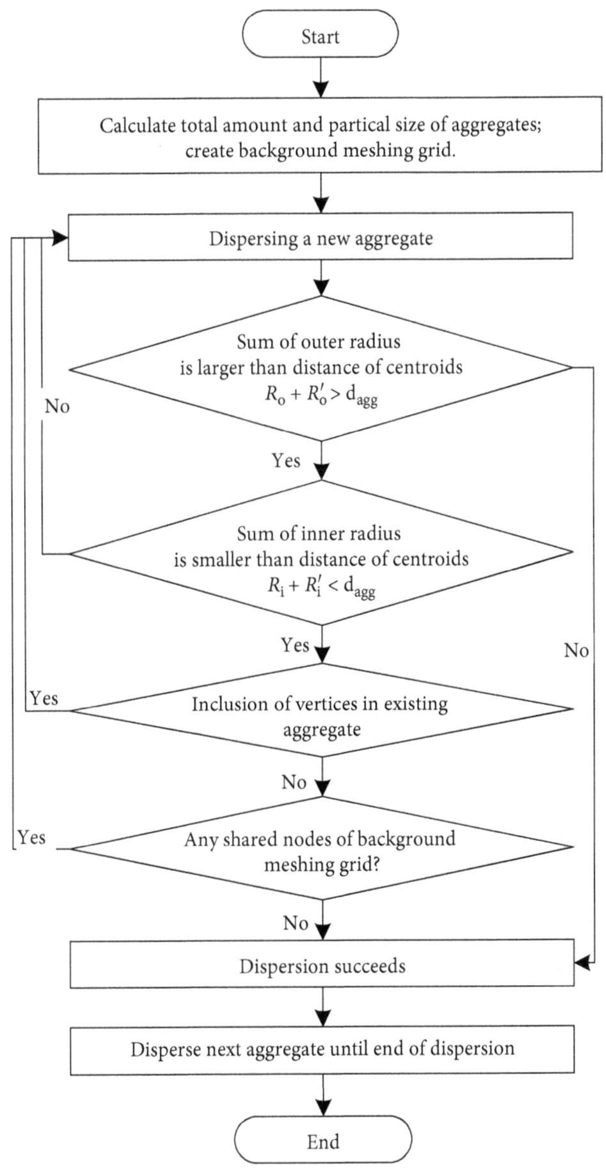

Figure 10: Process of generating random polygonal aggregates.

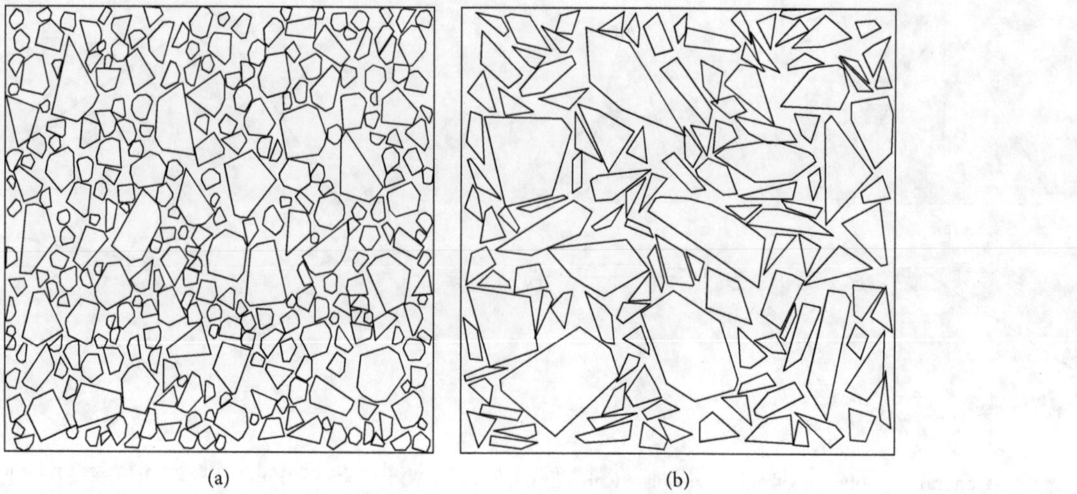

FIGURE 11: Generated aggregates with limitation on dimension ratio. (a) Allowable dimension ratio up to 2 : 1. (b) Allowable dimension ratio up to 5 : 1.

Additional process for modeling is meshing the geometry space including significantly irregular distribution of aggregate. Conventional commercial software was adopted for meshing process. However, low success rate of meshing in refined quadrilateral element cannot be avoided for most commercial software due to the complex model. Thus, an optimized way adopted in this paper was meshing the space in triangle element and subdividing one triangle into four small quadrilaterals, which was proved as high meshing success rate and high efficiency.

3.2. Modeling of ITZ Based on Meshed Space. Modeling approach for ITZ was determined by the mechanism of simulating chloride diffusion within ITZ. ITZ was recognized as an indeterminate zone around aggregates and bulk, in linearly decreasing porosity with increase of distance from aggregate surface. Existing approaches describe ITZ as an individual layer or compensation on equivalent diffusivity of aggregate [44–46]. In terms of aggregates in random polygon, the way of building actual elements of ITZ was adopted in this paper. Thus, more DOFs were introduced, which significantly increased time consumption of assembling global diffusion matrix and solving it.

Since ITZ was identified as the individual phase surround aggregate, additional uniform width of elements adhering on the edge of aggregates might induce unnecessary overlapping of ITZ elements in some cases that the pair of aggregate was too close to each other. In order to avoid the above circumstance, ITZ elements were assigned on the inner side of edge of aggregate instead of outer side. According to minor area proportion of ITZ, it was acceptable slightly decreasing the size of aggregate. There are totally three steps for this process.

(i) For the first step, for any aggregate, the lines on its edge were parallel offset inside by the designed width, h, shown as the dashed line in Figure 12(a).

(ii) For the second step, all intersection of the offset lines were found, the total number of which were equal to vertices of aggregate, shown as the black point in Figure 12(b). Additionally, a special subroutine called "func_L2L" was developed to identify the relationship of a pair of line segments. Its output result provided whether the two line segments intersect and the coordinate of their intersection.

(iii) For the third step, combining the previous available nodes of aggregate and newly generated nodes (intersection), a group of quadrilateral elements was built in counterclockwise order, which represented ITZ. Their physical properties were set individually being different from existing cement paste.

The above entire process of modeling FE model of concrete involving ITZ is shown in Figure 13.

4. Basis of Assessing Approach

4.1. Diffusion Model of Chloride. Transportation of chloride ion in porous medium such as cementitious materials is mainly a diffusive phenomenon due to concentration gradients [47]. The diffusion of chloride ion within concrete components was admitted as a complex process, which was influenced by many factors including water/cement ratio, porosity of cement, additive, aggregate, temperature, humidity, environmental chloride content, binding effect of binder, and hydration of cement. The above process could be simulated by numerical approach based on available environmental attacking condition and material properties. Fick's second law was widely adopted as the expression of principle differential equation [48]:

$$\omega_e \frac{\partial C_f}{\partial t} = \frac{\partial}{\partial x} D_e \omega_e \frac{\partial C_f}{\partial x} - \frac{\partial C_b}{\partial t},$$
$$C_t = C_b + C_f \cdot \omega_e, \tag{18}$$

where C_f is the free chloride content (in kg/m^3 of solution); C_b is the bound chloride content (in kg/m^3 of concrete); C_t is the total chloride content (in kg/m^3 of concrete); D_e is the

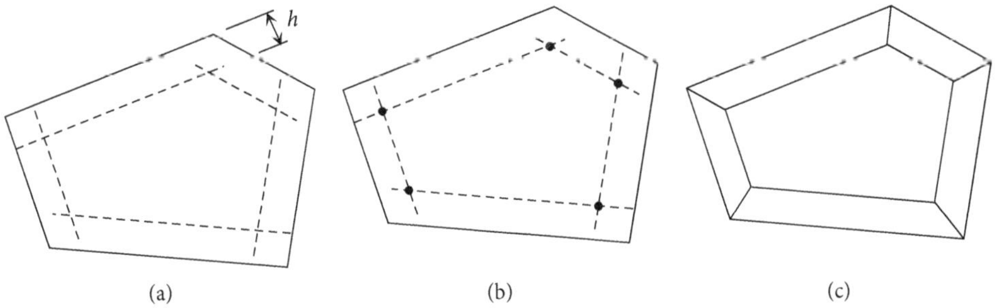

FIGURE 12: Process of generating new elements and nodes for ITZ. (a) Step 1: offset outline with ITZ width, h. (b) Step 2: find intersection (nodes) of outline. (c) Step 3: form elements of ITZ.

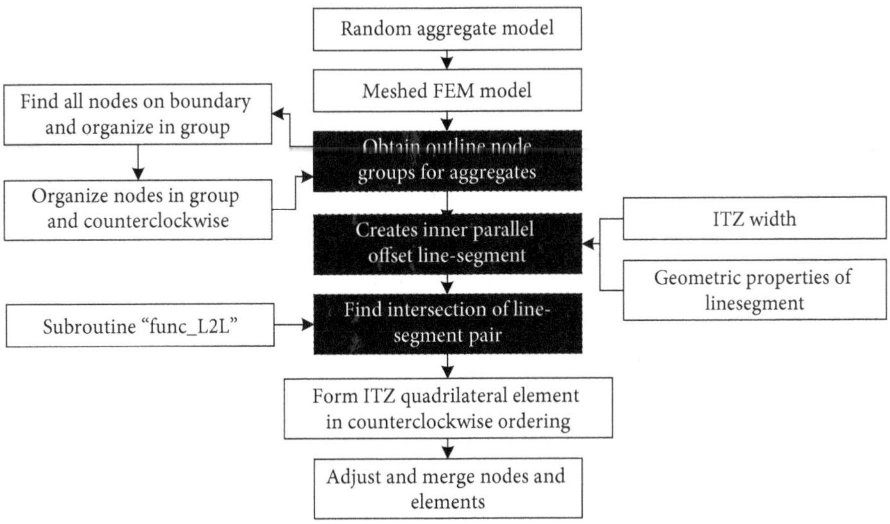

FIGURE 13: Flow chart for modeling of ITZ based on meshed FEM model.

effective diffusivity; and ω_e is the evaporable water content (in volume percentage of concrete). Considering complexity of cross sections of bridge pier, appropriate preliminary division and meshing grid are processed by means of some meshing tools or commercial FE software. Experimental data indicated that due to cement hydration and chloride content, the chloride diffusion coefficient is strongly dependent on the exposure period of concrete [49].

4.2. Corrosion Model of Rebar. Corrosion of rebar initiates in the form of galvanic reaction between ferrous and oxygen accelerated by the presence of active free chloride ion once the protection of high pH (>12.5) hydration products was depassivated. Corrosion continues with the cycle of electric current between cathode and anode, which was influenced by high chloride concentration at the level of the rebar, environmental temperature, electrical resistivity of concrete, and hydration of cement. The corrosion model regressed by Liu and Weyers [34] based on experimental data included the influence of chloride content, temperature, concrete cover resistance, and time of cement hydration, which was expressed as

$$\ln(1.08 i_{\text{corr}}) = 8.37 + 0.618 \cdot \ln(1.69\text{Cl}) - \frac{3034}{T} - 0.000105 R_c + 2.32 t^{-0.215}, \tag{19}$$

where i_{corr} is corrosion current intensity (μA/cm^2); Cl is chloride content (kg/m^3), which was obtained from numerical simulation result of chloride diffusion; T is temperature at the depth of steel surface (in degree Kelvin); R_c is the resistance of the cover concrete (ohms); and t is corrosion time duration (years). With the data chloride content obtained by numerical simulation and the other parameters assumed previously, the real-time corrosion current rate at each time step of numerical simulation was calculated by equation (19). Thus, the corrosion depth on surface of rebar was calculated as [50]

$$D_{t+\Delta t} = D_t - 0.023 \cdot i_{\text{corr}} \Delta t, \tag{20}$$

where D_t is remaining rebar diameter (mm) at t years, here 20 mm adopted; $D_{t+\Delta t}$ is reduced rebar diameter (mm) at $t + \Delta t$ years; and $\Delta D|_{\Delta t}$ is reduction of rebar diameter (mm) for the period of propagation for the duration time of Δt years.

5. Numerical Simulation for Diffusion of Chloride Including Compensation

5.1. Influence of ITZ Properties regarding Compensation. Considering that characteristics of ITZ significantly influence diffusion chloride within concrete, critical properties of

ITZ were detailed studied in this section, including diffusivity and width of ITZ, aggregate volume fraction, and grade of aggregate. Furthermore, chloride content on surface of rebar was also researched regarding presence of ITZ.

Another major consideration of this paper was the effect of compensation in multiscale modeling. A 100×100 mm two-dimensional square space was modeled for mesoscopic core part, and another 100×200 mm rectangle space was modeled for macroscopic compensation part (Figure 14). Width of ITL was set at 10 mm, which means 10% area of core part and 5% area of compensation part were overlapped. On the other hand, for the case without compensation, macroscopic compensation part was deleted and chloride ion can only diffuse within smaller mesoscopic core part.

For the other parameters, transportation mechanism of chloride ion between both parts was described in previous section of this paper. For all simplified cases studied in this section, the analytical apparent surface chloride content of 0.71 ($\%w_c$) and apparent concrete diffusivity of $1.84 \times 10^{-12}\,m^2/s$ were adopted for macroscopic model, which were regressed based on Zhao's data [51]. Meanwhile, the aggregate was considered as impermeable and corresponding modeling of FEM was ignored. Boundary condition and diffusivity of cement paste were calculated by these results. Regarding generating random aggregate of mesoscopic model, Fuller's curve was adopted as the grading of aggregate. Both types of cases with compensation and without compensation were studied.

Significant random distribution of aggregate was the major factor for uncertainty of chloride diffusion. Thus, totally, 100 specimens of aggregate model for each case were generated and analyzed considering reasonable accurate and acceptable time consumption. After all specimens were analyzed, mean results of these cases were calculated for further comparison.

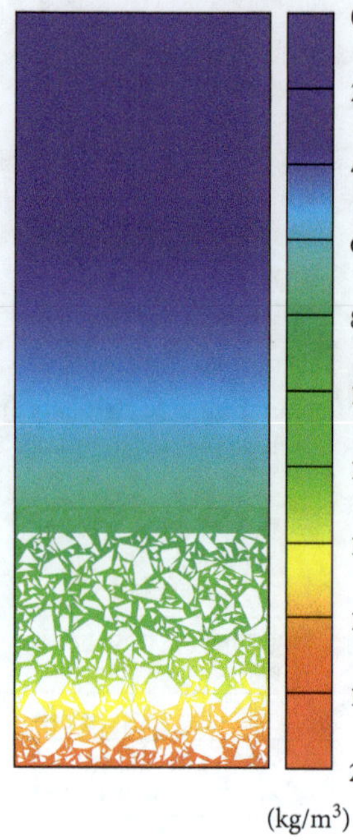

FIGURE 14: Contour of chloride content of multiscale model.

5.1.1. Diffusivity of ITZ. Firstly, diffusivity of ITZ was studied, which was represented by the ratio of D_{ITZ}/D_{CP}. Here, D_{ITZ} and D_{CP} denoted the diffusivity in ITZ and cement paste, respectively. As was discussed by Zheng et al. [4], D_{ITZ}/D_{CP} would vary within 1~5 considering different configuration of w/c, aggregate size, hydration time, and aggregate fraction theoretically, which were discussed by means of the approach introduced in this paper. 50 μm of ITZ width was modeled by expanding the edge of aggregate and rebar. Maximum aggregate size in 20 mm was considered.

Figure 15 showed typical distribution of chloride content regarding different diffusivity in ITZ. The major difference of contour line observed was the slightly higher chloride content due to the presence of ITZ in width of the micrometer scale. It means higher diffusivity in ITZ compared with cement paste accelerated the diffusion of chloride within concrete.

For mesoscopic numerical simulation, distribution of aggregate was recognized as the major influence factor for development of chloride [52]. Chloride content on surface of rebar significantly varied with different arrangement and grading of aggregate. Thus, both mean and maximum distribution of chloride ion in terms of random aggregate models became the major consideration. Figures 16 and 17 summarized average content and maximum versus depth regarding diffusivity of ITZ at 50 years and 100 years, respectively. It was clearly noted that chloride content increased as the diffusivity of ITZ increased. An interesting phenomenon observed was that the influence of compensation was larger in higher chloride content. It attributed to a larger diffusion space released in mixed model including compensation part. From the result of numerical simulation, another point should be noted that within the boundary layer in width of 10 mm from surface of concrete, both mean and maximum chloride content were closed to boundary condition, which was recognized as the lower aggregate volume fraction and smaller particle size in this region.

5.1.2. ITZ Width. The second major factor for chloride diffusion is ITZ width. A thin layer composed of quadrilateral elements in the width of 5 μm, 10 μm, 20 μm, 35 μm, and 50 μm adhering to the edge of aggregates was modeled (Figure 18). For the other parameters, diffusivity of ITZ in $D_{ITZ}/D_{CP} = 5$ was adopted. Maximum particle size of coarse aggregate was limited within $D_{max} = 20$ mm. Figure 18 showed ITZ elements in orange color of FE model in different width of ITZ.

Figures 19 and 20 showed average content and maximum versus depth regarding width of ITZ at 50 years and 100 years, respectively. A clear conclusion was drawn that thicker ITZ width induced higher volume fraction of ITZ and enhanced diffusion of chloride ion.

5.1.3. Aggregate Volume Fraction. Aggregate volume fraction determined total amount of aggregates within given space and accordingly total length of perimeter was also determined, which was the key factor for proportion of ITZ

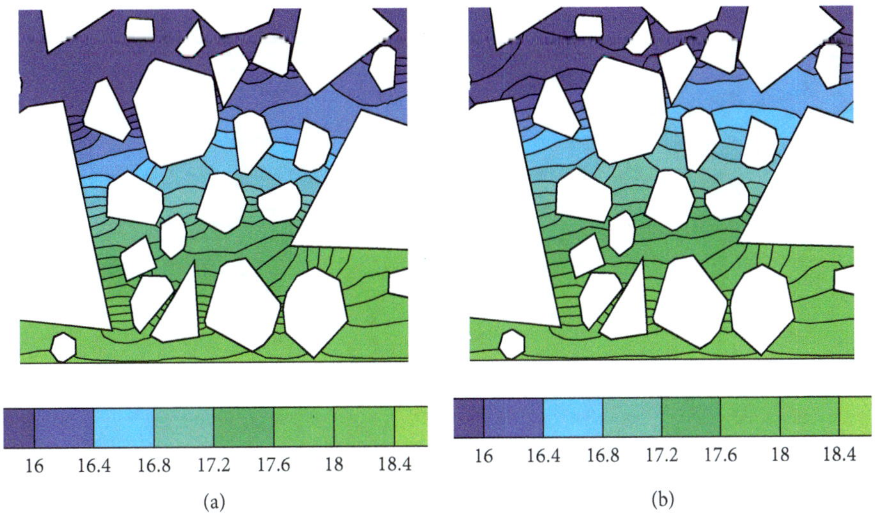

FIGURE 15: Typical distribution of chloride content regarding diffusivity in ITZ (solid line denotes isoline of chloride content). (a) $D_{ITZ}/D_{CP} = 1$. (b) $D_{ITZ}/D_{CP} = 5$.

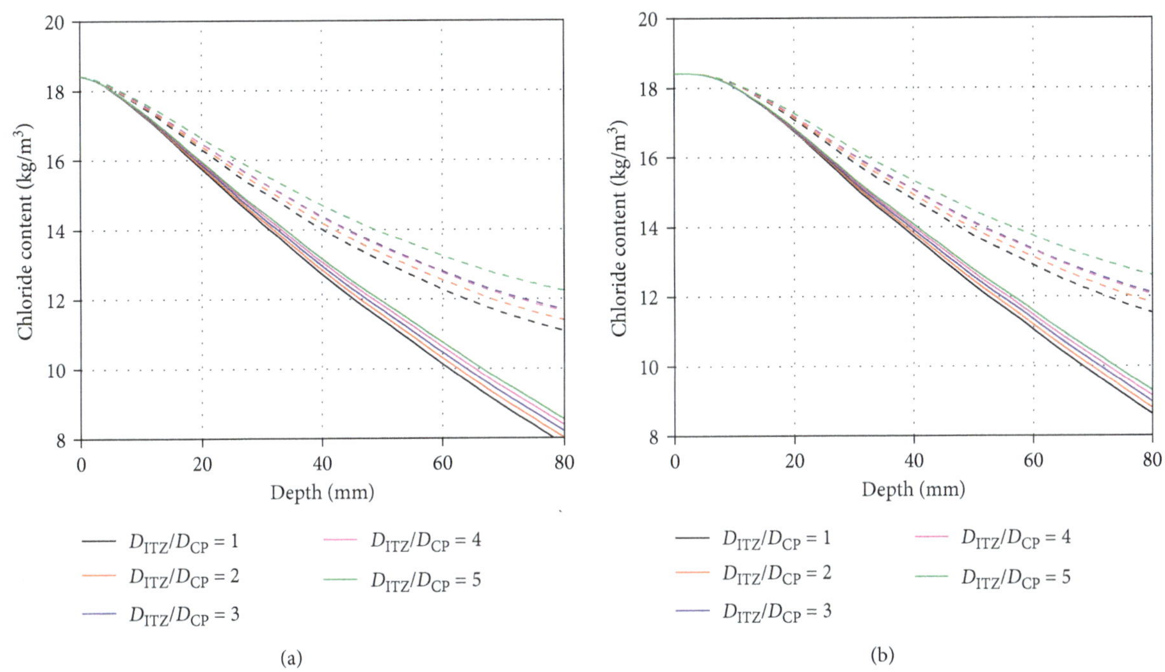

FIGURE 16: Average content and maximum vs. depth regarding compensation and ITZ D_{ITZ}/D_{CP} at 50 y of exposure (dashed: w/o compensation; solid: with compensation). (a) Mean chloride content. (b) Maximum chloride content.

in concrete. In this section, the cases of aggregate volume fraction in 20%, 30%, 40%, 50%, 60%, and 70% were discussed by means of the approach introduced in this paper. Diffusivity of ITZ in $D_{ITZ}/D_{CP} = 5$ and 50 μm of ITZ width were adopted. Maximum aggregate size in 20 mm and Fuller grading was adopted. Both types of cases with compensation and without compensation were studied. Figure 21 showed with aggregate volume fraction from 20% to 70%, the typical distribution of chloride content and aggregate with given analysis square space.

Figure 22 described mean content and maximum versus depth regarding aggregate volume fraction at 100 years. According to the result, presence of ITZ provided greater effect on higher aggregate volume fraction and thus reduced the difference of chloride content profile in cases with different aggregate volume fraction.

5.1.4. Grade of Aggregate. The cases for grade of aggregate with maximum aggregate size $D_{max} = 5$ mm, 10 mm, 15 mm, 20 mm, 25 mm, and 30 mm were discussed. Both cases with or without ITZ were included. Fuller grading was adopted. For the case with ITZ, diffusivity of ITZ in $D_{ITZ}/D_{CP} = 5$ was adopted. 50 μm of ITZ width was modeled by expanding the

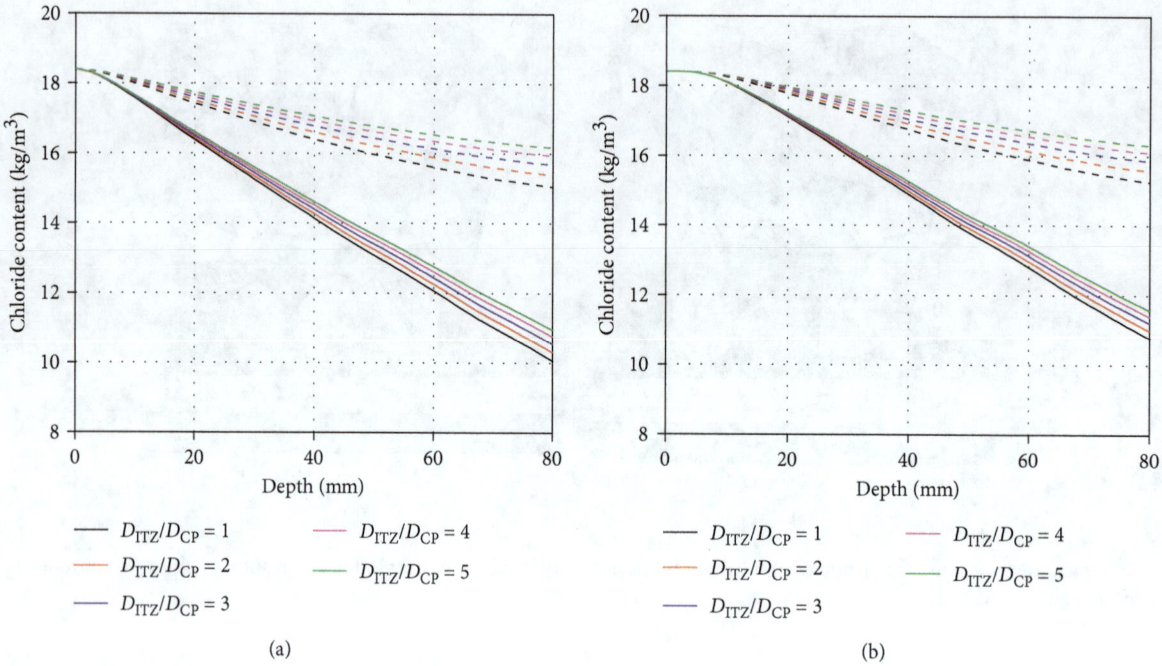

FIGURE 17: Average content and maximum vs. depth regarding compensation and ITZ D_{ITZ}/D_{CP} at 100 y of exposure (dashed: w/o compensation; solid: with compensation). (a) Mean chloride content. (b) Maximum chloride content.

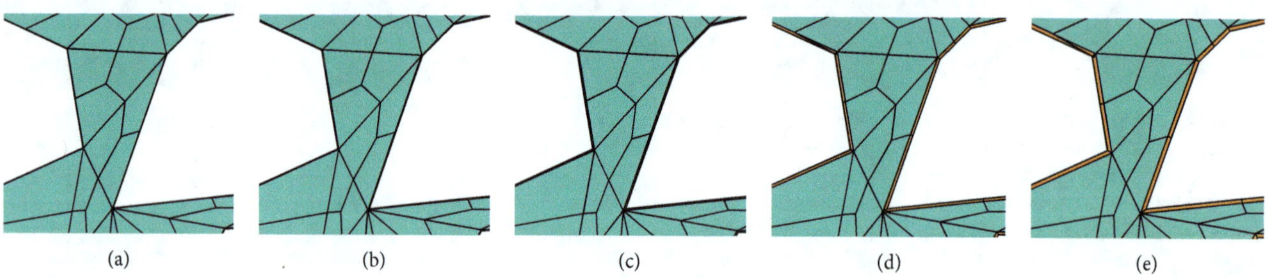

FIGURE 18: FE model for cement paste and ITZ regarding ITZ width, h (blue: cement paste; orange: ITZ). (a) $h = 5\,\mu$m. (b) $h = 10\,\mu$m. (c) $h = 20\,\mu$m. (d) $h = 35\,\mu$m. (e) $h = 50\,\mu$m.

edge of aggregate and rebar. Both types of cases with compensation and without compensation were studied. Figure 23 showed with maximum aggregate size from 5 mm to 30 mm, the typical distribution of chloride content and aggregate with given analysis square space. Figure 24 showed mean content and maximum versus depth regarding grade of aggregate at 100 years.

Figure 24 summarized mean and maximum chloride content vs. depth from surface of concrete. An interesting point should be noted that the lower aggregate volume fraction and smaller particle size in the region closed to boundary was not significant in the case with smaller maximum particle size. Moreover, presence of ITZ indeed enhanced diffusion of chloride ion especially in cases with smaller maximum particle size.

5.2. Chloride Content and Corrosion of Steel Rebar. In this section, chloride content on the front of rebar in terms of the configuration was studied by the multiscale numerical simulation introduced previously. According to general engineering design, a group of concrete cover thickness and rebar diameter was devised for detailed comparison of chloride diffusion and rebar corrosion, shown in Table 1. For the purpose of comparison, rebar in diameter of 16 mm was selected for the cases in different cover thickness and 50 mm cover thickness for the comparison of rebar diameters. By means of the random aggregate generation approach and considering reasonable computing loading, 100 two-dimensional samples for each combination of concrete cover thickness and rebar diameter was modeled considering random distribution of aggregate.

Considering three-phase composite of concrete, nodes and elements of ITZ were modeled not only surrounding aggregate, but also on the surface of steel rebar. It also justified to a certain extent that the formation of ITZ on surface of steel rebar in the light of the same mechanism of aggregate. Therefore the same width and diffusivity of ITZ on surface of steel were adopted. Chloride content on front of rebar was extracted from the newly created nodes on the edge of ITZ elements in terms of the neat surface of steel rebar.

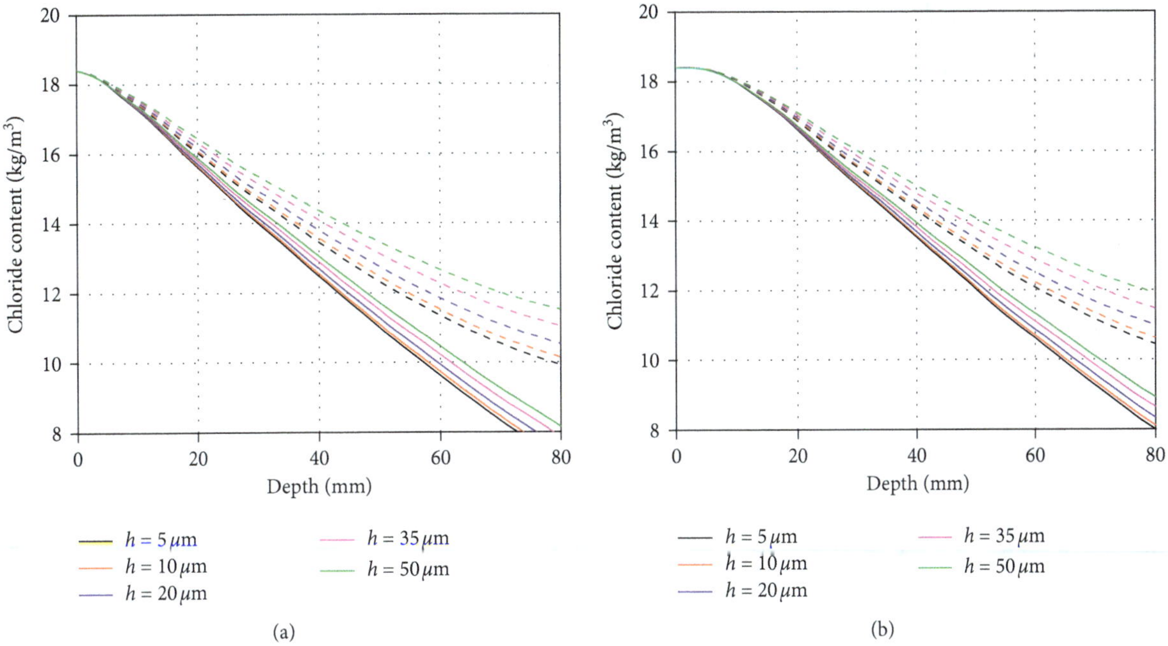

FIGURE 19: Average content and maximum vs. depth regarding compensation and ITZ width at 50 y of exposure (dashed: w/o compensation; solid: with compensation). (a) Mean chloride content. (b) Maximum chloride content.

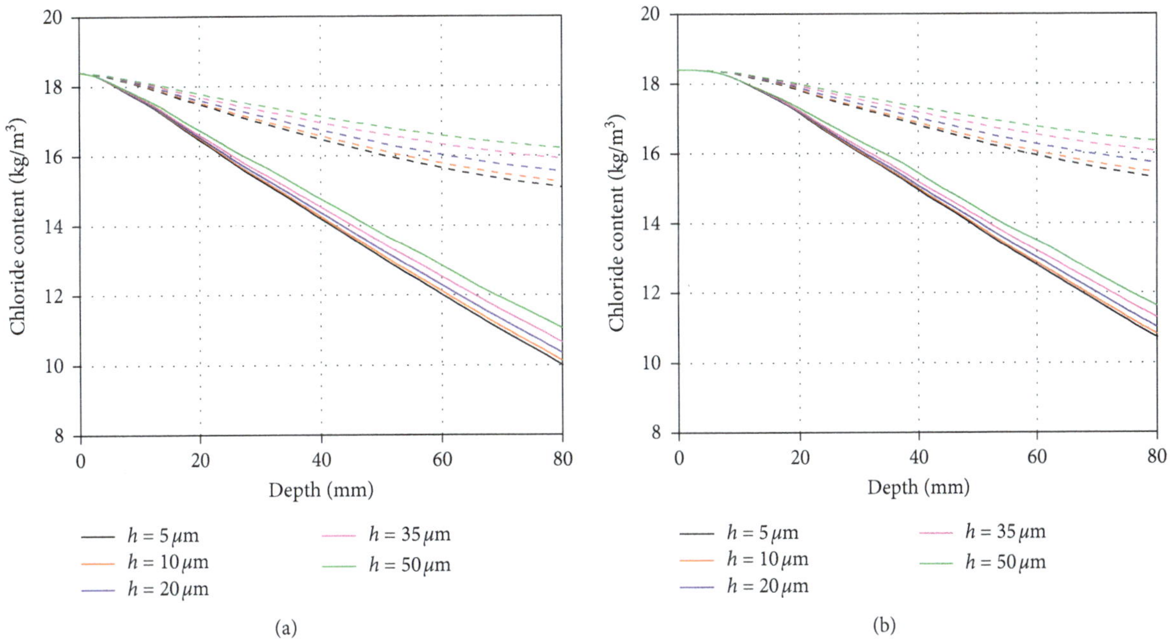

FIGURE 20: Average content and maximum vs. depth regarding compensation and ITZ width at 100 y of exposure (dashed: w/o compensation; solid: with compensation). (a) Mean chloride content. (b) Maximum chloride content.

The placement of steel rebar on the center of straight edge was considered. Rebar placed in the center of bottom side of concrete section (Figure 25(a) listed in Table 1 was tested. Term d denotes the cover thickness and the diameter; D of circle refers to diameter of rebar. A two-dimensional mesoscopic core model in the size of 100×100 mm was created. In order to guarantee the effect of compensation, 100×200 mm total size of compensation model was designed as two times of core model. 10 mm thickness of ITL was considered for compensation process. Apart from ITZ elements surrounding aggregates, new elements adjacent to steel rebar were also modeled by expanding outline. Therefore, chloride content on surface of rebar was extracted from the newly built nodes of ITZ. Both types of cases with compensation and without compensation were studied. Diffusivity of ITZ in $D_{ITZ}/$

Figure 21: FE model and typical distribution of chloride content for different aggregate volume fraction, fa. (a) fa = 20%. (b) fa = 30%. (c) fa = 40%. (d) fa = 50%. (e) fa = 60%. (f) fa = 70%.

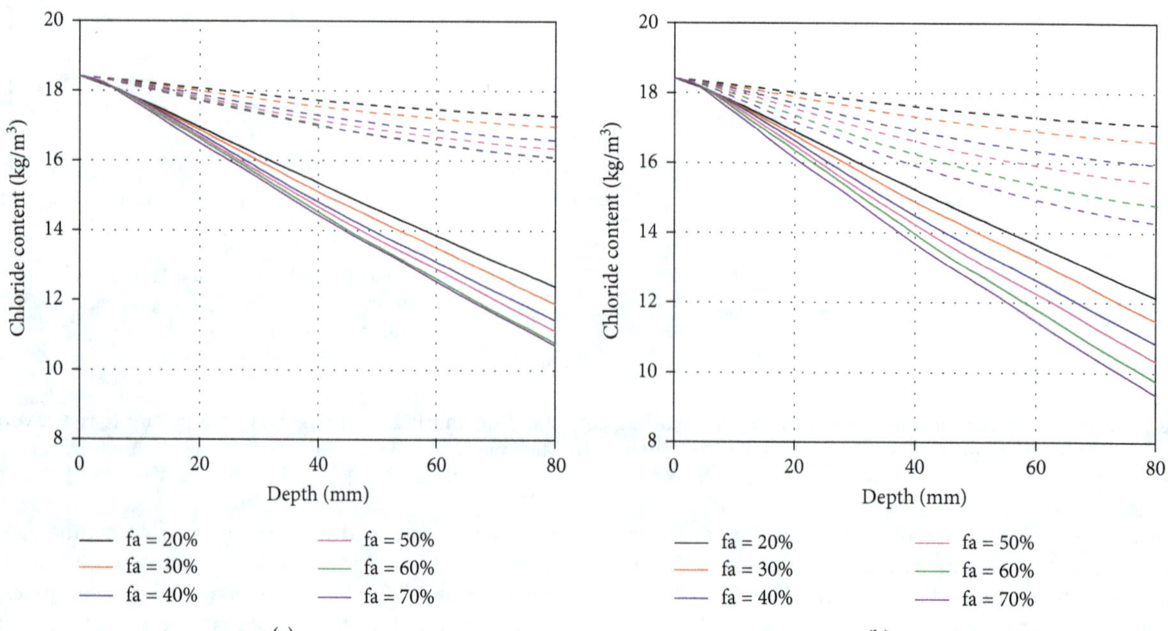

Figure 22: Continued.

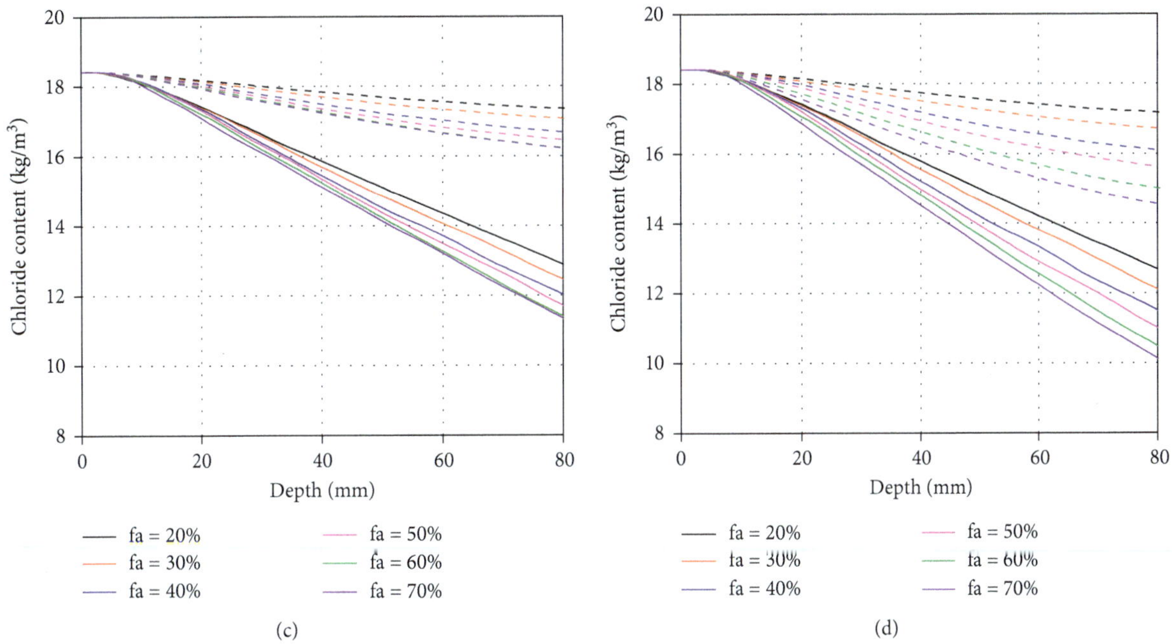

FIGURE 22: Average content and maximum vs. depth regarding aggregate volume fraction (fa) and presence of ITZ at 100 y of exposure (dashed: w/o compensation; solid: with compensation). (a) Mean chloride content with ITZ. (b) Mean chloride content w/o ITZ. (c) Max chloride content with ITZ. (d) Max chloride content w/o ITZ.

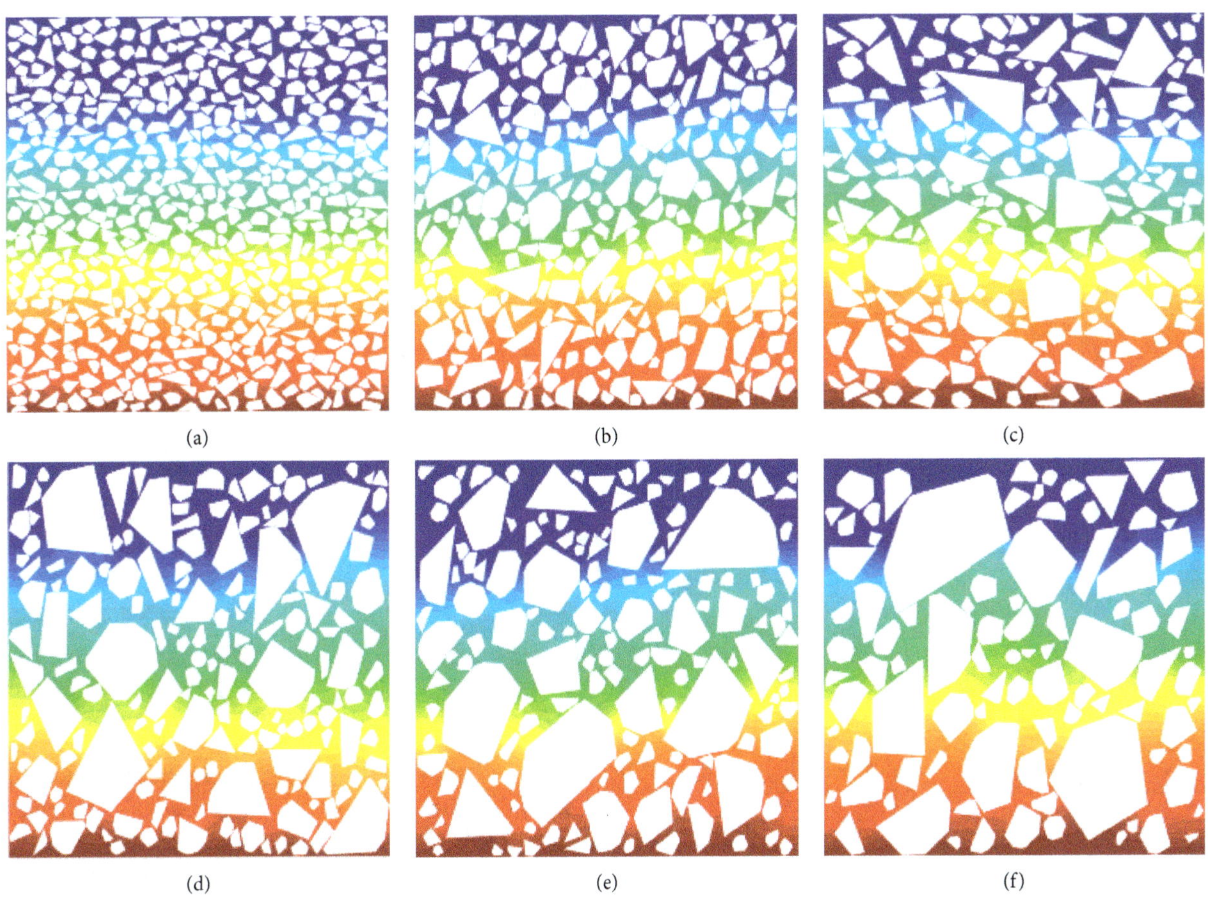

FIGURE 23: FE model and typical distribution of chloride content for different aggregate and gradation. (a) $D_{max} = 5$ mm. (b) $D_{max} = 10$ mm. (c) $D_{max} = 15$ mm. (d) $D_{max} = 20$ mm. (e) $D_{max} = 25$ mm. (f) $D_{max} = 30$ mm.

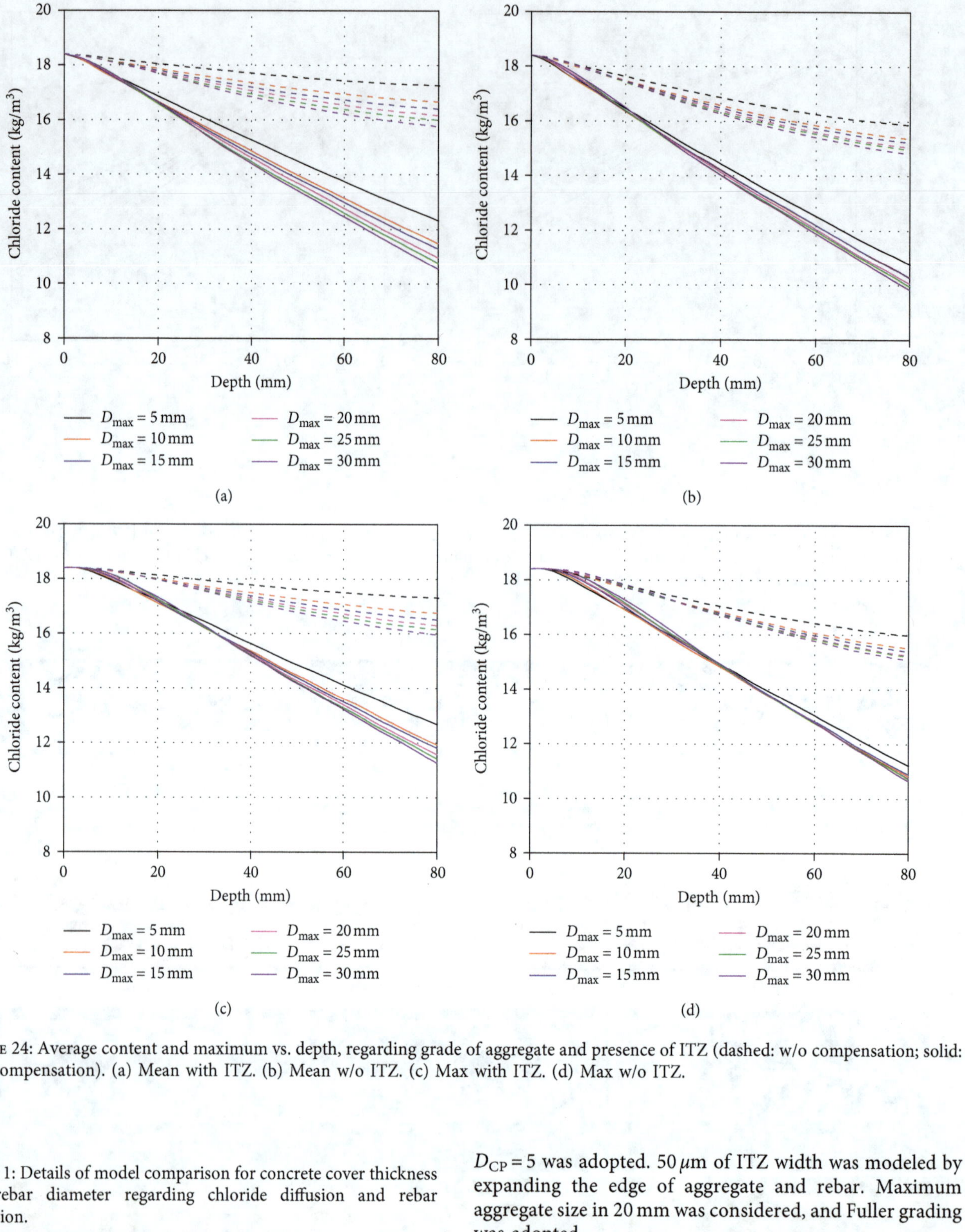

FIGURE 24: Average content and maximum vs. depth, regarding grade of aggregate and presence of ITZ (dashed: w/o compensation; solid: with compensation). (a) Mean with ITZ. (b) Mean w/o ITZ. (c) Max with ITZ. (d) Max w/o ITZ.

TABLE 1: Details of model comparison for concrete cover thickness and rebar diameter regarding chloride diffusion and rebar corrosion.

Cover thickness, C (mm)	Rebar diameter, D (mm)				
	12	16	20	25	28
20		×			
30		×			
40		×			
50	×	×	×	×	×
60		×			

D_{CP} = 5 was adopted. 50 μm of ITZ width was modeled by expanding the edge of aggregate and rebar. Maximum aggregate size in 20 mm was considered, and Fuller grading was adopted.

Figure 25(b) showed typical distribution of chloride content within core mesoscopic model and compensated macroscopic model at 100 years of exposure period for the cases of rebar placed in center of bottom. The contour clearly described sound transition of chloride content near ITL on both cases and proved the effect of compensation as well.

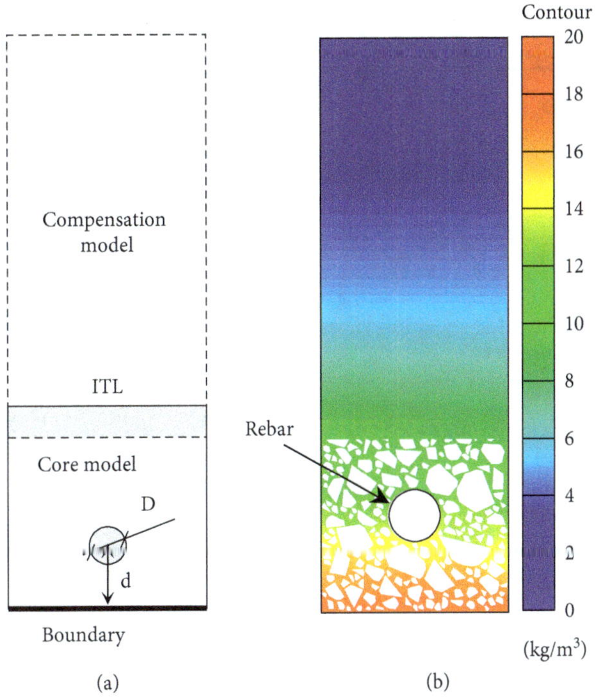

FIGURE 25: Geometrical configuration and chloride content distribution of compensation case. (a) Geometrical configuration of concrete analysis space and rebar. (b) Typical distribution of aggregate and chloride content in concrete, 100 y.

The circle curve of rebar was meshed into a series of node, and nodal chloride content was obtained from result of FE analysis. For all 100 samples of each research case, mean value of chloride content on each node around rebar was calculated and plotted in polar coordinate. Figure 26 showed the polar distribution of mean chloride content on cross section of rebar at 20 years (solid) and 100 years (dashed) of exposure period. Depending on whether compensation model or ITZ was considered, chloride content was plotted in different colors, shown in legend on the bottom. Distribution of chloride content developed towards lower-middle, which matched the configuration of boundary condition. Result indicated higher chloride content was obtained for longer period of exposure. Moreover, lower value was observed in the cases with compensation.

Mean chloride content vs. cover thickness for 16 mm rebar and mean chloride content vs. diameter for 50 mm cover are summarized in Figure 27. An obvious result showed lower chloride content in compensation case since solid lines were all under corresponding dashed lines in the figures, which attributed the larger diffusion space in compensation cases.

Another observation was the insignificant difference between the solid line of "ITZ" and "ITZ + Compensation" (20 years). Since diffusion of chloride did not exceed the border of core part and compensation part, there was no clear difference for both diffusion spaces of "ITZ" and "ITZ + Compensation," which is the major reason.

The corrosion model regressed by Liu and Weyers [34] based on experimental data included the influence of chloride content, temperature, concrete cover resistance, and time of cement hydration, which was expressed as

$$\ln(1.08 i_{\mathrm{corr}}) = 8.37 + 0.618 \cdot \ln(1.69\mathrm{Cl}) - \frac{3034}{T} - 0.000105 R_{\mathrm{c}} + 2.32 t^{-0.215}, \tag{21}$$

where i_{corr} is corrosion current intensity (μA/cm^2); Cl is chloride content (kg/m^3), which was obtained from numerical simulation result of chloride diffusion; T is temperature at the depth of steel surface (in degree Kelvin); R_{c} is the resistance of the cover concrete (ohms); and t is corrosion time duration (years). For corrosion calculation in the following discussion, environmental temperature 293 K was adopted. A generally adopted value of concrete resistance 1500 Ω was adopted. Besides, pitting of rebar was neglected. With the data chloride content obtained by numerical simulation, the real-time corrosion current rate at each time step of numerical simulation was calculated. The corrosion depth on surface of rebar was calculated as [50]

$$D_{t+\Delta t} = D_t - 0.023 \cdot i_{\mathrm{corr}} \Delta t, \tag{22}$$

where D_t is remaining rebar diameter (mm) at t years, here 20 mm adopted and $D_{t+\Delta t}$ is reduced rebar diameter (mm) at $t + \Delta t$ years.

A single mean chloride content for each time step was calculated as the attacking chloride content for rebar, and mean corrosion depth of rebar was calculated based on equation (19). Figure 28(a) (cover thickness) and Figure 28(b) (rebar diameter) showed corrosion process and

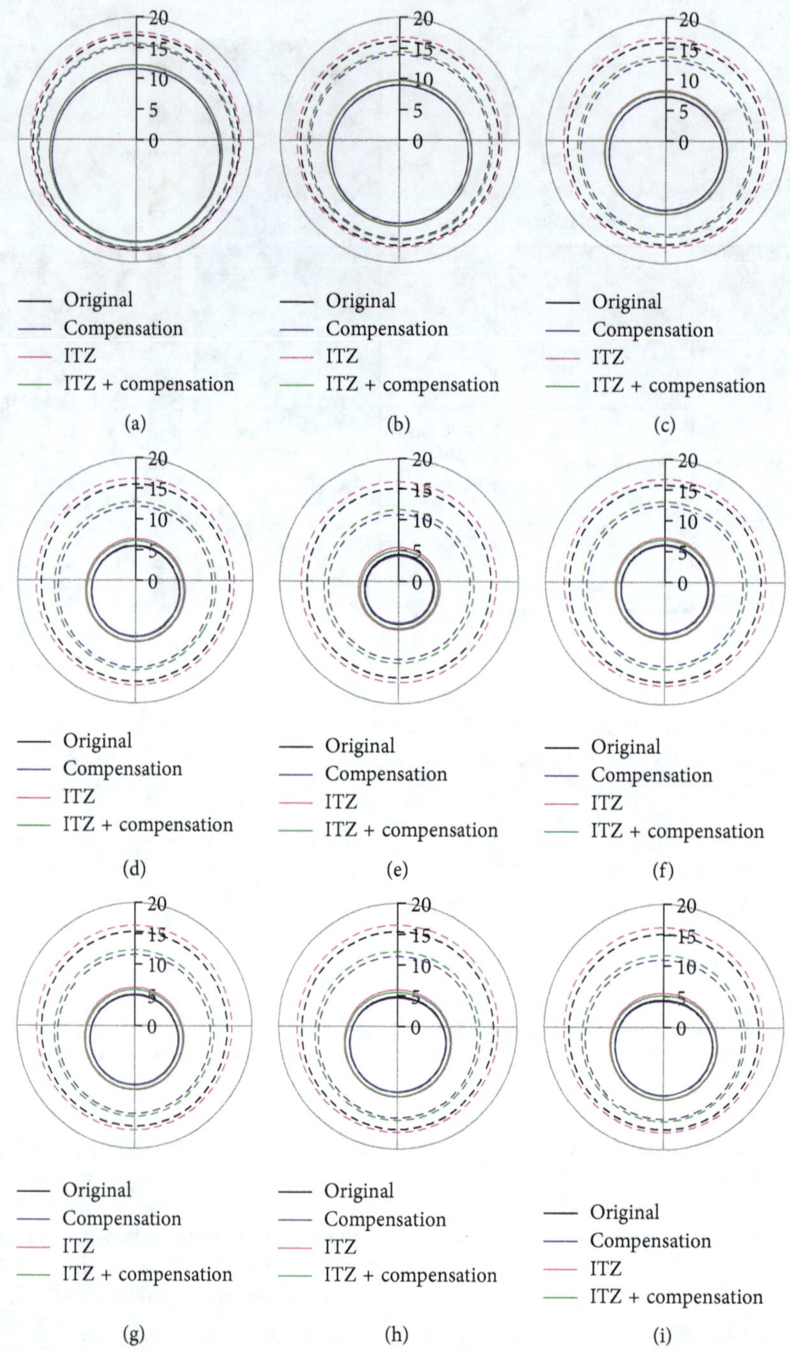

FIGURE 26: Polar distribution of chloride content on cross section of rebar on center of bottom side (solid: 20 y; dashed: 100 y). (a) $R = 16$ mm, $C = 20$ mm. (b) $R = 16$ mm, $C = 30$ mm. (c) $R = 16$ mm, $C = 40$ mm. (d) $R = 16$ mm, $C = 50$ mm. (e) $R = 16$ mm, $C = 60$ mm. (f) $R = 12$ mm, $C = 50$ mm. (g) $R = 20$ mm, $C = 50$ mm. (h) $R = 25$ mm, $C = 50$ mm. (i) $R = 28$ mm, $C = 50$ mm.

characteristics of rebar based on obtained distribution of chloride content around rebar. Ordinate in these figures indicated the ratio of diameter of corroded rebar to initial value.

Obtained diagrams showed less corrosion of steel rebar in the cases with compensation due to larger space for chloride diffusion and lower distributed chloride content. More obvious difference comparing between the result from the compensation model (solid) and the case without compensation (dashed) due to a longer exposure period was observed.

Moreover, it should be noted that increasing concrete cover provided slight effect on protecting steel rebar from corrosion. The higher diffusivity in ITZ adhering on steel rebar provided an equivalent express channel for diffusion of chloride ion around steel rebar. Thus, ITZ actually balanced the difference of depth from surface of concrete for these cases and insignificant difference was observed for cases in different covers. Similar phenomenon was noticed that compensation decreased chloride content on surface of rebar as well as its corrosion degree.

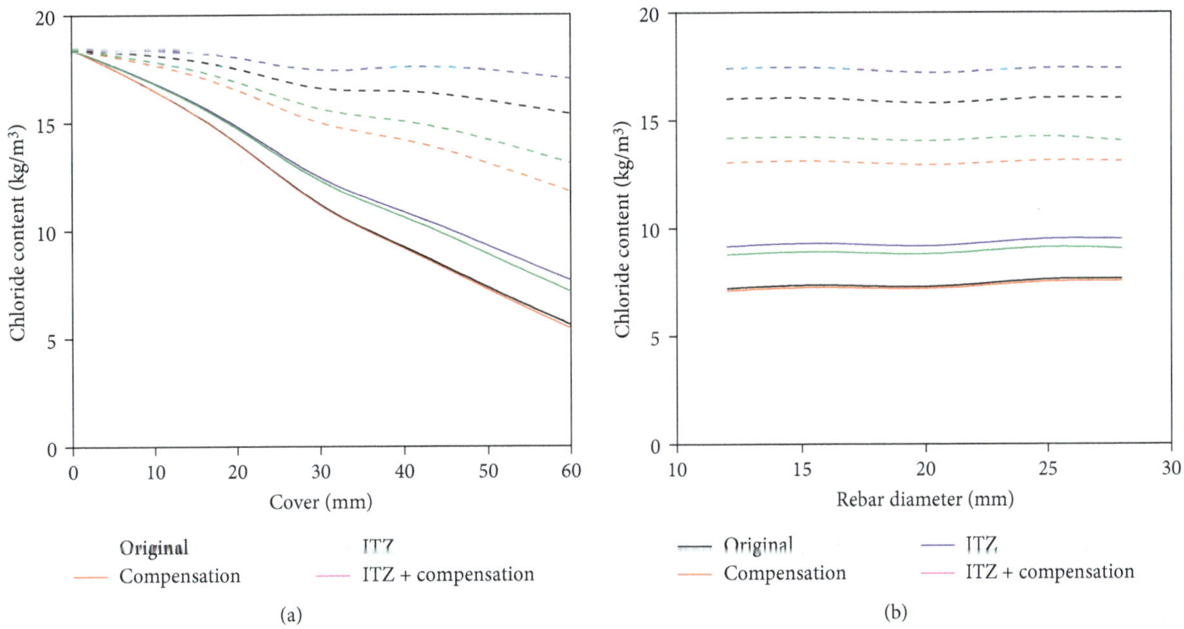

FIGURE 27: Mean chloride content of rebar on center of bottom side (solid: 20 y; dashed: 100 y). (a) Mean chloride content vs. cover thickness, 16 mm rebar. (b) Mean chloride content vs. diameter, 50 mm cover.

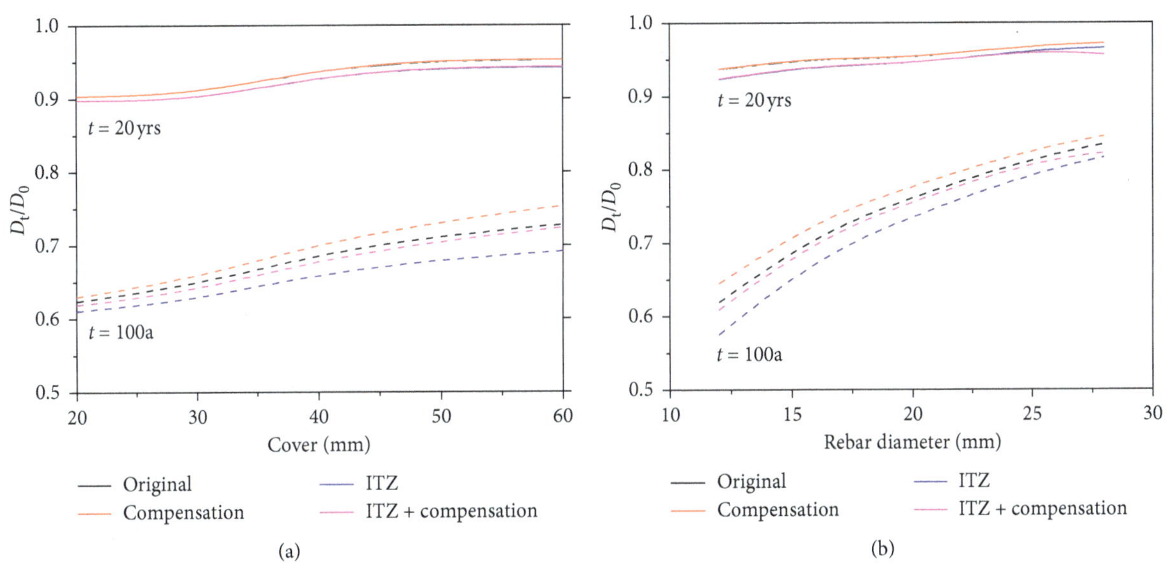

FIGURE 28: Mean chloride content of rebar on center of bottom side (solid: 20 y; dashed: 100 y). (a) Mean remaining diameter vs. cover thickness, 16 mm rebar. (b) Mean remaining diameter vs. diameter, 50 mm cover.

6. Conclusion

This paper presented a study on the approach of multiscale modeling for diffusion of chloride ion within concrete including ITZ. A two-phase integrated model including core part and compensation part was introduced, where both phases were connected by an overlapped zone called interfacial transition layer. Nodes and elements of ITZ in FE model were built by expanding outline of aggregates which were randomly generated. Cases considering multiscale and ITZ were studied for obtaining chloride content profile. Moreover, corrosion of rebar on straight edge was discussed by means of the multiscale approach. The conclusions of this study are summarized as follows:

(1) The multiscale approach based on the theory of compensation for mesoscopic numerical simulation of chloride diffusion within concrete introduced in this paper was applied by means of numerical simulation. The presented work showed the influence of compensated model on distribution of chloride content, which provided the basis for future large-scale structural durability analysis in optimized size, number of DOFs, and time consumption.

(2) Regarding diffusivity and width of ITZ, volume fraction, and grade of aggregate, effects of ITZ on distribution of chloride were compared in the presented work. The result showed significant difference on distribution of chloride content between different configuration of ITZ and compensation.

(3) In the light of corrosion model of steel rebar in terms of surrounding chloride content, presence of ITZ and multiscale compensation was discussed by numerical simulation. The result obtained is related to the configuration of rebar diameter, concrete cover, and exposure period. It was clearly concluded that consideration of both ITZ and compensation changed significantly the distribution of chloride content as well as corrosion process of steel rebar.

Conflicts of Interest

The authors declare that they have no conflicts of interest.

Acknowledgments

This study was sponsored by the National Key Research and Development Program of China (2016YFC0701202). This study was also supported by the National Natural Science Foundation of China (51508053), the National Natural Science Foundation of China (51808069), and the Fundamental Research Funds for the Central Universities (106112014CDJZR200016), which were greatly appreciated by the authors.

References

[1] W. Zhu, R. François, Q. Fang, and D. Zhang, "Influence of long-term chloride diffusion in concrete and the resulting corrosion of reinforcement on the serviceability of RC beams," *Cement and Concrete Composites*, vol. 71, pp. 144–152, 2016.

[2] Q.-F. Liu, G.-L. Feng, J. Xia, J. Yang, and L.-Y. Li, "Ionic transport features in concrete composites containing various shaped aggregates: a numerical study," *Composite Structures*, vol. 183, pp. 371–380, 2018.

[3] U. Angst, B. Elsener, C. K. Larsen, and Ø. Vennesland, "Critical chloride content in reinforced concrete–a review," *Cement and Concrete Research*, vol. 39, no. 12, pp. 1122–1138, 2009.

[4] J.-J. Zheng, H. S. Wong, and N. R. Buenfeld, "Assessing the influence of ITZ on the steady-state chloride diffusivity of concrete using a numerical model," *Cement and Concrete Research*, vol. 39, no. 9, pp. 805–813, 2009.

[5] L.-X. Mao, Z. Hu, J. Xia et al., "Multi-phase modelling of electrochemical rehabilitation for ASR and chloride affected concrete composites," *Composite Structures*, vol. 207, pp. 176–189, 2019.

[6] J. Xu and F. Li, "A meso-scale model for analyzing the chloride diffusion of concrete subjected to external stress," *Construction and Building Materials*, vol. 130, pp. 11–21, 2017.

[7] X. Du, L. Jin, and G. Ma, "A meso-scale numerical method for the simulation of chloride diffusivity in concrete," *Finite Elements in Analysis and Design*, vol. 85, pp. 87–100, 2014c.

[8] F. Bernard and S. Kamali-Bernard, "Numerical study of ITZ contribution on mechanical behavior and diffusivity of mortars," *Computational Materials Science*, vol. 102, pp. 250–257, 2015.

[9] J.-J. Zheng and X.-Z. Zhou, "Effective medium method for predicting the chloride diffusivity in concrete with ITZ percolation effect," *Construction and Building Materials*, vol. 47, pp. 1093–1098, 2013.

[10] K. Wu, L. Xu, G. D. Schutter, H. Shi, and G. Ye, "Influence of the interfacial transition zone and interconnection on chloride migration of portland cement mortar," *Journal of Advanced Concrete Technology*, vol. 13, no. 3, pp. 169–177, 2015.

[11] Z. Zhang and Q. Zhang, "Self-healing ability of engineered cementitious composites (ECC) under different exposure environments," *Construction and Building Materials*, vol. 156, pp. 142–151, 2017.

[12] Z. Zhang and Q. Zhang, "Matrix tailoring of engineered cementitious composites (ECC) with non-oil-coated, low tensile strength PVA fiber," *Construction and Building Materials*, vol. 161, pp. 420–431, 2018.

[13] H. Chen, Z. Zhu, J. Lin, W. Xu, and L. Liu, "Numerical modeling on the influence of particle shape on ITZ's microstructure and macro-properties of cementitious composites: a critical review," *Journal of Sustainable Cement-Based Materials*, vol. 7, no. 4, pp. 248–269, 2018.

[14] J. F. Unger and S. Eckardt, "Multiscale modeling of concrete," *Archives of Computational Methods in Engineering*, vol. 18, no. 3, pp. 341–393, 2011.

[15] G.-L. Feng, L.-Y. Li, B. Kim, and Q.-F. Liu, "Multiphase modelling of ionic transport in cementitious materials with surface charges," *Computational Materials Science*, vol. 111, pp. 339–349, 2016.

[16] K. Maekawa, T. Ishida, and T. Kishi, "Multi-scale modeling of concrete performance," *Journal of Advanced Concrete Technology*, vol. 1, no. 2, pp. 91–126, 2003.

[17] L. Chen, Z. Qian, and J. Wang, "Multiscale numerical modeling of steel bridge deck pavements considering vehicle–pavement interaction," *International Journal of Geomechanics*, vol. 16, no. 1, article B4015002, 2016.

[18] Y. Hiratsuka and K. Maekawa, "Multi-scale and multi-chemo-physics analysis applied to fatigue life assessment of strengthened bridge decks," in *Proceedings of XIII International Conference on Computational Plasticity: fundamentals and applications*, pp. 596–607, Barcelona, Spain, September 2015.

[19] X. Tu et al., "A multiscale numerical simulation approach for chloride diffusion and rebar corrosion with compensation model," *Computers and Concrete*, vol. 21, no. 4, pp. 471–484, 2018.

[20] X. Wang, M. Zhang, and A. P. Jivkov, "Computational technology for analysis of 3D meso-structure effects on damage and failure of concrete," *International Journal of Solids and Structures*, vol. 80, pp. 310–333, 2016.

[21] Z. M. Wang, A. K. H. Kwan, and H. C. Chan, "Mesoscopic study of concrete I: generation of random aggregate structure and finite element mesh," *Computers and Structures*, vol. 70, no. 5, pp. 533–544, 1999.

[22] M. Bailakanavar, Y. Liu, J. Fish, and Y. Zheng, "Automated modeling of random inclusion composites," *Engineering with Computers*, vol. 30, no. 4, pp. 609–625, 2012.

[23] Z. Y. Ren and J. Y. W., "Generation of spatial convex polyhedron random aggregate model," in *Proceedings of ICOME 2006*, Hefei, China, October 2006.

[24] A. Caballero, C. M. López, and I. Carol, "3D meso-structural analysis of concrete specimens under uniaxial tension," *Computer Methods in Applied Mechanics and Engineering*, vol. 195, no. 52, pp. 7182–7195, 2006.

[25] F. Biondini, F. Bontempi, D. M. Frangopol, and P. G. Malerba, "Cellular automata approach to durability analysis of concrete structures in aggressive environments," *Journal of Structural Engineering*, vol. 130, no. 11, pp. 1724–1737, 2004.

[26] B. Šavija, J. Pacheco, and E. Schlangen, "Lattice modeling of chloride diffusion in sound and cracked concrete," *Cement and Concrete Composites*, vol. 42, pp. 30–40, 2013.

[27] H. Ma, W. Xu, and Y. Li, "Random aggregate model for mesoscopic structures and mechanical analysis of fully-graded concrete," *Computers and Structures*, vol. 177, pp. 103–113, 2016.

[28] R. François and G. Arliguie, "Effect of microcracking and cracking on the development of corrosion in reinforced concrete members," *Magazine of Concrete Research*, vol. 51, no. 2, pp. 143–150, 1999.

[29] T. Vidal, A. Castel, and R. François, "Corrosion process and structural performance of a 17 year old reinforced concrete beam stored in chloride environment," *Cement and Concrete Research*, vol. 37, no. 11, pp. 1551–1561, 2007.

[30] T. Vidal, A. Castel, and R. François, "Analyzing crack width to predict corrosion in reinforced concrete," *Cement and Concrete Research*, vol. 34, no. 1, pp. 165–174, 2004.

[31] R. Zhang, A. Castel, and R. François, "Concrete cover cracking with reinforcement corrosion of RC beam during chloride-induced corrosion process," *Cement and Concrete Research*, vol. 40, no. 3, pp. 415–425, 2010.

[32] R. Zhang, A. Castel, and R. François, "The corrosion pattern of reinforcement and its influence on serviceability of reinforced concrete members in chloride environment," *Cement and Concrete Research*, vol. 39, no. 11, pp. 1077–1086, 2009.

[33] C. Andrade, C. Alonso, and F. J. Molina, "Cover cracking as a function of bar corrosion: Part I-Experimental test," *Materials and Structures*, vol. 26, no. 8, pp. 453–464, 1993.

[34] T. Liu and R. W. Weyers, "Modeling the dynamic corrosion process in chloride contaminated concrete structures," *Cement and Concrete Research*, vol. 28, no. 3, pp. 365–379, 1998.

[35] L. C. Qing, "Corrosion initiation of reinforcing steel in concrete under natural salt spray and service loading—results and analysis," *Materials Journal*, vol. 97, no. 6, 2000.

[36] S. M. Pritpal and S. E. Mahmoud, "Flexural strength of concrete beams with corroding reinforcement," *ACI Structural Journal*, vol. 96, no. 1, 1999.

[37] A. A. Torres-Acosta, S. Navarro-Gutierrez, and J. Terán-Guillén, "Residual flexure capacity of corroded reinforced concrete beams," *Engineering Structures*, vol. 29, no. 6, pp. 1145–1152, 2007.

[38] A. A. Torres-Acosta and M. Martnez-Madrid, "Residual life of corroding reinforced concrete structures in marine environment," *Journal of Materials in Civil Engineering*, vol. 15, no. 4, pp. 344–353, 2003.

[39] F. Biondini, F. Bontempi, D. M. Frangopol, and P. G. Malerba, "Probabilistic service life assessment and maintenance planning of concrete structures," *Journal of Structural Engineering*, vol. 132, no. 5, pp. 810–825, 2006.

[40] P. Zhang, D. Li, Y. Qiao, S. Zhang, C. Sun, and T. Zhao, "Effect of air entrainment on the mechanical properties, chloride migration, and microstructure of ordinary concrete and fly ash concrete," *Journal of Materials in Civil Engineering*, vol. 30, no. 10, article 04018265, 2018.

[41] P. Zhang, F. H. Wittmann, M. Vogel, H. S. Müller, and T. Zhao, "Influence of freeze-thaw cycles on capillary absorption and chloride penetration into concrete," *Cement and Concrete Research*, vol. 100, pp. 60–67, 2017.

[42] P. Zhang, D. Hou, Q. Liu, Z. Liu, and J. Yu, "Water and chloride ions migration in porous cementitious materials: an experimental and molecular dynamics investigation," *Cement and Concrete Research*, vol. 102, pp. 161–174, 2017.

[43] X. Li, F. Wu, and Z. Huang, "Analytical solution to chloride diffusion equation on concrete," *Concrete(Chinese)*, vol. 30, no. 10, pp. 30–33, 2009.

[44] W. Dridi, "Analysis of effective diffusivity of cement based materials by multi-scale modelling," *Materials and Structures*, vol. 46, no. 1-2, pp. 313–326, 2012.

[45] G. Sun, Y. Zhang, W. Sun, Z. Liu, and C. Wang, "Multi-scale prediction of the effective chloride diffusion coefficient of concrete," *Construction and Building Materials*, vol. 25, no. 10, pp. 3820–3831, 2011.

[46] Z. Zhu, H. Chen, L. Liu, and X. Li, "Multi-scale modelling for diffusivity based on practical estimation of interfacial properties in cementitious materials," *Powder Technology*, vol. 307, pp. 109–118, 2017.

[47] K. Maekawa, T. Ishida, and T. Kishi, *Multi Scale Modeling of Structural Concrete*, Taylor & Francis, Vol. 670, Taylor & Francis, Abingdon, UK, 2009.

[48] Q. Yuan, C. Shi, G. De Schutter, K. Audenaert, and D. Deng, "Chloride binding of cement-based materials subjected to external chloride environment: a review," *Construction and Building Materials*, vol. 23, no. 1, pp. 1–13, 2009.

[49] P. S. Mangat and B. T. Molloy, "Prediction of long term chloride concentration in concrete," *Materials and Structures*, vol. 27, no. 6, pp. 338–346, 1994.

[50] B. Šavija, M. Luković, J. Pacheco, and E. Schlangen, "Cracking of the concrete cover due to reinforcement corrosion: a two-dimensional lattice model study," *Construction and Building Materials*, vol. 44, pp. 626–638, 2013.

[51] Y. Zhao et al., "Concrete surface chloride ion concentration varying with seasons in marine environment (Chinese)," *Journal of Zhejiang University (Engineering Science)*, vol. 43, no. 11, pp. 2120–2124, 2009.

[52] Z. Pan, A. Chen, and X. Ruan, "Spatial variability of chloride and its influence on thickness of concrete cover: a two-dimensional mesoscopic numerical research," *Engineering Structures*, vol. 95, pp. 154–169, 2015.

4

Synthesis of Spatially Correlated Earthquake Ground Motions Based on Hilbert Transform

Erlei Yao,[1] Yu Miao,[1] and Guobo Wang[2]

[1]*School of Civil Engineering and Mechanics, Huazhong University of Science and Technology, Wuhan 430074, China*
[2]*Hubei Key Laboratory of Roadway Bridge & Structure Engineer, Wuhan University of Technology, Wuhan 430074, China*

Correspondence should be addressed to Yu Miao; miaoyu@hust.edu.cn

Academic Editor: Hongyi Li

A simplified method for synthesizing spatially correlated earthquake ground motions is developed based on Hilbert transform and a reference earthquake record. In this method, one reference earthquake record is treated as the original ground motion, based on a series of generated ground motions. This procedure uses the instantaneous amplitude and the instantaneous phase of the record obtained using Hilbert transform to achieve the nonstationarity of ground motion. To establish the coherency between generated ground motions, an incoherence model is employed to describe the relation between the instantaneous phase at the present station and the instantaneous phases at previous stations. This type of phase is defined as the instantaneous coherence phase. In addition, time lag is included in the instantaneous coherence phase to prescribe the wave passage effect. The proposed Hilbert-transform-based method is efficient and avoids cumbersome parameter estimations as well as other drawbacks involved in some traditional synthesizing methods. Applications of this method demonstrate that the generated ground motions are statistically analogous with the reference record.

1. Introduction

In the past decades, the methods of generating spatially variable ground motions were studied thoroughly. A number of studies demonstrate that the effect of spatially variable ground motions on the responses of long structures is not negligible [1–7]. Nowadays, the synthesizing methods [8–11] based on stochastic process theory are popular and are implemented as a tool in most of the finite-element packages [12]. Stochastic procedures usually include power spectral matrix/incoherence matrix decomposition [13, 14] or spectral factorization [15], which involves massive calculation, thus decreasing the synthesizing efficiency. In addition, the nonstationarity of ground motion is a significant factor in the process of generating ground motions, which includes intensity nonstationarity and frequency nonstationarity [16]. Intensity nonstationarity is usually achieved by multiplying the generated stationary process using an envelope function. However, this method may not be applicable for simulations based on a recorded accelerogram, because the applied envelop function may not completely present the intensity nonstationarity of the original ground motion. Moreover, studies by Ohsaki [17] showed that the seismic waveform was governed by the distribution of the phase difference spectrum; thus, phase difference spectrum could be used to achieve intensity nonstationarity. Zhu and Feng [18, 19] studied the distribution characteristic of the phase difference spectrum and proposed a method of generating a random phase. This method can achieve full nonstationarity; however, the effect of the randomness of the generated phase on the ground motions is not confirmed. Several methods have been proposed for frequency nonstationarity [20–26]. These methods involve complex processes and massive calculation. Conditional simulation is an alternative and several contributions have been made to the study of ground motion simulation [27, 28].

In the present paper, a simplified conditional simulation method of synthesizing spatially correlated earthquake ground motions is proposed based on Hilbert transform and earthquake record. In this method, one reference earthquake record is treated as the original ground motion, based on which a series of ground motions are generated. By performing Hilbert transform on the known earthquake record, the instantaneous amplitude and the instantaneous phase of the earthquake record can be obtained. The instantaneous amplitude is utilized as an envelope function to achieve the intensity nonstationarity of each simulated ground motion. To establish the coherency between generated ground motions, a coherence model is employed to describe the relation between the instantaneous phase at the present station and the instantaneous phases at previous stations. The instantaneous coherence phases at different stations are all statistically analogous with that of the known record. Thus every generated ground motion can present similar frequency nonstationarity. This type of phase is defined as an instantaneous coherence phase. In addition, time lag is included in the instantaneous coherence phase to prescribe the wave passage effect. The proposed Hilbert-transform-based method can efficiently achieve intensity nonstationarity and frequency nonstationarity, and this approach avoids cumbersome parameter estimations, as well as other drawbacks involved in some traditional synthesizing methods. The application of this method demonstrates its validity and practical value.

2. Hilbert Transform

For a real-valued function $x(t)$, its Hilbert transform $y(t)$ is defined as

$$y(t) = H[x(t)] = \frac{1}{\pi} P \int_{-\infty}^{\infty} \frac{x(\tau)}{t-\tau} d\tau, \tag{1}$$

where P denotes the Cauchy principal value of the integral. Thus, by using the Hilbert transform, an analytic signal $z(t)$ which is a complex-valued function can be obtained as follows [29]:

$$z(t) = x(t) + iy(t) \tag{2}$$

which can be further expressed as

$$z(t) = a(t) e^{i\theta(t)}, \tag{3}$$

where $i^2 = -1$.

$$a(t) = \sqrt{x^2(t) + y^2(t)}$$
$$\theta(t) = \arctan \frac{y(t)}{x(t)}. \tag{4}$$

Thus, the original function $x(t)$ can be expressed as

$$x(t) = a(t) \cos[\theta(t)], \tag{5}$$

where the independent variable t denotes time and time functions $a(t)$ and $\theta(t)$ are called the instantaneous amplitude and the instantaneous phase function or, by Bendat and Piersol [30], the envelop signal and the instantaneous phase signal of $x(t)$, respectively. In addition, $a(t)$ represents the amplitude modulation, and $\cos[\theta(t)]$ represents the frequency modulation mechanisms contained in the original signal and $\theta(t)$ is between $-\pi$ and π. In other words, $a(t)$ governs the intensity nonstationarity (or the temporal variation of amplitude), whereas $\theta(t)$ dominates the frequency nonstationarity (or the temporal variation of frequency content) of the signal $x(t)$.

3. Proposed Method

In this study, the spatial variation of ground motion is prescribed in terms of wave scattering as well as wave passage effects. The soil condition and geology of the field of interest were assumed to be uniform. The well-known north-south component of the natural ground motion recorded at the El Centro station during the 1940 earthquake in Imperial Valley, California, which had a magnitude of 6.95 Mw, is chosen as the reference accelerogram, as shown in Figure 1(a). The recorded ground motion exhibits a sampling rate of 50 Hz and the anterior 40.94 s time history of the record is extracted to be studied with a total of 2048 values. First, the instantaneous amplitude $a(t)$ and the instantaneous phase functions $\theta(t)$ can be obtained by performing the Hilbert transform on the known earthquake record, as indicated in Figures 1(b) and 1(c), respectively. The instantaneous amplitude is treated by the method as an envelop function, with a value that will be preserved into the synthetic samples.

According to (5), the original earthquake ground motion can be expressed, and the acceleration process at station n can be given as

$$x_n(t) = a(t) \cos\left[\theta(t) + \sum_{k=0}^{n-1} \arccos\left(\mu_{k,n} \cos\left(\theta_k(t)\right)\right) + \tau_n\right], \tag{6}$$

where τ_n is the time lag between the two stations given by [13]

$$\tau_n = \frac{\omega \cdot r_n}{V}, \tag{7}$$

where r_n denotes the separation distance between the two stations projected parallel to the dominant wave propagation direction and is assumed to be 200 m uniformly; V represents the apparent seismic wave velocity in the medium, which is 500 m/s; and $\mu_{k,n}$ indicates the coherence function between two instantaneous phases. The instantaneous phase at the present station is assumed to be only affected by these phases at previous stations. Thus, the effect of phases at previous stations on the present phase can be expressed by $\sum_{k=0}^{n-1} \arccos(\mu_{k,n} \cos(\theta_k(t)))$. The relevant verification is shown in appendix.

Many coherency models were proposed, such as the models proposed by Loh and Lin [31], Feng and Hu [32], Loh and Yeh [33], Hao et al. [34], and Harichandran and Vanmarcke [35]. In this study, the Sobczyk model [36] is selected to describe the coherency loss between the ground motions at points m and n as follows:

$$\mu_{m,n}(i\omega) = \exp\left(-\frac{\beta\omega d_{m,n}^2}{V}\right) \cdot \exp\left(-i\omega d_{m,n} \cos\frac{\alpha}{V}\right) \tag{8}$$

which reflects the level of coherency loss; $\beta = 0.0005$ is used in the present paper, which indicates highly correlated motions [37]; α is the incident angle of the incoming wave to the site, and it is assumed to be 0°.

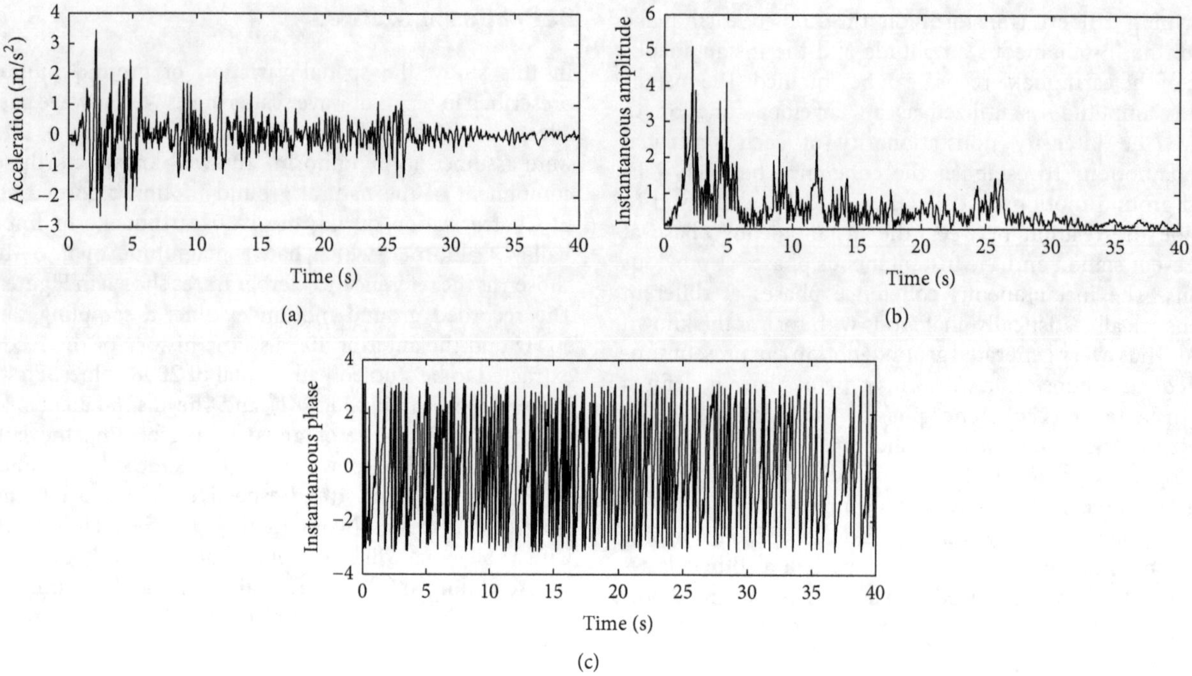

FIGURE 1: El Centro earthquake record: (a) acceleration; (b) instantaneous amplitude; (c) instantaneous phase angle.

4. Verification of the Proposed Method

To verify the proposed method, a set of comparisons are conducted, including comparison of (1) peak ground accelerations (PGA), peak ground velocity (PGV), and peak ground displacement (PGD); (2) Fourier amplitude and response spectra amplitude; and (3) coherence function of simulated motions to those of the known record.

4.1. Comparison of PGA, PGV and PGD. A series of accelerograms at three stations arranged at 200 m, 400 m, and 600 m from the original record are simulated using the proposed methodology. The acceleration, velocity, and displacement time histories at each point are presented in Figures 2–4, respectively. The waveform of time histories of acceleration, velocity, and displacement at each point are all consistent with the original record. The PGAs, PGVs, and PGDs of the simulated motions are compared with those of the original ground motion, which are presented in Table 1. The PGA, PGV, and PGD at every station are close to those of the known record. Besides, from those time histories figures, it can be seen that the generated ground motions have the identical intensity nonstationarity and frequency nonstationarity with the original record. These comparison results verify that the simulated ground motions are rational.

4.2. Comparison of Fourier Amplitude and Response Spectra. Figure 5 provides the results of Fourier amplitude of simulated motions compared with those of the original record. Figure 6 presents the response spectra amplitude of simulated motions, as well as that of the original record. The spectral values are close to that of the known record. The comparison of the time-frequency spectra of the generated ground motions to that of the known record is shown in Figure 7. The time-frequency spectra were obtained by using the short time Fourier transform. The figure shows that both spectral amplitude and tendency are similar to that of the known record.

TABLE 1: Peak values of the simulated and recorded ground motions.

Station	Acceleration (m/s^2)	Velocity (m/s)	Displacement (m)
Recorded motion	3.417	0.335	0.109
r = 200 m	3.426	0.351	0.098
r = 400 m	3.404	0.317	0.099
r = 600 m	3.804	0.342	0.085

4.3. Comparison of Coherence. Figure 8 indicates the coherency values between the generated time histories and the known record as well as the corresponding Sobczyk model. A good match can also be observed except for $\mu_{0,2}$ and $\mu_{0,3}$ in the high frequency range. However, this result is expected because the cross correlation between the generated motions or their coherency values decrease rapidly with frequency as the separation increases [37]. A previous study [38] indicated that the coherency value of approximately 0.3–0.4 is the threshold of the cross correlation between two signals because numerical calculations of the coherency function between any two white noise series result in a value of roughly 0.3–0.4. Thus the calculated coherency function between two simulated time histories is smaller than the target value.

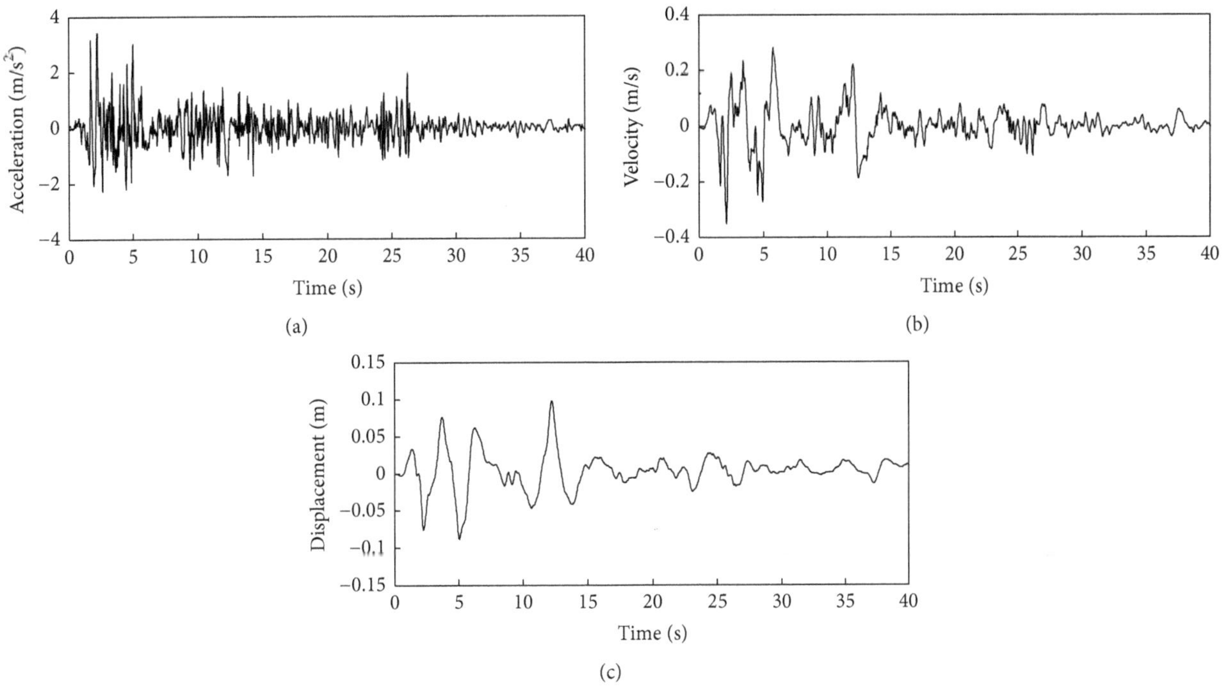

FIGURE 2: Simulated ground motion at station with r_n = 200 m: (a) acceleration; (b) velocity; (c) displacement.

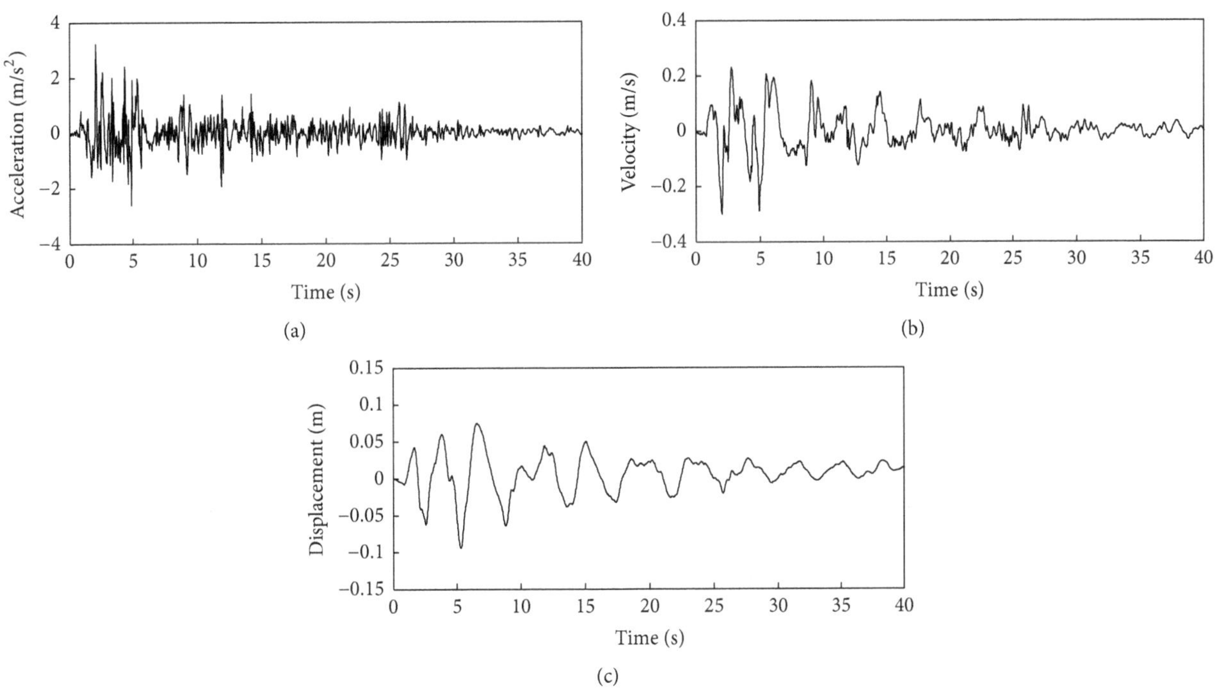

FIGURE 3: Simulated ground motion at station with r_n = 400 m: (a) acceleration; (b) velocity; (c) displacement.

5. Conclusion

In this study, an approach was developed to simulate spatially correlated earthquake ground motions based on Hilbert transform and earthquake records where instantaneous coherence phase is defined. In this method, one recorded accelerogram is treated as the original ground motion, based on which a series of ground motions are generated. By performing Hilbert transform on the known earthquake record, the instantaneous amplitude and the instantaneous phase of the earthquake record can be obtained. The instantaneous amplitude is utilized as an envelop function to achieve the intensity nonstationarity of each simulated ground motion. To establish the coherency between

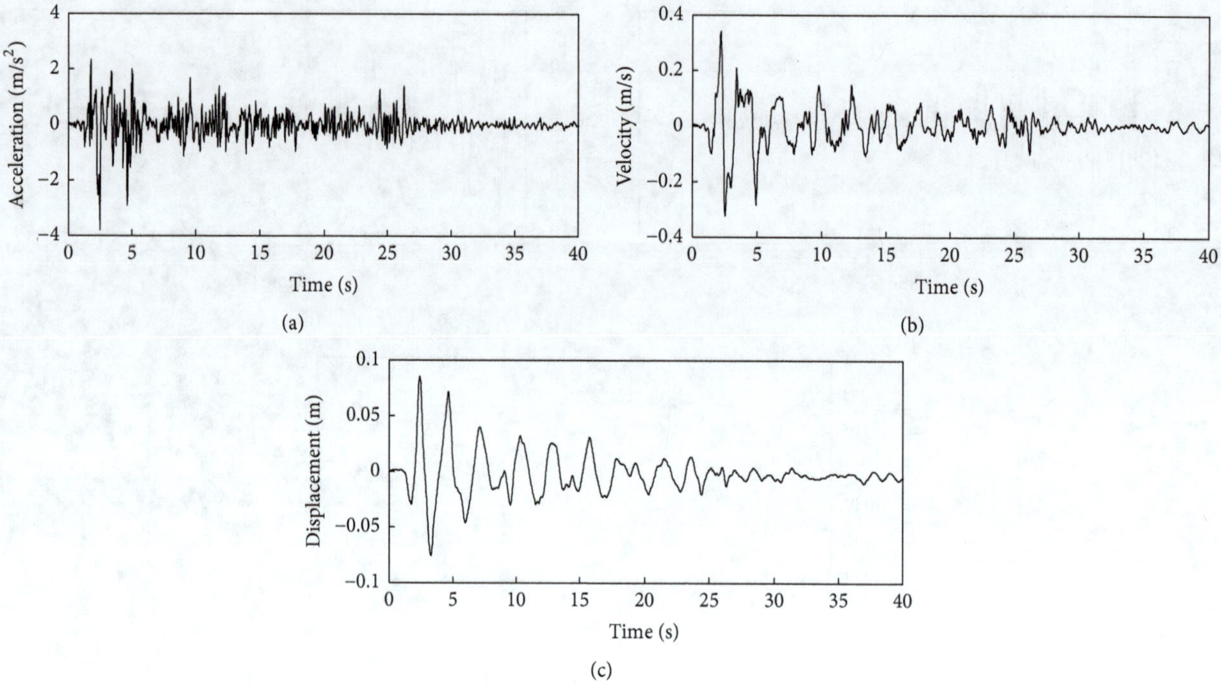

FIGURE 4: Simulated ground motion at station with r_n = 600 m: (a) acceleration; (b) velocity; (c) displacement.

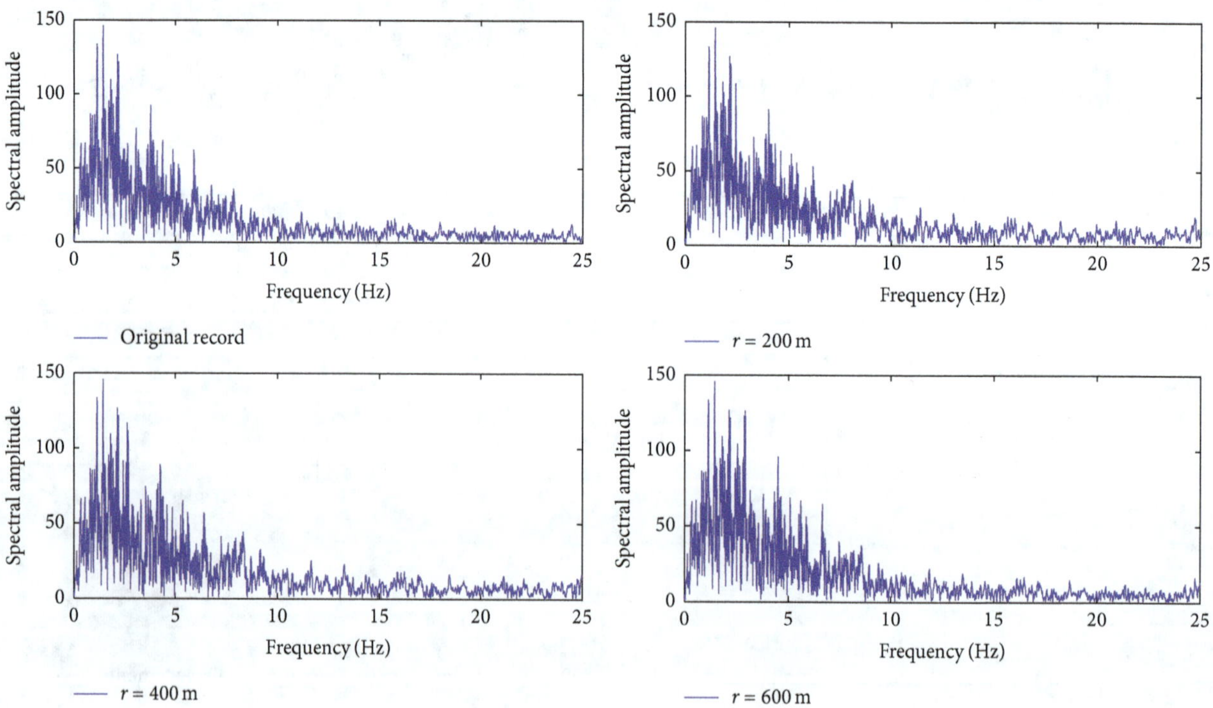

FIGURE 5: Comparison of Fourier amplitude of simulated motions to that of the original record.

generated ground motions, the Sobczyk model is employed in describing the relation between the instantaneous phase at the present station and the instantaneous phases at previous stations. The instantaneous coherence phases at different stations are all statistically analogous with that of the known record. Thus every generated ground motion can present similar frequency nonstationarity. In addition, time lag is included in the instantaneous coherence phase to prescribe the wave passage effect. The proposed Hilbert-transform-based method is concise and avoids many

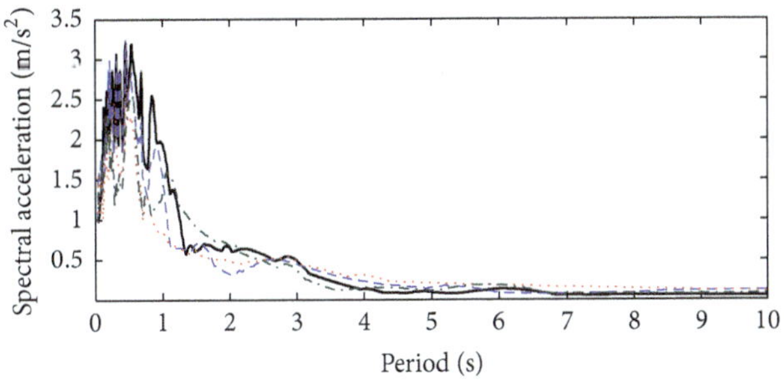

FIGURE 6: Comparison of response spectra with 2% damping ratio of simulated motions to that of the original record.

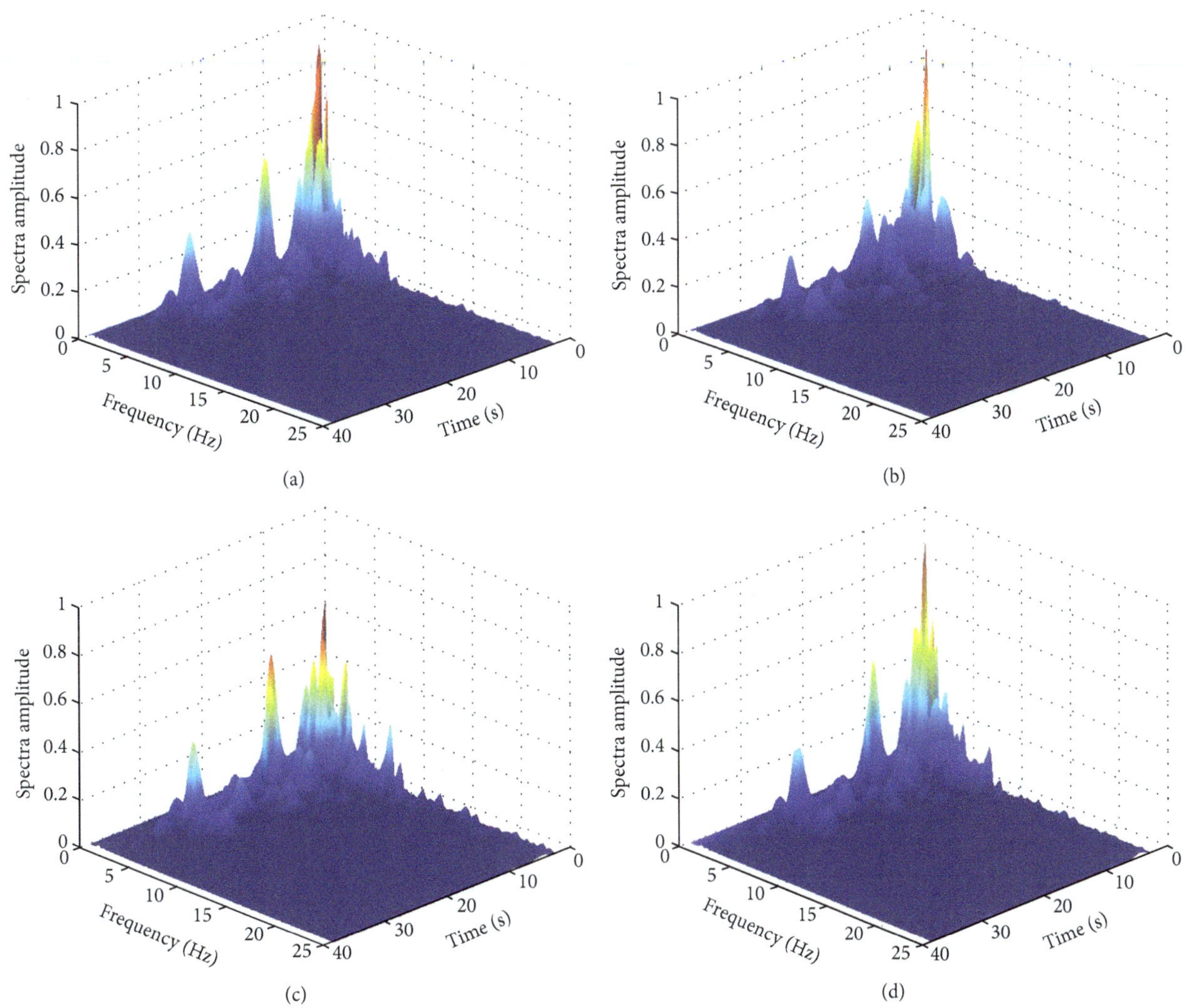

FIGURE 7: Comparison of time-frequency spectral at simulated stations to that of the original record: (a) the time-frequency spectrum of the known record; (b), (c), (d) the time-frequency spectrum of 1–3 simulated motion.

drawbacks involved in some traditional synthesizing methods. The power spectra density of every simulation point is considered to be identical in the proposed method, so it is applicable to the relatively small field only. Because for large-scale field, the power spectra density of each simulation point might be different. Thus, it has been considered that the local site effect should be incorporated in the synthesis method in the future.

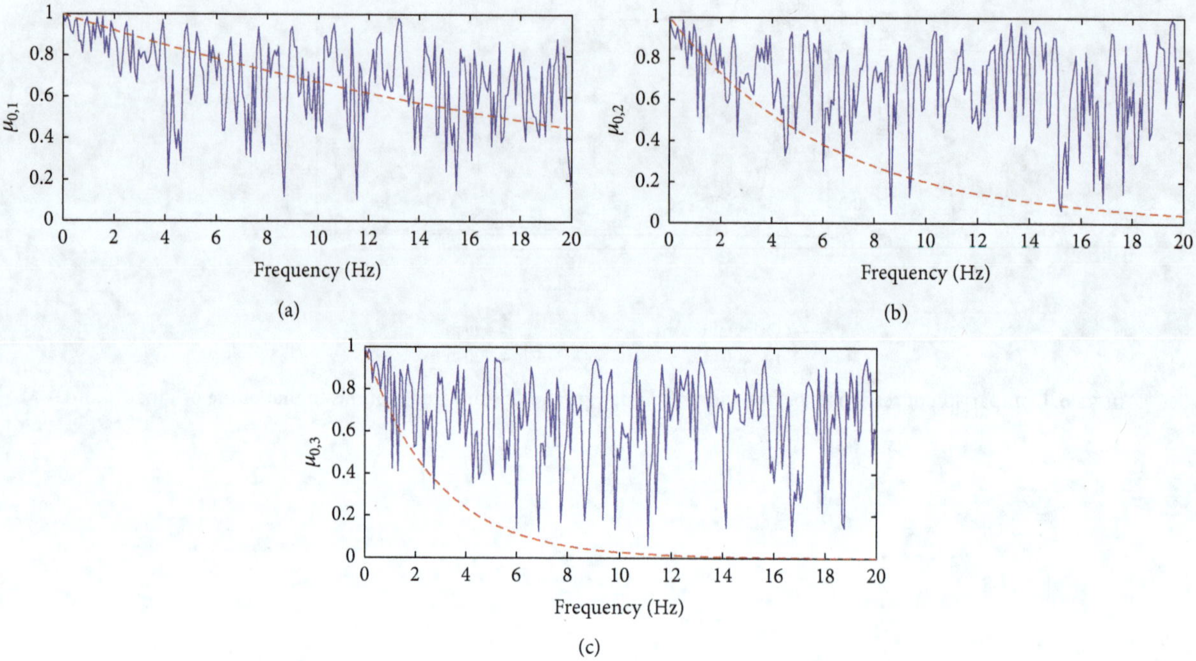

FIGURE 8: Comparison of coherency loss between the simulated accelerations with target value.

Appendix

Verification for the Instantaneous Coherency Phase

As mentioned above, it is assumed that the soil condition and geology of the field of interest were uniform, so the power spectra density at each point is uniform and the coherency function between mth and nth point can be given as follows:

$$\mu_{m,n}(\omega) = \frac{S_{mn}(\omega)}{\sqrt{S_{nn}(\omega) S_{mm}(\omega)}} = \frac{S_{mn}(\omega)}{S_{nn}(\omega)} \quad \text{(A.1)}$$

$$\mu_{m,n}(\omega) = \frac{\int_{-\infty}^{\infty}\left[\int_{-\infty}^{\infty} x_m(t)\, x_n(t-\tau)\, dt\right] e^{-i\omega\tau} d\tau}{\int_{-\infty}^{\infty}\left[\int_{-\infty}^{\infty} x_m(t)\, x_m(t-\tau)\, dt\right] e^{-i\omega\tau} d\tau}, \quad \text{(A.2)}$$

where τ denotes a time lag. By using Hilbert transform, $x_m(t)$ and $x_n(t)$ can be expressed as

$$\begin{aligned} x_m(t) &= a_0(t)\cos\left[\theta_m(t)\right] \\ x_n(t) &= a_0(t)\cos\left[\theta_n(t)\right]. \end{aligned} \quad \text{(A.3)}$$

Substituting (A.3) into (A.2), $\mu_{m,n}(\omega)$ can be expressed as

$$\mu_{m,n}(\omega) = \frac{\int_{-\infty}^{\infty}\left[\int_{-\infty}^{\infty} a_0(t)\, a_0(t-\tau)\cos\left[\theta_m(t)\right]\cos\left[\theta_n(t-\tau)\right] dt\right] e^{-i\omega\tau} d\tau}{\int_{-\infty}^{\infty}\left[\int_{-\infty}^{\infty} a_0(t)\, a_0(t-\tau)\cos\left[\theta_m(t)\right]\cos\left[\theta_m(t-\tau)\right] dt\right] e^{-i\omega\tau} d\tau} \quad \text{(A.4)}$$

and then

$$\begin{aligned} &\mu_{m,n}(\omega)\int_{-\infty}^{\infty}\left[\int_{-\infty}^{\infty} a_0(t)\, a_0(t-\tau)\cos\left[\theta_m(t)\right] \right. \\ &\left. \cdot \cos\left[\theta_m(t-\tau)\right] dt\right] e^{-i\omega\tau} d\tau = \int_{-\infty}^{\infty}\left[\int_{-\infty}^{\infty} a_0(t) \right. \\ &\left. \cdot\, a_0(t-\tau)\cos\left[\theta_m(t)\right]\cos\left[\theta_n(t-\tau)\right] dt\right] e^{-i\omega\tau} d\tau \end{aligned} \quad \text{(A.5)}$$

$$\begin{aligned} &\int_{-\infty}^{\infty}\int_{-\infty}^{\infty} a_0(t)\, a_0(t-\tau)\cos\left[\theta_m(t)\right] \\ &\cdot e^{-i\omega\tau}\left\{\mu_{m,n}\cos\left[\theta_m(t-\tau)\right] - \cos\left[\theta_n(t-\tau)\right]\right\} dt\, d\tau \\ &= 0. \end{aligned} \quad \text{(A.6)}$$

When $\forall\tau \in (-\infty, \infty)$, $\mu_{m,n}(\omega)\cos[\theta_m(t-\tau)] = \cos[\theta_n(t-\tau)]$, (A.6) is valid. Herein, τ is set to be 0 and then (A.8) can be expressed as

$$\mu_{m,n}(\omega)\cos\left[\theta_m(t)\right] = \cos\left[\theta_n(t)\right] \quad \text{(A.7)}$$

$$\theta_n(t) = \arccos\left[\mu_{m,n}(\omega)\cos\left[\theta_m(t)\right]\right]. \quad \text{(A.8)}$$

Conflicts of Interest

The authors declare no conflicts of interest.

Acknowledgments

Financial support for the project from the National Key Research and Development Program of China (no. 2016YFC0800206), the National Natural Science Foundation of China (no. 51378234 and no. 51678465), and the Fundamental Research Funds for the Central Universities (no. 2015MS060) is acknowledged.

References

[1] A. Hindy and M. Novak, "Pipeline response to random ground motion," *Journal of Engineering Mechanics ASCE*, vol. 106, no. 2, pp. 339–360, 1980.

[2] W. L. Liu, X. G. Wu, L. M. Zhang, J. Zheng, and J. Teng, "Global sensitivity analysis of tunnel-induced building movements by a precise metamodel," *Journal of Computing in Civil Engineering*, vol. 31, no. 5, p. 04017037, 2017.

[3] A. S. Nazmy and A. M. Abdel Ghaffar, "Effects of ground motion spatial variability on the response of cable-stayed bridges," *Earthquake Engineering & Structural Dynamics*, vol. 21, no. 1, pp. 1–20, 1992.

[4] R. S. Harichandran, A. Hawwari, and B. N. Sweidan, "Response of long-span bridges to spatially varying ground motion a," *Journal of Structural Engineering*, vol. 122, no. 5, pp. 476–484, 1996.

[5] M.-T. Chen and R. S. Harichandran, "Response of an earth dam to spatially varying earthquake ground motion," *Journal of Engineering Mechanics*, vol. 127, no. 9, pp. 932–939, 2001.

[6] A. A. Dumanogluid and K. Soyluk, "A stochastic analysis of long span structures subjected to spatially varying ground motions including the site-response effect," *Engineering Structures*, vol. 25, no. 10, pp. 1301–1310, 2003.

[7] K. Soyluk, "Comparison of random vibration methods for multi-support seismic excitation analysis of long-span bridges," *Engineering Structures*, vol. 26, no. 11, pp. 1573–1583, 2004.

[8] M. Shinozuka and G. Deodatis, "Stochastic process models for earthquake ground motion," *Probabilistic Engineering Mechanics*, vol. 3, no. 3, pp. 114–123, 1988.

[9] A. Zerva, "Seismic ground motion simulations from a class of spatial variability models," *Earthquake Engineering & Structural Dynamics*, vol. 21, no. 4, pp. 351–361, 1992.

[10] A. Zerva and M. Shinozuka, "Stochastic differential ground motion," *Structural Safety*, vol. 10, no. 1–3, pp. 129–143, 1991.

[11] G. Deodatis, "Non-stationary stochastic vector processes: seismic ground motion applications," *Probabilistic Engineering Mechanics*, vol. 11, no. 3, pp. 149–168, 1996.

[12] A. A. Shama, "Simplified procedure for simulating spatially correlated earthquake ground motions," *Engineering Structures*, vol. 29, no. 2, pp. 248–258, 2007.

[13] Y.-F. Gao, Y.-X. Wu, and B. Li, "Simplified method for simulation of ergodic spatially correlated seismic ground motion," *Applied Mathematics and Mechanics. English Edition*, vol. 32, no. 10, pp. 1297–1314, 2011.

[14] Y. Wu, Y. Gao, D. Li, C. Xu, and A. H. Mahfouz, "Study of the approximate approaches to the POD based spectral representation method," *Science China Technological Sciences*, vol. 56, no. 4, pp. 970–979, 2013.

[15] M. Shrikhande and V. K. Gupta, "Synthesizing ensembles of spatially correlated accelerograms," *Journal of Engineering Mechanics*, vol. 124, no. 11, pp. 1185–1192, 1998.

[16] S. Rezaeian and A. Der Kiureghian, "A stochastic ground motion model with separable temporal and spectral nonstationarities," *Earthquake Engineering and Structural Dynamics*, vol. 37, no. 13, pp. 1565–1584, 2008.

[17] Y. Ohsaki, "On the significance of phase content in earthquake ground motions," *Earthquake Engineering & Structural Dynamics*, vol. 7, no. 5, pp. 427–439, 1979.

[18] Zhu Y. and Feng Q. M., "Digital characteristics of phase difference spectrum distribution of earthquake accelerogram," *Earthquake Engineering & Engineering Vibration*, vol. 12, no. 1, pp. 37–44, 1992.

[19] Zhu Y. and Feng Q. M., "Distribution characteristic of phase difference spectrum and artificial accelerogram," *Earthquake Engineering & Engineering Vibration*, vol. 13, no. 2, pp. 30–37, 1993.

[20] M. B. Priestley, *SpEctral Analysis and Time Series*, Academic Press, 1981.

[21] K. Sarkar and V. K. Gupta, "Wavelet-based simulation of spatially correlated and spectrum-compatible accelerograms," in *Proceedings of the national symposium on structural dynamics, random vibrations earthquake engineering (NSSD '05)*, pp. 69–78, Department of Civil Engineering, Indian Institute of Science, Bangalore, 2005.

[22] P. Cacciola and G. Deodatis, "A method for generating fully non-stationary and spectrum-compatible ground motion vector processes," *Soil Dynamics and Earthquake Engineering*, vol. 31, no. 3, pp. 351–360, 2011.

[23] T. Kubo and J. Penzien, "Simulation of three-dimensional strong ground motions along principal axes, San Fernando earthquake," *Earthquake Engineering & Structural Dynamics*, vol. 7, no. 3, pp. 279–294, 1979.

[24] Y. K. Lin and Y. Yong, "Evolutionary Kanai-Tajimi earthquake models," *Journal of Engineering Mechanics*, vol. 113, no. 8, pp. 1119–1137, 1987.

[25] A. Der Kiureghian and J. Crempien, "An evolutionary model for earthquake ground motion," *Structural Safety*, vol. 6, no. 2, pp. 235–246, 1989.

[26] J. P. Conte, K. S. Pister, and S. A. Mahin, "Nonstationary ARMA modeling of seismic motions," *Soil Dynamics and Earthquake Engineering*, vol. 11, no. 7, pp. 411–426, 1992.

[27] L. Hu, Y. L. Xu, and Y. Zheng, "Conditional simulation of spatially variable seismic ground motions based on evolutionary spectra," *Earthquake Engineering and Structural Dynamics*, vol. 41, no. 15, pp. 2125–2139, 2012.

[28] Y. Wu, Y. Gao, and D. Li, "Error assessment of multivariate random processes simulated by a conditional-simulation method," *Journal of Engineering Mechanics*, vol. 141, no. 5, Article ID 04014155, 2015.

[29] L. Cohen, *Time-Frequency Analysis: Theory and Applications*, Prentice-Hall, Inc, Englewood Cliffs, New Jersey, 1995.

[30] J. S. Bendat and A. G. Piersol, *Random Data: Analysis and Measurement Procedures*, John & Wileys, New York, NY, USA, 1986.

[31] C.-H. Loh and S.-G. Lin, "Directionality and simulation in spatial variation of seismic waves," *Engineering Structures*, vol. 12, no. 2, pp. 134–143, 1990.

[32] Q. M. Feng and Y. X. Hu, "Spatial Correlation of Earthquake Motion and Its Effect on Structural Response, Bilateral Workshop on Earthquake Engineering, Beijing," 1982.

[33] C. Loh and Y. Yeh, "Spatial variation and stochastic modelling of seismic differential ground movement," *Earthquake Engineering & Structural Dynamics*, vol. 16, no. 4, pp. 583–596, 1988.

[34] H. Hao, C. S. Oliveira, and J. Penzien, "Multiple-station ground motion processing and simulation based on smart-1 array data," *Nuclear Engineering and Design*, vol. 111, no. 3, pp. 293–310, 1989.

[35] R. S. Harichandran and E. H. Vanmarcke, "Stochastic variation of earthquake ground motion in space and time," *Journal of Engineering Mechanics*, vol. 112, no. 2, pp. 154–174, 1986.

[36] K. Sobczyk, *StochasTic Wave Propagation*, Kluwer Academic Publishers, Netherlands, 1991.

[37] K. Bi and H. Hao, "Modelling and simulation of spatially varying earthquake ground motions at sites with varying conditions," *Probabilistic Engineering Mechanics*, vol. 29, pp. 92–104, 2012.

[38] H. Hao, EffEcts of Spatial Variation of Ground Motions on Large Multiple Supported Structures UCB/EERC-89-06, University of California, Berkeley, CA, USA, 1989.

5

Modelling of Creep and Stress Relaxation Test of a Polypropylene Microfibre by Using Fraction-Exponential Kernel

Andrea Sorzia

Dipartimento di Scienze e Metodi dell'Ingegneria (DISMI), Università di Modena e Reggio Emilia, 42122 Reggio Emilia, Italy

Correspondence should be addressed to Andrea Sorzia; andrea.sorzia@unimore.it

Academic Editor: Julius Kaplunov

A tensile test until breakage and a creep and relaxation test on a polypropylene fibre are carried out and the resulting creep and stress relaxation curves are fit by a model adopting a fraction-exponential kernel in the viscoelastic operator. The models using fraction-exponential functions are simpler than the complex ones obtained from combination of dashpots and springs and, furthermore, are suitable for fitting experimental data with good approximation allowing, at the same time, obtaining inverse Laplace transform in closed form. Therefore, the viscoelastic response of polypropylene fibres can be modelled straightforwardly through analytical methods. Addition of polypropylene fibres greatly improves the tensile strength of composite materials with concrete matrix. The proposed analytical model can be employed for simulating the mechanical behaviour of composite materials with embedded viscoelastic fibres.

1. Introduction

Fibre-reinforced composite materials consist of fibres with high strength and elastic modulus embedded in a matrix to produce a combination of properties that cannot be achieved by single constituents. Usually, fibres are the principal load-carrying members, while the surrounding matrix keeps them in the desired location and orientation. The matrix acts as a load transfer medium between fibres and plays a number of useful functions, for example, protecting the fibres from environmental damage. Fibres may be made up of various materials, such as steel, polymer, glass, or carbon, whereas polymer, metal, or ceramic can be chosen for the matrix material.

Fibre-reinforced polymer composites are probably the most important and widespread fibre-reinforced materials used for commercial and industrial applications. This is due to the combination of their low density, strength-weight ratios, and modulus-weight ratios that make them more attractive than many traditional metallic materials [1].

A classic example of fibre-reinforced composite used in civil engineering is Fibre-Reinforced Concrete (FRC), widely adopted for industrial pavements, tunnel linings, marine structures, earthquake-resistant structures, and plate and slab foundation [2]. In FRC composites, fibres are commonly added to concrete mixture in random disposition and short cuts in order to increase the cracking behaviour of concrete, thus transforming concrete from brittle into a ductile material [3]. However, there are composite materials in which fibres are aligned along particular directions, as it occurs in FRP frame elements (these composites can be studied by adopting, for instance, the approaches proposed in [4, 5]).

Recently, the use of macro synthetic fibres made of polymeric materials has been proposed for structural purposes [6, 7]. Experimental tests performed by Lanzoni et al. [8] show that the addition of polypropylene-based draw-wired fibres significantly improves crack resistance of the concrete mixture and enhances toughness and durability of FRC structural elements. Fibres could be found in interesting applications as an additive to improve concrete under aggressive environments (the mechanical behaviour has been studied under high thermal loads also [9, 10] and nuclear radiation [11]). However, particularly in such harsh conditions, a wide class of additives can be inserted at the mixing stage (like fly ash) to increase resistance of the concrete mixture [12].

Ductility and flexural strength of FRC are defined by energy-dissipation mechanisms during the pullout of the fibres that occurs in the opening propagating cracks [13, 14].

Pullout of the fibres begins after cracking the concrete matrix (for the stress and strain localizations at the crack tip, see, e.g., [14–17]) and ceases with the complete extraction of the fibres and its evolution depends on the bonding between fibre and matrix [18, 19]. Nevertheless due to their chemical inertness, polypropylene (PP) synthetic fibres have poor adhesion to the cement matrix with respect to other kinds of fibres. However, special surface nanotreatment can actually increase the adhesion properties between concrete and synthetic fibres, as shown by Di Maida et al. [3]. In this case, PP fibres may experience significant loading and consequently the viscous deformations of the composite material may considerably increase (for time-dependent effects in concrete structures see [20–22]). A single fibre can be studied as an embedded cylindrical body or a viscoelastic circular nanobeam in frictional contact with the cement matrix under axial tensile load (e.g., [23–25]).

Another application of synthetic fibres in civil engineering is the fibre-reinforced polymer (FRP) composites for reinforcement and retrofitting of concrete and masonry members, with applications in new buildings as well as for strengthening and/or rehabilitation of existing (prestressed as well as nonprestressed) structural members of both prefabricated and cast-in-place frames. FRP reinforcement consists in strengthening fibres applied to structural elements by a cementitious or polymeric-based layer. The mechanical performances of such systems can be assessed by solving the contact problem between two bounded layers [26–28] where the adhesive layer is a fibre-reinforced composite material [29] or, in a simpler way, as a fibre-reinforced Kirchhoff plate [30].

This work presents a creep and stress relaxation test performed over a PP synthetic fibre used for FRC. Creep is a time-dependent deformation of a viscoelastic material under the application of a constant stress at a constant temperature. Relaxation is the counterpart of creep, namely, a time-dependent stress of a viscoelastic material under the application of a constant deformation at a constant temperature. Both are complex phenomena for they depend on material properties (e.g., molecular orientation and crystallinity) and external conditions (e.g., applied stress, temperature, and moisture). Moreover, the viscoelastic behaviour of PP fibres embedded in an elastic matrix complicates the modelling of creep and stress relaxation response of the composite material, which depends on many additional parameters such as concentration, aspect ratio, orientation, and, obviously, mechanical properties of PP fibres.

Creep and stress relaxation tests best demonstrate the viscoelastic characteristics of a polymeric solid. In creep test, a constant stress is maintained on a specimen while its deformation is monitored as a function of time, and deformation increases with time. In stress relaxation test, a constant deformation is maintained while the stress on the specimen is monitored as a function of time, and stress decreases with time. Typical creep and stress relaxation diagrams exhibit an instantaneous elastic response followed by a delayed time-dependent response [1].

Over the experimental creep and stress relaxation tests performed, this paper proposes an analytical model to fit these experimental curves. The model uses fraction-exponential kernel in the viscoelastic operator and was proposed, for the first time, by Scott Blair and Coppen [31, 32] and Rabotnov [33] independently.

The classical viscoelastic constitutive models represented by a combination of dashpots and springs are usually adopted for simulating creep behaviour of composite materials. However, the simplest ones (Maxwell and Kelvin-Voight) are not sufficiently flexible to match experimental data for real materials. The more complex ones, obtained from combination of different Maxwell and Kelvin-Voight models, require instead many parameters and do not allow obtaining inverse Fourier or Laplace transforms in closed form [34, 35]. Therefore, in the 50s of the last century, a viscoelastic stress-strain relation based on fractional derivative has been proposed. The fractional derivative model (FDM) is more flexible and requires a smaller number of parameters, so that their calibration is considerably simpler. The flexibility is due to the order of derivatives which can vary to obtain a constitutive law suitable for the considered material. In 1948 Rabotnov [33] suggested to use fraction-exponential operators that can describe experimental data of real materials with sufficient accuracy and allows one to obtain inverse Laplace transforms analytically [36]. Scott Blair and Coppen [31, 32] used fraction-exponential operators for description of viscoelastic properties of materials experimentally (see Rogosin and Mainardi [37]). Fraction-exponential functions in viscoelastic operators were used by many authors last decade. Detailed discussion can be found in the book of Podlubny [38]. Published broad surveys are, for example, [39–41].

The model developed in the present paper is able to describe the creep curve of a PP fibre carefully and it allows obtaining the creep and stress relaxation test response of a fibre composite material in closed form. Moreover, the adopted fraction-exponential operators can be efficiently employed for the homogenization of synthetic FRC by extending the Maxwell scheme developed for elastic composites to viscoelastic behaviour of the constituents (e.g., [34, 42]).

2. Materials and Methods

2.1. Material and Breaking Test. The fibre consists of PP monofilament with a diameter of 0.78 mm and length of 200 mm. Since the cross section of the fibre is not perfectly circular, the diameter is an average of six measurements: Two measurements in two orthogonal directions in three points of the fibre, namely, at the middle and at both ends. To evaluate elastic Young modulus and tensile strength, tensile tests were performed on four specimens of fibres until their breakage. Each fibre was clamped at its ends and pulled by an electromechanic traction machine under displacement control.

The load cell is a GALDABINI 514262 TYPE TCA, with OUTPUT sensitivity of 2 mV/V. The machine uses a 20-bit A/D converter to acquire the analogical quantities. The resolution of the load cell is 0.002 N over the entire field

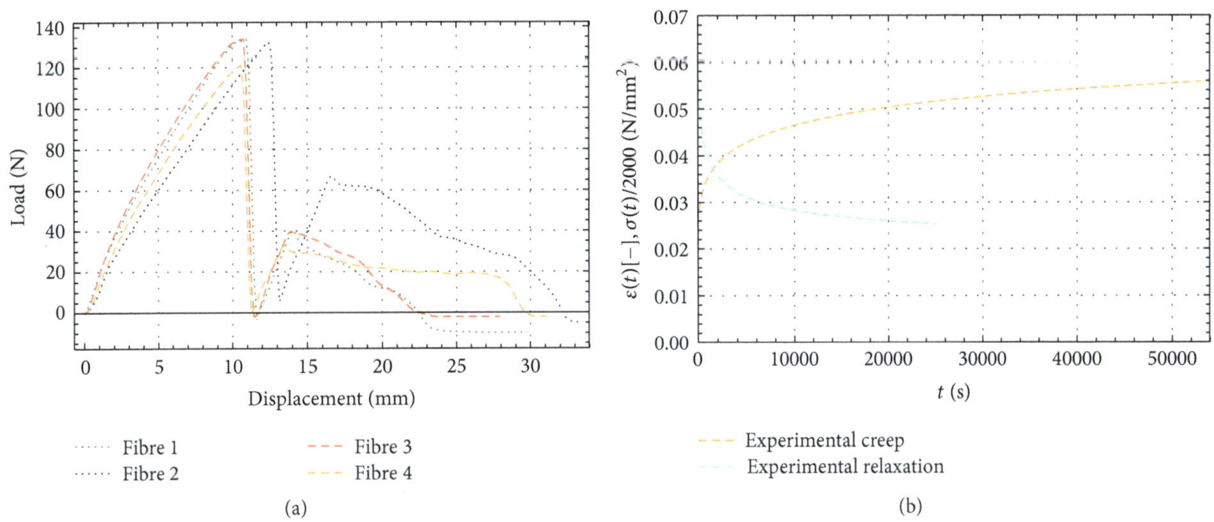

FIGURE 1: (a) Tensile tests until breakage on four specimens of fibres: load versus displacement. (b) Experimental curves: creep versus time (dashed orange curve) and stress relaxation versus time (dashed cyan curve: stress scaled by 1/2000).

TABLE 1: Properties of polypropylene fibre.

Diameter (mm)	0.78
Tensile strength (N/mm^2)	273.0
Elastic modulus E_0 at $t = 0$ (N/mm^2)	5.131×10^3
Elastic modulus E_∞ at $t \to \infty$ (N/mm^2)	1.959×10^3

of use, with a capacity of 250 N. The displacement control was performed by the actuator at a rate of 40 mm/min. The average value of the breakage tensile load is 130.5 N that occurs approximately at a displacement of 10 mm, namely, at 5% of strain.

The elastic Young modulus E_0 (Table 1) was determined from the ratio between stress and strain. The stresses and strains were calculated based on the early stage of the experimental load-displacement curve averaged on the four fibres.

The experiments were performed at 25°C. Since the glass transition temperature of the PP is approximately −20°C [43], at room temperature PP fibres exhibit a mechanical behaviour corresponding to that of a viscoelastic material in rubbery state, that is, a time-dependent response as shown in Figure 1(b).

Plots of breaking tests are reported in Figure 1(a). In particular, the first peak that can be observed in the curve is due to the breakage of the fibres that occurred on the clamped portion, at which followed a pullout stage until the complete detachment of the fibres. The properties adopted for PP fibre are listed in Table 1. The PP fibre was provided by the Italian company Fili&Forme® Ltd. and shown in Figure 3.

2.2. Creep and Relaxation Tests. The creep test was carried out on a sample of length 200 mm subject to a constant tensile force of 60 N by fixing between two clamps and measuring the displacement of the fibre over the time. The load cell is the same used for the breaking test (see Section 2.1) and the speed to gain 60 N was performed at a rate of 40 mm/min. The total duration of the creep test was 15 hours needed for the fibre to achieve a (temporary) stationary value of the deformation (plateau region of the creep curve). However, it should be remarked that the deformation could further increase after 15 hours, since the maximum level of deformation is theoretically achieved at $t \to \infty$.

It is worth noting that at time longer than 20,000 s the strain exceeds 5% but the fibre does not break (Figure 1(b)). Conversely, the fibre under tensile test falls at a strain level equal to 5% (Figure 1(a)). This is due to the fact that the creep test was performed at a constant load of 60 N, less than one-half the ultimate tensile load of 130.5 N.

Similarly, the stress relaxation test was carried out on a sample of length 200 mm subject to a constant displacement of 5 mm and measuring the tensile force over the time. The speed to gain 5 mm was performed at a rate of 40 mm/min. The total duration time was 7 hours, until the force of the fibre was almost stationary. Plots of creep and stress relaxation tests are reported in Figure 1(b).

Both creep and relaxation tests were performed on a single specimen.

The elastic Young modulus E_∞ (Table 1) was assessed based on the final stage of the experimental creep and relaxation tests by knowing the imposed constant load and displacement values at which the tests were performed. The assumed values were the averages of the obtained experimental results.

3. Results and Discussion

3.1. Analytical Model for Creep and Stress Relaxation Curves. We wanted to fit experimental creep and stress relaxation curves with an analytical model by using fraction-exponential kernel that on one side fits carefully experimental data and, at the same time, allows analytical expression for inverse Laplace transform. Let us consider the Boltzmann convolution integral that describes the creep strain $\varepsilon(t)$ of a material under a stress $\sigma(t)$ variable over time

$$\varepsilon(t) = \sigma_0 \psi(t) + \int_0^t \dot{\sigma}(\tau) \psi(t-\tau)\, d\tau, \tag{1}$$

where $\psi(t)$ is the creep function and $\sigma_0 = \sigma(0)$ the stress applied at $t = 0$. Note that the creep function $\psi(t)$ coincides with the axial strain produced by the constant stress $\sigma(t) = 1$. Integrating by parts (1) it follows

$$\varepsilon(t) = \sigma(t) \psi_0 - \int_0^t \sigma(\tau) \dot{\psi}(t-\tau)\, d\tau. \tag{2}$$

By following [34], we write (2) in the form

$$\varepsilon(t) = \frac{1}{E_0}\left[\sigma(t) - \lambda \int_0^t R_\alpha(\beta-\lambda, t-\tau)\sigma(\tau)\, d\tau\right], \tag{3}$$

where

$$\begin{aligned} R_\alpha(\beta-\lambda, t-\tau) &= (t-\tau)^\alpha \sum_{n=0}^{\infty} \frac{(\beta-\lambda)^n (t-\tau)^{n(1+\alpha)}}{\Gamma[(n+1)(1+\alpha)]}, \\ \psi_0 &= \frac{1}{E_0}, \\ \lambda &= \frac{E_0 - E_\infty}{E_0}\beta, \end{aligned} \tag{4}$$

where R_α is the Rabotnov function that allows using fraction-exponential kernel in viscoelastic operators, and E_0, E_∞ are the elastic Young modulus at $t = 0$, $t \to \infty$, respectively. A constant stress $\sigma(t) = \sigma_0$ is then assumed in order to simulate the creep test, so that

$$\begin{aligned} &\int_0^t R_\alpha(\beta-\lambda, t-\tau)\sigma(\tau)\, d\tau \\ &= \sigma_0 \sum_{n=0}^{\infty} \frac{(\beta-\lambda)^n}{\Gamma[(n+1)(1+\alpha)]} \int_0^t (t-\tau)^{n(1+\alpha)+\alpha} d\tau \\ &= \sigma_0 \sum_{n=0}^{\infty} \frac{(\beta-\lambda)^n (t-\tau)^{(n+1)(1+\alpha)}}{\Gamma[(n+1)(1+\alpha)+1]}, \end{aligned} \tag{5}$$

where the property of Euler gamma function $\Gamma(z+1) = z\Gamma(z)$ has been used.

By shifting the index $m = n + 1$, it follows

$$\begin{aligned} &\frac{\sigma_0}{\beta-\lambda} \sum_{n=0}^{\infty} \frac{(\beta-\lambda)^{n+1} t^{(n+1)(1+\alpha)}}{\Gamma[(n+1)(1+\alpha)+1]} \\ &= \frac{\sigma_0}{\beta-\lambda}\left\{\sum_{m=0}^{\infty} \frac{(\beta-\lambda)^m t^{m(1+\alpha)}}{\Gamma[m(1+\alpha)+1]} - 1\right\} \\ &= \frac{\sigma_0}{\beta-\lambda}\left\{M_{1+\alpha}\left[(\beta-\lambda) t^{1+\alpha}\right] - 1\right\}, \end{aligned} \tag{6}$$

where the Mittag-Leffler function,

$$M_a(z) = \sum_{m=0}^{\infty} \frac{z^m}{\Gamma[ma+1]}, \tag{7}$$

has been introduced. Therefore, (3) becomes

$$\varepsilon(t) = \frac{\sigma_0}{E_0}\left[1 - \frac{\lambda}{\beta-\lambda}\left\{M_{1+\alpha}\left[(\beta-\lambda) t^{1+\alpha}\right] - 1\right\}\right]. \tag{8}$$

By using the following properties of the Mittag-Leffler function

$$\begin{aligned} \lim_{t\to 0} M_{1+\alpha}\left[(\beta-\lambda) t^{1+\alpha}\right] &= 1, \\ \lim_{t\to \infty} M_{1+\alpha}\left[(\beta-\lambda) t^{1+\alpha}\right] &= 0, \end{aligned} \tag{9}$$

then, from (8) and (9), it follows

$$\begin{aligned} \varepsilon_0 &= \lim_{t\to 0} \varepsilon(t) = \frac{\sigma_0}{E_0}, \\ \varepsilon_\infty &= \lim_{t\to \infty} \varepsilon(t) = \frac{\sigma_0}{E_\infty}, \end{aligned} \tag{10}$$

where ε_0 and ε_∞ are the creep strains at $t = 0$, $t \to \infty$, respectively.

Again, let us consider the Boltzmann convolution integral that describes the relaxation stress $\sigma(t)$ of a material under a strain variable over time $\varepsilon(t)$.

$$\sigma(t) = \varepsilon_0 \phi(t) + \int_0^t \dot{\varepsilon}(\tau) \phi(t-\tau)\, d\tau, \tag{11}$$

where $\phi(t)$ is the relaxation function and $\varepsilon_0 = \varepsilon(0)$ the strain applied at $t = 0$. The relaxation function $\phi(t)$ coincides with the axial stress produced by the constant strain $\varepsilon(t) = 1$. Integrating by parts (11) it follows

$$\sigma(t) = \varepsilon(t) \phi_0 - \int_0^t \varepsilon(\tau) \dot{\phi}(t-\tau)\, d\tau. \tag{12}$$

By following [34], (12) can be written in the form

$$\begin{aligned} \sigma(t) &= E_0\left[\varepsilon(t) + \lambda \int_0^t R_\alpha(\beta, t-\tau)\varepsilon(\tau)\, d\tau\right], \\ \phi_0 &= E_0. \end{aligned} \tag{13}$$

Assuming a constant strain $\varepsilon(t) = \varepsilon_0$ in order to simulate the stress relaxation test and following the way to find (8), (13) becomes

$$\sigma(t) = E_0\varepsilon_0\left\{1 + \frac{E_0 - E_\infty}{E_0}\left[M_{1+\alpha}\left(\beta t^{1+\alpha}\right) - 1\right]\right\}. \tag{14}$$

For the properties of the Mittag-Leffler function, from (14) it follows

$$\begin{aligned} \sigma_0 &= \lim_{t\to 0} \sigma(t) = E_0\varepsilon_0, \\ \sigma_\infty &= \lim_{t\to \infty} \sigma(t) = E_\infty\varepsilon_0, \end{aligned} \tag{15}$$

where σ_0 and σ_∞ are the stress relaxation at $t = 0$, $t \to \infty$, respectively.

The equations (8) and (14) represent the model by means of the fact that it was possible to fit accurately the experimental creep and stress relaxation tests for suitable values of parameters α and β. The plots of experimental and analytical creep and stress relaxation curves are reported in Figure 2(a) and the corresponding parameters α and β are reported in Table 2.

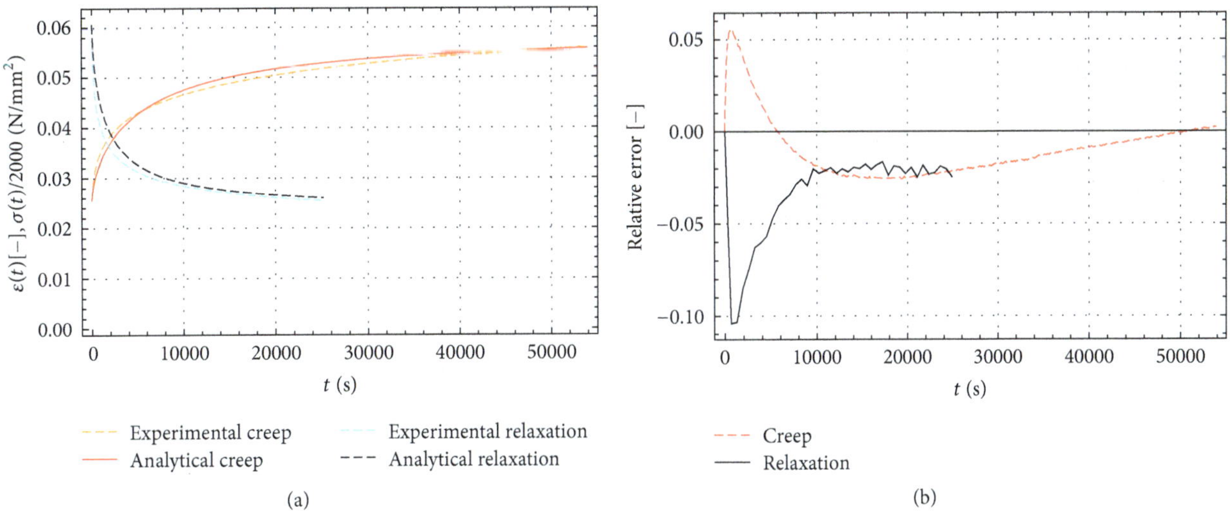

Figure 2: (a) Experimental creep curve (dashed orange curve) and analytical creep curve according to (8) (solid red curve). Experimental stress relaxation curve (dashed cyan curve) and analytical stress relaxation curve according to (14) (dashed blue curve) both scaled by 1/2000. (b) Relative error between experimental data on creep and relaxation and their model given by (8) and (14).

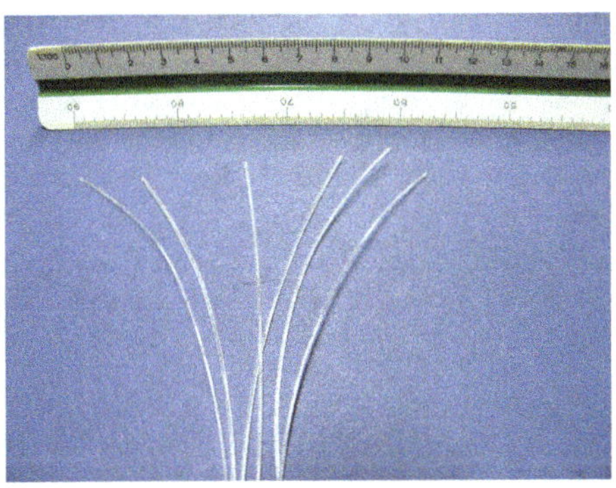

Figure 3: Polypropylene fibres used for tensile until breakage and creep and stress relaxation tests [3].

Table 2: Parameters α and β that fit experimental creep and relaxation curves using the model given by (8) and (14).

α (dimensionless)	$-1/3$
β ($s^{-(1+\alpha)}$)	$-1/(100\sqrt{3})$

4. Conclusions

In the present paper we carried out a tensile test until breakage and a creep and stress relaxation test on a PP fibre and fitted these experimental tests by using a viscoelastic model based on fractional-exponential kernel. The curves plotted in Figure 2(a) show that the theoretical model closely fits the experimental results. The relative error, showing the deviation between the experimental data and the theoretical prediction, is plotted in Figure 2(b). It shows a maximum relative error of 5% and 10%, respectively, for creep and relaxation, occurring in a very small portion of the curves.

Conversely to complex viscoelastic models based on combinations of Maxwell and Kelvin-Voight schemes, the proposed model requires the calibration of only two parameters (α and β) to simulate creep and relaxation phenomena [35]. Moreover, for simple load or displacement histories, it allows calculating straightforwardly the stress and strain fields based on the calculation of the inverse Laplace transform in closed form. Note also that the Rabotnov function generally allows fitting the experimental data better than other conventional schemes based on the combination of springs and dashpots [34].

It has been proved that the model here adopted is able to match carefully experimental data obtained for PP fibres. In a forthcoming work, the present investigation will be extended in order to take into account the effects induced by thermal variations acting on FRC elements (for the thermodynamic aspects of thermoelasticity, see, e.g., [44]).

Competing Interests

The author declares that they have no competing interests.

References

[1] P. K. Mallick, *Fiber Reinforced Composites—Materials, Manufacturing, and Design*, CRC Press, New York, NY, USA, 2008.

[2] L. Lanzoni, A. Nobili, E. Radi, and A. Sorzia, "Axisymmetric loading of an elastic-plastic plate on a general two-parameter foundation," *Journal of Mechanics of Materials and Structures*, vol. 10, no. 4, pp. 459–479, 2015.

[3] P. Di Maida, E. Radi, C. Sciancalepore, and F. Bondioli, "Pullout behavior of polypropylene macro-synthetic fibers treated with nano-silica," *Construction and Building Materials*, vol. 82, pp. 39–44, 2015.

[4] R. Barretta, L. Feo, and R. Luciano, "Some closed-form solutions of functionally graded beams undergoing nonuniform torsion," *Composite Structures*, vol. 123, pp. 132–136, 2015.

[5] R. Luciano and J. R. Willis, “Hashin-Shtrikman based FE analysis of the elastic behaviour of finite random composite bodies,” *International Journal of Fracture*, vol. 137, no. 1, pp. 261–273, 2006.

[6] A. Nobili, L. Lanzoni, and A. M. Tarantino, “Experimental investigation and monitoring of a polypropylene-based fiber reinforced concrete road pavement,” *Construction and Building Materials*, vol. 47, pp. 888–895, 2013.

[7] E. S. Bernard, “Design of fibre reinforced shotcrete linings with macro-synthetic fibres,” in *Proceedings of the 11th International Conference on Shotcrete for Underground Support*, F. Amberg and K. F. Garshol, Eds., vol. P11 of *Engineering Conferences International Symposium Series*, 2009, http://dc.engconfintl.org/shotcrete/14.

[8] L. Lanzoni, A. Nobili, and A. M. Tarantino, “Performance evaluation of a polypropylene-based draw-wired fibre for concrete structures,” *Construction and Building Materials*, vol. 28, no. 1, pp. 798–806, 2012.

[9] G. Xotta, G. Mazzucco, V. A. Salomoni, C. E. Majorana, and K. J. Willam, “Composite behavior of concrete materials under high temperatures,” *International Journal of Solids and Structures*, vol. 64-65, pp. 86–99, 2015.

[10] V. A. Salomoni, C. E. Majorana, G. M. Giannuzzi, and A. Miliozzi, “Thermal-fluid flow within innovative heat storage concrete systems for solar power plants,” *International Journal of Numerical Methods for Heat and Fluid Flow*, vol. 18, no. 7-8, pp. 969–999, 2008.

[11] V. A. Salomoni, C. E. Majorana, B. Pomaro, G. Xotta, and F. Gramegna, “Macroscale and mesoscale analysis of concrete as a multiphase material for biological shields against nuclear radiation,” *International Journal for Numerical and Analytical Methods in Geomechanics*, vol. 38, no. 5, pp. 518–535, 2014.

[12] G. Dinelli, G. Belz, C. E. Majorana, and B. A. Schrefler, “Experimental investigation on the use of fly ash for lightweight precast structural elements,” *Materials and Structures*, vol. 29, no. 194, pp. 632–638, 1996.

[13] A. M. Tarantino, “On the finite motions generated by a mode I propagating crack,” *Journal of Elasticity*, vol. 57, no. 2, pp. 85–103, 1999.

[14] A. M. Tarantino, “Crack propagation in finite elastodynamics,” *Mathematics and Mechanics of Solids*, vol. 10, no. 6, pp. 577–601, 2005.

[15] A. M. Tarantino, “The singular equilibrium field at the notch-tip of a compressible material in finite elastostatics,” *Zeitschrift für Angewandte Mathematik und Physik*, vol. 48, no. 3, pp. 370–388, 1997.

[16] A. M. Tarantino, “On extreme thinning at the notch tip of a neo-Hookean sheet,” *The Quarterly Journal of Mechanics and Applied Mathematics*, vol. 51, no. 2, pp. 179–190, 1998.

[17] A. M. Tarantino, “Nonlinear fracture mechanics for an elastic Bell material,” *The Quarterly Journal of Mechanics and Applied Mathematics*, vol. 50, no. 3, pp. 435–456, 1997.

[18] M. Di Prisco, G. Plizzari, and L. Vandewalle, “Fibre reinforced concrete: new design perspectives,” *Materials and Structures*, vol. 42, no. 9, pp. 1261–1281, 2009.

[19] E. Radi and P. Di Maida, “Analytical solution for ductile and FRC plates on elastic ground loaded on a small circular area,” *Journal of Mechanics of Materials and Structures*, vol. 9, no. 3, pp. 313–331, 2014.

[20] L. Dezi and A. M. Tarantino, “Time dependent analysis of concrete structures with variable structural system,” *ACI Materials Journal*, vol. 88, no. 3, pp. 320–324, 1991.

[21] L. Dezi, G. Menditto, and A. M. Tarantino, “Viscoelastic heterogeneous structures with variable structural system,” *Journal of Engineering Mechanics*, vol. 119, no. 2, pp. 238–250, 1993.

[22] L. Dezi, G. Menditto, and A. M. Tarantino, “Homogeneous structures subjected to repeated structural system changes,” *Journal of Engineering Mechanics*, vol. 116, no. 8, pp. 1723–1732, 1990.

[23] R. Barretta, L. Feo, and R. Luciano, “Torsion of functionally graded nonlocal viscoelastic circular nanobeams,” *Composites Part B: Engineering*, vol. 72, pp. 217–222, 2015.

[24] R. Barretta, L. Feo, R. Luciano, and F. Marotti de Sciarra, “A gradient Eringen model for functionally graded nanorods,” *Composite Structures*, vol. 131, pp. 1124–1131, 2015.

[25] R. Barretta, L. Feo, R. Luciano, and F. Marotti de Sciarra, “Variational formulations for functionally graded nonlocal Bernoulli-Euler nanobeams,” *Composite Structures*, vol. 129, pp. 80–89, 2015.

[26] V. Salomoni, G. Mazzucco, C. Pellegrino, and C. Majorana, “Three-dimensional modelling of bond behaviour between concrete and FRP reinforcement,” *Engineering Computations*, vol. 28, no. 1, pp. 5–29, 2011.

[27] A. Caporale, L. Feo, R. Luciano, and R. Penna, “Numerical collapse load of multi-span masonry arch structures with FRP reinforcement,” *Composites Part B: Engineering*, vol. 54, no. 1, pp. 71–84, 2013.

[28] A. Caporale and R. Luciano, “Limit analysis of masonry arches with finite compressive strength and externally bonded reinforcement,” *Composites Part B: Engineering*, vol. 43, no. 8, pp. 3131–3145, 2012.

[29] P. Bisegna and R. Luciano, “Bounds on the overall properties of composites with debonded frictionless interfaces,” *Mechanics of Materials*, vol. 28, no. 1–4, pp. 23–32, 1998.

[30] A. Apuzzo, R. Barretta, and R. Luciano, “Some analytical solutions of functionally graded Kirchhoff plates,” *Composites Part B: Engineering*, vol. 68, pp. 266–269, 2015.

[31] G. W. Scott-Blair and F. M. V. Coppen, “The subjective judgement of the elastic and plastic properties of soft bodies; the ‘differential thresholds’ for viscosities and compression moduli,” *Proceedings of the Royal Society B: Biological Sciences*, vol. 128, no. 850, pp. 109–125, 1939.

[32] G. W. Scott Blair and F. M. V. Coppen, “The estimation of firmness in soft materials,” *The American Journal of Psychology*, vol. 56, no. 2, pp. 234–246, 1943.

[33] Yu. N. Rabotnov, “Equilibrium of an elastic medium with after-effect,” *Journal of Applied Mathematics and Mechanics*, vol. 12, pp. 53–62, 1948 (Russian).

[34] I. Sevostianov, V. Levin, and E. Radi, “Effective properties of linear viscoelastic microcracked materials: application of Maxwell homogenization scheme,” *Mechanics of Materials*, vol. 84, pp. 28–43, 2015.

[35] F. Di Paola and F. P. Pinnola, *Calcolo Frazionario & Viscoelasticità*, Dipartimento di Ingegneria Civile Ambientale e Aerospaziale, Università degli Studi di Palermo, Palermo, Italy, http://www1.unipa.it/fabio.bagarello/didattica/Di%20Paola%20e%20Pinnola_calcolo%20frazionario.pdf.

[36] Yu. N. Rabotnov, *Elements of Hereditary Solid Mechanics*, Mir, Moscow, Russia, 1977.

[37] S. Rogosin and F. Mainardi, “George William Scott Blair—the pioneer of fractional calculus in rheology,” *Communications in Applied and Industrial Mathematics*, vol. 6, no. 1, article e-481, 2014.

[38] I. Podlubny, *Fractional Differential Equations*, Academic Press, New York, NY, USA, 1998.

[39] F. Mainardi, "Fractional calculus: some basic problems in continuum and statistical mechanics," in *Fractals and Fractional Calculus in Continuum Mechanics*, A. Carpinteri and F. Mainardi, Eds., Springer, Vienna, Austria, 1997.

[40] F. Mainardi, "Applications of fractional calculus in mechanics," in *Transform Methods and Special Functions, Varna '96*, P. Rusev, I. Dimovski, and V. Kiryakova, Eds., SCT, Singapore, 1997.

[41] Yu. A. Rossikhin and M. V. Shitikova, "Applications of fractional calculus to dynamic problems of linear and nonlinear hereditary mechanics of solids," *Applied Mechanics Reviews*, vol. 50, no. 1, pp. 15–67, 1997.

[42] F. Greco and R. Luciano, "A theoretical and numerical stability analysis for composite micro-structures by using homogenization theory," *Composites Part B: Engineering*, vol. 42, no. 3, pp. 382–401, 2011.

[43] B. Ellis and R. Smith, *Polymers A Property Database*, CRC Press, New York, NY, USA, 2nd edition, 2009.

[44] F. Marotti de Sciarra and M. Salerno, "On thermodynamic functions in thermoelasticity without energy dissipation," *European Journal of Mechanics A: Solids*, vol. 46, pp. 84–95, 2014.

6

Wind Load Reduction in Hollow Panel Arrayed Set

Michalina Markousi, Dimitrios K. Fytanidis, and Johannes V. Soulis

Fluid Mechanics/Hydraulics Division, Department of Civil Engineering, Democritus University of Thrace, 67100 Xanthi, Greece

Correspondence should be addressed to Michalina Markousi; mmarkous@civil.duth.gr

Academic Editor: Valerio Lo Brano

Reducing the wind loading of photovoltaic structures is crucial for their structural stability. In this study, two solar panel arrayed sets were numerically tested for load reduction purposes. All panel surface areas of the arrayed set are exposed to the wind similarly. The first set was comprised of conventional panels. The second one was fitted with square holes located right at the gravity center of each panel. Wind flow analysis on standalone arrayed set of panels at fixed inclination was carried out to calculate the wind loads at various flow velocities and directions. The panels which included holes reduced the velocity in the downwind flow region and extended the low velocity flow region when compared to the nonhole panels. The loading reduction, in the arrayed set of panels with holes ranged from 0.8% to 12.53%. The maximum load reduction occurred at 6.0 m/s upwind velocity and 120.0° approach angle. At 30.00 approach angle, wind load increased but marginally. Current research work findings suggest that the panel holes greatly affect the flow pattern and subsequently the wind load reduction. The computational analysis indicates that it is possible to considerably reduce the wind loading using panels with holes.

1. Introduction

Renewable energy resources trends show an increase in solar energy usage. Existing photovoltaic installation system methods may differ in their design but all are strongly influenced by wind loads. Extreme winds can cause damage and reduce the photovoltaic systems effectiveness. High wind loads effects can possibly create the need for costly maintenance or even panel replacement. Given the growing demand for solar energy, the studies that have been conducted in the field of photovoltaic systems aerodynamic load reduction analysis are relatively few.

Chevalien and Norton [1] conducted a study which addresses the wind load problem for roof-mounted solar collector panels and their support structures. Kopp et al. [2] have carried out experimental studies on the torque evaluation created due to wind on photovoltaic arrays arranged in series. The peak system torque was generally observed to occur at approach azimuthal angles near the diagonals of the panel at 45.0°, 135.0°, 225.0°, and 315.0° although large loads also occurred at 270.0°, where wind is in the plane of the panel perpendicular to the individual modules. Chung et al. [3] conducted experimental study of a scaled commercial system to investigate the mean surface pressure distributions and uplift forces at different wind speeds resulting in the following: the wind uplift is effectively reduced with a guide plate placed normal to wind direction. Chou et al. [4] investigated experimentally the effect of a crosswind on the aerodynamic characteristics of two different solar water heaters setups. The first one is a typical residential solar water heater and the second one is an inclined flat plate without the horizontal cylinder. The results revealed that at the near upwind corner the higher suction and fluctuating pressure are observed and that the presence of the horizontal cylinder increases the lift force.

Recent developments in computer technology and computing engineering make the application of Computational Fluid Dynamics (CFD) in wind engineering quite encouraging. Shademan and Hangan [5] carried out simulations to estimate the wind loads for various wind directions on standalone and arrayed panels. The study identified areas facing the greatest impact from wind loads and also observed that at a specific distance between two sets of panels the drag coefficient for the downstream sets of panels reaches a minimum. One year later, Shademan and Hangan [6] investigated wind loads on an array of panels for various wind directions using

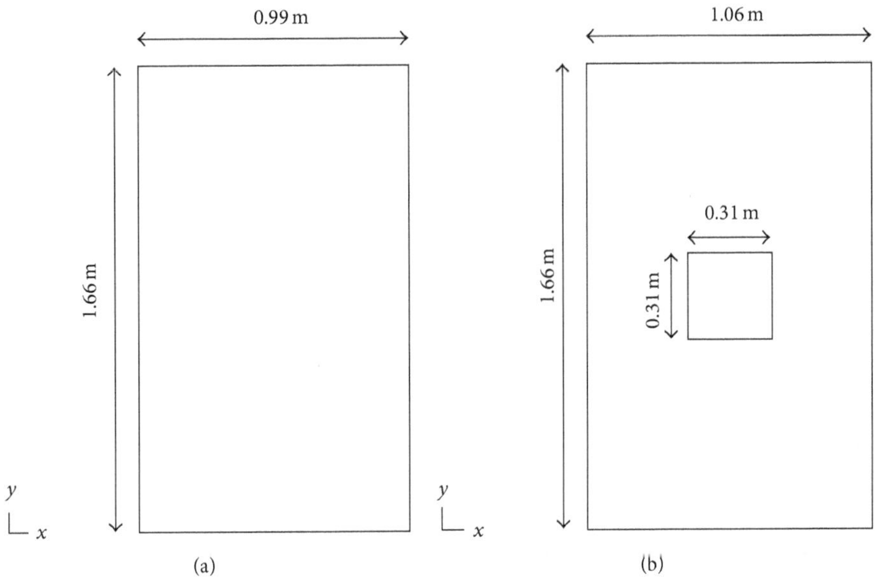

FIGURE 1: (a) Typical solar panel 0.99 × 1.66 m × m and (b) a 0.31 × 0.31 m × m rectangular shape hole cut at the panel gravity center. The solar panel with hole measures 1.06 × 1.66 m × m. Each panel type exposes to wind the same surface area.

CFD. The simulations identified the corner panels as the critical ones while more critical angles of wind incidence are 0.0° and 180.0°. Bitsuamlak et al. [7] conducting a similar study attempted to investigate the aerodynamic features of ground-mounted panels under atmospheric boundary layer flows using numerical techniques. The Reynolds Averaged Navier Stokes (RANS) equations turbulence modeling approach was adapted to obtain initial flow conditions for use by the more reliable Large Eddy Simulation (LES) technique. The calculated results have been compared and validated with full-scale experimental measurements concluding that LES underestimated the mean pressures compared to full-scale measurements. Meroney and Neff [8] used numerical analysis to calculate the wind loads on solar photovoltaic collector in order to estimate drag, lift, and overturning moments on different collector support systems. Results were compared with experimental measurements performed in wind tunnel. The comparison showed that the turbulence models that better approach the experimental measurements are the RNG k-ε and k-ω, whereas the k-ε model failed to replicate measurements. Jubayer and Hangan [9] used the shear stress transport (SST) k-ω turbulence closure for unsteady 3D numerical simulations in order to investigate the wind load and flow field around a ground-mounted stand-alone photovoltaic (PV) system. Simulation results compared with the wind tunnel measurements and an agreement within 46% is found.

In this study 3D numerical simulations were performed to investigate the wind load reduction on photovoltaic structures. Two arrayed sets of panels are considered to reduce the aerodynamic load: a standard photovoltaic structure (conventional) and a similar structure but with square holes evenly distributed exposing to the wind the same panel surface area. For each arrayed set of panels, a parametric analysis is performed relatively to approach angle and wind intensity. Purpose was to investigate the quantitative wind load reduction under the panels with holes presence.

TABLE 1: Panel dimensions and weight.

Length (mm)	1662.0 (±2.5)
Width (mm)	990.0 (±2.5)
Thickness (mm)	46.0
Weight (kg)	21.0

2. Methodology

2.1. Geometry and Computational Grid. The conventional panel dimensions are 1662.0 mm long, 990.0 mm wide, and 46.0 mm thick, Table 1. Figure 1 shows the two geometries of the solar panel types used for computational analysis. A rectangular shape hollow, measuring 0.31 × 0.31 m × m, is cut at the panel gravity center for the panel shown in Figure 1(b). The actually tested geometry is comprised of an arrayed set of 12 panels arranged in 3 rows, Figure 2. Each row encompasses 4 panels and the gap between panels measures 0.02 m. Each arrayed set of panels exposes to the wind the same panel surface area. This was on purpose done for wind loading comparison reasons and to ensure that the effective area, that is, the solar energy production, for both solar panels is equal.

The arrayed set of panels is placed at inclination angle β of 26.0° in relation to the flat ground surface, Figure 3(a). Longest panel axis is set parallel to north-south direction. However, a small gap 0.3 m was on purpose left between the lower panel end and the ground surface, Figure 3(a). This configuration closely approximates the reality. A schematic diagram showing the approach azimuthal angle is shown in Figure 3(b).

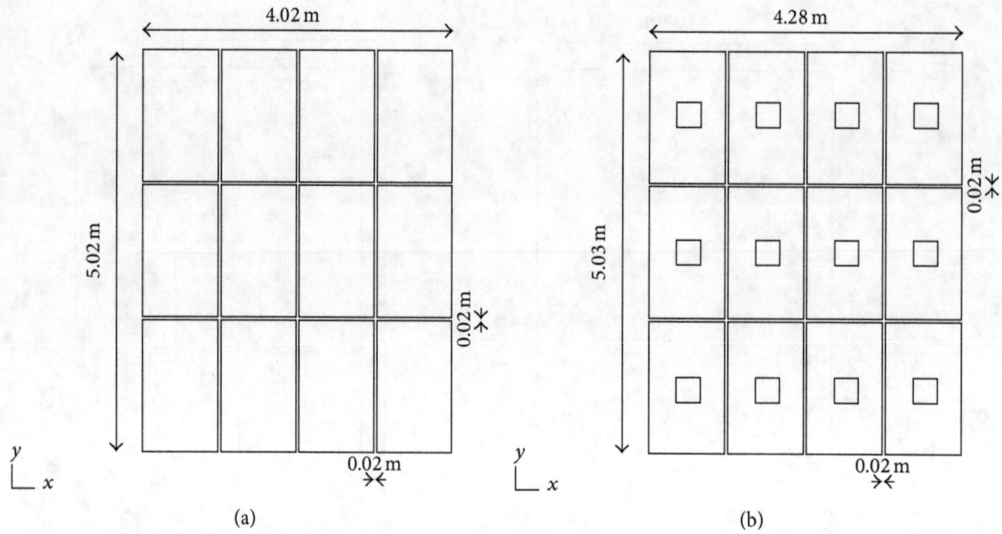

FIGURE 2: (a) Arrayed set of conventional solar panels 4.02 × 5.02 m × m with gaps and (b) arrayed set of holes with solar panels 4.28 × 5.03 m × m with gaps. Panel types expose to the wind the same surface area.

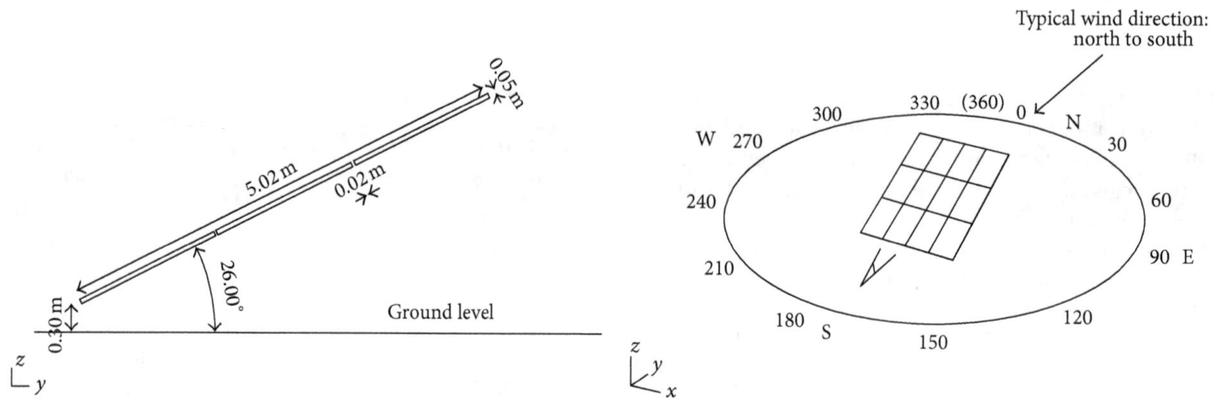

FIGURE 3: (a) Solar panels side view inclined at angle $\beta = 26.0°$ and (b) schematic diagram of the arrayed solar panels set. Longest panel axis is set parallel to north-south direction. The north is 0.0°. Azimuthal angles in degrees.

Figure 4 shows the utilized area for computational analysis. The geometry was constructed using the ANSYS-Design Modeler software and the grid generation was formed with the ANSYS-Meshing software package. The computational area is widely extended in all three dimensions to leave enough space for flow to be developed. A typical computational grid is shown in Figure 5. Grid nodes are densely located near to the panel flow regions. Some grid details in regions between panels (gaps) are shown in Figure 6. For a typical computational analysis, 1037091 grid nodes and 3524682 cells were used; see also Section 3.1. The number of the nonuniformly distributed grid nodes was considered to be satisfactory for subsequent analysis. The decision was made after numerical experimentation.

2.2. Governing Wind Flow Equations, Boundary Conditions, and Assumptions. The utilized numerical code utilizes the ANSYS-Fluent 14.0 solver. The wind flow is assumed to be 3D, steady, homogeneous, and incompressible with no external forces applied to it. To simulate the turbulence modeling the steady state RANS equations are solved. In their generality these equations are written as

$$\begin{aligned} &\frac{\partial}{\partial x_i}(\rho u_i) = S_m \quad \text{Continuity} \\ &\frac{\partial}{\partial x_j}(\rho u_i u_j) = -\frac{\partial p}{\partial x_i} + \frac{\partial \tau_{ij}}{\partial x_j} + \rho g_i + F_i \quad \text{Momenta} \end{aligned} \tag{1}$$

with

$$\tau_{ij} = \left[\mu\left(\frac{\partial u_i}{\partial x_i} + \frac{\partial u_i}{\partial x_i}\right)\right] - \frac{2}{3}\mu\frac{\partial u_i}{\partial x_i}\delta_{ij} \quad \text{Shear stress.} \tag{2}$$

x_i represent the Cartesian coordinate positions in the longitudinal, transverse, and vertical directions, respectively; ρ is the wind (air) density; S_m are sources terms (zero in our case);

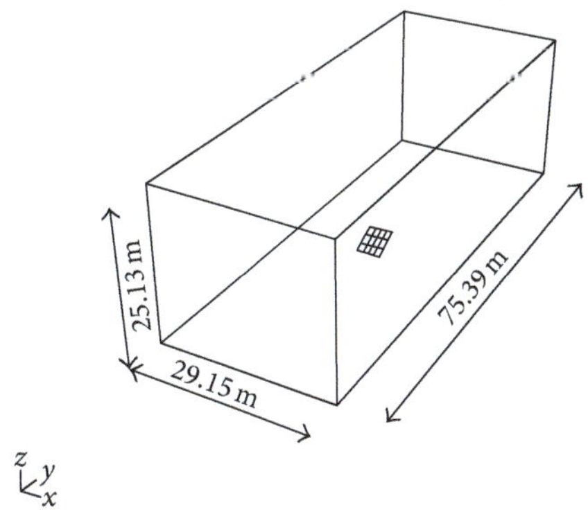

FIGURE 4: Arrayed set of solar panels and computational space used for analysis. The space geometry reads 75.39 m in length, 29.15 m in width, and 25.13 m in height.

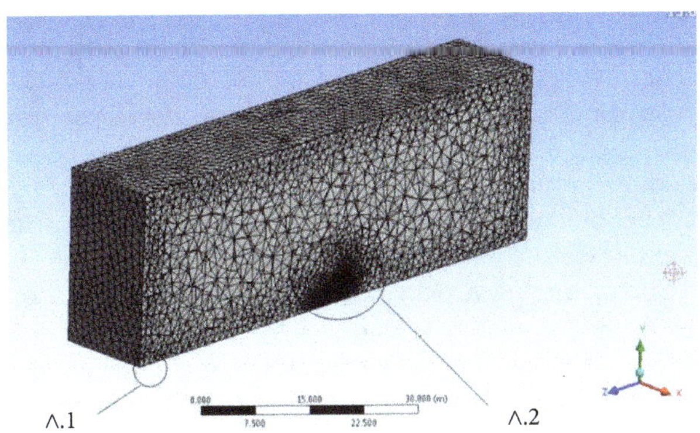

FIGURE 5: The computational grid is comprised of 1037091 grid nodes. Nodes are densely located close to solar panels.

u_i are the velocity components in x_i directions; p is the static pressure; g_i is the gravity acceleration along the i-axis; F_i are the applied external forces; μ is the molecular viscosity; δ_{ij} is the Kronecker symbol; and τ_{ij} are the Reynolds stresses.

For panel wall surfaces as well as ground surfaces the no-slip boundary condition was adopted; that is, velocity components at the wall face must be zero. The symmetry conditions are applied for all free-side walls (surfaces) as this boundary condition is widely applied, Stangroom [10]. For the downstream surface a "pressure outlet" boundary condition is adopted, which corresponds to Dirichlet boundary condition and Neumann boundary condition for the velocities field. Table 2 shows the inflow boundary conditions used for computational analysis.

Examined independent variables are the wind velocity and the approach azimuthal angle. Wind velocity ranges from 6.0 m/s to 14.0 m/s with 2.0 m/s step. The approach angles range from 360.0° (or 0.0°, north) to 180.0° with 30.0° step, excluding the 90.0° approach angle. The above range flow conditions cover typical applications regarding the north Hellenic State latitudes. The air density and molecular viscosity are set 1.225 kg/m^3 and 1.7894 × 10^{-5} kg/m-s, respectively. The air turbulence intensity is set 16.0% at the inlet flow plane, Bitsuamlak et al. [7], and the turbulent length scale is calculated as 3.5182 (= 0.7 × panels characteristic length of 5.03 m). For ground surface, the roughness height is set 0.03 m with roughness constant 0.5. For panels, the roughness height is set to 0.0000015 m with 0.5 roughness constant, Blocken et al. [11].

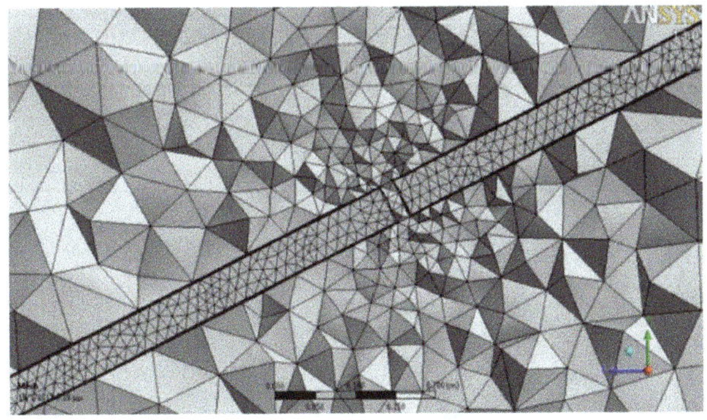

FIGURE 6: Computational grid gap details between solar panels.

TABLE 2: Approach azimuthal angles and velocity range for computational analysis.

Angle (degrees)	Upwind velocity (m/s)				
0.0	6.0	8.0	10.0	12.0	14.0
30.0	6.0	8.0	10.0	12.0	14.0
60.0	6.0	8.0	10.0	12.0	14.0
120.0	6.0	8.0	10.0	12.0	14.0
150.0	6.0	8.0	10.0	12.0	14.0
180.0	6.0	8.0	10.0	12.0	14.0

TABLE 3: Number of cells in million for the used meshes.

Cells	Coarse	Medium	Fine
	2.4 × 10^6	3.5 × 10^6	4.0 × 10^6

The aforementioned set of partial differential equations was solved using the numerical algorithms included in the commercial solver ANSYS-Fluent. For the spatial discretization, a first-order accuracy numerical scheme was utilized for all equations. The Semi-Implicit Method for Pressure-Linked Equations (SIMPLE) algorithm was applied for the velocity pressure coupling, Patankar [12], while the "RNG k-ε" turbulence model was used for the turbulence closure, with "standard wall functions" for the wall boundary treatment.

3. Results

3.1. Mesh Convergence Study. To investigate any differences in the simulation data resulting from mesh resolution, three unstructured Delaney tetrahedral grids were generated in ANSYS-Meshing. The number of cells used is shown in Table 3.

The mesh convergence analysis was performed at 30.0° approach angle and 14.0 m/s upwind velocity. The selected vertical profile location for mean velocities plotting is shown in Figure 7. Streamwise and lateral velocities profiles show good agreement with coarse, medium, and fine meshes, shown in Figure 8, with reference velocity U_{ref} equal to

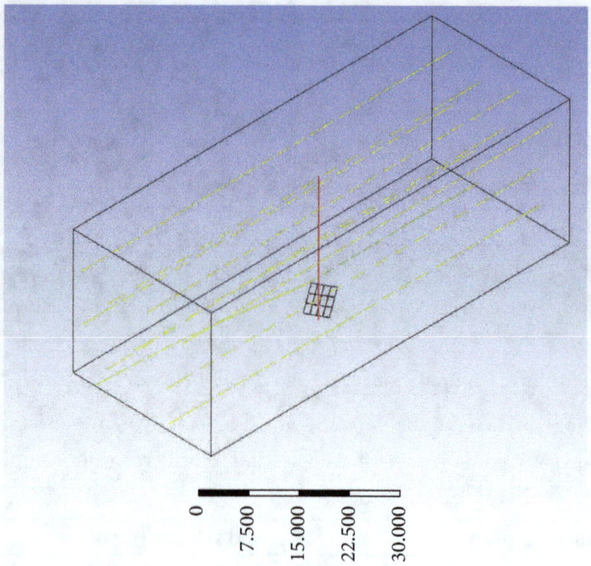

FIGURE 7: Profile location of the velocities plot for mesh convergence.

14.0 m/s. No significant differences were observed for the velocity profiles using the three meshes.

3.2. Static Pressure at the Upwind and Downwind Panel Sides. A schematic diagram over the inclined solar panel is shown in Figure 9. High and low pressure fields are developing up- and downwind of the panel flow regions, respectively.

Only a very limited amount of information results is given here. Contour graphics were performed using the CFD-POST software package. All currently reported research work results refer to 0.0° approach angle and 14.0 m/s (maximum available) upwind velocity unless otherwise noted. At this angle the arrayed longest panel axis is parallel to the north-south direction, Figure 3(a). The wind velocity magnitude is considered as a moderate one but realistic for analysis. Ogedengbe [13] in his analysis reports maximum wind velocities up to 23.0 m/s. At the same time, the approach angle was set 0.0° because this is the dominant wind direction in north Hellenic State latitudes.

Static pressure (N/m^2) contours for the up- and downwind conventional panel sides are shown in Figure 10. The upwind surface exhibits high wind pressure. The lower downwind panel surface part exhibits higher pressure than the upper one due to the escape corridor developing underneath. The downwind panel surface is less loaded. Panel gaps alter the pressure field mostly at the downwind panel region. Near symmetry of pressure flow distribution is preserved in either panel side. Static pressure contours at panels with holes for the up- and downwind panel sides are shown in Figure 11. Rectangular panel holes alter the static pressure flow field in the nearby regions. The near symmetry of the flow field is again preserved. At each hole, the upper side exhibits higher static pressure than the lower one. The overall static pressure pattern is altered in either panel side in comparison to the nonholes panel. The pressure differences across the panels with holes are now reduced. Again, the lower upwinding surfaces for either panel formation are highly loaded due to escape corridor between panels and ground surface.

3.3. Side View of Velocity and Static Pressure. To elucidate the flow velocity pattern, it was decided to present the results in two arrayed set panel cross sections, shown schematically in Figure 12. Section 1 plane is located in meridional direction (north to south). It divides the arrayed set of panels into two regions leaving two panel series on the left- and two on the right-hand side. Its actual location is adjacent to the central panel gap. Section 2 is a plane perpendicular to Section 1 (mid arrayed panel height). It divides the arrayed set of panels into two parts, namely, the upper half and the lower one.

Velocity magnitude (m/s) and static pressure (N/m^2) contours at conventional panels over Section 1 are shown in Figures 13(a) and 13(b), respectively. Top velocity magnitude reaches 20.0 m/s just above the panels tip at the downstream panel side. A wide rirculation area develops at the downstream flow region. Part of the wind escapes between the lower tip panel and ground surface, Figure 13(a). High pressure difference develops between up- and downwind panel surface and pushes it, Figure 13(b).

Corresponding velocity magnitude (m/s) and static pressure (N/m^2) contours at panels with holes are shown in Figures 14(a) and 14(b), respectively. Both velocity and pressure field patterns are drastically altered in comparison to the nonhole panels. In this case the maximum velocity magnitude falls to 17.56 m/s. No distinct recirculation region is present unlike panels with nonhole holes, Figure 13(a). The pressure differences between panel sides have now been reduced. Highest pressure differences occur at the panels tip. The pressure field at the downwind side is relatively uniform compared to the nonhole panel.

Velocity magnitude (m/s) and static pressure (N/m^2) contours for conventional panels at Section 2 (Figure 12) are shown in Figures 15(a) and 15(b), respectively. The flow field is symmetrical and the maximum velocity magnitude occurs at the far downwind flow region. Small strength vortices develop at the left and right panel tips. Gaps between panels disrupt the flow velocity field. High static pressures are uniformly present at the upwind surface. Low static pressure values are to be found in the near to downwind flow region, thus giving rise for pressure differences across the panel ~ 240.0 N/m^2, Figure 15(b).

Corresponding velocity magnitude (m/s) and static pressure (N/m^2) contours at Section 2 (Figure 12) for panels with holes are shown in Figures 16(a) and 16(b), respectively. It is again evident that the pressure differences between up- and downwind surfaces now stand at ~135.0 N/m^2 and are drastically altered (reduced) compared to the nonhole panels.

3.4. Side View Streamlines. Velocity magnitude (m/s) streamlined contours at conventional panels and panels with holes are shown in Figures 17(a) and 17(b), respectively. A large vortex has been developed at the downwind side, Figure 17(a). The vortex covers most of the adjacent downwind panel flow

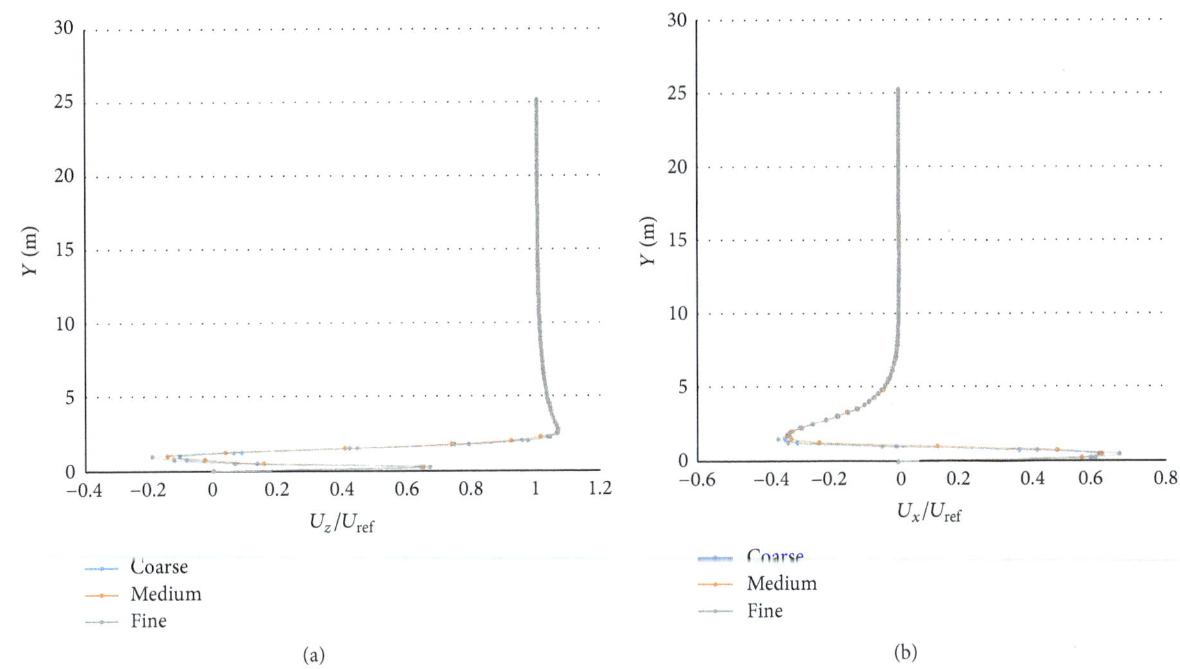

FIGURE 8: (a) Streamwise and (b) lateral wind velocity (m/s) profiles at location shown in Figure 7 using coarse, medium, and fine meshes.

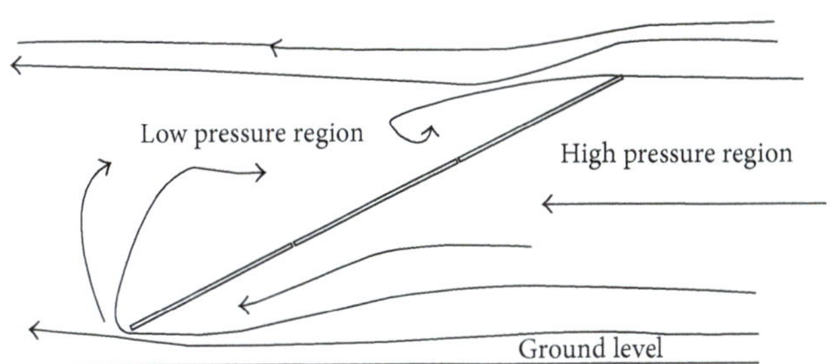

FIGURE 9: Schematic wind loading for inclined solar panel. The approach azimuthal angle is 0.0° (north to south).

region. In Figure 17(b) the vortex has been replaced with two smaller strength vortices. This phenomenon resulted from flow mixture through holes and flow over and under passing panels tips. The holes existence reduced the velocity magnitude downwind of panels flow region and extended the low velocity flow region.

3.5. *Wall Shear Stress.* Wall shear stress (N/m^2) contours at panels for up- and downwind sides are shown in Figures 18(a) and 18(b), respectively. For panels with holes, the wall shear stress contours at up- and downwind sides are shown in Figures 19(a) and 19(b), respectively. Maximum values reach 1.0 N/m^2 for either panel configuration. As it is expected and as it was previously mentioned, the holes drastically alter the flow field. Higher values appear on the upwind panel's side in either case. Panels with holes reduce the wall shear stress values. This particularly holds for the downwind side, Figure 19(b). At 120.0° approach angle and maximum upwind velocity the wall shear stress contours at up- and downwind side are shown in Figures 20(a) and 20(b), respectively. The flow field has now been completely altered with high wall shear stresses.

TABLE 4: Percentage wind load reduction. Negative sign indicates increase in load.

Velocity/angle	0.0°	30.0°	60.0°	120.0°	150.0°	180.0°
6.0 (m/s)	5.16	−0.32	2.86	12.53	6.28	0.96
8.0 (m/s)	6.46	−0.32	2.88	12.39	5.40	0.66
10.0 (m/s)	5.18	−0.41	2.80	11.83	4.60	2.80
12.0 (m/s)	5.86	−0.48	2.86	11.22	4.60	0.80
14.0 (m/s)	6.18	−0.44	2.79	12.43	4.60	1.72

3.6. *Loads.* The load calculation is based on the area-average pressure differences between up- and downwind panel side surfaces. In percentage terms the wind loading does not follow the increased upwind velocity and subsequently the increased loading. The maximum wind loading, in descending approach azimuthal angles, order is 30.0°, 0.0°, 180.0°, 60.0°, 120.0°, and 150.0°. The 0.0° approach azimuthal angle case, under all upwind velocities, yields the highest load reduction. Percentage wind load reduction between conventional panels and panels with holes is shown in Table 4. Percentage load reductions appear at all approach angles except at the 30.0%. Even in this case the load increase is marginal. The loading reduction, in the arrayed set of panels with holes, ranges from 0.8% to 12.53%. At 0.0° approach angle, the loading reduction for the upwind velocities ranges between 5.16% and 6.46%. Under all upwind velocities, the

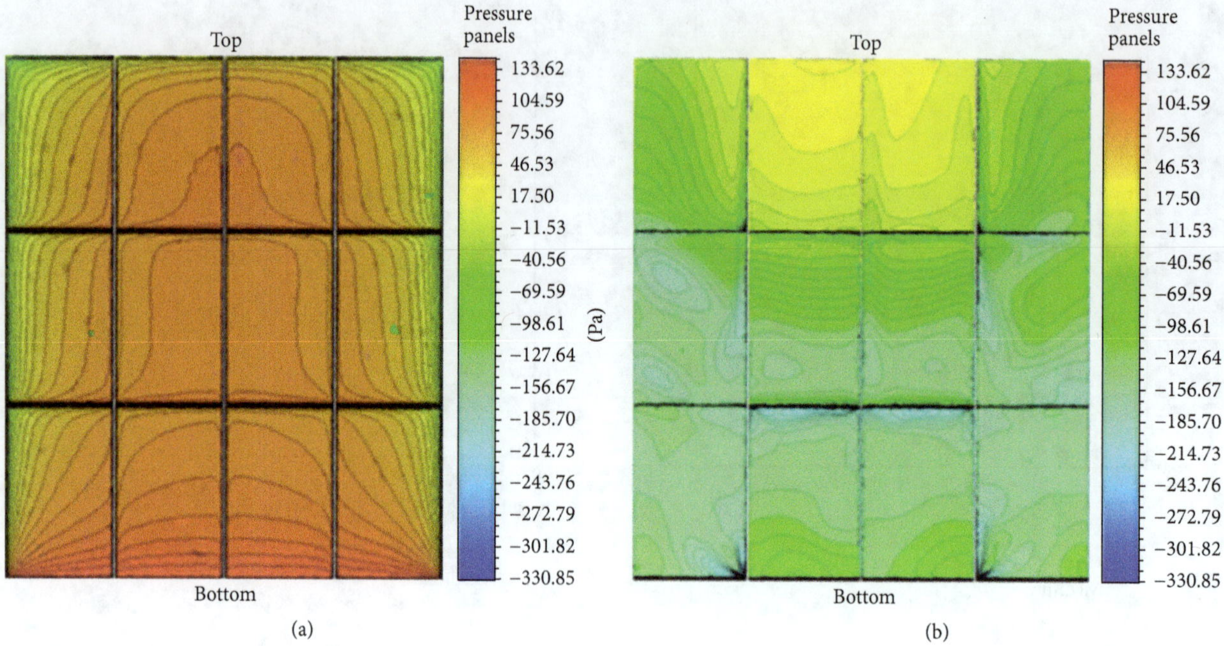

FIGURE 10: Static pressure (N/m^2) contours at conventional panels: (a) upwind side and (b) downwind side.

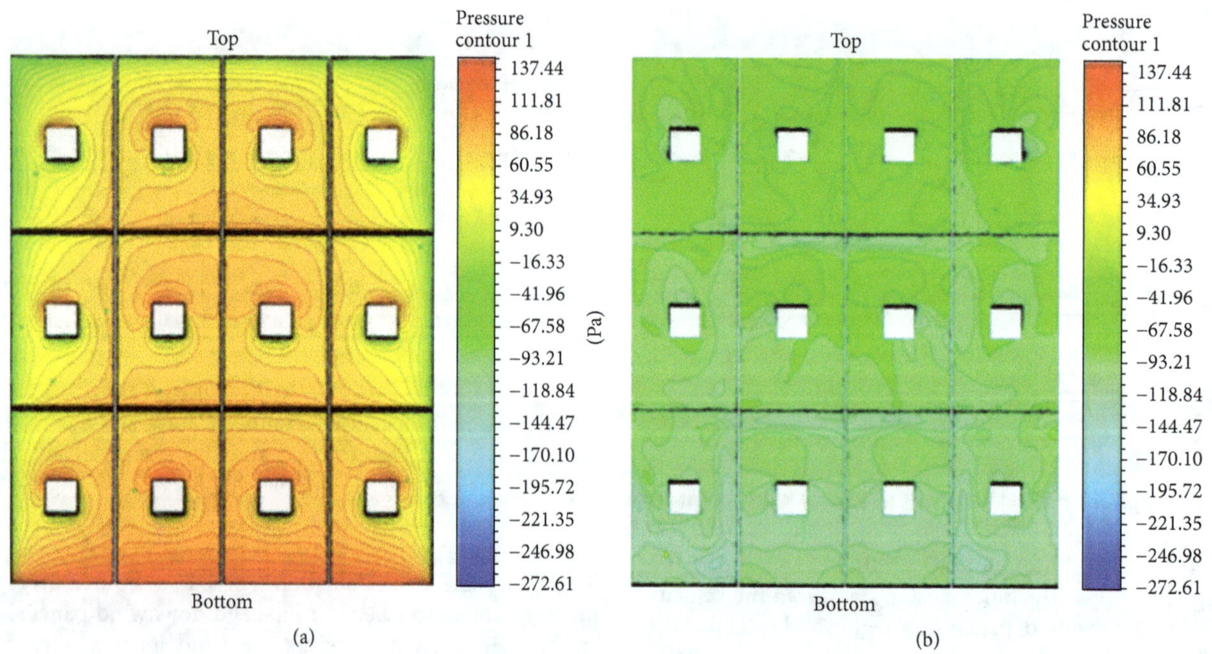

FIGURE 11: Static pressure (N/m^2) contours at panels with holes: (a) upwind side and (b) downwind side.

percentage load reduction is higher at 120.0°. The maximum load reduction occurs at 6.0 m/s upwind velocity and 120.0° approach angle.

3.7. Velocity and Approach Azimuthal Angle. Velocity magnitude (m/s) contours at maximum upwind velocity for panels with holes and (a) 0.0°, (b) 30.0°, (c) 60.0°, (d) 120.0°, (e) 150.0°, and (f) 180.0° approach angles (Section 2, Figure 12) are shown in Figure 21. The velocity flow field is drastically altered with the approach angle. At flow regions adjacent to upwind side, the velocity magnitude is relatively high in comparison to the adjacent downwind flow region. Maximum velocity magnitude is ~19.0 m/s. Within holes the velocity magnitude increases due to flow area reduction. Symmetrical flow field develops at 0.0° and 180.0° approach azimuthal angles. The velocity attains its maximum value at panel corners which the flow at first encounters. This is true for all approach azimuthal angles.

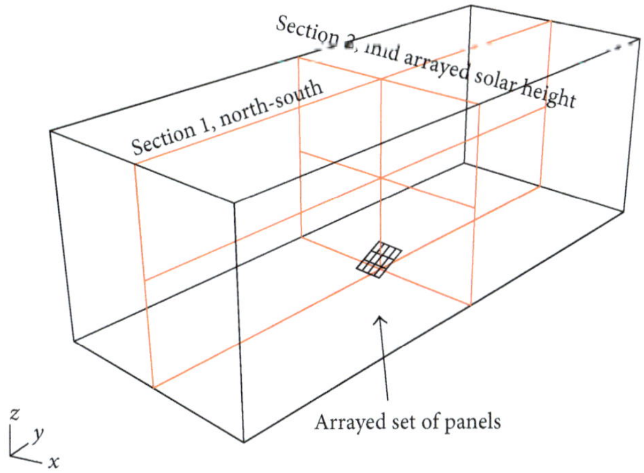

FIGURE 12: Schematic diagram showing two arrayed set solar panel cross sections used for results presentation.

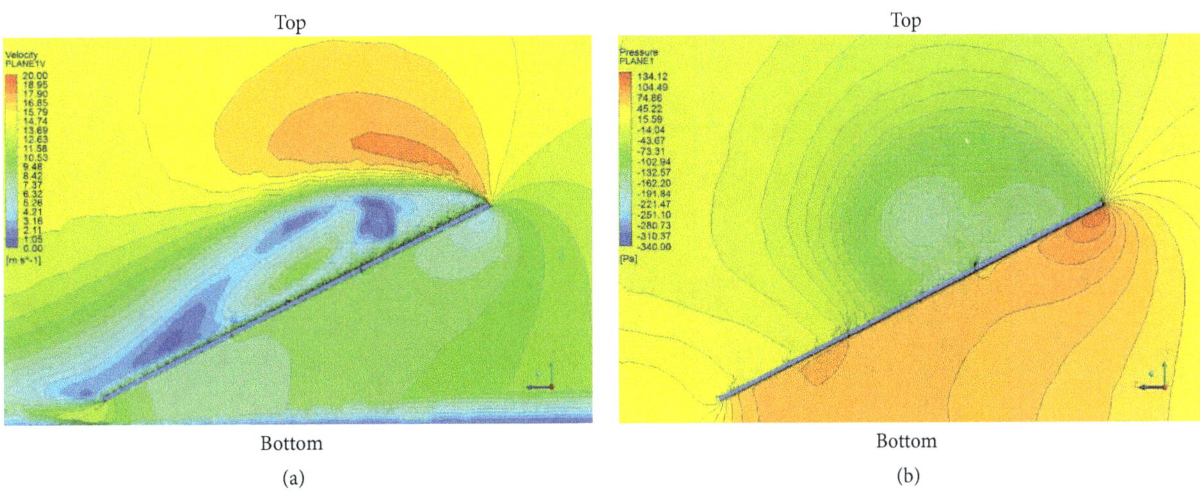

FIGURE 13: (a) Velocity magnitude (m/s) and (b) static pressure (N/m^2) contours at conventional panels, Section 1, Figure 12, wind from right to left.

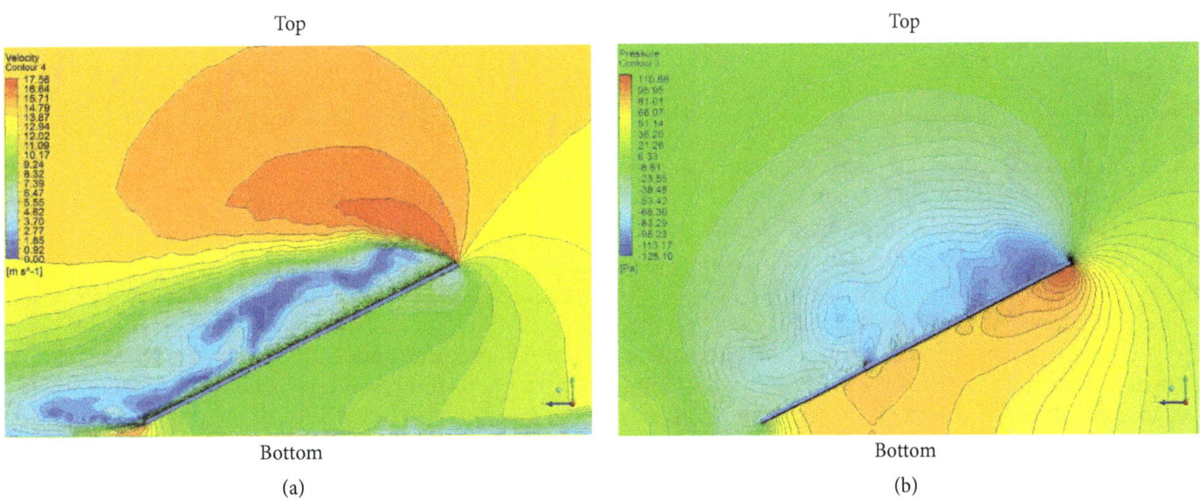

FIGURE 14: (a) Velocity magnitude (m/s) and (b) static pressure (N/m^2) contours at panels with holes, Section 1, Figure 12, wind from right to left.

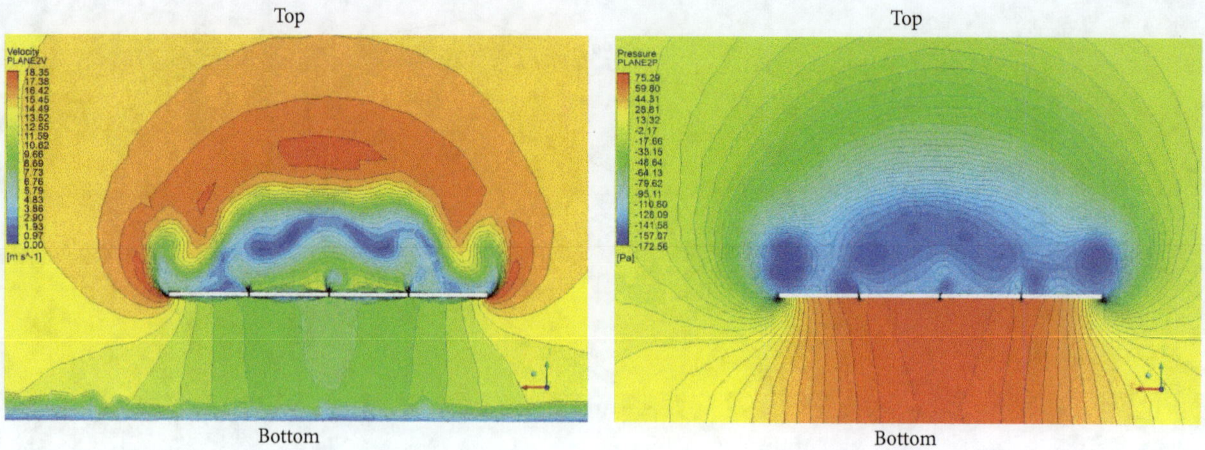

(a) (b)

FIGURE 15: (a) Velocity magnitude (m/s) and (b) static pressure (N/m^2) contours at conventional panels, Section 2, Figure 12, wind from bottom to top.

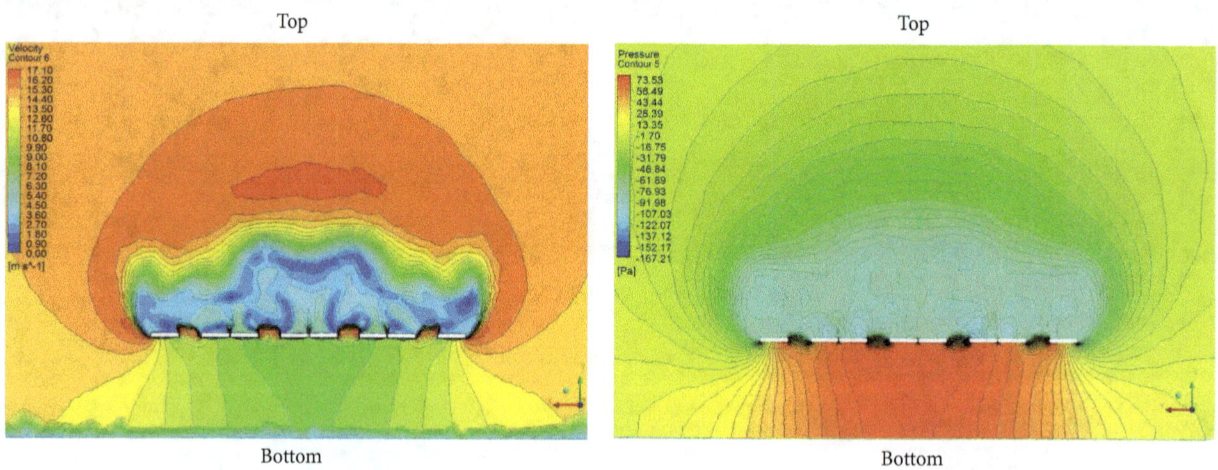

(a) (b)

FIGURE 16: (a) Velocity magnitude (m/s) and (b) static pressure (N/m^2) contours at panels with holes, Section 2, Figure 12, wind from bottom to top.

4. Discussion

The wind forces acting on panels can cause severe mechanical problems. Knowing the flow field around panels and its interaction with the structure is of paramount importance in minimizing the structural damaging effects. Panel loading depends on wind speed and its direction, height of the panel above the ground, and the relative inclination angle of the panel over the ground surface. An accurate determination of the wind loading is an important factor in deciding what degree of safety and economy can be achieved. Current analysis does not perform parametric analysis related either to panel height location above ground surface or to panel inclination effects. Furthermore, the ground roughness effects are not included in the analysis. Main purpose is to investigate the wind loading using panels with holes and to compare the load reduction from a conventional arrayed set of panels. Each panel type exposes to the wind the same panel surface area. Henceforth, the load reduction would be meaningful.

Higher wall shear stresses develop on the upwind side and mostly in regions close to the perimeter of the panel holes. Always wall shear stresses are considerably lower to static pressures (loads). Large values occur at maximum upwind velocity and 30.0° approach angle. High upwind velocities and not approach angles are responsible for large wall shear stresses. The developed flow pattern is a highly complex one and this is more evident at the downwind sides of the panels (fitted or not with holes).

High positive wind forces, that is, forces acting in the direction up- to downwind panel surface, develop with panels fitted with holes at 0.0°, 30.0°, and 60.0° approach angles. However, for the 120.0°, 150.0°, and 180.0° approach angles the emerged forces are negative in the sense that they act in the

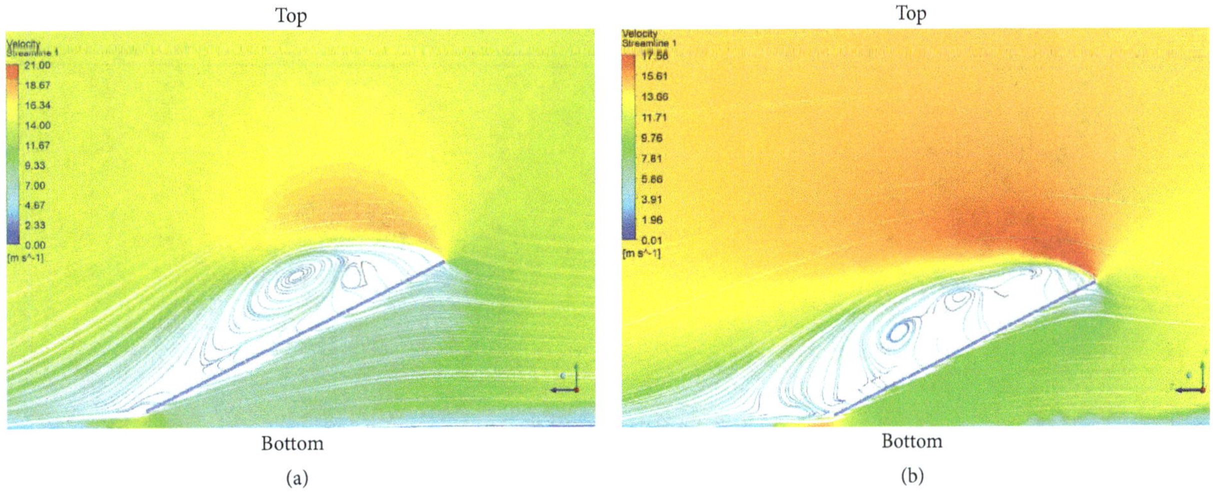

FIGURE 17: Streamlined velocity magnitude (m/s) contours at (a) conventional panels and (b) panels with holes, Section 1, Figure 12, wind from right to left.

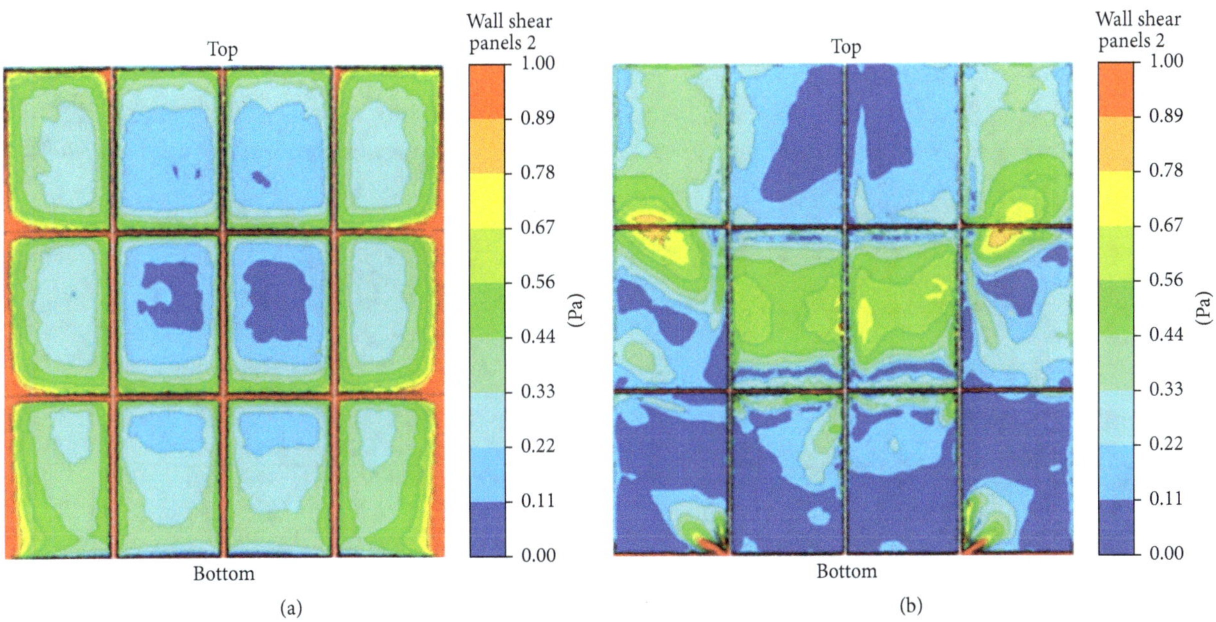

FIGURE 18: (a) Wall shear stress (N/m^2) contours at conventional panels: (a) upwind side and (b) downwind side.

direction down- to upwind panel surface. Force differences acting on panels fitted with holes are less powerful since the wind finds effective escape corridors through the panel holes and gaps. The holes within each panel and the gap between panels allow high momentum (velocity) wind to flow through and reduce the windward pressure, as well as the negative pressure acting on the back panels side. In Tables 5 and 6, the acting forces are appearing in descending order for conventional panels and for those fitted with holes, respectively. Number 1 indicates the highest possible wind force while 30 denotes the lowest one. Bold font values in Tables 5 and 6 indicate that the wind force ranking has been changed. As it is expected, the forces are higher at the maximum upwind velocity. However, the maximum wind load reduction in percentage terms does not follow the increased upwind velocity. At 90.0° approach angle the acting forces over the panel surface are minimal since the wind flows parallel to side panel walls (facing the panel width). The maximum upwind velocity at 30.0° approach angle yields the highest possible force on the panels with either conventional or solar panels fitted with holes. The maximum wind loading in descending order for the approach angles is 30.0°, 0.0°, 180.0°, 60.0°, 120.0°, and 150.0°.

The wind loading reduction using holes in the arrayed set of panels ranges from 0.8% to 12.53%. The maximum load reduction occurs at 6.0 m/s upwind velocity and 120.0° approach angle. Under all tested upwind velocities, the percentage load reduction is higher at this angle. For the

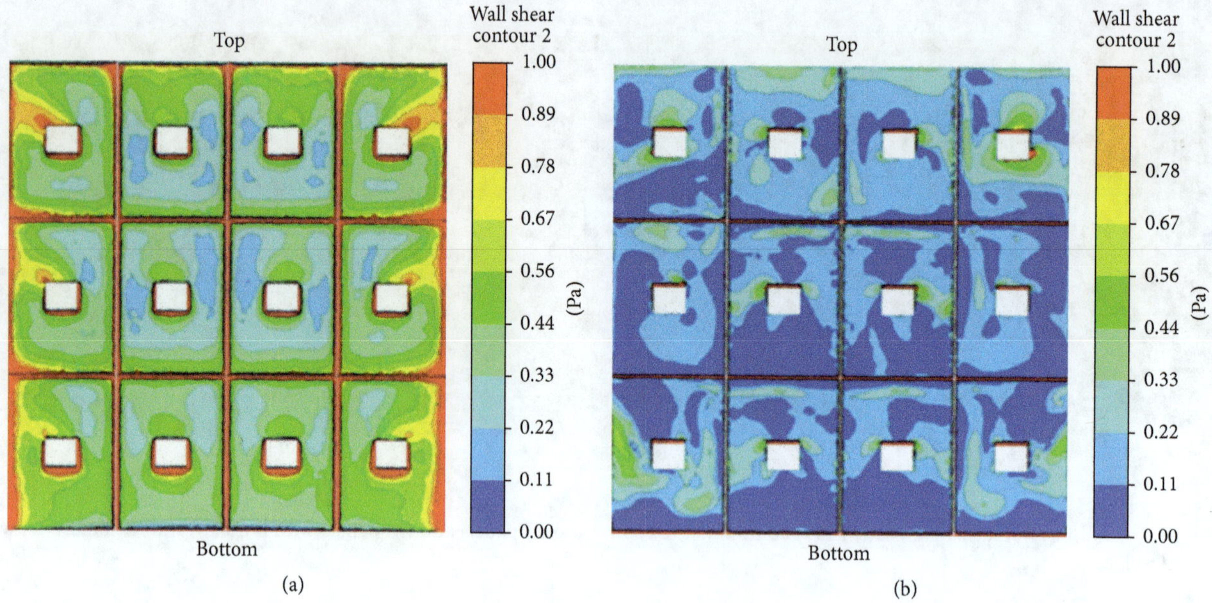

FIGURE 19: Wall shear stress (N/m^2) contours at panels with holes: (a) upwind side and (b) downwind side.

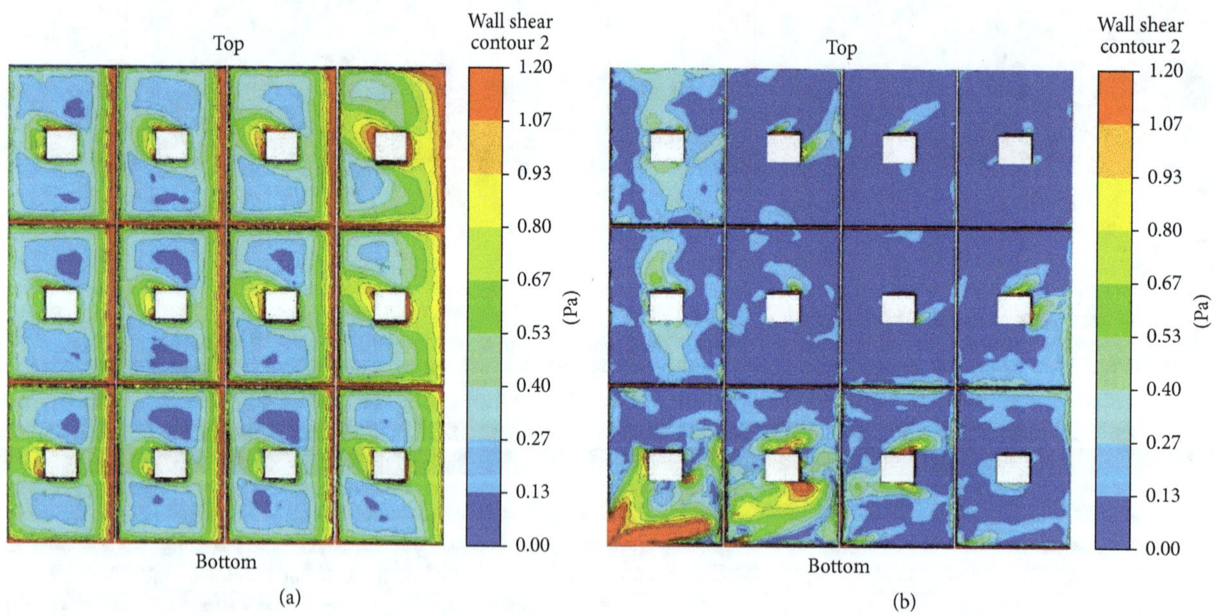

FIGURE 20: Wall shear stress (N/m^2) contours at panels with holes with 120.0° approach angle and maximum upwind velocity: (a) upwind side and (b) downwind side. Arrows show the wind direction (azimuthal angle).

0.0° approach angle the reduction in loading for all upwind velocities ranges between 5.16% and 6.46%. At 30.0° approach angle, under all upwind velocities, the percentage wind load increases.

Ogedengbe [13] conducted a series of pressure measurements (PIV technique) over panels on, similar to currently tested approach, azimuthal angles over panels inclined at 25.0° and 40.0° to the ground surface. They concluded that the gaps between each panel unit influenced the wind loading on the panel. Their research work findings closely approximate to the current numerically calculated research work results. However, they used experimental techniques and they were interested in panel gaps effects and reducing the wind loading with other means rather than panels with holes. Chung et al. [3] found that the wind uplift was effectively reduced with a guide plate inserted normal to the wind direction. Shademan and Hangan [6] also observed that at a specific distance between two sets of panels the drag coefficient for the downstream sets of panels reaches a minimum. Furthermore, Shademan et al. [14] numerically analyzed the effects of lateral gap spacing between subpanels and the ground clearance at different wind speeds and directions. The bottom solar

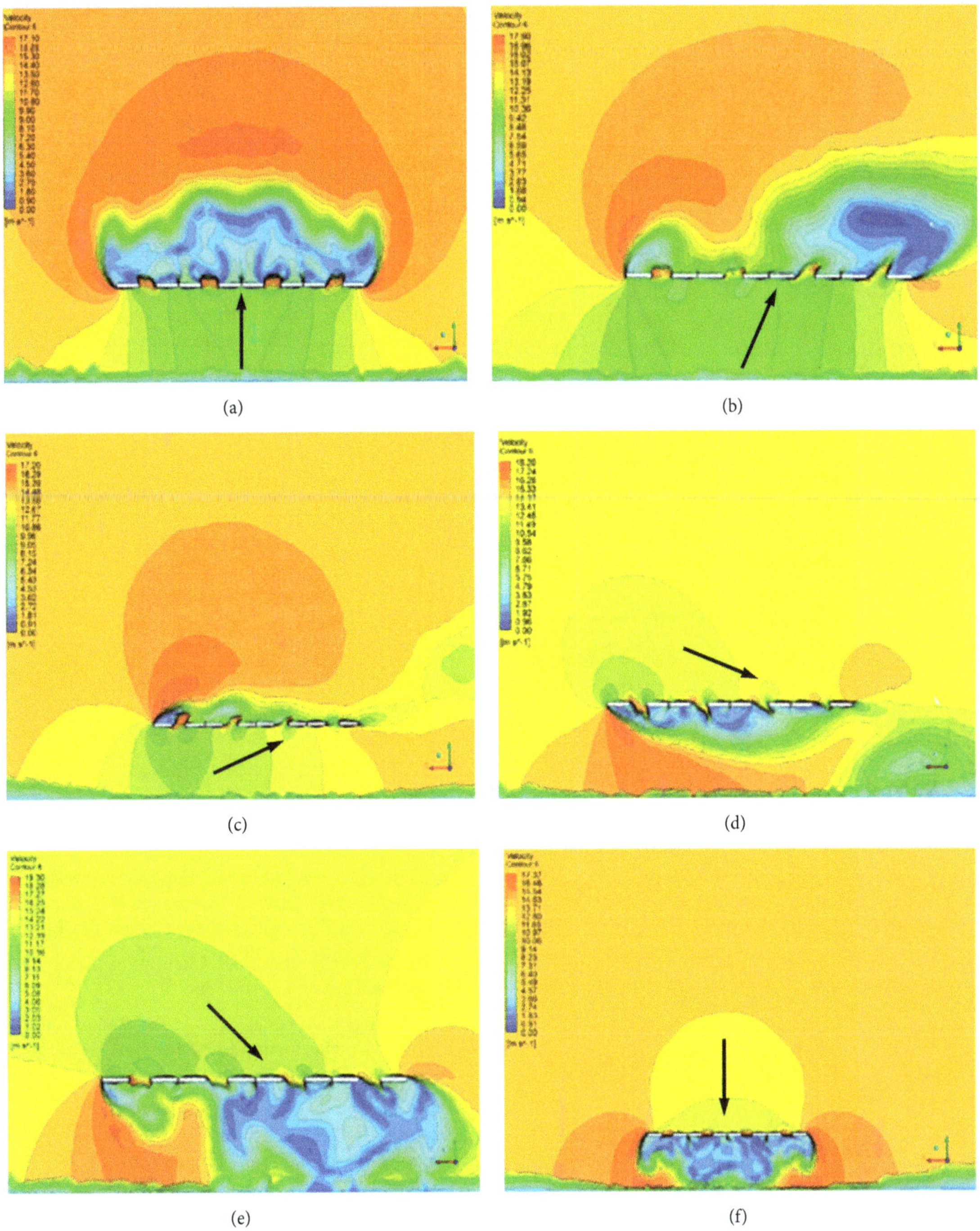

FIGURE 21: Velocity magnitude (m/s) contours at maximum upwind velocity 14.0 m/s. Panels with holes at (a) 0.0°, (b) 30.0°, (c) 60.0°, (d) 120.0°, (e) 150.0°, and (f) 180.0° approach angles, Section 2, Figure 12; arrows show the wind direction (azimuthal angle).

subpanels experience higher wind load compared to the top subpanels, results similar to ours, Figures 10 and 11. Jubayer and Hangan [9] concluded that the maximum wind load occurs close to the leading edge for all four different wind directions (0°, 45°, 135°, and 180°), a conclusion that can be also made by our analysis, Figures 14(a) and 14(b). Moreover for 0° and 180° wind directions, symmetry was observed in both mean C_p distributions and the wake structures, about the streamwise midline of the panel. Figures 10 and 11 of our analysis agree with this observation. Wind loads on solar panels were systematically investigated in wind tunnel, Peterka et al. [15]. They provided useful information, thus enriching the general literature, for the design of the solar panels mounted on buildings. Wind-induced loads on PV solar panels installed on roof tops were of main concern for Liu and Dragomirescu [16]. Measurements were conducted for a solar panel installed on a building. The maximum wind load of 1208.0 N was obtained on the northwest corner of

TABLE 5: Wind load ranking magnitude for conventional panels. Number 1 indicates the highest possible wind force and number 30 the smallest one. Bold font values indicate that the wind force ranking has been affected in comparison to panels with holes.

Angle/ velocity	6.0 (m/s)	8.0 (m/s)	10.0 (m/s)	12.0 (m/s)	14.0 (m/s)
0.0°	24	**17**	**9**	5	2
30.0°	23	**16**	8	**4**	1
60.0°	28	**21**	14	**10**	6
120.0°	29	**25**	**20**	**15**	11
150.0°	30	27	22	18	12
180.0°	**26**	19	13	7	**3**

TABLE 6: Wind load ranking magnitude for panels with holes. Number 1 indicates the highest possible wind force and number 30 the smallest one. Bold font values indicate that the wind force ranking has been affected in comparison to panels with holes.

Angle/ velocity	6.0 (m/s)	8.0 (m/s)	10.0 (m/s)	12.0 (m/s)	14.0 (m/s)
0.0°	24	**16**	**10**	5	2
30.0°	23	**15**	8	**3**	1
60.0°	28	**20**	14	**9**	6
120.0°	29	**26**	**21**	**17**	11
150.0°	30	27	22	18	12
180.0°	**25**	19	13	7	**4**

the PV solar panel arrays. The field measurements indicated that the highest wind-induced stress is on the lower edge of the PV panel. Upwind panel sides exhibit higher stresses than downwind sides. Furthermore, lower upwind panel side also exhibits higher stress than the center or upper panel side. In addition to the field measurements Liu and Dragomirescu [16] performed analysis with the Large Eddy Simulation technique. Our computational results are close to measured and calculated results of theirs. Sensors located on edge of the downwind side exhibit stresses lower than the center line. This is also an interesting finding and it is verified by our analysis, Figure 15(b). The crosswind effects on the aerodynamic characteristics of residential inclined flat plate mounted with a horizontal cylinder are experimentally investigated, Wood et al. [17]. They conducted low wind tunnel velocity with 0.0° to 135.0° wind direction. High suction and fluctuating pressure were observed near the upwind corner of the structure. These findings are also observed in our research for the tested solar panels (conventional and fitted with holes). Current research work findings suggest that the gaps and the panel holes greatly affect the flow pattern and subsequently the wind load reduction under upwind velocities and approach angles. To the best of our knowledge there was not similar research work undertaken and subsequently published on panel wind loading reduction using holes.

5. Conclusion

Current research work findings suggest that the panel holes greatly affect the flow pattern and subsequently the wind load reduction. These panels reduced the velocity magnitude downwind of panels flow region and extended the low velocity flow region. Wind forces acting on panels fitted with holes are less powerful since the wind finds effective escape corridors through the panels. The maximum wind load reduction in percentage terms does not follow the increased upwind velocity. The maximum wind loading in descending order for the approach angles is 30.0°, 0.0°, 180.0°, 60.0°, 120.0°, and 150.0°. The loading reduction in the arrayed set of panels ranges from 0.8% to 12.53%. The maximum load reduction occurs at 6.0 m/s upwind velocity and 120.0° approach angle. Under all upwind velocities, the percentage load reduction is higher at 120.0°. The only case where there is a small increase of wind loading refers to 30.0° approach angle. For the 0.0° approach angle, the reduction in wind loading for the upwind velocities ranges between 5.16% and 6.46%. Both the wind velocity and the angle of approach affect the wind loading on panels with or without holes. The current computational analysis indicates that it is possible to reduce wind loading considerably using panels with holes.

Competing Interests

The authors declare that they have no competing interests.

References

[1] L. Chevalien and J. Norton, *Wind Loads on Solar Collector Panels and Support Structure*, Aerospace Engineering Department, Texas A&M University, College Station, Tex, USA, 1979.

[2] G. A. Kopp, D. Surry, and K. Chen, "Wind loads on a solar array," *Wind and Structures*, vol. 5, no. 5, pp. 393–406, 2002.

[3] K. Chung, K. Chang, and Y. Liu, "Reduction of wind uplift of a solar collector model," *Journal of Wind Engineering and Industrial Aerodynamics*, vol. 96, no. 8-9, pp. 1294–1306, 2008.

[4] C.-C. Chou, K.-M. Chung, and K.-C. Chang, "Wind loads of solar water heaters: wind incidence effect," *Journal of Aerodynamics*, vol. 2014, Article ID 835091, 10 pages, 2014.

[5] M. Shademan and H. Hangan, "Wind loading on solar panels at different inclination angles," in *Proceedings of the 11th Americas Conference on Wind Engineering*, San Juan, Puerto Rico, June 2009.

[6] M. Shademan and H. Hangan, "Wind loading on panels at different azimuthal and inclination angles," in *Proceedings of the 5th International Symposium on Computational Wind Engineering (CWE '10)*, Chapel Hill, NC, USA, May 2010.

[7] G. Bitsuamlak, A. Dagnew, and J. Erwin, "Evaluation of wind loads on panel modules using CFD," in *Proceedings of the 5th International Symposium on Computational Wind Engineering*, Chapel Hill, NC, USA, 2010.

[8] R. Meroney and D. Neff, "Wind effects on roof-mounted solar photovoltaic arrays: CFD and wind-tunnel evaluation," in *Proceedings of the The 5th International Symposium on Computational Wind Engineering*, Chapel Hill, NC, USA, 2010.

[9] C. M. Jubayer and H. Hangan, "Numerical simulation of wind effects on a stand alone ground mounted photovoltaic (PV) system," *Journal of Wind Engineering and Industrial Aerodynamics*, vol. 134, pp. 56–64, 2015.

[10] P. Stangroom, *CFD Modelling of Wind Flow*, University of Nottingham, Nottingham, UK, 2004.

[11] B. Blocken, T. Defraeye, D. Derome, and J. Carmeliet, "High-resolution CFD simulations for forced convective heat transfer coefficients at the facade of a low-rise building," *Building and Environment*, vol. 44, no. 12, pp. 2396–2412, 2009.

[12] S. V. Patankar, *Numerical Heat Transfer and Fluid Flow*, Taylor & Francis, Abingdon, UK, 1980.

[13] A. A. Ogedengbe, *Experimental Investigation of Wind Effect on Solar Panels*, The University of Western Ontario, London, Canada, 2013.

[14] M. Shademan, R. M. Barron, R. Balachandar, and H. Hangan, "Numerical simulation of wind loading on ground-mounted solar panels at different flow configurations," *Canadian Journal of Civil Engineering*, vol. 41, no. 8, pp. 728–738, 2014.

[15] J. A. Peterka, B. Bienkiewicz, N. Hosoya, and J. E. Cermak, "Heliostat mean wind load reduction," *Energy*, vol. 12, no. 3-4, pp. 261–267, 1987.

[16] Y. Liu and E. Dragomirescu, "Field measurements and CFD simulation for wind effect on PV solar panels mounted on flat roofs," in *Proceedings of the World Congress on Advances in Civil Engineering and Materials Research (ACEM '14)*, Busan, South Korea, 2014.

[17] G. S. Wood, R. O. Denoon, and K. C. S. Kwok, "Wind loads on industrial solar panel arrays and supporting roof structure," *Wind & Structures*, vol. 4, no. 6, pp. 481–494, 2001.

7

Numerical Simulation of Dam Break Flows Using a Radial Basis Function Meshless Method with Artificial Viscosity

Elmiloud Chaabelasri[1,2]

[1]*National School of Applied Sciences, First Mohammed University, Al-Hociema, Morocco*
[2]*LME, Faculty of Sciences, First Mohammed University, Oujda, Morocco*

Correspondence should be addressed to Elmiloud Chaabelasri; chaabelasri@gmail.com

Academic Editor: Dimitrios E. Manolakos

A simple radial basis function (RBF) meshless method is used to solve the two-dimensional shallow water equations (SWEs) for simulation of dam break flows over irregular, frictional topography involving wetting and drying. At first, we construct the RBF interpolation corresponding to space derivative operators. Next, we obtain numerical schemes to solve the SWEs, by using the gradient of the interpolant to approximate the spatial derivative of the differential equation and a third-order explicit Runge–Kutta scheme to approximate the temporal derivative of the differential equation. For the problems involving shock or discontinuity solutions, we use an artificial viscosity for shock capturing. Then, we apply our scheme for several theoretical two-dimensional numerical experiments involving dam break flows over nonuniform beds and moving wet-dry fronts over irregular bed topography. Promising results are obtained.

1. Introduction

Development of robust meshless methods for the numerical solution of partial differential equations has attracted considerable interest over the past twenty years, for example, [1–3]. There are three types of meshless techniques: meshless techniques based on weak forms such as the element-free Galerkin method [4], meshless techniques based on collocation techniques such as the meshless collocation technique based on radial basis functions (RBFs) [5], and meshless techniques based on the combination of weak forms and collocation technique. In the literature, several meshless weak form techniques have been reported; among others, the smooth particle hydrodynamic method [6] and boundary point interpolation method are worth noticing [7]. The weak forms are used to derive a set of algebraic equations through a numerical integration process using a set of quadrature domain that might be constructed globally or locally in the domain of the problem. In this topic, Liu et al. [8] applied the notion of meshless local Petrov–Galerkin and developed the meshless local radial point interpolation method. This method was studied and used on a class of three-dimensional wave equations later by Shivanian [9]. A combination of the natural neighbour finite element method with the radial point interpolation method, using the multiquadric radial basis function, is developed by Dinis et al. [10] and Belinha et al. [11] to analyse a three-dimensional solid.

Initially, the radial basis function meshless method was developed for data surface fitting, and later, with the work developed by Kansa [5], the radial basis function was used for solving partial differential equations. Fedoseyev et al. [12] and Cheng et al. [13] have shown that meshless radial basis functions (RBFs) are attractive options because of the exponential convergence of certain RBFs. Various RBFs have been successfully applied to obtain accurate, efficient solutions of partial differential equations of engineering interest. This method has been applied to solve inviscid compressible flows [14], natural convection [15], heat conduction [16], three-dimensional incompressible viscous flows [17], and long waves in shallow water [18].

The Saint-Venant equations, also called shallow water equations, have often been preferred for the propagation of the flow in the open channel, and they exhibit a simplified mathematical structure, with an ability to take into account the smoothly varying flow conditions and flow discontinuities such as hydraulic jumps, moving bores, and propagation over dry beds, despite it still poses many theoretical and practical problems [19]. The dam break flow problem is an ideal theoretical example that involves these hydraulic challenges. In this paper, we examine the application of the radial basis function meshless method to the numerical solution of the shallow equations for dam break flow problems involving wetting/drying over complicated, frictional bed topography. In the formulation of RBFs, the radial basis functions represented by the multiquadric functions $\sqrt{1+(\varepsilon\|\mathbf{x}-\mathbf{x}_i\|)^2}$ are employed as basis functions to compute weighting coefficients for space differential operators, over a global set of computational collocation points [20]. The friction term is included in the momentum equations and discretized by a splitting implicit scheme [21]. A third-order Runge–Kutta algorithm is used for time integration [22].

It is known that the system of shallow water equations admits nonsmooth solutions that may contain shocks and rarefaction waves and, in the case of nonsmooth bottom topography, also contain contact discontinuities. To perform properly, a numerical method must be nonlinearly stable because linearly stable methods may develop large spurious oscillations and even blow up. To stabilize the proposed RBF model for slowly varying flows, as well as rapidly varying flows involving shocks or discontinuities such as dam breaks and hydraulic jumps, an artificial viscosity technique is used, following [23]. Furthermore, to avoid numerical instability caused by negative water depth near wet/dry fronts, the local flow variables are modified, imposing zero discharge when the water height became very small. As a consequence, the present numerical scheme ensures preservation of nonnegative water depth, and there is no need to limit the fluxes during simulation.

This paper is organized as follows: Section 2 outlines the shallow water equations. Section 3 presents the general formulation of partial differential problem interpolation using RBFs. Section 4 describes the application of RBFs to generate the discrete form of shallow water equations. Details are given of the numerical methods used to represent the friction term and carry out time integration. Section 4 presents the way to use the artificial viscosity for shock capturing. Section 6 discusses validation and application of the method; several numerical experiments are carried out for previously published benchmark cases in order to confirm the potential of the proposed scheme. Section 7 summarizes the main findings.

2. Two-Dimensional Shallow Water Equations

In conservation form, the two-dimensional nonlinear shallow water equations are given by (1) below [24]. The depth-averaged continuity and momentum equations in the x- and y-directions are

$$\begin{gathered}\frac{\partial h}{\partial t}+\frac{\partial(hu)}{\partial x}+\frac{\partial(hv)}{\partial y}=0,\\ \frac{\partial(hu)}{\partial t}+\frac{\partial(hu^2+(gh^2/2))}{\partial x}+\frac{\partial(huv)}{\partial y}=-gh\frac{\partial z_b}{\partial x}-\frac{S_{fx}}{\rho},\\ \frac{\partial(hv)}{\partial t}+\frac{\partial(huv)}{\partial x}+\frac{\partial(hv^2+(gh^2/2))}{\partial y}=-gh\frac{\partial z_b}{\partial y}-\frac{S_{fy}}{\rho},\end{gathered}\tag{1}$$

where h is the total depth from the sea bed to the free surface, u and v are the depth-averaged velocity components in the Cartesian x and y directions, z_b is the bed elevation above a fixed horizontal datum, g is the acceleration due to gravity, and S_{fx} and S_{fy} are the bed shear stress components, which are defined as

$$\begin{aligned}S_{fx}&=\rho C_b u\sqrt{u^2+v^2},\\ S_{fy}&=\rho C_b v\sqrt{u^2+v^2},\end{aligned}\tag{2}$$

where ρ is the water density and C_b is the bed friction coefficient, which may be estimated from $C_b=(gn_M^2/h^{1/3})$, where n_M is the Manning coefficient.

$$\frac{\partial \mathbf{W}}{\partial t}+\frac{\partial \mathbf{F}_x(\mathbf{W})}{\partial x}+\frac{\partial \mathbf{F}_y(\mathbf{W})}{\partial y}=\mathbf{S}(\mathbf{W}),\tag{3}$$

where $\mathbf{W}$ is the vector of dependent variables, $\mathbf{F}_x$ and $\mathbf{F}_y$ are the inviscid flux vectors, and $\mathbf{S}$ is the vector of source terms. In full, the vectors are

$$\begin{gathered}\mathbf{W}=\begin{pmatrix}h\\ hu\\ hv\end{pmatrix},\\ \mathbf{F}_x=\begin{pmatrix}hu\\ hu^2+\dfrac{gh^2}{2}\\ huv\end{pmatrix},\\ \mathbf{F}_y=\begin{pmatrix}hv\\ huv\\ hv^2+\dfrac{gh^2}{2}\end{pmatrix},\\ \mathbf{S}=\begin{pmatrix}0\\ -ghS_{fx}-gh\dfrac{\partial z_b}{\partial x}\\ -ghS_{fy}-gh\dfrac{\partial z_b}{\partial y}\end{pmatrix}.\end{gathered}\tag{4}$$

3. The Radial Basis Function Meshless Method

Let Ω be an open domain of R^2. Suppose $U=U(\mathbf{x},t)$ is a function to be approximated in a set of N pairwise distinct nodes $\mathbf{x}=\{x_1,x_2,\ldots,x_N\}$. In the RBF meshless scheme, the approximation of Uat the node $\mathrm{x}=(x_i,y_i)$ can be written as a linear combination of N RBFs:

$$U(\mathbf{x},t)=\sum_{j=1}^{N}\lambda_j(t)\varphi\left(\left\|\mathbf{x}-\mathbf{x}_j\right\|,\varepsilon\right),\tag{5}$$

where $U(\mathbf{x},t) = [U(\mathbf{x}_1,t), U(\mathbf{x}_2,t),\ldots,U(\mathbf{x}_N,t)]^T$ are the function values at node $\mathbf{x}$, $\Phi(\mathbf{x}) = [\varphi(\|\mathbf{x}-\mathbf{x}_1\|,\varepsilon), \varphi(\|\mathbf{x}-\mathbf{x}_2\|,\varepsilon), \ldots, \varphi(\|\mathbf{x}-\mathbf{x}_N\|,\varepsilon)]^T$ is a RBF centered at $\mathbf{x}$, $\|.\|$ denotes the Euclidean norm between nodes $\mathbf{x}$ and $\mathbf{x}_j$, and $\Lambda = [\lambda_1, \lambda_2, \ldots, \lambda_N]^T$ are the coefficients to be determined.

One of the most commonly used RBFs is the multiquadric (MQ) RBFs [25]. Here, we use the infinitely smooth multiquadric radial basis function defined as

$$\varphi\left(\left\|\mathbf{x}-\mathbf{x}_j\right\|,\varepsilon\right) = \sqrt{1+\varepsilon^2\left(\mathbf{x}-\mathbf{x}_j\right)^2}, \tag{6}$$

where ε is a shape parameter that controls the fit of a smooth surface to the data. In the present work, we used the following selection [26]:

$$\varepsilon = 0.8\frac{\sqrt{N}}{d_{\mathrm{m}}}, \tag{7}$$

where d_{m} denotes the smallest nodal distance. The expansion coefficients $\mathbf{\Lambda}$ in (5) are obtained by solving the following linear system of $N \times N$ algebraic equations of $U(\mathbf{x}_i,t) = U_i$:

$$\mathbf{\Phi\Lambda} = \mathbf{U}. \tag{8}$$

The expansion coefficients are then calculated by

$$\mathbf{\Lambda} = \mathbf{\Phi}^{-1}\mathbf{U}, \tag{9}$$

where $\mathbf{U} = [U_1, U_2, \ldots, U_N]^T$ is the vector of approximate solutions and $\Phi = (\varphi(\|x_i - x_j\|, \varepsilon))$, $1 \le i, j \le N$ is an $N \times N$ matrix of RBFs given as

$$\begin{pmatrix} \varphi\left(\left\|x_1-x_1\right\|,\varepsilon\right) & \cdots & \varphi\left(\left\|x_1-x_N\right\|,\varepsilon\right) \\ \vdots & & \vdots \\ \varphi\left(\left\|x_N-x_1\right\|,\varepsilon\right) & \cdots & \varphi\left(\left\|x_N-x_N\right\|,\varepsilon\right) \end{pmatrix}_{N\times N}. \tag{10}$$

Space derivatives of the interpolant (5) may be readily calculated, due to its linearity. Generally, the first- and second-order spatial derivatives at the point $\mathbf{x}$ can be calculated as

$$\frac{\partial^{(k)}\mathbf{U}}{\partial x^k} = \sum_{j=1}^{N}\lambda_j(t)\frac{\partial^{(k)}\varphi}{\partial x^k}\left(\left\|x_i - x_j\right\|,\varepsilon\right), \tag{11}$$

where $k = [1, 2]$ denotes the first- and second-order derivatives. In a compact matrix form, using (9), (11) can be written as

$$\frac{\partial^{(k)}\mathbf{U}}{\partial x^k} = \mathbf{\Phi}x^{(k)}\mathbf{\Lambda} = \mathbf{\Phi}x^{(k)}\mathbf{\Phi}\mathbf{U}, \tag{12}$$

where

$$\begin{aligned} \frac{\partial^{(k)}\mathbf{U}}{\partial x^k} &= \left[\frac{\partial^{(k)}U(x_1,t)}{\partial x^k}, \frac{\partial^{(k)}U(x_2,t)}{\partial x^k}, \ldots, \frac{\partial^{(k)}U(x_N,t)}{\partial x^k}\right]^T, \\ \mathbf{\Phi}x^{(k)} &= \left[\frac{\partial^{(k)}\mathbf{\Phi}(x_1)}{\partial x^k}, \frac{\partial^{(k)}\mathbf{\Phi}(x_2)}{\partial x^k}, \ldots, \frac{\partial^{(k)}\mathbf{\Phi}(x_N)}{\partial x^k}\right]^T. \end{aligned} \tag{13}$$

4. Discretized Form of Shallow Water Equations by the RBF Meshless Method

4. 1. Convective Flux and Bottom Topography Term Approximations. Let us assume that $F_\xi^n(\mathbf{W}(x))$ are known convective fluxes along the axes x and y at time $t = n\Delta t$, where $\xi = (x, y)$. Using (5), it can be approximated by

$$F_\xi^n(W) = \sum_{j=1}^{N}\lambda_{\xi,j}(t)\varphi\left(\left\|x - x_j\right\|,\varepsilon\right). \tag{14}$$

In the matrix form, this equation becomes

$$F_\xi^n = \Lambda_\xi^n\mathbf{\Phi}. \tag{15}$$

However, the expansion coefficients are then calculated by

$$\mathbf{\Lambda}_\xi^n = \mathbf{\Phi}^{-1}F_\xi^n. \tag{16}$$

From (14), the partial flux derivative $(\partial F_\xi^n(W)/\partial\xi)$ can be written as

$$\frac{\partial F_\xi^n(\mathbf{W})}{\partial\xi} = \sum_{j=1}^{N}\lambda_{\xi,j}^n(t)\frac{\partial\varphi\left(\left\|\mathbf{x}-\mathbf{x}_j\right\|,\varepsilon\right)}{\partial\xi}. \tag{17}$$

Then, the compact matrix form of the partial derivative of the flux vector is

$$\frac{\partial F_\xi^n(\mathbf{W})}{\partial\xi} = \mathbf{\Phi}_\xi\mathbf{\Lambda}_\xi^n = \mathbf{\Phi}_\xi\mathbf{\Phi}^{-1}F_\xi^n(\mathbf{W}). \tag{18}$$

Applying the same procedure described above for the bottom topography function $z_{\mathrm{b}}(\mathbf{x})$, it is easy to obtain its discrete form. The topography function can be approximated by

$$z_b^n(\mathbf{x}) = \sum_{j=1}^{N}\alpha_j(t)\varphi\left(\left\|\mathbf{x}-\mathbf{x}_j\right\|,\varepsilon\right). \tag{19}$$

Its partial derivative is then expressed as

$$\frac{\partial z_b^n(\mathbf{x})}{\partial\xi} = \sum_{j=1}^{N}\alpha_j(t)\frac{\partial\varphi\left(\left\|\mathbf{x}-\mathbf{x}_j\right\|,\varepsilon\right)}{\partial\xi}. \tag{20}$$

In the same way, the compact matrix form of the bottom topography term $S_{z_b}^n$ along the axis ξ is given as

$$S_{z_b}^n(\xi) = -gh^n\frac{\partial z_b^n(x,t)}{\partial\xi} = -gh^n\mathbf{\Phi}_\xi\mathbf{\Phi}^{-1}z_b^n(x,t). \tag{21}$$

4.2. Addition of Friction Effects. To incorporate friction into the present numerical scheme, the friction term is discretized using an operator splitting procedure described by Boushaba et al. [27], which splits the shallow water equations (2) into two equations:

$$\begin{aligned} \frac{\partial h\mathbf{U}}{\partial t} &= S_b^{n+1}(\mathbf{W}) = \left(-gn^2\|\mathbf{U}\|\mathbf{U}h^{-1/3}\right)^{n+1}, \\ \frac{\partial h\mathbf{U}}{\partial t} &+ \mathbf{L}(\mathbf{W}, Z_{\mathrm{b}}) = 0, \end{aligned} \tag{22}$$

```
For i = 1, 2, 3 do
        W(i) = W^0(i)
For t = Δt, 2Δt, 3Δt, ... do
    For i = 1, 2, 3 do
        K_1 = W(i) + Δt RHS(W(i))
        K_2 = (3/4)W(i) + (1/4)K_1 + (1/4)Δt RHS(K_1)
        W(i) = (1/3)W(i) + (2/3)K_2 + (2/3)Δt RHS(K_2)
    For k = 1, 2, ..., N_Boundary do   {For all boundary points}
        If x(k) ∈ Γ_open Then   {Boundary conditions for inlet and outlet depending on the case}
          W(k) = W_knowed||∂W(k)/∂n = 0
      Else   {Boundary conditions for solid walls}
        hu(k) = hv(k) = 0 & ∂h(k)/∂n = 0
```

Algorithm 1: Time integration procedure.

```
F_x(1) = W(2), F_x(2) = W(2)^2/W(1) + 0.5gW(1)^2, F_x(3) = W(2) × W(3)/W(1)
F_y(1) = W(3), F_y(2) = W(2) × W(3)/W(1), F_y(3) = W(3)^2/W(1) + 0.5gW(1)^2
For i = 1, 2, 3 do
    RSH(i) = M_x M^-1 F_x(i) + M_y M^-1 F_y(i)
For i = 1, 2, 3 do
    RSH(i) = RSH(i) − (max(||U(i)|| + sqrt(gh(i)))/P_e) d_min M_xx M^-1 W(i)
```

Algorithm 2: Calculation of flux vector and right-hand side.

where $U = (u, v)$. In the first step of the calculation, the upper ordinary differential equation in (23) is approximated by an implicit method as described in [21]:

$$\frac{\overline{h\mathbf{U}} - (h\mathbf{U})^n}{\Delta t} = S_b^{n+1}(\mathrm{W}), \tag{23}$$

where the friction term S_b^{n+1} may be expressed using a Taylor series as

$$S_{\mathrm{b}}^{n+1}(\mathbf{W}) = S_{\mathrm{b}}^{n}(\mathbf{W}) + \left(\frac{\partial S_{\mathrm{b}}}{\partial h\mathbf{U}}\right)^{n}\left((\overline{h\mathbf{U}}) - (h\mathbf{U})^{n} + 0\left(\Delta (h\mathbf{U})^{2}\right)\right. \tag{24}$$

Rearranging the above equation leads to the following formula for updating water discharge $\overline{hu}$ at the new time step:

$$\overline{h\mathbf{U}} = (h\mathbf{U})^n + \Delta t \frac{S_b^n(\mathbf{W})}{1 + 2\Delta t C_{\mathrm{f}} \|\mathbf{U}\| h^{-1}}. \tag{25}$$

In the second step, the value $\overline{h\mathbf{U}}$ is taken to be the initial condition when solving the second equation in (23).

4.3. Time Integration and Stability Condition. To date, the forward Euler method has been mainly used as the preferred time-stepping scheme for RBF methods. However, the forward Euler method is only the first-order accurate in time and so may introduce excessive numerical dissipation in the computed RBF solutions. To achieve a higher order of accuracy, we use the explicit Runge–Kutta method recommended by [22]. The procedure to advance the solution from the time t^n to the next time t^{n+1} is carried out as

$$\begin{aligned}
\mathbf{W}^{(1)} &= \mathbf{W}^n + \Delta t \mathbf{L}(\mathbf{W}^n), \\
\mathbf{W}^{(2)} &= \frac{3}{4}\mathbf{W}^n + \frac{1}{4}\mathbf{W}^{(1)} + \frac{1}{4}\Delta t \mathbf{L}\left(\mathbf{W}^{(1)}\right), \\
\mathbf{W}^{n+1} &= \frac{1}{3}\mathbf{W}^n + \frac{2}{3}\mathbf{W}^{(2)} + \frac{2}{3}\Delta t \mathbf{L}\left(\mathbf{W}^{(2)}\right),
\end{aligned} \tag{26}$$

where n represents the time level and Δt is the time step. To achieve stability, for this explicit scheme, the time step must meet the following criterion:

$$\Delta t = \mathrm{CFL} \frac{d_{\min}}{\max\left(\|u_i\| + \sqrt{gh_i}\right)}, \tag{27}$$

where CFL is the Courant number, such that $0 < \mathrm{CFL} < 1$, and $d_{\min}$ denotes the smallest nodal distance between collocation points.

4.4. Boundary Conditions. In this work, transmissive boundary conditions for open inflow/outflow and reflective boundary conditions for solid walls are applied in the simulations. At a transmissive boundary, the flow variables at a collocation point on the boundary are set to the same values as at the interior point normal to the boundary. At a reflective boundary, the value at a collocation point is simply the mirror image of that at the associated boundary point so that the normal velocity component is zero at the boundary. However, the representation of boundary conditions within an RBF method is less trivial.

4.5. Implementation Algorithms. For a given initial condition $\mathbf{W}^0 = \mathbf{W}(\mathbf{X}, t = 0)$, the time integration procedure is

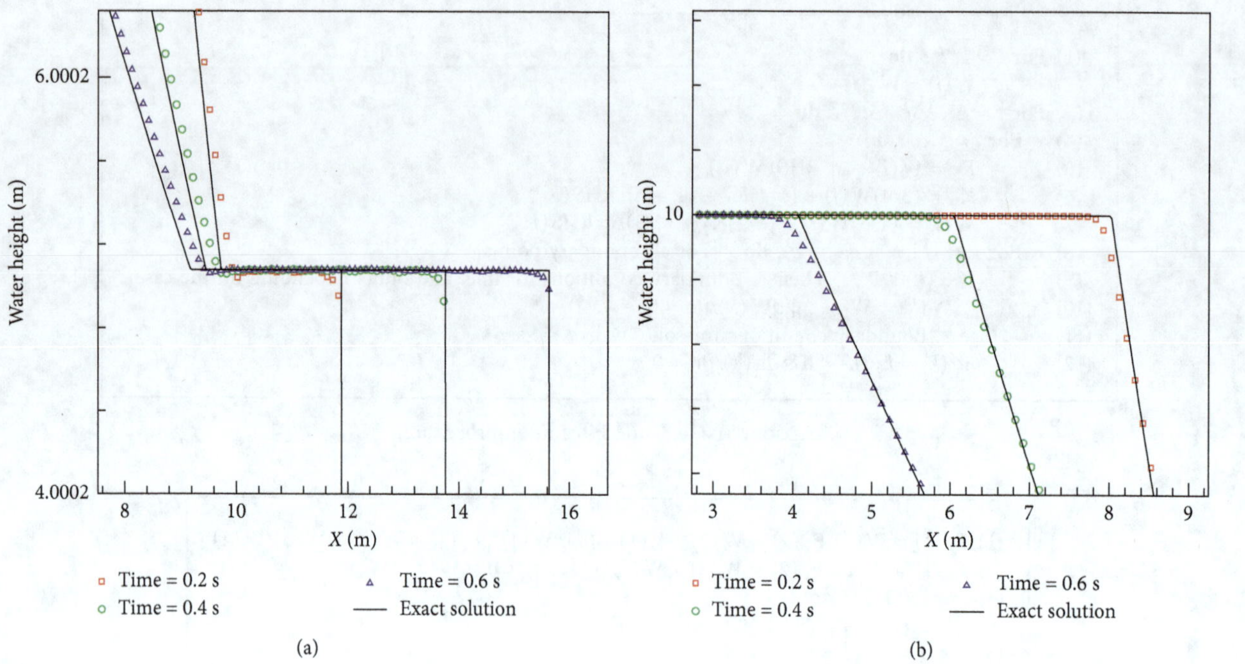

FIGURE 1: Numerical and exact solutions of a dam break on a wet bed, at different times, using a grid of 800 points: (a) free surface elevation profiles; (b) profiles of water discharge per unit breadth.

described in Algorithm 1. The right-hand side (RHS) used in the algorithm represents the convective flux and bottom topography that are calculated in Algorithm 2.

5. Artificial Viscosity for Shock Capturing

For particular hydraulic problems involving shock waves, such as a hydraulic jump or dam break flows, the numerical model is required to represent a steady or unsteady discontinuity, where the presence of oscillations in the solution is expected, and it sometimes may grow over time. However, by introducing a small amount of artificial diffusion, it is possible to damp these oscillations [23, 28]. We therefore augment the right-hand side of the hyperbolic system (2) with the artificial diffusion term:

$$\mathbf{D}(\mathbf{W}) = D_{\mathrm{h}} \frac{\partial^2 \mathbf{W}}{\partial x^2}, \tag{28}$$

where D_{h} is a tunable viscosity coefficient. The Peclet number is defined as follows:

$$P_e = \frac{\max\left(\|\mathbf{U}_i\| + \sqrt{g h_i}\right)}{\mathbf{D}_{\mathbf{h}}} \mathbf{d}_{\min}. \tag{29}$$

It often controls the stability of the numerical solutions [23, 29]. In the case of convection-dominated flow, the Peclet number is large but finite, and the effect of the diffusion term (29) becomes negligible. Therefore, a natural and simple way to stabilize the solution is to reduce the Peclet number.

Using the RBF interpolation of an arbitrary function (12), the augmented term (29) can be obtained as

$$\mathbf{D}(\mathbf{W}) = \frac{\max\left(\|\mathbf{U}_i\| + \sqrt{g h_i}\right)}{P_e} d_{\min} \mathbf{\Phi}_{xx} \mathbf{\Phi} \mathbf{W}. \tag{30}$$

6. Numerical Results

6.1. One-Dimensional Dam Break over a Wet Bed. A one-dimensional dam break over a wet, horizontal bed is first simulated to demonstrate the shock-capturing capability of the corrected radial basis function method using artificial viscosity. The channel is of length 20 m. The initial water discharge and depth are given by

$$hu(x,0) = 0 \text{ and } h(x,0) = \begin{cases} 10, & \text{if } x \le 10, \\ 2, & \text{otherwise.} \end{cases} \tag{31}$$

As boundary conditions, a zero discharge and a free boundary are considered at the left and right ends of the channel. The analytical solution for this simple dam break test consists of a backward-propagating rarefaction and a forward-moving shock wave. Figure 1 shows the numerical results of flow after the dam fails, at three different times in terms of water surface elevation and discharge. Here, the model domain consists of 800 points. An artificial viscosity of 0.001 $\mathrm{m^2 s^{-1}}$ is applied corresponding to a Peclet number of 300. Figure 2 shows close-up views of the evolving free surface elevation and discharge profiles along the channel at different times. In general, satisfactory agreement is achieved between the numerical and analytical solutions, although a very small amount of numerical diffusion occurs where the surface gradient is steepest.

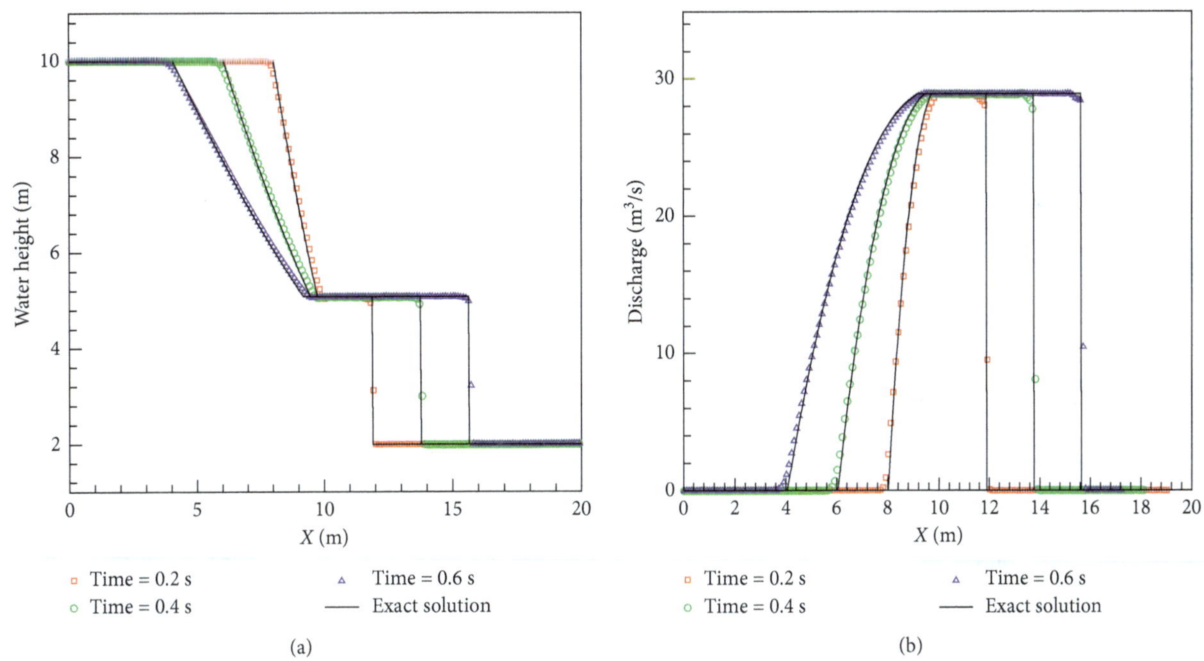

FIGURE 2: Zoom-in view of Figure 1, focusing on discontinuities in the water level.

6.1. Two-Dimensional Partial Dam Break Simulation. For this second test case, we consider a hypothetical two-dimensional dam break problem with nonsymmetric breach that is a typical validation made in many presented papers, for example, [30–33]. An illustration of this problem is shown in Figure 3, in which the domain is defined by a 200 m × 200 m channel with the horizontal bed. A dam is located in the middle of the domain, and the nonsymmetric breach is 75 m wide with negligible thickness and is located 95 m from the left side of the domain. The initial water discharge and depth are given by

$$hu(t = 0, x, y) = hv(t = 0, x, y) = 0,$$

$$h(t = 0, x, y) = \begin{cases} 10, & \text{if} \quad x \le 95, \\ 5, & \text{otherwise}. \end{cases} \qquad (32)$$

As boundary conditions, at the left and right ends of the channel, a zero discharge and a free boundary are considered, and a solid wall is considered for others. Figure 4 shows the numerical results of flow after the dam fails, at three different times in terms of water depth and velocity. As we can show, the upstream water is released into the downstream side through the breach, creating the surge wave propagating to downstream and the neglective wave to upstream. Here, the model domain consists of 2000 points. An artificial viscosity of $0.001\,m^2s^{-1}$ is applied corresponding to a Peclet number of 300. No analytical solution is available for this case, but the results can be compared with those of other numerical schemes. Generally, the results are in agreement with the results obtained by other numerical methods in aforementioned literatures.

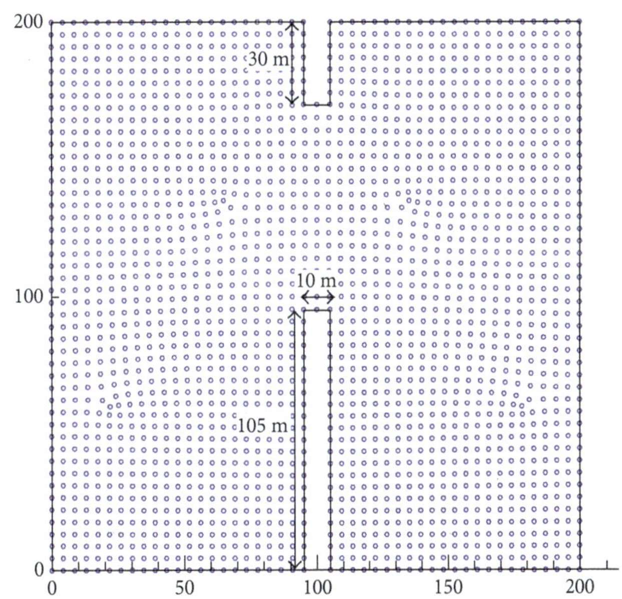

FIGURE 3: The computational domain and RBF collocation points of the 2D dam break problem.

6.2. Circular Dam Break. This second test case consists of the instantaneous breaking of a cylindrical tank of diameter 20 m (Figure 5), initially filled with 2 m of water at rest. The wave generated by the breaking of the tank propagates into still water with an initial depth of 0.5 m. This classic problem is widely used to test the shock capture capability of numerical schemes [34, 35].

Figure 6 illustrates the wave propagation on computational 8000 collocation points. An artificial viscosity of $0.04\,m^2s^{-1}$ is applied corresponding to a Peclet number of 100. At the beginning, that is, at $t = 0$ s, the dam is broken instantaneously and the column of water is released, and the

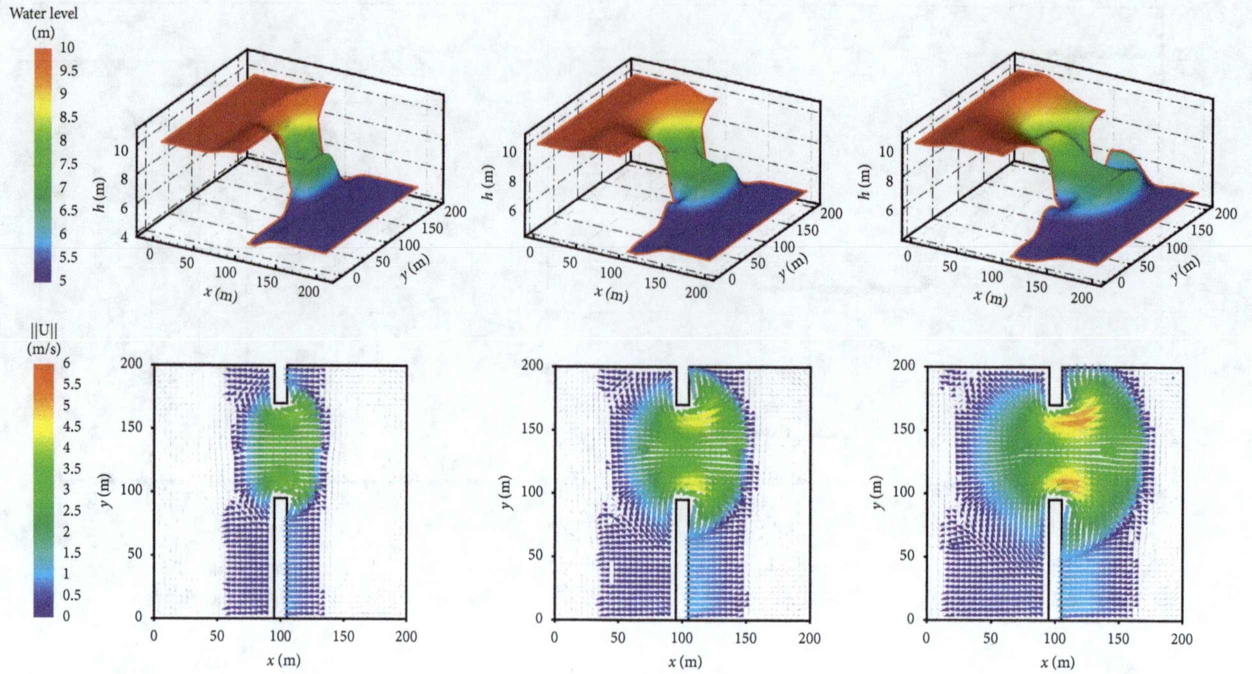

FIGURE 4: The profile of water depth and velocity vectors of the 2D dam break at different output times: $t = 3$ s, $t = 5$ s, and $t = 7.2$ s.

shock wave results in the increase of water depth in the lower depth region, propagating in the radial direction. Our state solutions obtained at time $t = 1$ and 2.5 s by different methods, that is, the RBF meshless method and the triangular finite volume method, as a based mesh method and a 1D reference solution are in an agreement (Figure 7). No singular corner effects on smoothness of the solution can be noticed.

6.3. Dam Break over Three Humps. The nonlinear shallow water equations are not too many analytical solutions for the problem of the flow over wetting and drying topography. In this context, Antuono et al. [36] studied an analytical solution of the shallow water flow over uneven topography, and this is for analysis of wave run-up on a beach. The current shallow water model is applied to simulate a dam break over an initially dry floodplain with three humps, recommended by Kawahara and Umetsu [37] for this challenging problem that involves complex flow hydrodynamics, bottom topography, and wetting and drying. The simulated setup is sketched in Figure 8, where the dam break occurs in a 75×30 m rectangular domain with the dam located 16 m away from the upstream end. A reservoir with a still water surface elevation of 1.875 m is assumed to be upstream of the dam. The bed topography of the domain is defined as

$$z(x, y) = \max\Big[0, 1 - \frac{1}{8}\sqrt{(x-30)^2 + (y-6)^2}, \quad 1 - \frac{1}{8}\sqrt{(x-30)^2 + (y-24)^2}, \quad 1 - \frac{3}{10}\sqrt{(x-47.5)^2 + (y-15)^2}\Big]. \tag{33}$$

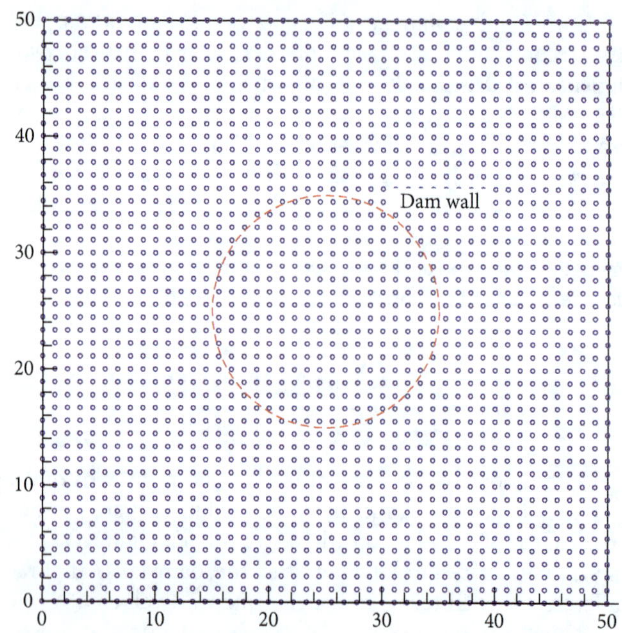

FIGURE 5: The computational domain, RBF collocation points, and location of the circular dam break problem.

The floodplain is initially dry, and a constant Manning coefficient of $0.018\ \mathrm{s/m^{1/3}}$ is used throughout the domain. The domain walls are assumed to be solid. The dam collapses instantly at $t = 0$ s, and simulation is run for 300 s on 2000 collocation points.

After the dam fails, the initial still water in the reservoir rushes onto the downstream floodplain. In Figure 9, we present the profile of water depth and contours at different

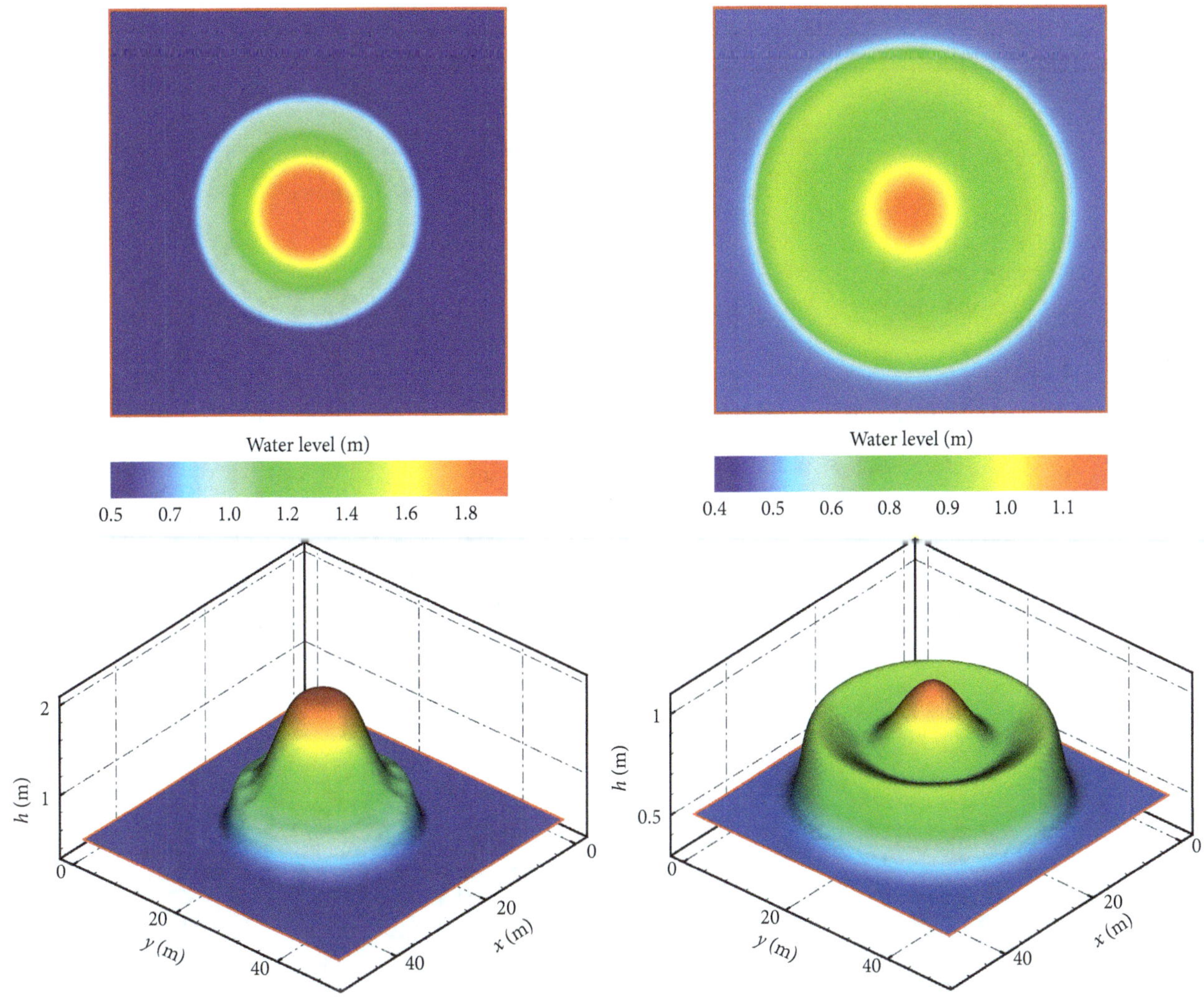

FIGURE 6: The water depth contours and profile of the circular dam break at different output times: $t = 1$ s and $t = 2.5$ s.

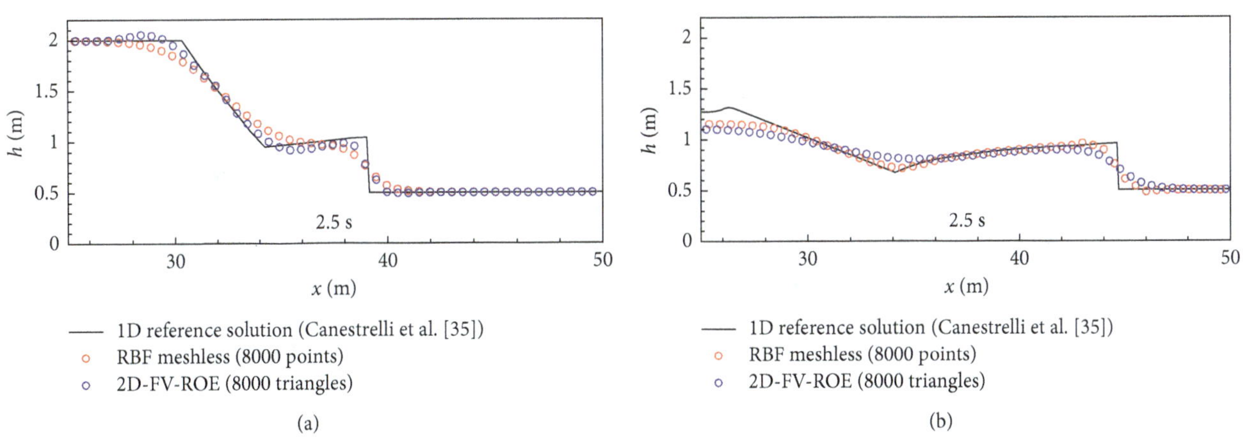

FIGURE 7: Cross section comparison between the numerical solution using the RBF meshless method and 2D FV-Roe scheme and the 1D reference solution.

output times $t = 1$, 6, 12, 30, and 300 s. As can be observed from these results, after about 1 s, the wet-dry front reaches the two small humps and begins to climb over them. At $t = 6.0$ s, the two small humps are entirely submerged and the wave front hits the large hump. At $t = 12$ s, the wave front passes through both sides of the large hump and begins to submerge the lee of the hump. Finally, the flow becomes steady due to the dissipation caused by bed friction, as shown at $t = 300$ s when the flow is nearly motionless and the peaks of the humps are no longer submerged.

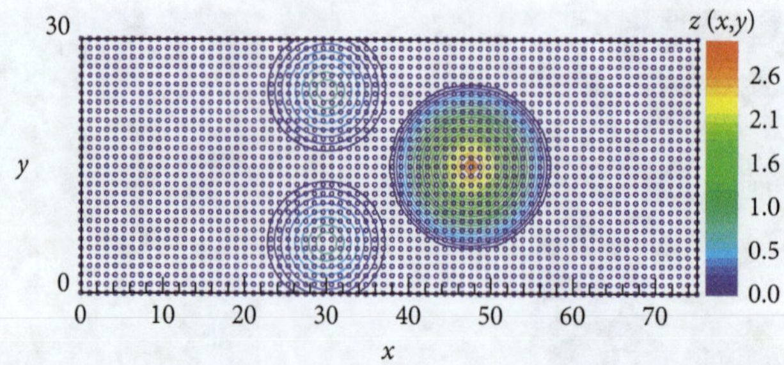

FIGURE 8: The computational domain, RBF collocation points, and bed contours of the dam break problem over humps.

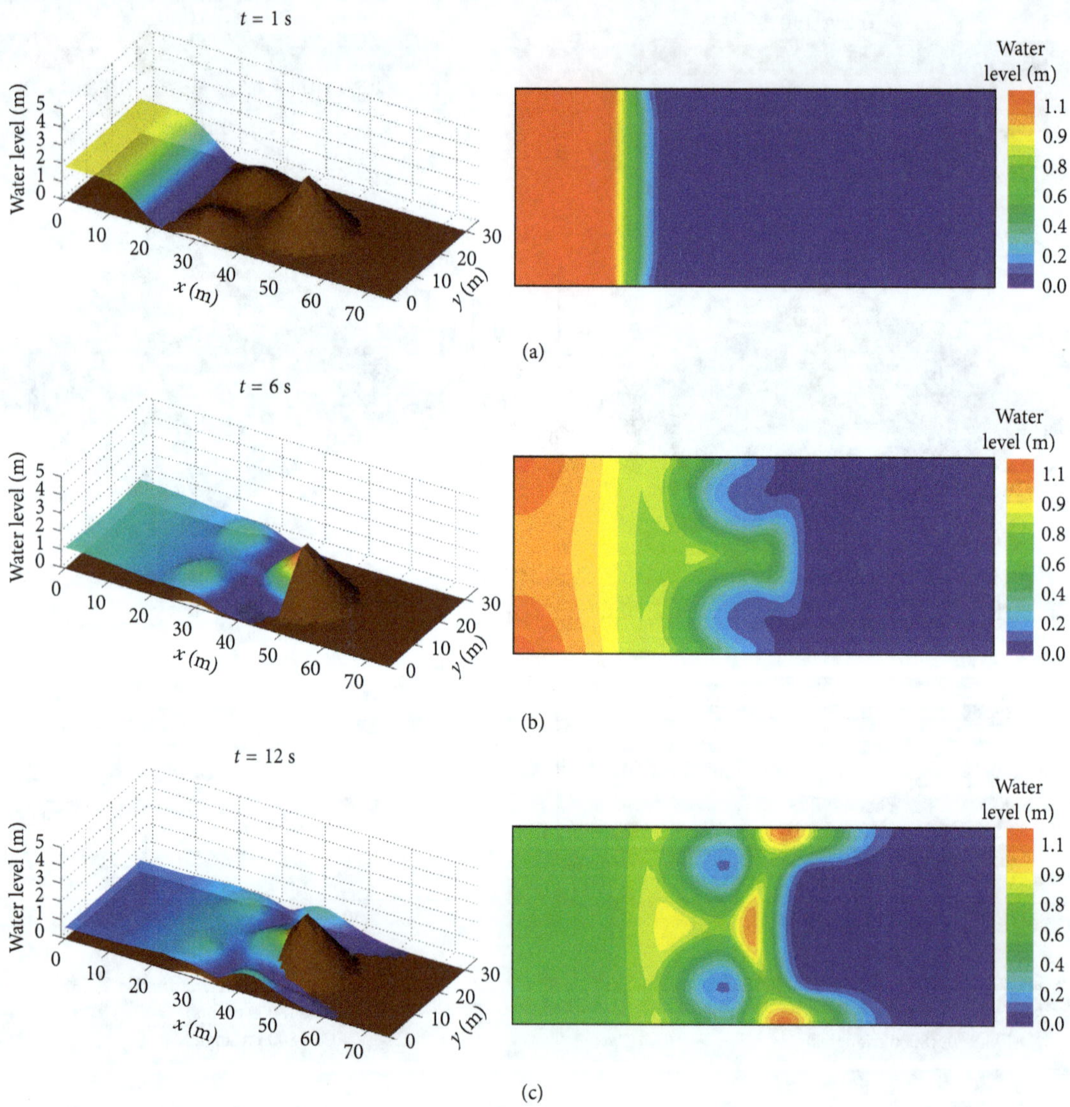

FIGURE 9: Continued.

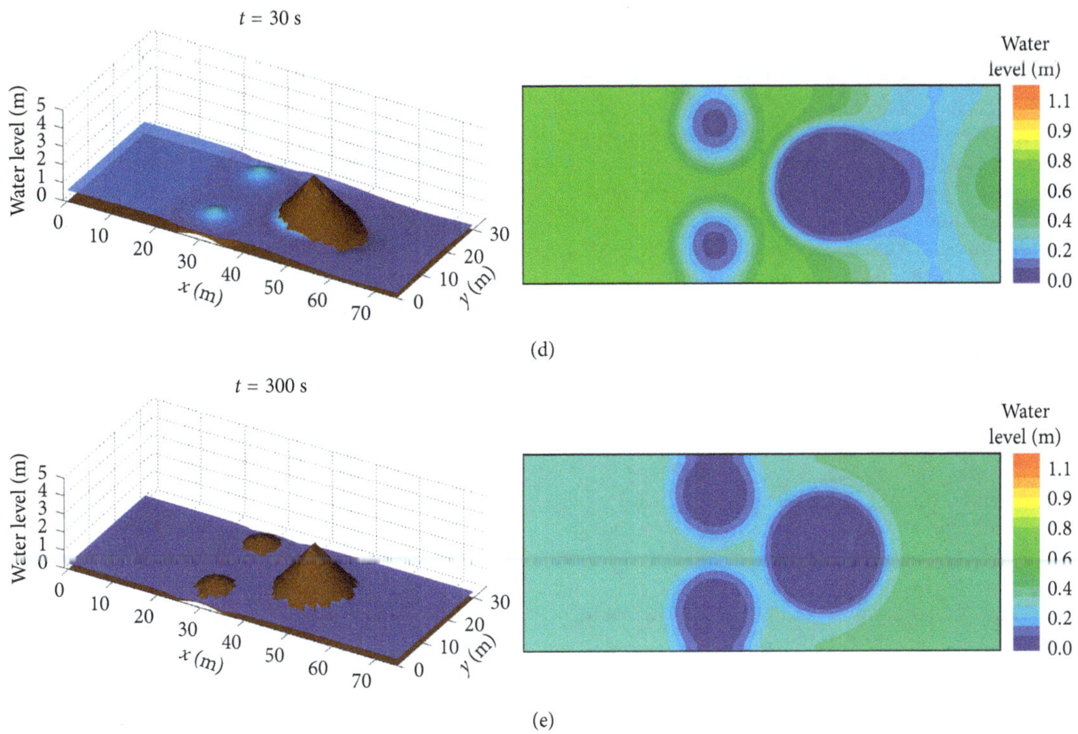

FIGURE 9: The profile of water depth and depth contours of the dam break over three humps at different output times: $t = 1$ s, $t = 6$ s, $t = 12$ s, $t = 30$ s, and $t = 300$ s.

The overall flow pattern for this example is preserved with no spurious oscillations appearing in the results by the RBF meshless method. Obviously, the obtained results verify the stability and the shock-capturing properties of the proposed meshless method. The RBF meshless method performs well for this test problem since it does not diffuse the moving fronts, and no spurious oscillations have been observed when the water flows over the humps.

7. Conclusions

This paper has investigated a simple RBF meshless method for numerical simulation of dam break flows by solving the shallow water equations including bed friction and nonuniform bed elevation terms. The space derivative has been approximated by radial basis functions on global collocation points. An explicit third-order Runge–Kutta scheme was used for time integration. The RBF meshless method has been found to be flexible and straightforward to implement. The method is particularly attractive for the shallow water equations in comparison with classical mesh-based methods because it is inherently mesh-free and requires no special treatment of the wet-dry interface. The method was tested against several theoretical dam break problems, including a dam break with a nonsymmetric breach, a circular dam break, and a dam break on an irregular bed involving the wet/dry interface. The numerical model gives promising predictions in comparison with previously references and published results.

Conflicts of Interest

The author declares that there are no conflicts of interest regarding the publication of this paper.

References

[1] T. Belytschko, Y. Krongauz, D. Organ, M. Fleming, and P. Krysl, "Meshless methods: an overview and recent developments," *Computer Methods in Applied Mechanics and Engineering*, vol. 139, no. 1–4, pp. 3–47, 1996.

[2] G. Liu, *Mesh Free Methods: Moving Beyond the Finite Element Method*, CRC Press, Boca Raton, FL, USA, 2002.

[3] V. Nguyen, T. Rabczuk, S. Bordas, and M. Duflot, "Meshless methods: a review and computer implementation aspects," *Mathematics and Computers in Simulation*, vol. 79, no. 3, pp. 763–813, 2008.

[4] A. Fili, A. Naji, and Y. Duan, "Coupling three-field formulation and meshless mixed Galerkin methods using radial basis functions," *Journal of Computational and Applied Mathematics*, vol. 234, no. 8, pp. 2456–2468, 2010.

[5] E. J. Kansa, "A scattered data approximation scheme with applications to computational fluid-dynamics–I and II," *Computers and Mathematics with Applications*, vol. 19, no. 8-9, pp. 127–161, 1990.

[6] A. Bratsos, "An improved numerical scheme for the sine-Gordon equation in 2+1 dimensions," *International Journal for Numerical Methods in Engineering*, vol. 75, no. 7, pp. 787–799, 2008.

[7] Y. Gu and G. Liu, "A boundary point interpolation method for stress analysis of solids," *Computational Mechanics*, vol. 28, no. 1, pp. 47–54, 2002.

[8] G. Liu, L. Yan, J. Wang, and Y. Gu, "Point interpolation method based on local residual formulation using radial basis functions," *Structural Engineering and Mechanics*, vol. 14, no. 6, pp. 713–732, 2002.

[9] E. Shivanian, "On the convergence analysis, stability, and implementation of meshless local radial point interpolation on a class of three-dimensional wave equations," *International*

Journal for Numerical Methods in Engineering, vol. 105, no. 2, pp. 83–110, 2016.

[10] L. M. J. S. Dinis, R. M. Natal Jorge, and J. Belinha, "Analysis of 3D solids using the natural neighbour radial point interpolation method," *Computer Methods in Applied Mechanics and Engineering*, vol. 196, no. 13–16, pp. 2009–2028, 2007.

[11] J. Belinha, L. M. J. S. Dinis, and R. M. Natal Jorge, "The natural radial element method," *International Journal for Numerical Methods in Engineering*, vol. 93, no. 12, pp. 1286–1313, 2013.

[12] A. Fedoseyev, M. Friedman, and E. Kansa, "Improved multiquadric method for elliptic partial differential equations via PDE collocation on the boundary," *Computers and Mathematics with Applications*, vol. 43, no. 3–5, pp. 491–500, 2002.

[13] A. Cheng, M. Golberg, E. Kansa, and T. Zammito, "Exponential convergence and H-c multiquadric collocation method for partial differential equations," *Numerical Methods for Partial Differential Equations*, vol. 19, no. 5, pp. 571–594, 2003.

[14] C. Shu, H. Ding, H. Chen, and T. Wang, "An upwind local RBF-DQ method for simulation of inviscid compressible flows," *Computer Methods in Applied Mechanics and Engineering*, vol. 194, no. 18–20, pp. 2001–2017, 2005.

[15] H. Ding, C. Shu, K. Yeo, and Z. Lu, "Simulation of natural convection in eccentric annuli between a square outer cylinder and a circular inner cylinder using local MQ-DQ method," *Numerical Heat Transfer, Part A: Applications*, vol. 47, no. 3, pp. 291–313, 2005.

[16] S. Soleimani, D. Ganji, E. Ghasemi, M. Jalaal, and H. Bararnia, "Meshless local RBF-DQ for 2-D heat conduction: a comparative study," *Thermal Science*, vol. 15, no. 1, pp. 117–121, 2011.

[17] H. Ding, C. Shu, K. Yeo, and D. Xu, "Numerical computation of three-dimensional incompressible viscous flows in the primitive variable form by local multiquadric differential quadrature method," *Computer Methods in Applied Mechanics and Engineering*, vol. 195, no. 7-8, pp. 516–533, 2006.

[18] A. Khoshfetrat and M. Abedini, "Numerical modeling of long waves in shallow water using LRBF-DQ and hybrid DQ// LRBF-DQ," *Ocean Model*, vol. 65, pp. 1–10, 2013.

[19] M. Antuono, L. Soldini, and M. Brocchini, "On the role of the Chezy frictional term near the shoreline," *Theoretical and Computational Fluid Dynamics*, vol. 26, no. 1–4, pp. 105–116, 2012.

[20] E. Kansa, "Exact explicit time integration of hyperbolic partial differential equations with mesh free radial basis functions," *Engineering Analysis with Boundary Elements*, vol. 31, no. 7, pp. 577–585, 2007.

[21] G. Kesserwani and Q. Liang, "Locally limited and fully conserved RKDG shallow water solutions with wetting and drying," *Journal of Scientific Computing*, vol. 50, no. 1, pp. 120–144, 2012.

[22] P. Ullrich, C. Jablonowski, and B. van Leer, "High-order finite-volume methods for the shallow water equations on the sphere," *Journal of Computational Physics*, vol. 229, no. 17, pp. 6104–6134, 2010.

[23] D. Stevens and H. Power, "The radial basis function finite collocation approach for capturing sharp fronts in time dependent advection problems," *Journal of Computational Physics*, vol. 298, pp. 423–445, 2015.

[24] E. Chaabelasri, A. Amahmouj, M. Jeyar, A. G. L. Borthwick, N. Salhi, and I. Elmahi, "Numerical survey of contaminant transport and self cleansing of water in Nador Lagoon, Morocco," *Modelling and Simulation in Engineering*, vol. 2014, Article ID 179504, 8 pages, 2014.

[25] S. Sarra, "Integrated multiquadric radial basis function approximation methods," *Computers and Mathematics with Applications*, vol. 51, no. 8, pp. 1283–1296, 2006.

[26] F. Benkhaldoun, A. Halassi, D. Ouazar, M. Seaid, and A. Taik, "A stabilized meshless method for time-dependent convection-dominated flow problems," *Mathematics and Computers in Simulation*, vol. 137, pp. 159–176, 2017.

[27] F. Boushaba, E. M. Chaabelasri, N. Salhi, I. Elmahi, F. Benkhaldoun, and A. G. L. Borthwick, "A comparative study of finite volume and finite element on some transcritical free surface flow problems," *International Journal of Computational Methods*, vol. 5, no. 3, pp. 413–431, 2008.

[28] X. Xia, Q. Liang, M. Pastor, W. Zou, and Y. Zhuang, "Balancing the source terms in a SPH model for solving the shallow water equations," *Advances in Water Resources*, vol. 59, pp. 25–38, 2013.

[29] S. J. Jackson, D. Stevens, D. Giddings, and H. Power, "An adaptive RBF finite collocation approach to track transport processes across moving fronts," *Computers and Mathematics with Applications*, vol. 71, no. 1, pp. 278–300, 2016.

[30] D. Liang, B. Lin, and R. A. Falconer, "Simulation of rapidly varying flow using an efficient TVD–MacCormack scheme," *International Journal for Numerical Methods in Fluids*, vol. 53, no. 5, pp. 811–826, 2007.

[31] H. Shirkhania, A. Mohammadian, O. Seidou, and A. Kurganov, "A well balanced positivity preserving central upwind scheme for shallow water equations on unstructured quadrilateral grids," *Computers and Fluids*, vol. 126, pp. 25–40, 2016.

[32] I. K. Nikolos and A. I. Delis, "An unstructured node-centered finite volume scheme for shallow water flows with wet/dry fronts over complex topography," *Computer Methods in Applied Mechanics and Engineering*, vol. 198, no. 4748, pp. 3723–50, 2009.

[33] C. K. Chou, C. P. Sun, D. L. Young, J. Sladek, and V. Sladek, "Extrapolated local radial basis function collocation method for shallow water problems," *Engineering Analysis with Boundary Elements*, vol. 50, pp. 275–290, 2015.

[34] A. Baghlani, "Simulation of dam-break problem by a robust flux-vector splitting approach in Cartesian grid," *Scientia Iranica A*, vol. 18, no. 5, pp. 1061–1068, 2011.

[35] A. Canestrelli, M. Dumbser, A. Siviglia, and E. F. Toro, "Well-balanced high-order centered schemes on unstructured meshes for shallow water equations with fixed and mobile bed," *Advances in Water Resources*, vol. 33, no. 2010, pp. 291–303, 2010.

[36] M. Antuono and M. Brocchini, "Analysis of the nonlinear shallow water equations over non-planar topography," *Studies in Applied Mathematics*, vol. 124, no. 1, pp. 85–103, 2010.

[37] M. Kawahara and T. Umetsu, "Finite element method for moving boundary problems in river flow," *International Journal for Numerical Methods in Fluids*, vol. 6, no. 6, pp. 365–86, 1986.

8

Traffic Multiresolution Modeling and Consistency Analysis of Urban Expressway Based on Asynchronous Integration Strategy

Liyan Zhang,[1,2] Jian Ma,[1,2] Bin Ran,[2] and Lixin Yan[3]

[1]*School of Civil Engineering, Suzhou University of Science and Technology, 1701 Binhe Road, New District, Suzhou, Jiangsu 215011, China*
[2]*Department of Civil and Environmental Engineering, University of Wisconsin, 2312 Engineering Hall, 1415 Engineering Drive, Madison, WI 53706, USA*
[3]*College of Transportation and Logistics, East China Jiaotong University, Changbei Open and Developing District, Nanchang, Jiangxi 330013, China*

Correspondence should be addressed to Jian Ma; 9764634@qq.com

Academic Editor: Jing-song Hong

The paper studies multiresolution traffic flow simulation model of urban expressway. Firstly, compared with two-level hybrid model, three-level multiresolution hybrid model has been chosen. Then, multiresolution simulation framework and integration strategies are introduced. Thirdly, the paper proposes an urban expressway multiresolution traffic simulation model by asynchronous integration strategy based on Set Theory, which includes three submodels: macromodel, mesomodel, and micromodel. After that, the applicable conditions and derivation process of the three submodels are discussed in detail. In addition, in order to simulate and evaluate the multiresolution model, "simple simulation scenario" of North-South Elevated Expressway in Shanghai has been established. The simulation results showed the following. (1) Volume-density relationships of three submodels are unanimous with detector data. (2) When traffic density is high, macromodel has a high precision and smaller error and the dispersion of results is smaller. Compared with macromodel, simulation accuracies of micromodel and mesomodel are lower but errors are bigger. (3) Multiresolution model can simulate characteristics of traffic flow, capture traffic wave, and keep the consistency of traffic state transition. Finally, the results showed that the novel multiresolution model can have higher simulation accuracy and it is feasible and effective in the real traffic simulation scenario.

1. Introduction

With the rapid development of social economy and the accelerating process of urbanization, traffic demands increase sharply and traffic-related problems have become the common challenges for countries in the world. In order to solve the above questions, it depends on the comprehension and its application of the traffic control and operating rules. According to the size of the granularity, traffic simulation system is usually divided into three levels, that is, macroscopic traffic simulation, mesoscopic traffic simulation, and microscopic traffic simulation.

In the early 1960s, many macroscopic traffic simulation model emerged in the traffic simulation field [1], and a certain application of the results was achieved. But the macrosimulation model has difficulty in expressing the real dynamic behavior of the crowded state of traffic flow properly and each system performance change cannot be described which is caused by the random phenomenon of traffic flow. Since the 1980s, mesoscopic traffic models have been proposed and have been part of the application. Mesoscopic model cannot reflect the single driver's reaction to information and effects of different control systems to road network can only be expressed in the form of the capacity [2]. With the development of computer technology and system simulation technology, microscopic traffic simulation has been widely used. However, due to the complexity of microscopic modeling, uncertainty of parameter calibration, and constraints of computing resource, the network range of microsimulation is generally small [3]. It can be seen that single resolution model cannot simultaneously solve a series of issues, such as an actual operating state description of traffic flow, operating state description of a single vehicle, simulation accuracy, simulation complexity, and limited resources [4].

With the expansion of the scale of the traffic network simulation and simulation accuracy, a single fixed resolution simulation model cannot effectively solve the contradictions between network complexities and limited resources [5]. Thus, ideas and methods of multiresolution traffic simulation were proposed. Multilevel traffic hybrid simulation model system can be formed by using multiresolution modeling technology, which can organically integrate all levels of traffic simulation models and play their respective advantages of each model. At the same time, the original intricate problem will be solved by using multigranularity analysis methodology. Therefore, the paper studies the traffic multiresolution modeling and its consistency analysis of urban expressway main line and its ramp. It provides a feasible and effective method in the way of multiresolution theory applied in the field of traffic. And it points out a new direction for the further study of multiresolution theory.

2. Overview of Multiresolution Traffic Simulation Model

According to different granularity level of the hybrid model, multiresolution traffic simulation model can be divided into two aspects, "two-level hybrid model" and "three-level or above hybrid model." They are introduced as follows.

2.1. Two-Level Hybrid Models

2.1.1. Macroscopic and Microscopic Hybrid Traffic Models. The main idea of macro- and microhybrid traffic model is to simulate the entire road network by macromodel and only parts of the roads or subnetwork by microsimulation model and to perform polymerization and depolymerization at the boundary of microscopic model. The main representative studies of such models are described below. Lerner et al. [6] (2000) achieved a hybrid model which combined PELOPS microscopic model and macroscopic models SIMONE. In the paper, he discussed the random depolymerization method from the macromodel to the microscopic model and the polymerization process from micro- to macromodels. Magne et al. [7] (2000) integrated SITRAB + (microscopic model) with SIMRES (macroscopic model) and focused on the five limit items which can be simultaneously met by the car-following models and macroscopic models, so as to explore the compatibility of macromodel and micromodel. Leclercq [8] (2007) proposed macro- and microhybrid model based on the principles of Lighthill–Whitham–Richards (LWR) model. In the paper, consistency and conversion issues were discussed between macroindicators and microindicators. On the basis of the Asymmetric Cell Transmission Model (ACTM), improvement of Traditional Cell Transmission Model (CTM), Torné et al. [15] (2014) proposed an elegant analytical relationship based on traffic flow theory, which can ensure the physical consistency of the dynamics involved in queue processes in the proximity of a noncontrolled merge junction on freeway. Such models focused on theoretical consistency between macro- and micromodel but did not pay attention to advantages of different resolution model.

2.1.2. Mesoscopic and Microscopic Hybrid Traffic Models. The main idea of mesoscopic and microscopic hybrid traffic model is to simulate part of road network by mesomodel and micromodel and achieve conversion between mesomodel and micromodel by specific means at boundary nodes. Study of such models is the current hotspot of traffic hybrid simulation system.

Early research results included the hybrid model about PARAMICS microscopic model and DYNASMATR mesoscopic model, implemented by Sahraoui et al. [9] (2001), which emphasized that the microscopic model is not suitable to describe the route choice behavior. Nizard [10] (2002) focused on calibrating the mesoscopic model by using the results of microscopic simulation model and implemented a hybrid model of microscopic MITSIMLab and Metropolis mesoscopic model. Burghout [11] (2004) achieved the compound of microscopic model MITSIMLab and mesoscopic model Messo, which discussed not only the boundary interface problem of two kinds of model but also the dynamic path selection method based on virtual road sections. Ziliaskopoulos et al. [12] (2006) implemented the mixing of mesoscopic CTM model and microscopic discrete car-following model. Nava et al. [16] (2012) documented the modeling efforts in regard to a dynamic lane closure and reversible lane operation by integrating microscopic and mesoscopic model. Joueiai et al. [14] (2015) provided a brief overview of mesoscopic LWR model and its solution by the Variational theory. It is the central issue of current traffic hybrid simulation system but it has some disadvantages, such as being based on the interaction of cross-resolution, poor scalability, and instability caused by high coupling of system.

At present, traffic hybrid models have received more and more attention by scholars and there are some achievements in all aspects of the traffic field. The main representative studies of two-level hybrid model are described in Table 1.

Two-level hybrid traffic model, usually integrated with two different granularity simulation models, which can take full advantage of different models, overcome or reduce the disadvantages of single model and maximize to meet the needs of applications. However, with the development of the actual traffic demand, a simple two-level hybrid model cannot meet the demands of reality. In order to meet the practical applications and increasing theoretical research, multiresolution traffic hybrid model is the future direction of development and research focus, which uses the more rational modeling idea and method.

2.2. Three-Level and above Hybrid Traffic Models. The traditional integrated model approach is suitable for the two-level hybrid models, but it is a little insufficient to build three-level or above hybrid models. Multiresolution modeling (MRM) is an international research focus in the field of modeling and simulation since the 1990s. MRM has become one of the key

TABLE 1: Two-level hybrid simulation models and its characteristics.

Number	Model name	Level	Time	Characteristics
1	Lerner et al. [6]	Integration of macro- and micromodels	2000	Integration of micromodel PELOPS and macromodel SIMONE
2	Magne et al. [7]	Integration of macro- and micromodels	2000	Integration of micromodel SITRAB+ and macromodel SIMRES. Focus on compatibility of microscopic and macroscopic model
3	Leclercq [8]	Integration of macro- and micromodels	2007	Integration of macromodel LWR and its micromodel. Consistency and conversion of parameters
4	Sahraoui et al. [9]	Integration of meso- and micromodels	2001	Integration of micromodel PARAMICS and mesomodel DYNASMATR
5	Nizard [10]	Integration of meso- and micromodels	2002	Integration of micromodel MITSIMLab and mesomodel Metropolis
6	Burghout [11]	Integration of meso- and micromodels	2004	Integration of micromodel MITSIMLab and mesomodel Messo
7	Ziliaskopoulos et al. [12]	Integration of meso- and micromodels	2006	Integration of mesoscopic CTM model and microscopic discrete car-following model
8	Xu et al. [13]	Integration of macro- and micromodels	2010	Traffic flow analysis of dynamic network; real-time traffic prediction
9	Joueiai et al. [14]	Integration of meso- and micromodels	2015	Investigate the influence of various boundary conditions on traffic features

technologies for large-scale simulation of the United States Navy. United States National Research Council consider that MRM is one of the fundamental challenges of the modern modeling and simulation techniques [17].

2.2.1. Multiresolution Modeling Method. In the study of MRM, the modeling method is one of the key contents and the method provides a methodological basis for the multiresolution traffic simulation. Among modeling methods, there are some influential theories such as Aggregation/Disaggregation, Select Viewing (SV), UNIFY method, Integrated Hierarchical Variable-Resolution Modeling (IHVR), and Cross-Resolution Interaction. They are not going to be described here in detail.

Aggregation and Disaggregation method is easy to understand, but there are some problems, such as transient characteristic inconsistencies and chain disaggregation [18, 19]. Select Viewing method is relatively simple. It is easy to maintain consistency between the models but its calculation cost is too large and its flexibility is poor [20, 21]. UNIFY method can avoid disadvantages of Aggregation and Disaggregation and ensure consistency between the different resolution models; however, the efficiency of UNIFY is very low and the resource consumption is huge [21, 22].

IHVR method is a process-oriented approach and is easy to understand, but it is difficult to choose the variables because the hierarchical relationship between variables is not strict [23]. Alternate Submodels need higher interface requirements for the models and its reusability and transformation are poor because of its fixed resolution [24]. Only one submodel is active in the course of simulation at a time, which leads multimodel formalism to have status recovery and synchronization problems [25]. Cross-Resolution Interaction method can use independent simulation models on different resolution levels, which is not only to retain the advantages of the original model but also to maintain loose coupling between the models [26]. It reduces system cost while improving the stability of the simulation system scalability.

Therefore, the article will use Cross-Resolution Interaction method in the study of multiresolution traffic modeling.

2.2.2. Multiresolution Traffic Model. Currently, multiresolution modeling theory is not very mature, but there have been some successful applications, which highlights the fact that the multiresolution modeling has broad prospect and research value. In the field of multiresolution modeling applications, the representative work includes the following. Davis and Hillestad (1998) presented an exploratory analysis based on multiresolution model by using multiresolution modeling theory [27]. Lee and Fishwick (1999) studied the multiresolution modeling applications in real-time systems [28]. Sekine et al. (2001) studied the multiresolution modeling in Traffic System Simulation [29]. Schaffer and Kukolich (1998), Schaffer and Marion (1999), and Butler (2002) studied the multiresolution modeling in comprehensive natural environment [30–32]. Reynolds et al. [33] (2002) studied the meteorological and oceanographic multiresolution modeling problems and so on.

Since the middle of 1990s, China began to study multiresolution modeling theory and its application in the field of combat simulation aspect, but the research is not systematic. In recent years, traffic multiresolution simulation models have been studied widely, but all of them are in their first step. Currently, representative research in the field of multiresolution traffic simulation is Transmodeler [34] (Yang and Slavin, 2002), which is a package of a famous traffic simulation/planning software—TransCAD. Transmodeler integrated the three granularity traffic models—macroscopic traffic model, mesoscopic traffic model, and microscopic traffic model. User can freely select the granularity of sections, simultaneously choose different granularity simulation model in different sections, and keep the simulation system in different granularity running concurrently. Transmodeler can maintain balance among simulation granularity, computing speed, evaluation indexes, and system overall performance. It can meet the specific requirements of simulation program very well. However, multiresolution modeling research is just emerging, especially in the field of transport; there is no relevant systematical academic literature. It also illustrates the importance and feasibility of the multiresolution traffic simulation study.

2.3. Comprehensive Analysis and Commentary. First of all, in a single resolution traffic model, the macroscopic model has a high computational efficiency and low computing requirements, but it is not suitable to simulate this traffic behavior, which needs to consider the interaction between adjacent vehicles. Microscopic model can simulate the movement behavior of a vehicle and visual simulation process, which is in favor of the simulation results analysis. However, road network modeling of microscopic model is so complex, the model is highly sensitive, and it requires a high computing ability. Mesoscopic model is studied on vehicle fleet and it can describe the interaction between the vehicles to some extent. Computing ability requirement of mesoscopic model is lower than microscopic models', but it needs a good balance between width and depth of simulation.

Secondly, the two-level hybrid model based on different granularity is generally achieved by parameters conversion method, which combines the advantages of a single model to some extent. However, the applicability of two-level hybrid model is rather poor, basic data processing is complex, and its expansibility is relatively weak. Three-level or above multiresolution mixed traffic model is based on multiresolution modeling theory and it not only integrates the advantages of multiple models, but also provides a shared basic database in order to save data maintenance cost for a single model, which ensures the consistency between models. Features comparison between the different models is shown in Table 2 [5, 35].

In summary, traffic simulation theory and models have made many achievements, but there are still some shortcomings, mainly reflected in the following aspects.

(1) In terms of defining the concept, macroscopic model, mesoscopic model, and microscopic model of multiresolution modeling are a relative concept. For example, models can be distinguished obviously when establishing multiresolution model by Aggregation and Disaggregation method. It can be distinguished clearly between macroscopic model and microscopic model when modeling by other methods; however, mesoscopic model is difficult to distinguish from granularity. Therefore, how to define different resolution models is worthy of deep researching.

(2) In the study objects aspect, most models are aimed at freeway or highway, less at urban expressway. Meanwhile, models proposed by foreigners need to be verified in the Chinese traffic situation because of different transport infrastructure, traffic composition, traffic control means, and characteristics of traffic participants.

(3) In terms of granularity, most models focus on single resolution model, which is modeling on a certain level of granularity, while less studies are focused on multiresolution model, especially three-level or above multiresolution models. At the same time, the study of integrated manner and type and directions of hybrid models have been less mentioned.

(4) In terms of theoretical foundations, different types of models have different theoretical basis; at the same time, the hybrid model has a number of different theoretical bases. Without a systematic theory, most hybrid models are based on conventional parameter transformation or partial data sharing. However, multiresolution simulation theory provides a new idea and theoretical basis for the hybrid model. But all the multiresolution simulation theories and their application in traffic are still in their infancy.

(5) In research methods, most previous hybrid models are based on interface parameters fusion method, whose case study is often an ideal or specific scene, and it is difficult to apply to practical problems. Aggregation and Disaggregation method, a common multiresolution modeling method, is more suitable for nonmemory system. However, Cross-Resolution Interaction is suitable for multiresolution traffic simulation system with memory function. In addition, Cross-Resolution Interaction can be integrated with integration simulation theory, providing a consistent methodology basis for mixing different resolution models, and it can make better use of existing research results and save costs.

(6) In terms of model implementation, existing models have too much control parameters and do not consider the actual traffic road conditions, which require adjusting, modifying, and improving the models under ideal conditions.

(7) In model optimization aspect, existing models' solving processes are too simple and the results accuracy is not high; the algorithm efficiency is relatively low, which cannot meet traffic simulation real-time and accuracy requirements. In maintaining the accuracy of the algorithm, parallel algorithm can greatly improve the efficiency of computing, which provides a feasible way to achieve real-time simulation.

(8) In model validation aspect, most models are based on purely mathematical derivation and lack of effective actual data validation and verification, which results in the model having a certain difference with the actual traffic situation.

3. Multiresolution Traffic Simulation Model and Its Framework

3.1. Definition of Model Resolution. Model resolution still does not have a very clear definition. International Simulation Interoperability Standards Organization defined resolution as detailed and accurate representation of the real world level in the models and simulation, which can be also called granularity [36]. Resolution's definition by Roza et al. was to describe the accuracy and precision of some models and simulation applications in the real world [37]. Granularity and level of detail are equivalent to the concept of resolution in the field of modeling and simulation [17].

Table 2: Comparative analysis of different resolution traffic simulation models.

Item	Single resolution traffic simulation model			Multiresolution traffic simulation model	
	Microscopic model	Mesoscopic model	Macroscopic model	Two-level hybrid traffic model	Three-level and above hybrid traffic model
Theoretical background	Discrete event system modeling, fixed-step	Discrete event system modeling, fixed-step	Continuous system modeling, numerical simulation	Similar to the distributed simulation	Multiresolution modeling and simulation theory
Simulation object	Single vehicle	Single vehicle, driven by queue	Vehicle is regarded as fluid or gas	Depending on the hybrid model	Objects can be dynamically changed on demand
Core model	Car-following model, lane changing model, cellular model	Queuing model, flow-velocity-density model, allocation model	To describe the system evolution rule in differential equations derived from the flow conservation	Mathematical derivation or interfaces defined in original model	Integration of consistency maintenance of different resolution models
Implementation	Time series method, event-based method	Numerical methods, time series method, event-based method	Numerical methods	Parameter conversion, partial data sharing	Integrated simulation, multiresolution modeling method
Application	Microdesign analysis such as intersection channeling and signal timing	Dynamic network simulation, but limited to the scope of the network	Planning analysis of large-scale road network	Depends on hybrid models	Gradual application from network to regional to intersection
Typical system	VISSIM, Aimsun, PARAMICS, CORSIM	Dynasmart, DynaMIT, Dynameq, Transims	NETFLO2, FREFLO	DynaCHINA	Transmodeler

According to the range scope of granularity, multiresolution model can be divided into low-resolution model (macroscopic model), normal resolution model (mesoscopic model), and high-resolution model (microscopic model) [36]. Under normal circumstances, you can use a high-resolution model to describe the microscopic properties of the system, the low-resolution model to describe the system of macroscopic properties, and normal resolution model to describe application scenarios that need to balance simulation accuracy and breadth, so that you can effectively reduce computing costs and increase operating efficiency.

3.2. Multiresolution Model. In theory, multiresolution traffic model is a mathematical description of the same object under different granularity. From the view of practical application, different people pay attention to different aspects for the same issue, and different granularity models are built according to different needs. In addition, the establishment of different resolution models' family of transportation research objects can enable the user to choose different resolution models based on specific needs.

In this paper, transportation system multiresolution model is created by Set Theory. The model describes the characteristics of the system, including behavioral characteristics, time characteristics, spatial characteristics, performance characteristics, granularity level, and internal structure. In the model formalization process, set structure of research objects is built through sufficiently abstracting the transportation system characteristics, so that the model has a more extensive description and scope. The paper draws seven-tuple structure to build a multiresolution model of ten-tuple structure of transportation system and discusses the state transition equation and the mechanism of traffic state [38, 39]. So, multiresolution traffic simulation model based on Set Theory is described as follows:

$$M = \langle X, Y, T, \Omega, S, \theta, Q, \delta, \Gamma, \lambda \rangle . \quad (1)$$

Among them, M represents a traffic system.

X indicates the input set of the system, which describes the effect of the external environment on the system. Generally, X represents input variables of real value. If external input is a discrete event, X can be described as $X_e \cup \{\phi\}$, in which X_e is a set of external events and ϕ is a set of empty events. In a specific traffic scene, X_e may be represented as speed, density, and so on.

Y indicates the output set of the system, which describes the sum of the output of the system to the external environment. In a specific traffic scene, it may represent the parameter, such as traffic flow and location.

T is a time coordinate system to describe the system time change. According to T, system is divided into discrete-time system or continuous-time system.

Ω is time input segment set, which describes the system input mode in a time interval. It can be described as $\omega : \langle t_0, t_1 \rangle \to X$. $\langle t_0, t_1 \rangle$ is an interval of time base from the initial time t_0 to terminate time t_1. The set of all input fragments in $\langle t_0, t_1 \rangle$ are denoted by (X, T). Ω is a subset of (X, T). (1) When T is a set of real numbers, Ω is divided into two cases. ① Ω is piecewise continuous segment set, which can be described as $\omega : \langle t_0, t_1 \rangle \to X,\ X = R^n$. ② Ω is discrete events section set, which can be described as $\omega : \langle t_0, t_1 \rangle \to X_e \cup \{\phi\}$. If limited event time collection $\{\tau_1 \cdots \tau_n\} \not\subset \langle t_0, t_1 \rangle$, then $\omega(t) = \phi$. (2) When T is a set of integers, Ω is finite set of time series.

S is a spatial coordinate to describe the system spatial location. According to S, system is divided into discrete space system or continuous space system.

θ is spatial input segment set, which describes the system input mode in a spatial interval. It can be described as $\varphi : \langle s_0, s_1 \rangle \to X$. $\langle s_0, s_1 \rangle$ is an interval of space base from the initial position s_0 to terminate position s_1. The set of all input fragments in $\langle s_0, s_1 \rangle$ are denoted by (X, S). θ is a subset of (X, S). (1) When S is a set of continuous values, θ is divided into two cases. ① θ is piecewise continuous segment set, which can be described as $\varphi : \langle s_0, s_1 \rangle \to X,\ X = R^n$. ② θ is discrete events section set, which can be described as $\varphi : \langle s_0, s_1 \rangle \to X_e \cup \{\phi\}$. If limited event spatial collection $\{\gamma_1 \cdots \gamma_n\} \not\subset \langle s_0, s_1 \rangle$, then $\varphi(s) = \phi$. (2) When the values of S are not continuous, θ is finite set of space series.

Q is the collection of internal state, which is the core of the internal structure modeling. It is a collection of information under a certain state of the system, and it affects the system behavior response of the present and the future. In a specific traffic scene, it may characterize all parameters of the traffic operation.

δ is state transition function, which defines how the internal state of the system changes. It can be described as $\delta : Q \times \Omega \times \theta \to Q$. System state is q when it is in position s_0 at time t_0. When an input segment $\omega : \langle t_0, t_1 \rangle \cup \varphi : \langle s_0, s_1 \rangle \to X$ is applied to q, $\delta(q, \omega, \varphi)$ represents the state of the system in position s_1 at time t_1. Thus, the internal state at any moment, the time input section, and space input section from that moment all determine the status of the position of the segment end time. δ can be simply divided into macroscopic traffic model, mesoscopic traffic model, and microscopic traffic model in multiresolution traffic simulation model. Different resolution traffic model controls the system internal state changing of different granularity and output the corresponding results. In the paper, different resolution model is referred to as traffic model. Submodel is defined as one-part model of the set of multiresolution models. Specific forms of each submodel are described below.

Γ is granularity function, which describes the level of granularity of the system's internal state. It can be described as $\Gamma : Q \to \delta$. Granularity function gives the standard of state transition function, which is $\Gamma : Q \times X \times T \times S \to \delta$. In a specific traffic scene, it may be the parameters of the model at different resolutions or mouse operation.

λ is output function, which can be described as $\lambda : Q \to Y$. Output function gives an output section, which is $\lambda : Q \times X \times T \times S \to Y$. In a specific traffic scene, it may be the formatted input operation of database.

The set structure based on ten dimensions can describe the properties and behavior characteristics of the object by time, space, and corresponding parameters. The multiresolution traffic model can be effectively described by Set Theory and it provides a basis for the study of the establishment and evaluation of complex models of different resolutions transportation system.

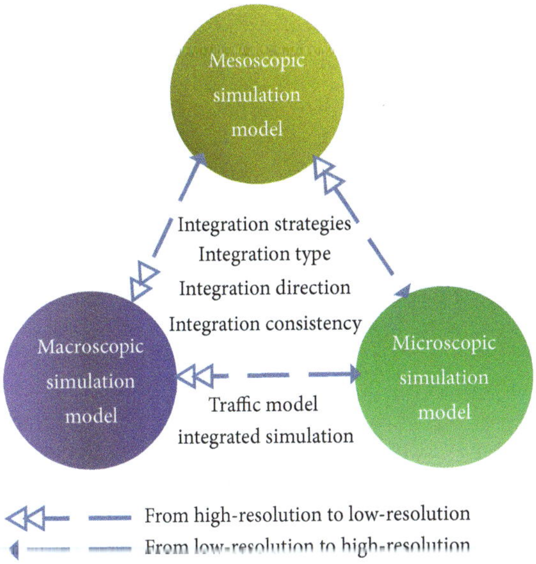

FIGURE 1: Triangular framework of multiresolution traffic simulation model.

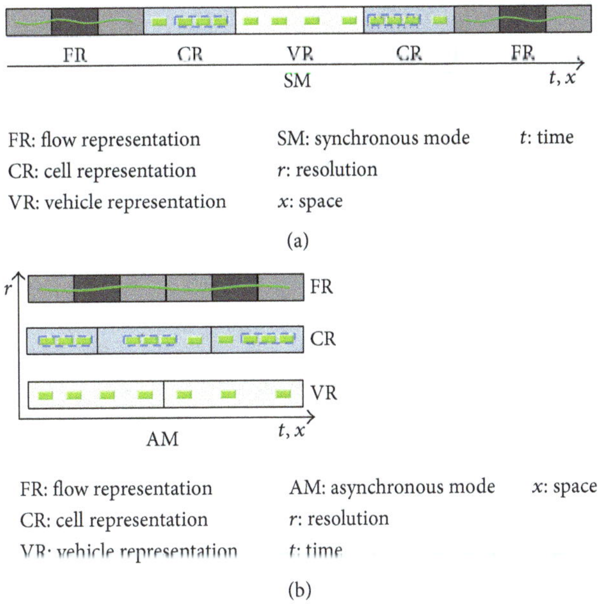

FIGURE 2: Multiresolution traffic simulation integration strategies schematic diagram.

3.3. Multiresolution Traffic Simulation Framework. Using the integrated simulation methods to achieve multiresolution traffic simulation system is the feasible and optimum method, both from the point of view of an implementation convenience or the cost. On the one hand, it can make full use of existing resources and systems, to avoid resources waste and to maintain the continuity of the existing application system. On the other hand, it can greatly reduce development and testing costs, meanwhile meeting the target needs of system.

Multiresolution traffic simulation model can be established and analyzed from the macromodel, mesomodel, and micromodel three granularity levels. The three granularity level models can be fused, analyzed, simulated, and experimented by integrated simulation methods from aspect of integration type, integration strategy, integration direction, and integration consistent, and so on. All of these build a multiresolution traffic simulation triangular framework, shown in Figure 1.

Multiresolution traffic simulation triangular framework makes the macro-, meso-, and micromodel as its vertexes, which are the basis and cornerstone of multiresolution traffic simulation. Traffic model integrated simulation method in the triangular framework, as a means, is fusion technology basis of different resolution simulation models. Integration type, integration strategy, integration direction, and integration consistent are the cores of different resolution simulation model in optimization process. All the above parts are the foundation of multiresolution traffic simulation in verification and validation process.

3.4. Integration Strategies of Multiresolution Simulation. There are different integration strategies when integrating macroscopic, mesoscopic, and microscopic simulation models. According to time standard, it can be divided into Synchronous Mode (SM), Asynchronous Mode (AM), and Mixing Mode (MM). SM is a serial execution policy in essence, which is a process that is different from granularity models combined in a certain order that can run sequentially. AM is that different granularity models that can run simultaneously in a certain space or time and it is a parallel execution strategy. MM is a mixed application of SM and AM. SM is used in part of the space or a period of time, while AM is adopted at other space or time. MM can maximally meet the actual simulation needs and make full use of the advantages of both integration strategies.

Figure 2 is a schematic view of the multiresolution traffic simulation integration strategies. Figure 2(a) shows a serial operation mode of macromodel (Flow Representation, FR), mesomodel (Cell Representation, CR), and micromodel (Vehicle Representation, VR), which is an implementation of SM. Figure 2(b) shows a parallel operating mode of macroscopic, mesoscopic, and microscopic model, which is an implementation of AM. In this paper, AM is used. Please refer to another article literature of the author [40] if studying SM.

4. Macroscopic Traffic Flow Modeling and Simulation Analysis

4.1. Urban Expressway Traffic Characteristics and the Target of Macroscopic Modeling. The main purpose of urban expressway is to build fast connecting channels between the various functional areas within the city, between functional areas and periurban satellite town, and between neighboring cities and functional areas. Urban expressway can separate vehicles from nonmotor vehicle in travel space [41]. Compared with other roads, urban expressway has its own features in the aspect of road environment, traffic flow characteristics, and other aspects.

(1) Main line of urban freeway is fully enclosed in the form of grade separation. So its traffic flow of main line is rapid and continuous, while relief road, affected by intersection traffic control, is typical intermittent traffic flow and queuing and fleet dissipate phenomenon occurs regularly [42].

(2) Urban freeway mainly services the vehicles in the city or a short travel.

(3) In general, design speed of freeway is 60–80 km/h, no more than 100 km/h, in China.

When vehicles number is small on urban freeway, the vehicle can run in expected speed limits as a free flow. With the increase of traffic flow and traffic density, traffic driving speed slows down and results in queuing.

(4) Urban expressway has plenty of entrances and exits, the distance between the entrances and exits is small, and its density is a little high.

(5) Operating speed of urban expressway is basically in normal distribution. At the entrance of ramp, the vehicle speed is relatively concentrated, which is general in skewed distribution [43].

(6) Compared with urban roads, capacity of urban freeway is larger and the average operating speed is bigger. Urban freeway has obviously tidal characteristics.

In this paper, urban expressway has been studied. According to analyzing the properties of urban expressway, macroscopic model should describe traffic flow characteristics of the following aspects. ① Describe the viscous and compression properties of traffic flow. ② Describe the characteristics of the external environment and driving characteristics of the drivers. ③ Simulate various actual traffic flow phenomena, such as traffic congestion caused by accident or construction. ④ Reflect the ramp signal control and no ramp metering impact on transportation state. ⑤ Describe the traffic flow operating state variation with time in the basic segment road, entrance and exit of ramps, and weaving section.

4.2. Macroscopic Mathematical Model of Traffic Flow

4.2.1. Hypothesis of Macromodel. For macroscopic model, make the following basic assumptions for the relationship between speed and density of traffic flow:

(1) Speed ranges from zero to free-flow speed $u_f(x,t)$; density ranges from zero to jam density $k_{\text{jam}}(x,t)$.

(2) When the space between vehicles is infinity, vehicle speed is limited to free-flow speed; that is, $\lim_{k\to 0} u = u_f(x,t)$, wherein k is traffic density and u is velocity.

(3) When the density is jam density $k_{\text{jam}}(x,t)$, the speed is zero; that is, $u(k_{\text{jam}}) = 0$.

(4) The vehicle speed decreases with the increase of density; $u'(k) < 0$ $(0 < k \le k_{\text{jam}}(x,t))$.

(5) Equilibrium speed function $u_e(k)$ satisfies the following formula. $u_e'(k) < 0$, $(ku_e(k))'' < 0$, $u_e(0,x,t) = u_f(x,t)$, $u_e(k_{\text{jam}}(x,t)) = 0$. According to the actual situation, different road sections have different free-flow speed $u_f(x,t)$ and different jam density $k_{\text{jam}}(x,t)$.

(6) The driver's response to stimulation has a time delay T_r, which can be constant or function.

(7) Equilibrium speed is $u_e(k,x)$ and speed reaction time is $(t+T_r)$ at $(x+\Delta x)$ position.

(8) The state change of rear vehicle depends on the relative speed of adjacent vehicles, the space between adjacent vehicles, viscous resistance caused by vehicle type difference, compression properties of traffic flow, and drivers' psychological vehicle safety distance.

4.2.2. The Derivation of Macromodel. Based on the above assumptions, combined with the derivation of the classical car-following model [44–46] and the observation of car-following phenomena [47], the macromodel should consider acceleration caused by relative speed difference of the adjacent two vehicles, viscous resistance caused by braking distance, difference of the adjacent two vehicles, compression properties of traffic flow caused by drivers' psychological vehicle safety distance, and actual transportation environment. And state of rear vehicle depends on the space between the adjacent vehicles.

For the two adjacent car-following vehicles, $x_n(t)$ is the position of vehicle n at the time t. $x_{n+1}(t)$ is the position of vehicle $(n+1)$ at the time t. $s(t)$ is the space headway between vehicles at time t. T_r is driver's reaction time, which is called delay time or relaxation time. Vehicle $(n+1)$ travels d_1 distance in the reaction time T_r. d_2 is braking distance of vehicle $(n+1)$. d_3 is braking distance of vehicle n. L is a safe stop distance.

As can be seen from Figure 3,

$$s(t) = x_n(t) - x_{n+1}(t) = d_1 + d_2 - d_3 + L, \tag{2}$$

$$d_1 = T_r \cdot u_{n+1}(t) = T_r \cdot u_{n+1}(t+T_r) = T_r \cdot x'_{n+1}(t). \tag{3}$$

Formula (2) can be rewritten as

$$\begin{gathered} s(t) = x_n(t) - x_{n+1}(t) = d_1 + L + W, \\ W = w_1 + w_2 = d_2 - d_3. \end{gathered} \tag{4}$$

W is a function about the proportion of the car, time t, road geometry features, and weather, which considered vehicle type difference, road conditions, weather conditions, and other factors. They are traffic characteristics items.

w_1 is viscous properties factor.

w_2 is compression properties factor.

Detailed definitions are formulas (10) and (11).

After above derivation transformation, formula (4) can be rewritten as

$$\begin{gathered} x''_{n+1}(t+T_r) = \lambda\left[x'_n(t) - x'_{n+1}(t)\right] - \lambda W' \\ n = 1,2,3,\ldots, \\ \lambda = \frac{1}{T_r}. \end{gathered} \tag{5}$$

It can be seen from formula (5) that vehicle traveling state is affected not only by speed difference between adjacent vehicles but also traffic flow composition, road, weather, and other factors. The latter in formula (5) is a traffic characteristics item.

Jiang et al. model thinks that the relative speed of two vehicles should be considered; that model evolved into [48]

$$x''_{n+1}(t+T_r) = \lambda\left[U(\Delta x) - u_{n+1}(t)\right] + \kappa\Delta u. \tag{6}$$

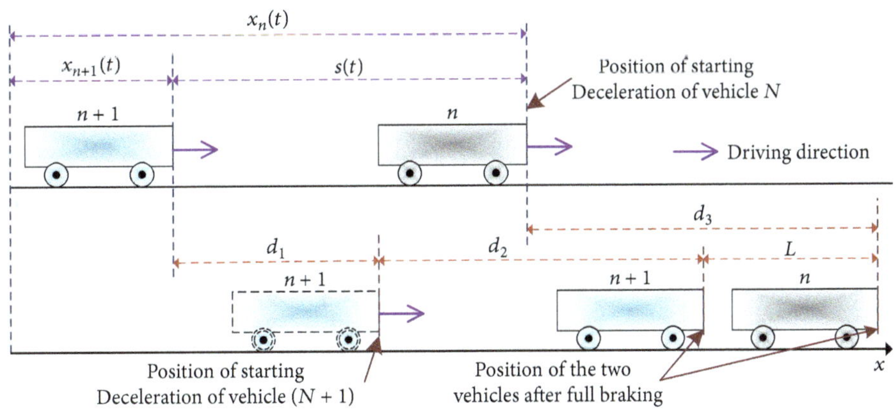

FIGURE 3: Car-following model diagram.

So formulas (5) and (6) could be expressed as follows:

$$\begin{aligned} x''_{n+1}(t+T_r) &= \lambda\left[U(\Delta x) - u_{n+1}(t)\right] + \kappa\Delta u - \lambda W', \\ W' &= w'_1 + w'_2, \\ \Delta u &= u_n(t) - u_{n+1}(t), \\ \Delta x &= x_n(t) - x_{n+1}(t). \end{aligned} \tag{7}$$

Among them, u_n is speed of vehicle n. u_{n+1} is speed of following vehicle $(n+1)$. λ is response coefficient and κ is reaction coefficient. x_n and x_{n+1}, respectively, are the position of vehicle n and the following vehicle $(n+1)$. U is the function of Δx, while Δx is displacement difference between front and following vehicles.

By converting the microscopic parameters into macroscopic parameters:

$$\begin{aligned} u_{n+1}(t) &: u(x,t), \\ u_n(t) &: u(x+\Delta x, t), \\ U(\Delta x) &: u_e(k), \\ \lambda &= \frac{1}{T_r}, \\ \kappa &= \frac{1}{\tau}. \end{aligned} \tag{8}$$

T_r, as described above, is driver's reaction time, which is called delay time or relaxation time. It takes time τ when disturbance propagates backwards the distance Δx.

Make $\lambda W' = \delta_{w1} + \delta_{w2}$; then

$$\begin{aligned} \frac{du(x,t)}{dt} = \frac{u_e(k) - u(x,t)}{T_r} + \frac{u(x+\Delta x,t) - u(x,t)}{\tau} \\ - (\delta_{w1} + \delta_{w2}), \end{aligned} \tag{9}$$

$$\delta_{w1} = \mu \cdot k \cdot \Delta u \cdot u \cdot \frac{S_z}{S_j}, \tag{10}$$

$$\delta_{w2} = \alpha \cdot k \cdot \Delta u \cdot u \cdot \frac{S_z}{S_j}, \tag{11}$$

$$\Delta u = u_i - u_{i+1} \begin{cases} \Delta u, & \Delta u > 0, \\ 0, & \Delta u = 0 \text{ or } \Delta u = \varepsilon, \\ 0, & \Delta u < 0. \end{cases} \tag{12}$$

μ is coefficient of viscosity, its unit is m/(veh·s), which varies with the proportion of the car, and generally its value is in line with the normal distribution [49].

α is the compression coefficient.

u is vehicle speed.

S_z is braking distance of vehicle; the unit is m.

S_j is line of sight; the unit is m.

The left side of (9) is written as a fully differential form, the Taylor expansion on its right, ignoring the high-order terms; then

$$\begin{aligned} \frac{du(x,t)}{dt} &= \frac{\partial u}{\partial t} + u\frac{\partial u}{\partial x} \\ &= \frac{u_e(k) - u(x,t)}{T_r} + \frac{\Delta x}{\tau}\frac{\partial u}{\partial x} - \delta_{w1} - \delta_{w2}. \end{aligned} \tag{13}$$

In formula (13), $\Delta x/\tau = -ku'_e(k) \equiv a(k) > 0$; it is also called the speed of sound.

Substituting (10) and (11) into (13), at the same time, the conservation of traffic flow should still be satisfied. Then, macroscopic traffic flow model of multiresolution simulation model combined with viscous and compression properties can be collectively represented as follows:

$$\frac{du(x,t)}{dt} = \frac{\partial u}{\partial t} + u\frac{\partial u}{\partial x} = \frac{u_e(k) - u(x,t)}{T_r} + \frac{\Delta x}{\tau}\frac{\partial u}{\partial x} - \mu \cdot k \cdot \Delta u \cdot u \cdot \frac{S_z}{S_j} - \alpha \cdot k \cdot \Delta u \cdot u \cdot \frac{S_z}{S_j}, \tag{14}$$

$$\frac{\partial k}{\partial t} + \frac{\partial q}{\partial x} = g(x,t). \tag{15}$$

k is traffic density. q is traffic volume. x and t, respectively, represent space and time. $g(x,t)$ is source-sink item, which means input or output volume. u is average velocity of traffic flow. $u_e(k)$ expresses the relationship between u and k under the equilibrium state.

5. Mesoscopic Traffic Flow Model

5.1. Hypothesis of Mesomodel. (1) Assume an approximate solution, which is expressed as a linear combination of a set of (formally) simple functions ψ_i and the coefficient of the linear combination is a set of undetermined coefficient C_i.

(2) Considering differential equations and boundary conditions, an objective function F is established to account for the difference between true solution φ and approximate solution $\overline{\varphi}$.

(3) The objective function is minimized by a suitable algorithm—the process of minimization determines the undetermined coefficient, and thus the approximate solution of the problem is achieved.

5.2. The Derivation of Mesomodel. The control equations of the anisotropic kinetic model can be expressed as follows:

$$\frac{du}{dt} = \frac{\partial u}{\partial t} + u\frac{\partial u}{\partial x} = F(k,u)\frac{\partial u}{\partial x} + H \cdot \frac{u_e - u}{T_r} + G(k,u). \tag{16}$$

$F(k,u)$ is velocity gradient coefficient in relaxation term. H is expected term coefficient of the equation. $G(k,u)$ is equation viscous term or nonhomogeneous term.

According to (13), (15), and (16), we can get the following equation:

$$G(k,u) = -\delta_{w1} - \delta_{w2}, \tag{17}$$

$$F(k,u) = \frac{\Delta x}{\tau}, \tag{18}$$

$$H = 1. \tag{19}$$

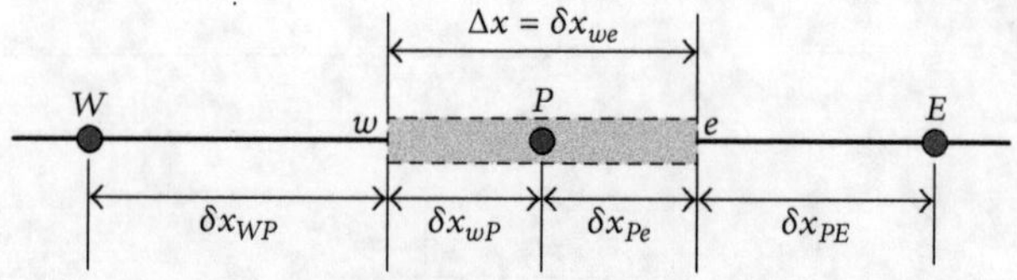

FIGURE 4: Traffic flow control volume diagram.

According to Figure 4 Traffic flow control volume diagram, within a control volume, is integral to (15) during intervals time from t to $(t + \Delta t)$:

$$\int_t^{t+\Delta t}\int_{\Delta v}\frac{\partial k}{\partial t}dv\,dt + \int_t^{t+\Delta t}\int_{\Delta v}\frac{\partial q}{\partial x}dv\,dt = \int_t^{t+\Delta t}\int_{\Delta v} g(x,t)\,dv\,dt. \tag{20}$$

Then,

$$\int_{\Delta v}\left(\int_t^{t+\Delta t}\frac{\partial k}{\partial t}dt\right)dv + \int_t^{t+\Delta t}\left[(Aq)_e - (Aq)_w\right]dt = \int_t^{t+\Delta t}\overline{g}(x,t)\,\Delta v\,dt. \tag{21}$$

In the formula, A is surface area of control volume. A is 1 at one dimension. Δv is volume of control volume. $\Delta v = A \cdot \Delta x$, in which Δx is length of control volume Δx_{we}. $\overline{g}(x,t)$ is average intensity of source-sink item.

For one-dimensional control volume, intervals time is from t to $(t + \Delta t)$; integrate (16) within the control volume and at the boundary:

$$\int_t^{t+\Delta t}\int_{\Delta v}\frac{\partial u}{\partial t}dv\,dt + \int_t^{t+\Delta t}\int_{\Delta v} u\frac{\partial u}{\partial x}dv\,dt = \int_t^{t+\Delta t}\int_{\Delta v} F(k,u)\frac{\partial u}{\partial x}dv\,dt + \int_t^{t+\Delta t}\int_{\Delta v}\frac{u_e - u}{T_r}dv\,dt + \int_t^{t+\Delta t}\int_{\Delta v} G(k,u)\,dv\,dt. \tag{22}$$

Then,

$$\int_{\Delta v}\left(\int_t^{t+\Delta t}\frac{\partial u}{\partial t}dt\right)dv + \int_t^{t+\Delta t} a\left[(Au)_e - (Au)_w\right]dt = \int_t^{t+\Delta t} F(k,u)\left[(Au)_e - (Au)_w\right]dt + \int_t^{t+\Delta t}\int_{\Delta v}\frac{u_e - u}{T_r}dv\,dt + \int_t^{t+\Delta t}\int_{\Delta v} G(k,u)\,dv\,dt. \tag{23}$$

In the formula, subscripts w and e, respectively, represent the initial and terminal section position of control volume. a is upwind scheme parameter.

Equations (17), (18), (21), and (23) compose mesoscopic traffic flow model, which is established by using Finite Volume method. Mesomodel not only has nature of macroscopic traffic model but also has conservation characteristics of integral equations; in particular, its control parameters may not be continuous. In addition, mesoscopic traffic model of the integral form is more suitable to describe traffic flow that contains real discontinuity. Since the integral equation may be intermittently in any position, which has laid a theoretical foundation for travel behavior research of small vehicles fleet, it can be seen as mesoscopic traffic flow model.

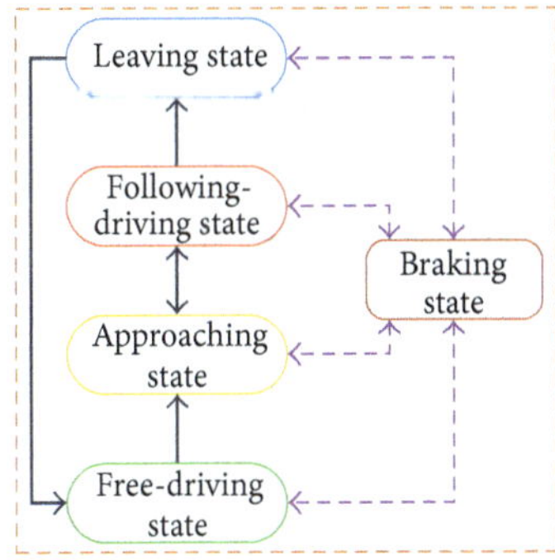

Figure 5: Basic state transition schematic diagram of the car-following driving.

6. Microscopic Traffic Flow Model

Microscopic simulation model of traffic multiresolution model is achieved by using COM methods integrated VISSIM system of German PTV company. VISSIM is a microscopic road traffic modeling tool by simulating traffic operation states to evaluate the program, which takes full account of the traffic characteristics of the motor vehicle, pedestrians, and bicycles. Its core simulation models include ① vehicle longitudinal motion adopted psychophysical car-following model that is put forward by professor Wiedemann of Germany University of Karlsruhe; ② vehicle lateral movement selected Sparmann's model of the lane changing behavior based on psychology and VISSIM used dynamic traffic allocation in path selection [50, 51].

6.1. Wiedemann Conceptual Model Based on Psychological Car-Following Model. Based on the visual psychological hypothesis, Wiedemann et al. proposed a concept model based on driving psychological car-following principle. A basic assumption of Wiedemann model is that the driver adjusts the following speed according to the relative movement between the front and following car, such as the variety of speed and distance. But these stimuli can only be sensed and reacted to by the driver beyond a certain threshold [52, 53]. At the same time, Wiedemann model divides driving status into free-driving state, approaching state, following-driving state, braking state, and leaving state [52, 54].

This paper argues that drivers in following-driving state generally follow these driving patterns: vehicles are usually from free-driving state to approaching state and then into following-driving state and then leaving state, and then become in free-driving state. In these states, there is a possibility that an abrupt state occurs and the vehicle is brought into a braking state. In addition, the approaching state and the following-driving state are often converted into each other. For each driving mode, the acceleration of the rear vehicle is determined by the vehicle speed, speed difference and distance between the front and rear cars, and characteristics of the vehicle. When the driver reaches a certain threshold expressed by the speed difference and distance, he will switch from one driving state to another driving state [50]. Such a change in driving state can be seen in Figure 5.

6.2. The Sparmann Conceptual Model Based on Psychological Lane Changing Rule. The lane changing behavior refers to the behavior of the general vehicle from the current lane to the adjacent lane. The Sparmann lane changing rule is a well-known lane changing model based on psychology [55], whose theoretical basis is the decision-making process model proposed by Willmann for lane changing behavior. Sparmann lane changing process is a typical hierarchical decision; decision-making order of the driver is considered as follows [52]: ① Is there a willingness to change lanes? ② Is driving in the adjacent lane better? ③ Is it feasible to change to the adjacent lane? The specific flow chart is shown in Figure 6.

7. Multiresolution Simulation and Evaluation

7.1. Traffic Simulation Scenario and Parameter Settings. In this paper, "simple simulation scenario" has been established, which is about 2 km from Yan'an East Road interchange to Tianmu Road rotary interchange of North-South Elevated Expressway in Shanghai. Weihai road north entrance ramp and Beijing road south exit ramp are ignored because they are too close to each other. But Xinzha road north entrance ramp is reserved in the simulation scenario. In order to reduce the impact of interchange, the start and end points of simulation scenario have some distance from the interchange of Yan'an East Road and Tianmu Road. Considering the characteristics of the traffic problems, grid cannot be divided too small. In the "simple simulation scenario," the grid is equally divided.

Figure 7 is a schematic diagram of the road model. The left part of Figure 7 is the real road from Yan'an East Road interchange to Tianmu Road rotary interchange. The middle part is the schematic diagram of macroscopic and mesoscopic road model, which identifies road segments and detector locations. The right part of Figure 7 is grid partitioning schematic diagram simulated traffic flow from south to north. Microscopic simulation model is established by VISSIM. Intersection between trunk road (simplified to a single lane) and ramp in grid partitioning schematic diagram corresponds to sections 8 of macroscopic and mesoscopic model. Arrow of sections 8 represents input volume, which actually shows Xinzha road north entrance ramp. To optimize the parameters, simulation test program is designed by using

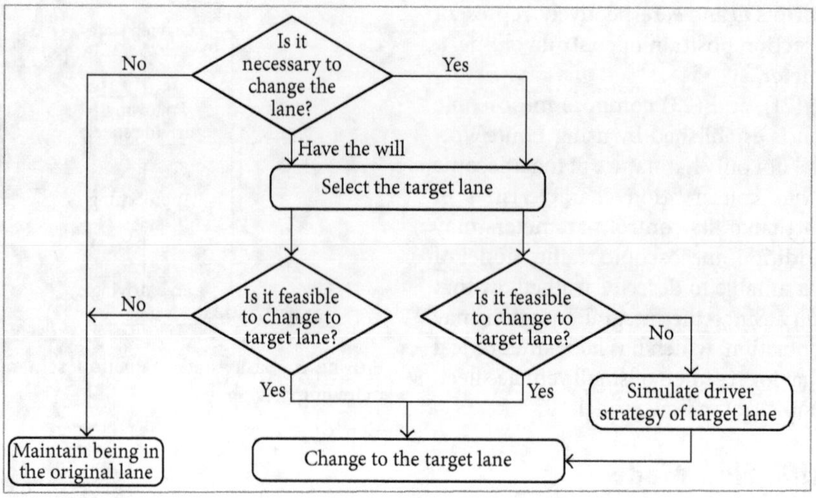

FIGURE 6: Schematic diagram of changing lane decision-making process.

TABLE 3: Ramp name, ID, and detector information table of part of North-South Elevated Road.

Link ID	Length of the link (m)	Ramp ID	Ramp name	Detector ID
7	595	Entrance ramp 7 Entrance ramp 8 Exit ramp 7 Exit ramp 8	Yan'an East Road Interchange left turn from west to north Yan'an East Road Interchange right turn from east to north Left turn-Yan'an elevated Right-Yan'an elevated	DX08 DX09
8	595	—	—	DX10
9	565	Entrance ramp 9 Exit ramp 10	Entrance ramp on the north of Weihai road Exit ramp on the south of Beijing road	DX11
10	565	—	—	DX12
11	552	Entrance ramp 11	Entrance ramp on the north of Xinzha road	DX13 DX14
12	552	Entrance ramp 12 Entrance ramp 13 Exit ramp 12 Exit ramp 13	Entrance ramp on the west of Tianmu Road rotary interchange Entrance ramp on the east of Tianmu Road rotary interchange Exit ramp on the west of Tianmu Road rotary interchange Exit ramp on the east of Tianmu road rotary interchange	DX15

Latin orthogonal test method in the simulation scenario. This figure is not drawn to scale for clarity of presentation.

In this paper, the actual test data is used, September 22, 2009 detector data, and sampling frequency is 1 minute. Table 3 shows the Ramp ID, Name, and Detector ID. Research scenario is part of North-South Elevated Road and detector IDs are from DX08 to DX15.

In this paper, upwind scheme is used to numerically solve the macromodel and mesomodel. In "simple simulation scenario," parameter initial values of the macroscopic and mesoscopic model are set as follows:

Free-flow speed $u_f(x) = \{21, 21, 21, 21, 21, 21, 21, 21, 21, 21\}$ m/s.

Jam density $k_j(x) = \{0.16, 0.16, 0.16, 0.16, 0.16, 0.16, 0.16, 0.16, 0.16, 0.16\}$ veh/m.

Initial density $k(i) = \{0.035, 1 \le i \le 5; 0.12, 6 \le i \le 7; 0.014, 8 \le i \le 10\}$ (unit is veh/m).

Input volume constant $Q_{\text{in}} = \{0.12, 0.12, 0.24, 0.24, 0.24, 0.48, 0.12, 0.12, 0.12, 0.12\}$ (unit is veh/s).

Output volume constant $Q_{\text{out,val}} = 0.16$ (unit is veh/s); output volume $Q_{\text{out}} = (k_{10}u(k_{10}), Q_{\text{out,val}})$.

Simulation step $t = 1$ s. T is 300 simulation steps. Link length L is 2000 m. Grid is equally divided and length of each grid is 200 m. Input volume, S8, of macromodel is 0.48 veh/s, while input volume, S8, of mesomodel is 0.44 veh/s. In the micromodel, parameters include the following. Free-flow speed of section is 80 km/h. Traffic capacity of expressway and ramp are, respectively, 2400 veh/h and 1200 veh/h.

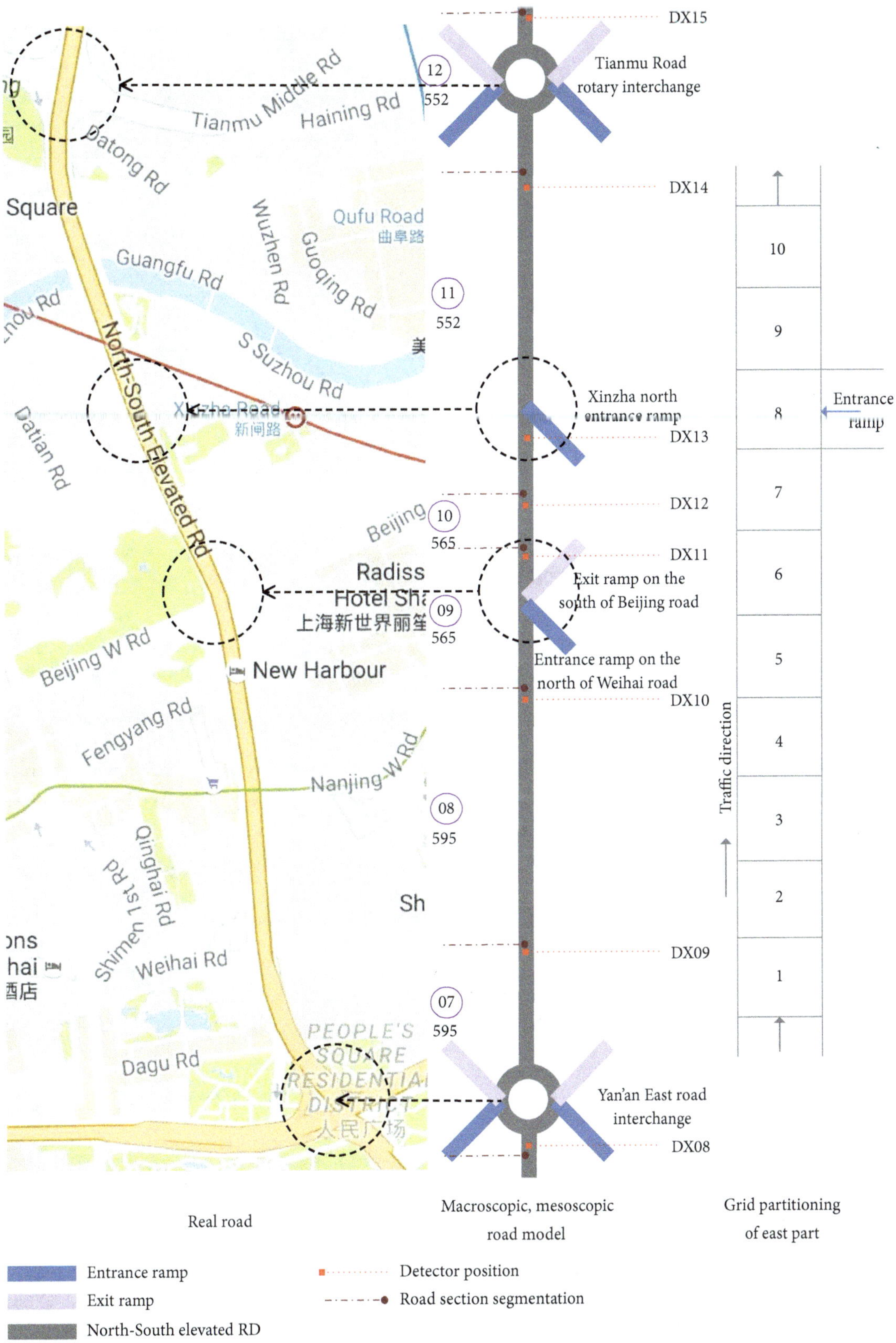

FIGURE 7: Basic road schematic diagram of multiresolution traffic simulation model.

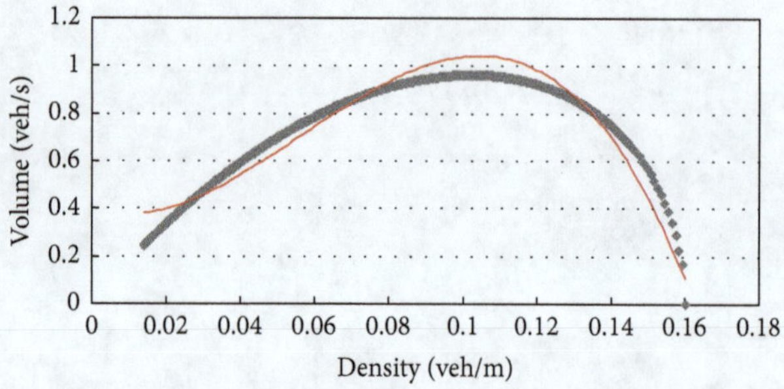

FIGURE 8: Volume-density relationship diagram of macromodel.

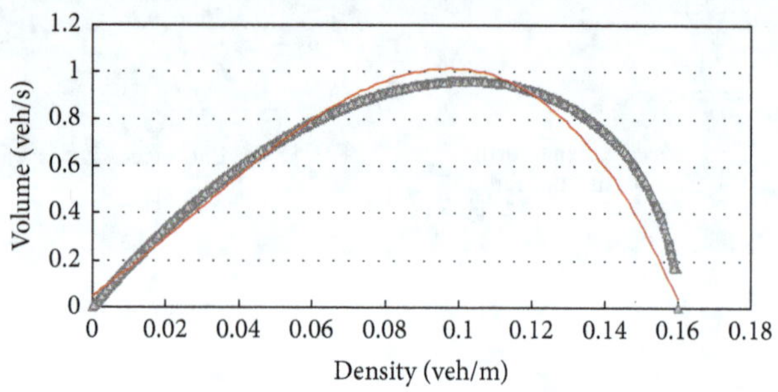

FIGURE 9: Volume-density relationship diagram of mesomodel.

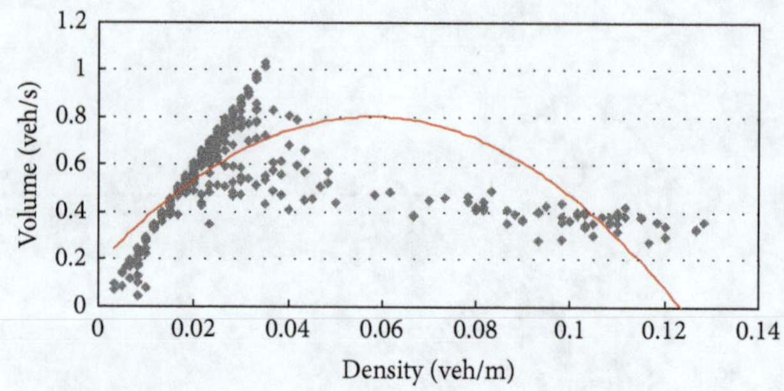

FIGURE 10: Volume-density relationship diagram of micromodel.

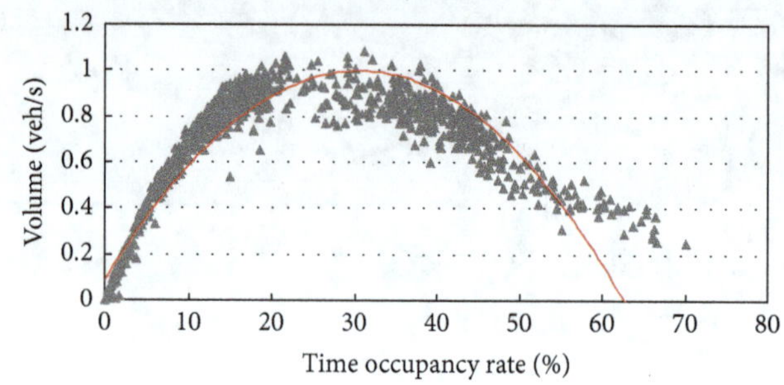

FIGURE 11: Volume-time occupancy rate relationship graph of detector data.

7.2. Volume Consistency Analysis of Multiresolution Model

7.2.1. Consistency Analysis of Road Section Volume-Density Relationship. Figures 8, 9, and 10, are, respectively, the volume-density relationship diagram of macromodel, mesomodel, and micromodel. Figure 11 is a volume-time occupancy rate relationship graph of detector data. The grey points are simulation data and red line is polynomial trend line of data.

As can be seen from the figure, oscillation of data points and trend line in the volume-density diagram of macromodel and mesomodel are small. The whole trend line is shaped as a parabolic, the maximum volume in trend line is about 1 veh/s, and jam density is about 0.16 veh/m. Data of volume-density diagram of micromodel distributed around the trend line and the whole trend line is shaped as a parabolic, but oscillation of data is bigger. Simultaneously, the maximum volume in micromodel trend line is about 0.8 veh/s and jam density is about 0.125 veh/m. Data in volume-time occupancy rate relationship graph of detector data distributed around the trend line, the whole trend line is shaped as a parabolic, and oscillation of data is small. The maximum volume in trend line of actual detector data is about 1 veh/s and its time occupancy ratio in traffic jam state is about 63%.

From a qualitative point of view, volume-density relationships of macroscopic, mesoscopic, and microscopic models are unanimous with detector data. Among them, fluctuations of simulation data of macromodel and mesomodel are relatively smaller and their consistencies are relatively higher. However, fluctuation of simulation data of micromodel is relatively bigger and its consistency is relatively lower. Maybe it is because of that fact that micromodel's description is more detailed and its randomness among vehicles is bigger.

7.2.2. Volume Error Analysis of Different Models. In order to quantitatively evaluate the validity and applicability of multiresolution transport model, the following several indicators are used: Mean Absolute Error (MAE), Relative Error (RE), Mean square Error (MSE), and Standard Deviation (St.D). For RE, the result considers the effect of "maximum relative error" and "gross error" has been processed according to Dixon guidelines [56]. Table 4 is the volume analysis of each model. In order to comparatively analyze several models, Cell Transmission Model (CTM) is as a reference model.

$$\begin{aligned}
\text{MAE} &= \frac{\left(\sum_1^n \left|\text{true value} - \text{prediction value}\right|\right)}{n}, \\
\text{RE} &= \frac{\left(\sum_1^n \left(\left(\left|\text{true value} - \text{prediction value}\right|\right) / \text{true value}\right)\right)}{n}, \\
\text{MSE} &= \frac{\left(\sum_1^n \left(\text{true value} - \text{prediction value}\right)^2\right)}{n}, \\
\text{St.D} &= \left(\frac{\left(\sum_1^n \left(\text{true value} - \text{prediction value}\right)^2\right)}{(n-1)}\right)^{1/2}.
\end{aligned} \tag{24}$$

When the traffic flow density is higher, MAE of mesomodel is the biggest one, macromodel's MAE is the smallest one, MAE of CTM and micromodel are in the between. Considering RE, Relative Error of macromodel is the smallest, its value of CTM is the biggest one, RE of mesomodel and

Table 4: Volume analysis of each model.

Model	Average density (veh/m)	MAE	RE	MSE	St.D
CTM	0.1154	0.24	0.25	0.09	0.29
Macromodel	0.1225	0.21	0.20	0.08	0.28
Mesomodel	0.1250	0.31	0.24	0.15	0.39
Micromodel	0.1033	0.26	0.23	0.13	0.36

micromodel are close, which are in between the minimum and maximum. Considering MSE and St.D, the values of macromodel and CTM are close, which are smaller. However, MSE and St.D of mesomodel and micromodel are close, which are bigger.

It can be seen, when the traffic density is high, because macromodel takes into account viscous and compression properties, that the simulation result of macromodel has a high precision, the error is small, and the dispersion of results is smaller. Micromodel can reflect the characteristics of individual vehicles, its error is larger, and its dispersion of simulation results is bigger; because randomness of micromodel is relatively large, time offset and position offset are high. The situation is more significant when traffic volume is small and its density is low. Considering mesomodel, it can combine partial characteristics of macromodel and micromodel, compared with CTM; their RE values have 4% differences. But mesomodel can reflect some random nature of traffic, which leads to its MSE and St.D being larger than CTM. Compared with micromodel, mesomodel greatly improves the simulation efficiency and application scope. Compared with macromodel, simulation accuracy of microscopic and mesoscopic model are lower but errors are bigger, because of random nature and its time offset and location offset in statistical analysis.

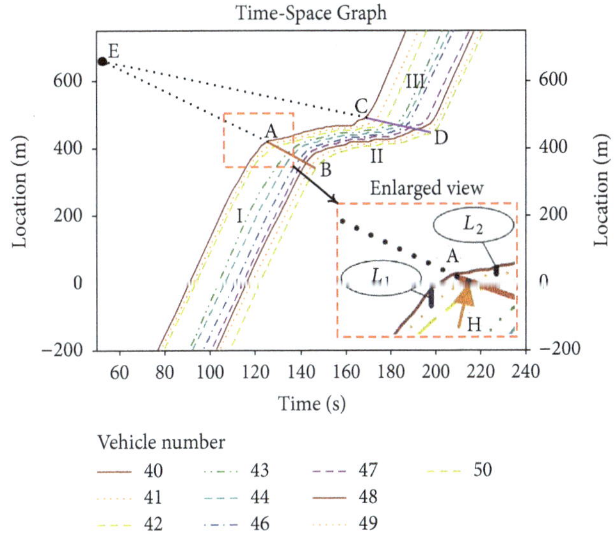

Figure 12: Vehicle time-space trajectories diagram of micromodel.

7.3. Consistency Analysis of Multiresolution Model

7.3.1. Traffic Wave Analysis of Microscopic Model. The paper compares each submodel of multiresolution model based on the same real scenario to evaluate consistency of multiresolution model, which is expressed by consistency analysis of traffic wave. Figure 12 is vehicle time-space trajectories diagram of micromodel, which can analyze traffic state transition process in microscopic simulation. Traffic waves verification data correspond to grids 6–10 of macroscopic and mesoscopic model in the paper. In Figure 12, horizontal axis is simulation time (second), while the actual time is simulation time multiplied by simulation speed coefficient; vertical axis represents location of the vehicle (meter). Each curve represents time-space trajectories of each vehicle. Horizontal distance of curves indicates time headway; vertical distance represents space headway. Solid lines AB and CD represent distributed wave, which reflects a shift interface of density state; slope of solid line is wave velocity; positive and negative slope, respectively, represent a forward or backward characteristic of fluctuations.

7.3.2. Traffic Wave Analysis of Macroscopic and Mesoscopic Model. (a) and (b) in Figure 13, respectively, are grid 8 traffic volume-density relationship of macromodel and mesomodel. In the volume-density relationship curve, velocity of distributed wave is slope of secant line, such as secants AB, BC, DE, and EF; velocity of weak wave is the slope of the tangent. Moreover, in the range of free flow, the fluctuation velocity is positive; namely, wave propagation direction is the same with the traveling direction of traffic flow. But in the range of jam traffic flow, the fluctuation velocity is negative; namely, wave propagation direction is the converse of the traveling direction of traffic flow.

When the traffic flow low-density and low volume state, A (D), transform to the higher density higher volume state, B (E), the velocity of distributed wave is positive and its propagation direction is along the road. When the traffic flow high-density and low volume state, C (F), transform to the lower density higher volume state, B (E), the velocity of distributed wave is negative and its propagation direction is back along the road. The wave from the state A (D) to the state B (E) is concentrated wave, and the wave from the state B (E) to the state A (D) is evanescent wave, both of which are progressive waves. The wave from the state C (F) to the state B (E) is evanescent wave, and the wave from the state B (E) to the state C (F) is concentrated wave, both of which are retrograde waves.

7.3.3. Traffic Wave Consistency Analysis of Multiresolution Model. Figure 14 is traffic wave consistency analysis of multiresolution model, and Figure 14(a) shows transportation

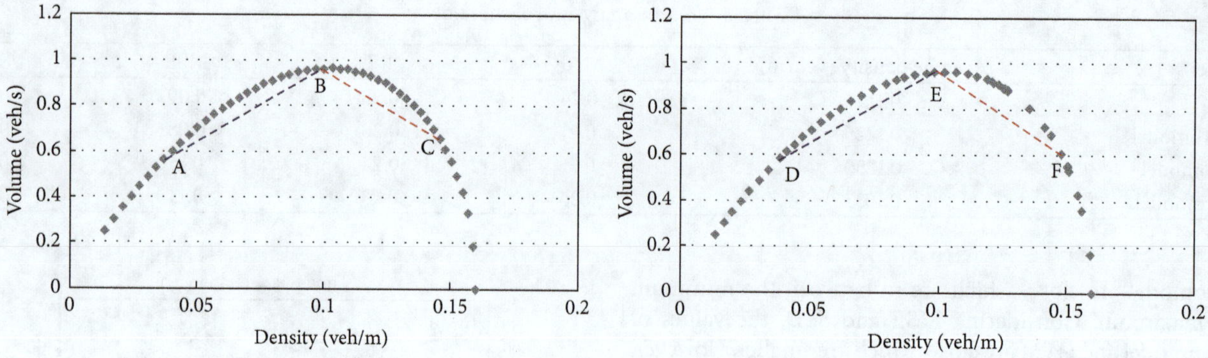

(a) Macromodel (b) Mesomodel

Figure 13: Grid 8 traffic volume-density relationship diagram.

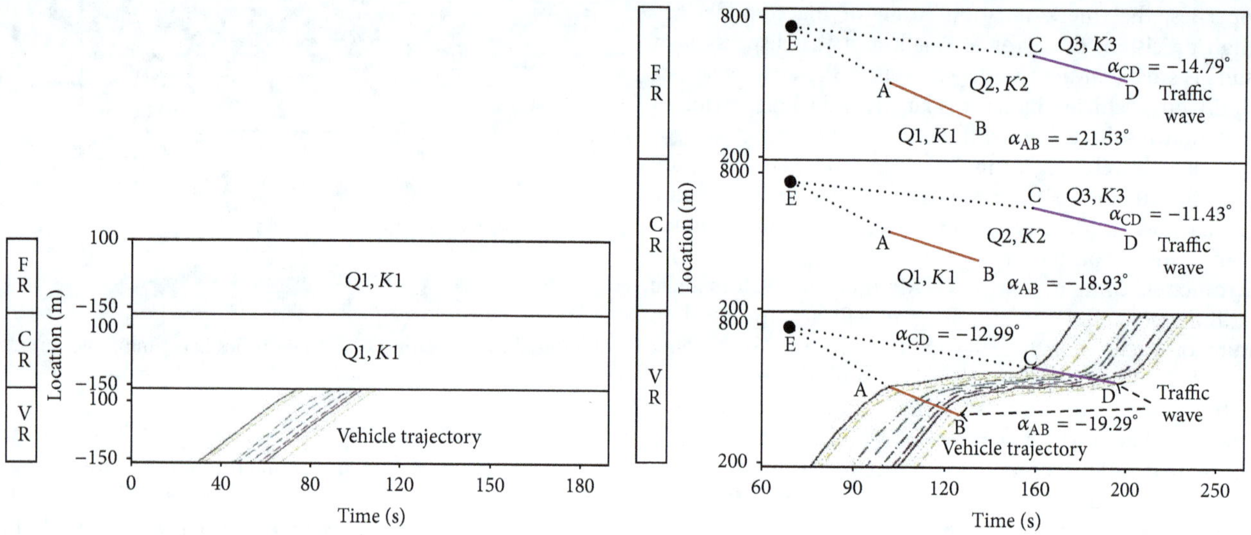

(a) (b)

Figure 14: Traffic wave consistency analysis schematic diagram of multiresolution model.

performance when free moving. That is to say, the transportation is free moving during 25~100 simulation seconds at road section −150~100 m. Meanwhile, time-space trajectories of vehicles are described in micromodel schematic diagram; however, there are no curves in macroscopic and mesoscopic model schematic diagram. As shown in the figure, traffic keeps in state Q_1, K_1 without changing during the period at this section.

Figure 14(b) is the schematic diagram of traffic wave consistency analysis of multiresolution model. It shows the traffic operating status when the traffic density changes, which describes the traffic state transition process from concentrated wave to evanescent wave during 60~250 s at road section 200~800 m. In micromodel results, it describes time-space trajectories of vehicles and concentrated wave AB and evanescent wave CD. In macromodel and mesomodel results, they describe the curves of traffic waves with respect to the micromodel waves. As shown in Figure 14, the area is divided into three traffic states according to solid lines AB and CD and the extended line, which includes $Q_1, K_1, Q_2, K_2, Q_3, K_3$. Boundaries AB and CD represent thresholds of the traffic state transition, and the solid line AB represents the concentrate wave, which reflects the phenomenon of traffic state transition boundaries from the low-density to the high-density. Simultaneously, the wave is retrograde wave, while the solid line CD represents the evanescent wave, which reflects the phenomenon of traffic state transition boundaries from the high-density to the low-density. Simultaneously, the wave is retrograde wave. In addition, depending on the angle between the lines AB and CD, we can know that the values of wave velocity of different models are different.

7.3.4. Traffic Wave Consistency Error Analysis of Multiresolution Model. Traffic wave qualitative analyses of different models have been studied above; traffic wave quantitative analyses will be discussed in the following, using MAE, RE, MSE, and St.D indexes. Table 5 is grid 8 traffic wave velocity Error Analysis of multiresolution model. About the traffic wave velocity, positive sign in the table represents the forward propagation wave; negative sign indicates backward

Table 5: Grid 8 traffic wave velocity Error Analysis of multiresolution model.

Model	Wave number	Wave velocity (m/s)	Angle (degree)	Index			
				MAE	RE	MSE	St.D
Macromodel	Wave AB	−3.95	−21.53	0.4858	0.1404	0.2360	0.4858
	Wave CD	−2.64	−14.79	0.0825	0.0323	0.0068	0.0825
	Mean value	—	—	0.2841	0.0863	0.1214	0.2841
Mesomodel	Wave AB	−3.43	−18.93	0.0303	0.0088	0.0009	0.0303
	Wave CD	−2.02	−11.43	0.5359	0.2096	0.2872	0.5359
	Mean value	—	—	0.2831	0.1092	0.1441	0.2831
CTM	Wave AB	−2.96	−16.51	0.4954	0.1432	0.2455	0.4954
	Wave CD	−3.26	−18.06	0.7030	0.2749	0.4942	0.7030
	Mean value	—	—	0.5992	0.2091	0.3698	0.5992
Micromodel	Wave AB	−3.50	−19.29	0.0400	0.0116	0.0016	0.0400
	Wave CD	−2.31	−12.99	0.2495	0.0976	0.0623	0.2495
	Mean value	—	—	0.1448	0.0546	0.0319	0.1448

propagation wave. Corresponding angle of wave velocity can be found in Figure 14. In the table, the wave velocity values correspond to the result based on time consistency, spatial consistency, and interface consistency of each model. Speed and solving process of wave velocity in microscopic model have considered the simulation accuracy parameter impact of VISSIM. Simulation accuracy is 10 time steps per simulation seconds here. Error Analysis in Table 5 is based on the mean of all model simulation results.

As shown in Table 5, for traffic wave of each model, RE of macromodel and micromodel are close, both of which are at a lower level; RE of CTM is bigger and RE of mesomodel is between them. For evaluation index MSE, MSE of microscopic model is in lower level; MSE of CTM is relatively larger, and MSE of macromodel and its value of mesomodel are in between.

As can be seen, these four models can simulate changes of traffic state, capture traffic wave, and have higher simulation accuracy, wherein micromodel's simulation accuracy is the highest and result's dispersion degree is the lowest, thanks to the micromodels' sophisticated simulation capabilities. The macromodel simulation accuracy is relatively higher because of the detailed description of traffic impedance (viscous and compression properties). The mesomodel simulation accuracy is higher than CTM and is lower than macromodel. Compared with CTM, mesomodel considers dynamics of traffic and can describe traffic state transition process more delicately than CTM. Compared with macromodel, mesomodel can describe a certain discrete stochastic state, but it also resulted in low accuracy and higher dispersion degree.

Meanwhile, the simulation results of different resolution models can have differences because of the time offset, position offset, inconsistency of interface, and control parameters. In order to obtain better results, they can be achieved by parameter optimization of different models, including unified interface and consistent optimization of time and space.

8. Conclusion

By comparing the characteristics of single resolution model, the paper has determined the multiresolution technology to study the traffic flow model of urban expressway main line and its ramp. Firstly, the idea of multiresolution modeling is expounded and the multiresolution traffic simulation triangular frame is constructed by using the integrated simulation method. Secondly, integration strategy issues, the key problem of multiresolution traffic simulation, are elaborated. The paper proposes a novel multiresolution traffic simulation model by asynchronous integration strategy based on Set Theory, which can describe the characteristics of traffic system, such as behavioral characteristics, time characteristics, spatial characteristics, performance characteristics, granularity level, and internal structure. Then, according to the characteristics of urban expressway traffic flow and macromodel goal, the conceptual model and mathematical model of macroscopic traffic flow, combining viscosity characteristic and compression characteristic, are established in detail. Meanwhile, based on the macroscopic traffic flow model, a new mesoscopic traffic flow model is established by using the weighted residual method and the finite difference method. Moreover, the microscopic model is discussed by using VISSIM of German PTV company. In addition, in order to simulate and evaluate the multiresolution model, the paper established "simple simulation scenario," which is about 2 km from Yan'an East Road interchange to Tianmu Road rotary interchange of North-South Elevated Expressway in Shanghai. Finally, simulation results of multiresolution model have been analyzed from several aspects, volume consistency, traffic wave consistency, and Error Analysis, and so on. The paper used MAE, RE, MSE, and St.D to quantitatively evaluate the validity and applicability of multiresolution transport model. The results show that the novel multiresolution model can simulate characteristics of traffic flow and keep the consistency of traffic state transition. It can capture traffic wave and have higher simulation accuracy. And it is feasible and effective in traffic simulation.

Conflicts of Interest

The authors declare that they have no financial and personal relationships with other people or organizations that can inappropriately influence their work; there is no professional or other personal interest of any nature or kind in any product, service, or company that could be construed as influencing the position presented in, or the review of, the manuscript.

Acknowledgments

This work is partly supported by the Natural Foundation of Jiangsu Province (BK20151201 and BK20160357), Industrial Technology Innovation Projects of Suzhou (SS201525), Ministry of Housing and Urban-Rural Development of the People's Republic of China (nos. 2015-K5-027 and 2013-K5-27), the Construction System of Jiangsu Province (no. 2014ZD86), and Suzhou University of Science and Technology Youth Fund (nos. XKQ201403 and XKQ201508).

References

[1] M. Wei, F.-T. Yang, and Zh.-Q. Cao, "A Review of Development and Study on the Traffic Simulation," *Journal of System Simulation*, vol. 15, no. 8, pp. 1179–1183, 2003.

[2] Q. Yang, "Comments on development and applications of traffic simulation software in China," *Urban Transport of China*, vol. 4, no. 3, pp. 77–81, 2006.

[3] B.-H. Liu and K.-D. Huang, "Multi-resolution modeling: present status and trends," *Journal of System Simulation*, vol. 16, no. 6, pp. 1150–1154, 2004.

[4] Ch.-F. Shao, Ch.-Z. Xiao, B.-B. Wang, and M. Meng, "Speed-density relation model of congested traffic flow under minimum safety distance constraint," *Journal of Traffic and Transportation Engineering*, vol. 15, no. 1, pp. 92–99, 2015.

[5] W.-N. Huang, *Study of multi-resolution traffic simulation [M.S. thesis]*, Tongji University, Shanghai, China, 2009.

[6] G. Lerner, A. Hochstaedter, R. Kates et al., "The interplay of multiple scales in traffic flow: coupling of microscopic, mesoscopic and macroscopic simulation," in *Proceedings of the World Congress on Intelligent Transportation Systems*, Torino, Italy, 2000.

[7] L. Magne, S. Rabut, and J. F. Gabard, "Towards a hybrid macro-micro traffic simulation model," in *Proceedings of the INFORMS*, Salt Lake City, Utah, USA, 2000.

[8] L. Leclercq, "Hybrid approaches to the solutions of the "Lighthill-Whitham-Richards" model," *Transportation Research Part B: Methodological*, vol. 41, no. 7, pp. 701–709, 2007.

[9] A.-E.-K. Sahraoui, J.-S. Oh, and R. Jayakrishnan, "Calibration and path dynamics issues in microscopic simulation for advanced traffic management and information systems," *Transportation Research Record Journal of the Transportation Research Board*, vol. 1771, no. 1, pp. 89–99, 2001.

[10] L. Nizard, *Combining Microscopic And Mesoscopic Traffic Simulators*, Ecole Polytechnique, Paris, France, 2002.

[11] W. Burghout, *Hybrid Microscopic-Mesoscopic Traffic Simulation*, Royal Institute of Technology, Sweden, 2004.

[12] A. Ziliaskopoulos, J. Zhang, and H.-J. Shi, "A hybrid mesoscopic-microscopic traffic simulation model: design, implementation and computational analysis," in *Proceedings of the TRB 2006 Annual Meeting CD-ROM*, Washington, Wash, USA, 2006.

[13] Zh.-X. Xu, Y. Lin, Sh.-B. Li, and L. Wang, "A mesoscopic traffic simulation model and its case applications," *Shandong Science*, vol. 23, no. 3, pp. 62–65, 2010.

[14] M. Joueiai, L. Leclercq, H. van Lint, and S. Hoogendoorn, "A multi-scale traffic flow model based on the mesoscopic LWR mode," in *Proceedings of the TRB 94th Annual Meeting 2015*, Washington, Wash, USA, January 2015.

[15] J. M. Torné, F. Soriguera, and N. Geroliminis, "On the consistency of freeway macroscopic merging models," *Transportation Research Record*, vol. 2422, pp. 34–41, 2014.

[16] E. Nava, J. Shelton, and Y.-Ch. Chiu, "Analyzing impacts of dynamic reversible lane systems using a multi-resolution modeling approach," in *Proceedings of the TRB 91st Annual Meeting*, Washington, Wash, USA, January 2012.

[17] B.-H. Liu, *Research on the Theory and Key Techniques of Multi-Resolution Modeling (MRM) [Ph.D. thesis]*, National University of Defense Technology, Changsha, China, 2003.

[18] M. A. Biddle, "Proposed scheme for implementing aggregation and disaggregation in HLA," in *Proceedings of 2000 Fall Simulation Interoperability Workshop*, 2000.

[19] A. Natrajan and A. Nguyen-Tuong, *To Disaggregate or Not to Disaggregate, That Is Not The Question*, Department of Computer Science, University of Virginia, Charlottesville, Va, USA, 1995.

[20] P. K. Davis and R. Hillestad, "Families of models that cross levels of resolution: issues for design, calibration and management," in *Proceedings of the 25th Conference on Winter Simulation (WSC '93)*, pp. 1003–1012, Los Angeles, Calif, USA, December 1993.

[21] A. Natrajan, *Consistency Maintenance in Concurrent Representations [Ph.D. thesis]*, School of Engineering and Applied Science at the University of Virginia, Charlottesville, Va, USA, 2000.

[22] P. F. Reynolds Jr., A. Natrajan, and S. Srinivasan, "Consistency maintenance in multiresolution simulation," *ACM Transactions on Modeling and Computer Simulation*, vol. 7, no. 3, pp. 368–392, 1997.

[23] B. D. Kornman and P. B. Marion, "Cross-model consistency in JSIMS," Tech. Rep., Lockheed Martin Information Systems Advanced Simulation Laboratory, Norfolk, Va, USA, 1998.

[24] P. K. Davis and R. K. Huber, "Variable-resolution combat modeling: motivations, issues, and principles," RAND N-3400-DARPA, 1992.

[25] F. J. Barros, B. P. Zeigler, and P. A. Fishwick, "Multimodels and dynamic structure models: an integration of DSDE/DEVS and OOPM," in *Proceedings of the 30th conference on Winter simulation*, pp. 413–419, Washington, Wash, USA, December 1998.

[26] D. R. Powell, "Control of entity interactions in a hierarchical variable resolution simulation," in *Proceedings of 1997 Fall Simulation Interoperability Workshop*, Orlando, Fla, USA, 1997.

[27] P. K. Davis and R. J. Hillestad, "Experiments on Multiresolution Modeling," RAND MR1004, RAND Corp, 1998.

[28] K. Lee and P. A. Fishwick, "OOPM/RT: a multimodeling methodology for real-time simulation," *ACM Transactions on Modeling and Computer Simulation*, vol. 9, no. 2, pp. 141–170, 1999.

[29] S. Sekine, M. Kanou, M. Ogata et al., "Advanced technique for MRM (multi-resolution models)," in *Proceedings of the 2001 Spring Simulation Interoperability Workshop*, Orlando, Fla, USA, 2001.

[30] R. Schaffer and S. Kukolich, "Multi-resolution modeling summary and discussion paper," Systems Modeling and Simulation, Lockheed Martin Information Systems Advanced Simulation Center, 1998.

[31] R. Schaffer and P. B. Marion, "Multi-resolution modeling final report," Tech. Rep., Advanced Simulation Center, Burlington, Vt, USA, 1999.

[32] B. Butler, "Deign-strategies for multi-resolution synthetic environment representation with examples from one SAFSNE," in *Proceedings of the 2002 Fall Simulation Interoperability Workshop*, Orlando, Fla, USA, 2002.

[33] R. A. Reynolds, H. Iskenderian, and S. O. Ouzts, "Using multiple representations and resolutions to compose simulated METOC environments," in *Proceedings of 2002 Spring Simulation Interoperability Workshop*, Orlando, Fla, USA, 2002.

[34] Q. Yang and H. Slavin, *High Fidelity, Wide Area Traffic Simulation Model*, Caliper Corporation, Boston, Mass, USA, 2002.

[35] J. Sun and W.-N. Huang, "A framework for multi-resolution traffic simulation modeling [DB/EB]," 2009, http://www.paper.edu.cn.

[36] I. T. Cameron, "Modern process modeling: multi-scale and goal-directed," in *Proceedings of the 14th International Drying Symposium (IDS '04)*, pp. 3–17, São Paulo City, Brazil, 2004.

[37] M. Roza, P. V. Gool, and H. Jense, "A fidelity management process overlay onto the FEDEP model," in *Proceedings of the Fall 1998 Simulation Interoperability Workshop*, Orlando, Fla, USA, 1998.

[38] M. Arbib, "A mathematical theory of systems engineering: The elements," *IEEE Transactions on Automatic Control*, vol. 15, no. 3, pp. 407-408, 1970.

[39] Y.-P. Feng, *Multi-resolution modeling of material flow for petrochemical production processes [Ph.D. thesis]*, Zhejiang University, Hangzhou, China, 2008.

[40] J. Ma, J. Sun, K.-P. Li, and L.-Y. Zhang, "A study on multi-resolution scheme of macroscopic-microscopic traffic simulation model," in *Proceedings of the 14th IEEE International Intelligent Transportation Systems Conference (ITSC '11)*, pp. 1421–1426, Washington, Wash, USA, October 2011.

[41] L. Wang, *Study on traffic flow collection and control of urban express ways [Ph.D. thesis]*, Tianjin University, Tianjin, China, 2005.

[42] H.-Ch. Gan and L.-J. Sun, "Modeling and simulation of urban expressway dynamic network traffic flow," *Computer and Communications*, vol. 23, no. 2, pp. 24–26, 2005.

[43] T. Qiao, *Research on the theory and technology of urban expressway traffic street [M.S. thesis]*, Southwest Jiaotong University, Chengdu, China, 2007.

[44] M. Bando, K. Hasebe, A. Nakayama et al., "Dynamical model of traffic congestion and numerical simulation," *Physical Review E*, vol. 51, pp. 1035–1042, 1955.

[45] P. G. Gipps, "A behavioural car-following model for computer simulation," *Transportation Research Part B: Methodological*, vol. 15, no. 2, pp. 105–111, 1981.

[46] Sh.-W. Zhou, *Traffic Engineering [M.S. thesis]*, Tongji University Press, Shanghai, China, 2007.

[47] W.-Sh. An, *Study on dynamic modeling of mixed Traffic Flow [Ph.D. thesis]*, Southwest Jiaotong University, Chengdu, China, 2010.

[48] R. Jiang, Q.-S. Wu, and Z.-J. Zhu, "A new dynamics model for traffic flow," *Chinese Science Bulletin*, vol. 45, pp. 1895–1899, 2000.

[49] X. Luo, J.-Y. Du, and Y.-W. Chen, "Analyses on multi-traffic flow's characteristics," *Journal of Southwest Jiaotong University*, vol. 6, pp. 297–300, 2000.

[50] PTV GROUP, "VISSIM 5.00 User's manual," Tech. Rep., PTV AG, Karlscruhe, Germany, 2008.

[51] L.-D. Zhang, Y.-L. Wang et al., "A review of traffic simulation," *Computer Simulation*, vol. 23, no. 6, pp. 255–258, 2006.

[52] L. Li and R. Jiang, *Modern traffic flow theory and application [M.S. thesis]*, Tsinghua University Press, Beijing, China, 2011.

[53] R. Wiedemann and U. Reiter, "Microscopic traffic simulation/the simulation system-mission," Tech. Rep. Germany, University Karlsruhe, 1992.

[54] R. Wiedemann, "Simulation of road traffic flow," Reports of the Institute for Transport and Communication Germany, University of Karlsruhe, 1974.

[55] U. Sparmann, "The importance of lane-changing on motorways," *Traffic Engineering and Control*, vol. 20, no. 6, pp. 320–323, 1979.

[56] Y.-T. Fei, *Error Theory and Data Processing*, Machinery Industry Press, Beijing, China, 2008.

9

A Note on Torsion of Nonlocal Composite Nanobeams

Luciano Feo and Rosa Penna

Department of Civil Engineering, University of Salerno, Via Ponte don Melillo, 84084 Fisciano, Italy

Correspondence should be addressed to Luciano Feo; l.feo@unisa.it

Academic Editor: Theodoros C. Rousakis

The ERINGEN elastic constitutive relation is used in this paper in order to assess small-scale effects in nanobeams. Structural behavior is studied for functionally graded materials in the cross-sectional plane and torsional loading conditions. The governing boundary value problem has been formulated in a mixed framework. Torsional rotations and equilibrated moments are evaluated by solving a first-order differential equation of elastic equilibrium with boundary conditions of kinematic-type. Benchmarks examples are briefly discussed, enlightening thus effectiveness of the proposed methodology.

1. Introduction

Assessments of stress and displacement fields in continuous media are a subject of special interest in the theory of structures. Numerous case studies have been examined in the current literature with reference to beams [1–6], half-spaces [7, 8], thin plates [9, 10], compressible cubes [11], and concrete [12, 13]. Several methodologies of analysis have been developed in the research field of geometric continuum mechanics [14, 15], limit analysis [16–19], homogenization [20], elastodynamics [21–25], thermal problems [26–28], random composites [29–32], and nonlocal and gradient formulations [33–38]. A comprehensive analysis of classical and generalized models of elastic structures, with special emphasis on rods, can be found in the interesting book by IEŞAN [39]. In particular, IEŞAN [40–42] formulated a method for the solution of SAINT-VENANT problems in micropolar beams with arbitrary cross-section. Detailed solution of the torsion problem for an isotropic micropolar beam with circular cross-section is given in [43, 44]. Experimental investigations are required for the evaluation of the behavior of composite structures [45].

In the context of the present research, particular attention is devoted to the investigation of scale effects in nanostructures; see, for example, [46–55] and the reviews [56, 57]. Recent contributions on functionally graded materials have been developed for nanobeams under flexure [58, 59] and torsion [60].

Unlike previous treatments on torsion of gradient elastic bars (see, e.g., [61]) in which higher-order boundary conditions have to be enforced, this paper is concerned with the analysis of composite nanobeams with nonlocal constitutive behavior conceived by ERINGEN in [62]. Basic equations governing the ERINGEN model are preliminarily recalled in Section 2. The corresponding elastic equilibrium problem of torsion of an ERINGEN circular nanobeam is then formulated in Section 3. It is worth noting that only classical boundary conditions are involved in the present study. Small-scale effects are detected in Section 4 for two static schemes of applicative interest. Some concluding remarks are delineated in Section 5.

2. Eringen Nonlocal Elastic Model

Before formulating the elastostatic problem of a nonlocal nanobeam subjected to torsion, we shortly recall in the sequel some notions of nonlocal elasticity. To this end, let us consider a body $\mathscr{B}$ made of a material, possibly composite, characterized by the following integral relation between the stress $\mathbf{t}_{ij}$ at a point $\mathbf{x}$ and the elastic strain field $\mathbf{e}_{ij}$ in $\mathscr{B}$ [62]:

$$\mathbf{t}_{ij}(\mathbf{x}) = \int_{\mathscr{B}} K\left(\left|\mathbf{x}' - \mathbf{x}\right|, \tau\right) \mathbb{E}_{ijhk}\left(\mathbf{x}'\right) \mathbf{e}_{hk}\left(\mathbf{x}'\right) dV. \quad (1)$$

The fourth-order tensor $\mathbb{E}_{ijhk}(\mathbf{x}')$, symmetric and positive definite, describes the material elastic stiffness at the point $\mathbf{x}' \in \mathscr{B}$.

The attenuation function K depends on the Euclidean distance $|\mathbf{x}' - \mathbf{x}|$ and on a nonlocal dimensionless parameter defined by

$$\tau = \frac{e_0 a}{l}, \tag{2}$$

where e_0 is a material constant, a is an internal characteristic length, and l stands for external characteristic length.

Assuming a suitable expression of the nonlocal modulus K in terms of a variant of the Bessel function, we get the inverse differential relationship of (1) between the nonlocal stress and the elastic deformation

$$\left(1 - (e_0 a)^2 \nabla^2\right) \mathbf{t}_{ij} = \mathbb{E}_{ijhk} \mathbf{e}_{hk} \tag{3}$$

with ∇^2 the Laplacian. Note that (3) can be conveniently used in order to describe the law between the nonlocal shear stress field $\boldsymbol{\tau}_i$ on the cross-section of a nanobeam and the elastic shear strain $\boldsymbol{\gamma}_i$ as follows:

$$\boldsymbol{\tau}_i - (e_0 a)^2 \frac{d^2 \boldsymbol{\tau}_i}{dx^2} = \mu \boldsymbol{\gamma}_i, \tag{4}$$

where x is the axial direction and μ is the shear modulus.

3. Torsion of Nonlocal Circular Nanobeams

Let Ω be the cross-section of a circular nanobeam, of length L, subjected to the following loading conditions depicted in Figure 1:

$$\begin{aligned} &m_t, \\ &\text{distributed couples per unit length in the interval } [0, L], \\ &\mathcal{M}_t, \\ &\text{concentrated couples at the end cross-sections } \{0, L\}. \end{aligned} \tag{5}$$

The triplet (x, y, z) describes a set of Cartesian axes originating at the left cross-section centre $\mathbf{O}$.

Equilibrium equations are expressed by

$$\frac{dM_t}{dx} = -m_t, \quad \text{in } [0, L], \tag{6a}$$

$$M_t = \mathcal{M}_t, \quad \text{at } \{0, L\}, \tag{6b}$$

where M_t is the twisting moment.

Components of the displacement field, up to a rigid body motion, of a circular nanobeam under torsion write as

$$\begin{aligned} s_x (x, yz) &= 0, \\ s_y (x, yz) &= -\theta (x) z, \\ s_z (x, yz) &= \theta (x) y, \end{aligned} \tag{7}$$

where $\theta(x)$ is the torsional rotation of the cross-section at the abscissa x. Shear strains, compatible with the displacement field equation (7), are given by

$$\begin{aligned} \gamma_{yx} (x, y, z) &= -\frac{d\theta}{dx} xz, \\ \gamma_{zx} (x, y, z) &= \frac{d\theta}{dx} xy, \end{aligned} \tag{8}$$

where

$$\chi_t (x) = \frac{d\theta}{dx} (x) \tag{9}$$

is the twisting curvature. The shear modulus μ is assumed to be functionally graded only in the cross-sectional plane (y, z).

The elastic twisting stiffness is provided by

$$k_t = \int_\Omega \mu \left(y^2 + z^2\right) dA, \tag{10}$$

where the symbol $\cdot$ is the inner product between vectors.

The differential equation of nonlocal elastic equilibrium of a nanobeam under torsion is formulated as follows. Let us preliminarily multiply (4) by $\mathbf{Rr} := \{-z, y\}$ and integrate on Ω

$$\int_\Omega \mathbf{Rr} \cdot \boldsymbol{\tau} dA - (e_0 a)^2 \int_\Omega \mathbf{Rr} \cdot \frac{d\boldsymbol{\tau}}{dx^2} dA = \int_\Omega \mu \mathbf{Rr} \cdot \boldsymbol{\gamma} dA, \tag{11}$$

with the vector $\boldsymbol{\gamma} := \{\gamma_{yx}, \gamma_{zx}\}$ given by (8).

Enforcing (6a) and imposing the static equivalence condition

$$M_t = \int_\Omega \left(\tau_{zx} y - \tau_{yx} z\right) dA, \tag{12}$$

we get the relation

$$M_t (x) + (e_0 a)^2 \frac{dm_t}{dx} (x) = k_t \frac{d\theta}{dx} (x). \tag{13}$$

This equation can be interpreted as decomposition formula of the twisting curvature χ_t into elastic $(\chi_t)_{\mathrm{EL}}$ and inelastic $(\chi_t)_{\mathrm{IN}}$ parts

$$\chi_t = (\chi_t)_{\mathrm{EL}} + (\chi_t)_{\mathrm{IN}}, \tag{14}$$

with

$$(\chi_t)_{\mathrm{EL}} = \frac{M_t}{k_t}, \tag{15a}$$

$$(\chi_t)_{\mathrm{IN}} = \frac{(e_0 a)^2}{k_t} \frac{dm_t}{dx}. \tag{15b}$$

Accordingly, the scale effect exhibited by the torsional rotation function of a nonlocal nanobeam can be evaluated by solving a corresponding linearly elastic beam subjected to the twisting curvature distortion $(\chi_t)_{\mathrm{IN}}$ (15b).

4. Benchmark Examples

Let us consider a nanocantilever and a fully clamped nanobeam of length L subjected to the following quadratic distribution of couples per unit length:

$$m_t = \frac{m}{L^2} x^2. \tag{16}$$

The cross-sectional torsional rotation is evaluated by following the methodology illustrated in the previous section.

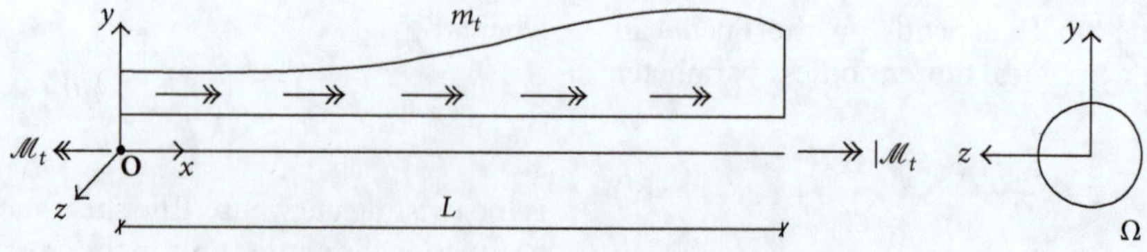

FIGURE 1: Sketch of a nanobeam under torsion.

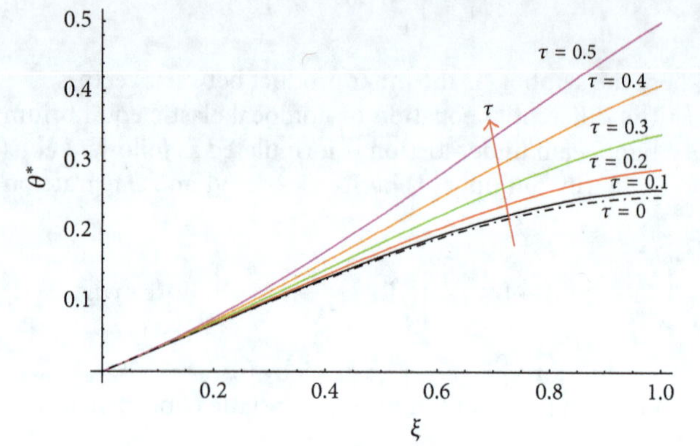

FIGURE 2: Torsional rotation θ^* for the nanocantilever versus ξ for increasing values of τ.

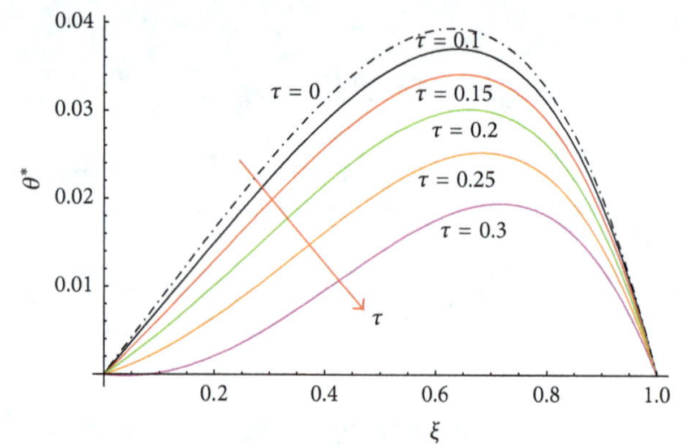

FIGURE 3: Dimensionless torsional rotation θ^* for the fully clamped nanobeam versus ξ for increasing values of τ.

The nonlocality effect is assessed by prescribing the twisting curvature distortion equation (15b)

$$(\chi_t)_{\mathrm{IN}} = \frac{(e_0 a)^2}{k_t} \frac{2m}{L^2} x \qquad (17)$$

on corresponding (cantilever and fully clamped) local nanobeams. Let us set $\xi = x/L$ and $\theta^*(\xi) = k_t/(mL^2)\theta(\xi)$. Torsional rotations θ^* versus ξ of both the nanobeams are displayed in Figures 2 and 3 for selected values of the nonlocal parameter $\tau := e_0 a/L$.

5. Concluding Remarks

The basic outcomes contributed in the present paper are listed as follows:

(1) Size-effects in nanobeams under torsion have been evaluated by resorting to the nonlocal theory of elasticity.

(2) Exact torsional rotations solutions of cross-sections of functionally graded nanobeams have been established for nanocantilevers and fully clamped nanobeams under a quadratic distribution of couples per unit length.

(3) It has been observed that the stiffness of a nanobeam under torsional loadings is affected by the scale parameter and depends on the boundary kinematic constraints. Indeed, as shown in Figures 2 and 3, contrary to the nanocantilever structural behavior, the fully clamped nanobeam becomes stiffer for increasing values of the nonlocal parameter.

Competing Interests

There is no conflict of interests related to this paper.

References

[1] R. Barretta, "On Cesàro-Volterra method in orthotropic Saint-Venant beam," *Journal of Elasticity*, vol. 112, no. 2, pp. 233–253, 2013.

[2] R. Barretta and M. Diaco, "On the shear centre in Saint-Venant beam theory," *Mechanics Research Communications*, vol. 52, pp. 52–56, 2013.

[3] G. Romano, A. Barretta, and R. Barretta, "On torsion and shear of Saint-Venant beams," *European Journal of Mechanics. A. Solids*, vol. 35, pp. 47–60, 2012.

[4] R. Barretta, "On stress function in Saint-Venant beams," *Meccanica*, vol. 48, no. 7, pp. 1811–1816, 2013.

[5] F. Marmo and L. Rosati, "Analytical integration of elasto-plastic uniaxial constitutive laws over arbitrary sections," *International Journal for Numerical Methods in Engineering*, vol. 91, no. 9, pp. 990–1022, 2012.

[6] F. Marmo and L. Rosati, "The fiber-free approach in the evaluation of the tangent stiffness matrix for elastoplastic uniaxial constitutive laws," *International Journal for Numerical Methods in Engineering*, vol. 94, no. 9, pp. 868–894, 2013.

[7] L. Rosati and F. Marmo, "Closed-form expressions of the thermo-mechanical fields induced by a uniform heat source acting over an isotropic half-space," *International Journal of Heat and Mass Transfer*, vol. 75, pp. 272–283, 2014.

[8] F. Marmo and L. Rosati, "A general approach to the solution of Boussinesq's problem for polynomial pressures acting over polygonal domains," *Journal of Elasticity*, vol. 122, no. 1, pp. 75–112, 2016.

[9] R. Barretta, "Analogies between Kirchhoff plates and Saint-Venant beams under torsion," *Acta Mechanica*, vol. 224, no. 12, pp. 2955–2964, 2013.

[10] R. Barretta, "Analogies between Kirchhoff plates and Saint-Venant beams under flexure," *Acta Mechanica*, vol. 225, no. 7, pp. 2075–2083, 2014.

[11] A. M. Tarantino, "Homogeneous equilibrium configurations of a hyperelastic compressible cube under equitriaxial dead-load tractions," *Journal of Elasticity*, vol. 92, no. 3, pp. 227–254, 2008.

[12] V. A. Salomoni, G. Mazzucco, C. Pellegrino, and C. E. Majorana, "Three-dimensional modelling of bond behaviour between concrete and FRP reinforcement," *Engineering Computations*, vol. 28, no. 1, pp. 5–29, 2011.

[13] V. A. Salomoni, C. E. Majorana, B. Pomaro, G. Xotta, and F. Gramegna, "Macroscale and mesoscale analysis of concrete as a multiphase material for biological shields against nuclear radiation," *International Journal for Numerical and Analytical Methods in Geomechanics*, vol. 38, no. 5, pp. 518–535, 2014.

[14] G. Romano, R. Barretta, and M. Diaco, "The geometry of nonlinear elasticity," *Acta Mechanica*, vol. 225, no. 11, pp. 3199–3235, 2014.

[15] G. Romano, R. Barretta, and M. Diaco, "Geometric continuum mechanics," *Meccanica*, vol. 49, no. 1, pp. 111–133, 2014.

[16] A. Caporale, R. Luciano, and L. Rosati, "Limit analysis of masonry arches with externally bonded FRP reinforcements," *Computer Methods in Applied Mechanics and Engineering*, vol. 196, no. 1–3, pp. 247–260, 2006.

[17] A. Caporale and R. Luciano, "Limit analysis of masonry arches with finite compressive strength and externally bonded reinforcement," *Composites Part B: Engineering*, vol. 43, no. 8, pp. 3131–3145, 2012.

[18] A. Caporale, L. Feo, R. Luciano, and R. Penna, "Numerical collapse load of multi-span masonry arch structures with FRP reinforcement," *Composites Part B: Engineering*, vol. 54, no. 1, pp. 71–84, 2013.

[19] A. Caporale, L. Feo, D. Hui, and R. Luciano, "Debonding of FRP in multi-span masonry arch structures via limit analysis," *Composite Structures*, vol. 108, no. 1, pp. 856–865, 2014.

[20] F. Greco and R. Luciano, "A theoretical and numerical stability analysis for composite micro-structures by using homogenization theory," *Composites Part B: Engineering*, vol. 42, no. 3, pp. 382–401, 2011.

[21] A. M. Tarantino, "Nonlinear fracture mechanics for an elastic Bell material," *The Quarterly Journal of Mechanics and Applied Mathematics*, vol. 50, no. 3, pp. 435–456, 1997.

[22] A. M. Tarantino, "The singular equilibrium field at the notch-tip of a compressible material in finite elastostatics," *Journal of Applied Mathematics and Physics*, vol. 48, no. 3, pp. 370–388, 1997.

[23] A. M. Tarantino, "On extreme thinning at the notch tip of a neo-Hookean sheet," *The Quarterly Journal of Mechanics and Applied Mathematics*, vol. 51, no. 2, pp. 179–190, 1998.

[24] A. M. Tarantino, "On the finite motions generated by a mode I propagating crack," *Journal of Elasticity*, vol. 57, no. 2, pp. 85–103, 1999.

[25] A. M. Tarantino, "Crack propagation in finite elastodynamics," *Mathematics and Mechanics of Solids*, vol. 10, no. 6, pp. 577–601, 2005.

[26] V. A. Salomoni, C. E. Majorana, G. M. Giannuzzi, and A. Miliozzi, "Thermal-fluid flow within innovative heat storage concrete systems for solar power plants," *International Journal of Numerical Methods for Heat and Fluid Flow*, vol. 18, no. 7-8, pp. 969–999, 2008.

[27] F. Marotti de Sciarra and M. Salerno, "On thermodynamic functions in thermoelasticity without energy dissipation," *European Journal of Mechanics. A. Solids*, vol. 46, pp. 84–95, 2014.

[28] G. Xotta, G. Mazzucco, V. A. Salomoni, C. E. Majorana, and K. J. Willam, "Composite behavior of concrete materials under high temperatures," *International Journal of Solids and Structures*, vol. 64, pp. 86–99, 2015.

[29] R. Luciano and J. R. Willis, "Bounds on non-local effective relations for random composites loaded by configuration-dependent body force," *Journal of the Mechanics and Physics of Solids*, vol. 48, no. 9, pp. 1827–1849, 2000.

[30] R. Luciano and J. R. Willis, "Non-local effective relations for fibre-reinforced composites loaded by configuration-dependent body forces," *Journal of the Mechanics and Physics of Solids*, vol. 49, no. 11, pp. 2705–2717, 2001.

[31] R. Luciano and J. R. Willis, "Boundary-layer corrections for stress and strain fields in randomly heterogeneous materials," *Journal of the Mechanics and Physics of Solids*, vol. 51, no. 6, pp. 1075–1088, 2003.

[32] R. Luciano and J. R. Willis, "Hashin-Shtrikman based FE analysis of the elastic behaviour of finite random composite bodies," *International Journal of Fracture*, vol. 137, no. 1–4, pp. 261–273, 2006.

[33] R. De Borst and H.-B. Muehlhaus, "Gradient-dependent plasticity: formulation and algorithmic aspects," *International Journal for Numerical Methods in Engineering*, vol. 35, no. 3, pp. 521–539, 1992.

[34] E. C. Aifantis, "Strain gradient interpretation of size effects," *International Journal of Fracture*, vol. 95, no. 1–4, pp. 299–314, 1999.

[35] E. C. Aifantis, "Gradient deformation models at nano, micro, and macro scales," *Journal of Engineering Materials and Technology, Transactions of the ASME*, vol. 121, no. 2, pp. 189–202, 1999.

[36] R. H. J. Peerlings, R. De Borst, W. A. M. Brekelmans, and J. H. P. De Vree, "Gradient enhanced damage for quasi-brittle materials," *International Journal for Numerical Methods in Engineering*, vol. 39, no. 19, pp. 3391–3403, 1996.

[37] R. H. J. Peerlings, M. G. D. Geers, R. de Borst, and W. A. M. Brekelmans, "A critical comparison of nonlocal and gradient-enhanced softening continua," *International Journal of Solids and Structures*, vol. 38, no. 44-45, pp. 7723–7746, 2001.

[38] H. Askes and E. C. Aifantis, "Gradient elasticity in statics and dynamics: an overview of formulations, length scale identification procedures, finite element implementations and new results," *International Journal of Solids and Structures*, vol. 48, no. 13, pp. 1962–1990, 2011.

[39] D. Ieşan, "Classical and generalized models of elastic rods," in *Modern Mechanics and Mathematics*, D. Gao and R. W. Ogden, Eds., CRC Series, CRC Press, 2008.

[40] D. Ieşan, "Saint-Venant's problem in micropolar elasticity," in *Mechanics of Micropolar Media*, O. Brulin and R. K. T. Hsieh, Eds., pp. 281–390, World Scientific, Singapore, 1982.

[41] D. Ieşan, "Generalized twist for the torsion of micropolar cylinders," *Meccanica*, vol. 21, no. 2, pp. 94–96, 1986.

[42] D. Ieşan, "Generalized torsion of elastic cylinders with microstructure," *Note di Matematica*, vol. 27, no. 2, pp. 121–130, 2007.

[43] R. D. Gauthier and W. E. Jahsman, "Quest for micropolar elastic constants," *Journal of Applied Mechanics, Transactions ASME*, vol. 42, no. 2, pp. 369–374, 1975.

[44] G. V. K. Reddy and N. K. Venkatasubramanian, "Saint-Venant's problem for a micropolar elastic circular cylinder," *International Journal of Engineering Science*, vol. 14, no. 11, pp. 1047–1057, 1976.

[45] G. Dinelli, G. Belz, C. E. Majorana, and B. A. Schrefler, "Experimental investigation on the use of fly ash for lightweight precast structural elements," *Materials and Structures/Materiaux et Constructions*, vol. 29, no. 194, pp. 632–638, 1996.

[46] J. Peddieson, G. R. Buchanan, and R. P. McNitt, "Application of nonlocal continuum models to nanotechnology," *International Journal of Engineering Science*, vol. 41, no. 3–5, pp. 305–312, 2003.

[47] J. N. Reddy, "Nonlocal theories for bending, buckling and vibration of beams," *International Journal of Engineering Science*, vol. 45, no. 2–8, pp. 288–307, 2007.

[48] Q. Wang and K. M. Liew, "Application of nonlocal continuum mechanics to static analysis of micro- and nano-structures," *Physics Letters, Section A: General, Atomic and Solid State Physics*, vol. 363, no. 3, pp. 236–242, 2007.

[49] M. Aydogdu, "A general nonlocal beam theory: its application to nanobeam bending, buckling and vibration," *Physica E: Low-Dimensional Systems and Nanostructures*, vol. 41, no. 9, pp. 1651–1655, 2009.

[50] O. Civalek and Ç. Demir, "Bending analysis of microtubules using nonlocal Euler-Bernoulli beam theory," *Applied Mathematical Modelling*, vol. 35, no. 5, pp. 2053–2067, 2011.

[51] H.-T. Thai and T. P. Vo, "A nonlocal sinusoidal shear deformation beam theory with application to bending, buckling, and vibration of nanobeams," *International Journal of Engineering Science*, vol. 54, pp. 58–66, 2012.

[52] R. Barretta and F. Marotti De Sciarra, "A nonlocal model for carbon nanotubes under axial loads," *Advances in Materials Science and Engineering*, vol. 2013, Article ID 360935, 6 pages, 2013.

[53] F. Marotti de Sciarra and R. Barretta, "A gradient model for Timoshenko nanobeams," *Physica E: Low-Dimensional Systems and Nanostructures*, vol. 62, pp. 1–9, 2014.

[54] R. Barretta, L. Feo, R. Luciano, and F. Marotti de Sciarra, "Application of an enhanced version of the Eringen differential model to nanotechnology," *Composites B*, vol. 96, pp. 274–280, 2016.

[55] R. Barretta, L. Feo, R. Luciano, F. Marotti de Sciarra, and R. Penna, "Functionally graded Timoshenko nanobeams: a novel nonlocal gradient formulation," *Composites B*, vol. 100, pp. 208–219, 2016.

[56] B. Arash and Q. Wang, "A review on the application of nonlocal elastic models in modeling of carbon nanotubes and graphenes," *Computational Materials Science*, vol. 51, no. 1, pp. 303–313, 2012.

[57] R. Rafiee and R. M. Moghadam, "On the modeling of carbon nanotubes: a critical review," *Composites Part B: Engineering*, vol. 56, pp. 435–449, 2014.

[58] R. Barretta, R. Luciano, and F. M. de Sciarra, "A fully gradient model for Euler-Bernoulli nanobeams," *Mathematical Problems in Engineering*, vol. 2015, Article ID 495095, 8 pages, 2015.

[59] M. A. Eltaher, S. A. Emam, and F. F. Mahmoud, "Static and stability analysis of nonlocal functionally graded nanobeams," *Composite Structures*, vol. 96, pp. 82–88, 2013.

[60] F. Marotti de Sciarra, M. Čanadija, and R. Barretta, "A gradient model for torsion of nanobeams," *Comptes Rendus—Mecanique*, vol. 343, no. 4, pp. 289–300, 2015.

[61] K. A. Lazopoulos and A. K. Lazopoulos, "On the torsion problem of strain gradient elastic bars," *Mechanics Research Communications*, vol. 45, pp. 42–47, 2012.

[62] A. C. Eringen, "On differential equations of nonlocal elasticity and solutions of screw dislocation and surface waves," *Journal of Applied Physics*, vol. 54, no. 9, pp. 4703–4710, 1983.

10

Collapse Analysis of Transmission Tower Subjected to Earthquake Ground Motion

Xiaohong Long, [1,2] **Wei Wang**, [1,2] **and Jian Fan** [1,2]

[1] *School of Civil Engineering and Mechanics, Huazhong University of Science and Technology, Wuhan 430074, China*
[2] *Hubei Key Laboratory of Control Structure, Huazhong University of Science and Technology, Wuhan 430074, China*

Correspondence should be addressed to Xiaohong Long; xhlong@hust.edu.cn

Academic Editor: Zhiping Qiu

The collapse of transmission towers involves a series of complex problems, including geometric nonlinearity, material nonlinearity, dynamic nonlinearity, and the failure of members. Simulation of the process of collapse is difficult using traditional finite element method (FEM), which is generated from continuum and variation principle, whereas the finite particle method (FPM) enforces equilibrium on each point. Particles are free to separate from one another, which is advantageous in the simulation of the structural collapse. This paper employs the finite particle method (FPM) to simulate the collapse of a transmission steel tower under earthquake ground motions; the three-dimensional (3D) finite particle model using MATLAB and the 3D finite element model using ANSYS of the transmission steel tower are established, respectively. And the static and elastic seismic response analyses indicate that the results of the FPM agree well with those of the FEM. To simulate the collapse of the transmission steel tower, a failure criterion based on the ideal elastic-plastic model and a failure mode are proposed. Finally, the collapse simulation of the transmission steel towers subjected to unidirectional earthquake ground motion and the collapse seismic fragility analysis can be successfully carried out using the finite particle method. The result indicates that the transmission steel tower has better seismic safety performance and anticollapse ability.

1. Introduction

The breakdown or collapse of transmission towers is common during an earthquake. In the United States, the Landers earthquake in 1992 and the Northridge earthquake in 1994 severely damaged the transmission system, respectively [1]. The Kobe earthquake in Japan in 1955 resulted in the destruction of a large number of transmission towers whose main failure modes were the subsidence of foundation, the tilt of tower, and compressive yield of structural member [2]. The Kocaeli earthquake in Turkey in 1999 also caused landslides, faulting, and ruptures of earth's surface, thereby damaging a considerable number of transmission towers [3]. Moreover, transmission lines also broke down in the Wenchuan earthquake in China in 2008 [4–6]. Figure 1 depicts the collapse of transmission towers during the earthquake.

The seismic dynamic responses of transmission towers involve initially solving a mechanical model, and its analysis involves many nonlinear problems, such as dynamic nonlinearity, geometric nonlinearity, and material nonlinearity. Kempner et al. [7] analyzed the structure of transmission towers by using the space frame model or the space truss model and discovered that the damping ratio of a 45 m high transmission tower varied between 0.015 and 0.04. Meanwhile, Kitipornchai et al. [8, 9] and Meek et al. [10, 11] established a space rigid frame model to analyze the correlations of transmission towers. They deduced the element stiffness matrix by considering initial stress, initial deformation, and geometry. The results of numerical simulation were highly consistent with those of structural tests, and they significantly improved the computational efficiency. Yasui et al. [12] established a finite element model by using the beam element and the truss element and conducted a dynamic response analysis of self-supporting and guyed transmission towers induced by wind loads, respectively, and the quality of conductors and insulators

FIGURE 1: Collapse of transmission steel towers during Wenchuan earthquake in China (2008).

was concentrated to arms or nodes of the finite element model.

Lin et al. [13] used the finite element software ABAQUS, to develop a subroutine based on the idea of the birth-death element to simulate the progressive collapse of transmission towers subjected to one-dimensional and three-dimensional earthquake, respectively. The weak parts of transmission towers were analyzed. They concluded that the transmission tower was more prone to collapsing under three-dimensional earthquake action. Three main methods can be used to analyze structural progressive collapse: theoretical analysis, experimental study, and numerical simulation. Given that the collapse process of structures involves many complicated nonlinear cases, theoretical analysis is difficult and cannot accurately evaluate the structural failure mechanisms. In addition, conducting structural collapse tests is expensive and dangerous, and the experimental process is difficult to control. Therefore, the numerical simulation method is widely adopted for the collapse process of structures. At present, the main methods of simulating structural collapse can be summarized as follows. (1) The finite element method (FEM): firstly, the structure is discretized according to the theory of the variational principle and continuum mechanics. Next, the element stiffness matrix of each discrete element is established. Then the global stiffness matrix is formed. Finally, the structural analysis is conducted by solving motion equations. If large deformation and nonlinear problems are involved, then the FEM can lead to the convergence problem of the numerical results. Thus, special treatment measures are needed to control the strong nonlinear behavior in the collapse process of structures. Other methods are needed to solve problems of discontinuity, such as those of fractures and collisions [14–16]. Numerical simulation of the collapse process of structures is usually based on explicit integration of the finite element program. Currently, some finite element software packages, such as LS-DYNA, ABAQUS/Explicit and MSC-MARC, can be used to simulate structural collapse. (2) The discrete element method (DEM): according to the Newtonian mechanics, the explicit time integration is adopted to solve governing equations and discontinuous problems involving the contact and collision between elements. In recent years, scholars have applied the DEM to simulate the structural collapse of brittle materials, such as concrete structures [17, 18]. However, the DEM is mainly used in the simulation of particulate materials. When the DEM is applied to simulate a large 3D space structure, the modeling process is more complicated and the analysis requires an excessively large amount of calculations; furthermore, the calculation results are also inaccurate. (3) The finite particle method (FPM): the FPM was proposed by Ting et al. [19, 20]. In space functions, the structure is discretized into finite number of particles connected to one another with a massless element. In time functions, the structural motion process is divided into a limited time element called the path element, in which motion equations can be established and solved by applying Newton's second law to each particle. Therefore, the entire structural mechanical behavior can be described by this method. As a particle method, it is possible to add or delete particles and elements in the FPM, which is important in the simulation of structural collapse. Particularly, no iterations are necessary to follow nonlinear laws, and no global matrices are formed.

The objective of this work is to establish a general finite particle method framework for simulating the collapse process of the transmission steel tower, including the discretization of the structure, motion equation of particles, the internal force calculation, and the explicit time integration and solution. This study conducts the collapse simulation of the transmission steel tower model based on the FPM. The three-dimensional finite particle model of transmission steel tower is developed in MATLAB. The static and elastic seismic response analysis of transmission tower is carried out and calculation results of the FPM agree well with those of the FEM, respectively. The failure criterion based on the ideal elastic-plastic model and the failure mode are proposed. Finally, the collapse simulation of the transmission steel towers subjected to unidirectional and three-dimensional earthquake ground motions can be successfully carried out using the finite particle method.

2. Description of Finite Particle Method

The FPM has three basic concepts: the point value description, the path element, and the virtual reversed movement. In space, the structure is discretized into certain number of particles by point value description. In time, the motion process is divided into certain number of periods through the path element. Moreover, the internal force of elements can

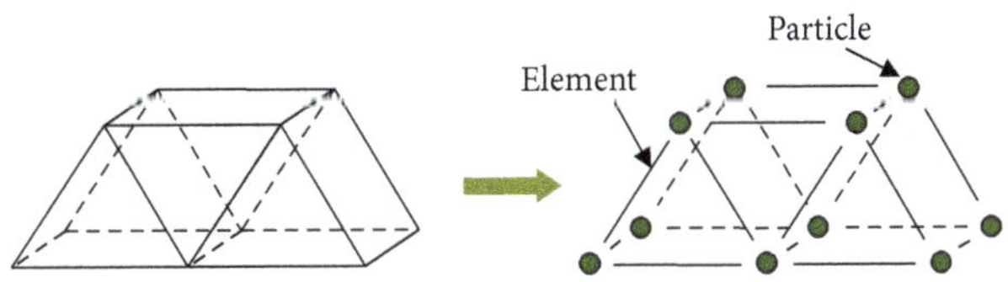

FIGURE 2: 3D truss structure and discretization in the FPM.

be calculated through virtual reversed movement. Finally, the mechanical behavior of structural systems can be accurately described by the FPM.

2.1. Discretization of the Truss Structure. Structural analysis is primarily used to study structural properties such as deformation, internal force, and displacement when structures are subjected to external action. The basic element of structure in the FPM is the particle, and the mass, space position, internal force, and deformation of the particle are described by the property of particle (Figure 2). However, elements are not used to describe the structure but to represent the topological relationship between two particles. And elements do not have mass, which is equivalent to the particles.

2.2. Establishment of Motion Equations for Truss Structure. The motion process of structures is the function of space and time. In the FPM, the motion process of particle is divided into certain number of motions in the small time step and particle simultaneously meets the control equations. Motion of particle in the small time step is called a path element, which is applied to simulate actual motion of structures. During the path element, the number and mass of particles in structural space, the connection, and mechanical properties of elements, the displacement constraints will remain unchanged. However, it can be changed only between path elements such as fracture, yield, and collision of structural members.

By the point value description and the path element, the structural motion in real time and space can be equivalent to the motion of particles at the small time step, and this motion follows Newton's second law. Therefore, the motion equation of particle in every path element can be established using Newton's second law. For space truss structures, the displacement of a particle can be decomposed into three translation degrees of freedom. In order to conduct the seismic response analysis using FPM, the acceleration time history record is imposed on particles of the upper structure. So, the motion equation of particles subjected to the earthquake can be expressed as

$$m\frac{d^2}{dt^2}\begin{bmatrix} d_x \\ d_y \\ d_z \end{bmatrix} = \begin{bmatrix} f_x^{\text{ext}} \\ f_y^{\text{ext}} \\ f_z^{\text{ext}} \end{bmatrix} + \begin{bmatrix} f_x^{\text{int}} \\ f_y^{\text{int}} \\ f_z^{\text{int}} \end{bmatrix} + \begin{bmatrix} f_x^{\text{damp}} \\ f_y^{\text{damp}} \\ f_z^{\text{damp}} \end{bmatrix} + \begin{bmatrix} f_x^{\text{eaq}} \\ f_y^{\text{eaq}} \\ f_z^{\text{eaq}} \end{bmatrix}, \tag{1}$$

where m is the equivalent mass of the particle, $[d_x \; d_y \; d_z]^T$ is the displacement vector of the particle, $F^{\text{ext}} = [f_x^{\text{ext}} \; f_y^{\text{ext}} \; f_z^{\text{ext}}]^T$ is the equivalent external force vector acting on the particle, $F^{\text{int}} = [f_x^{\text{int}} \; f_y^{\text{int}} \; f_z^{\text{int}}]^T$ is the equivalent internal force vector exerted by the elements connected with the particle, $F^{\text{damp}} = [f_x^{\text{damp}} \; f_y^{\text{damp}} \; f_z^{\text{damp}}]^T$ is the damping force, and $F^{\text{eaq}} = [f_x^{\text{eaq}} \; f_y^{\text{eaq}} \; f_z^{\text{eaq}}]^T$ is the earthquake action when structures are subjected to the earthquake ground motions.

In (1), the damping force is expressed as

$$\begin{bmatrix} f_x^{\text{damp}} \\ f_y^{\text{damp}} \\ f_z^{\text{damp}} \end{bmatrix} = -\mu m \begin{bmatrix} \dot{d}_x \\ \dot{d}_y \\ \dot{d}_z \end{bmatrix}, \tag{2}$$

where μ is damping factor which is the same as the definition in the dynamic relaxation method [21], and the structural damping is considered. $[\dot{d}_x \; \dot{d}_y \; \dot{d}_z]^T$ is the velocity vector of the particle.

The effect of earthquake action is similar to that of the external forces, and earthquake action can be calculated in

$$\begin{bmatrix} f_x^{\text{eaq}} \\ f_y^{\text{eaq}} \\ f_z^{\text{eaq}} \end{bmatrix} = -m \begin{bmatrix} \ddot{x}_{gx} \\ \ddot{x}_{gy} \\ \ddot{x}_{gz} \end{bmatrix}, \tag{3}$$

where $[\ddot{x}_{gx} \; \ddot{x}_{gy} \; \ddot{x}_{gz}]^T$ is the three-dimensional earthquake ground acceleration.

2.3. Internal Force Calculation of the Bar Element. The relationship between the internal force of elements and particle position should be solved in order to accurately depict structural deformation. However, the internal force of the bar element is only related to pure deformation in the relative position. In the FPM, virtual reversed movement is utilized to calculate internal forces of bar elements. Because the internal forces (axial force) are related only to the deformation of the bar element, it is necessary to remove rigid body translations and rotations from the relative displacement. The path element assumes that the bar element has virtual reversed translations and virtual reversed rotations to obtain its pure deformation (Figure 3).

The time segment is assumed to be very small, and the deformation during the time step is infinitesimal. So, the engineering stress and strain could be used to calculate the

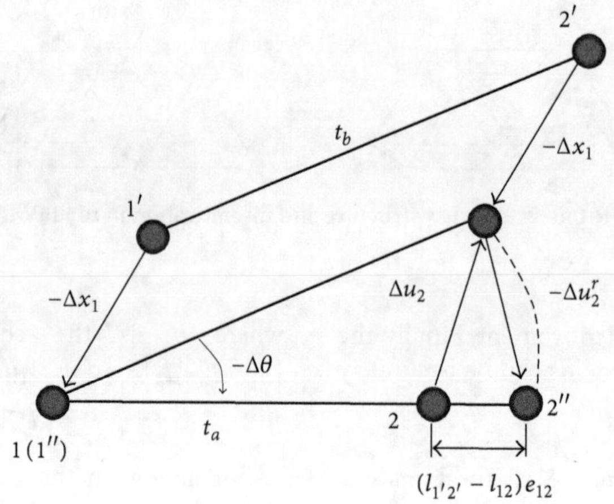

FIGURE 3: Reversed translation and rotation.

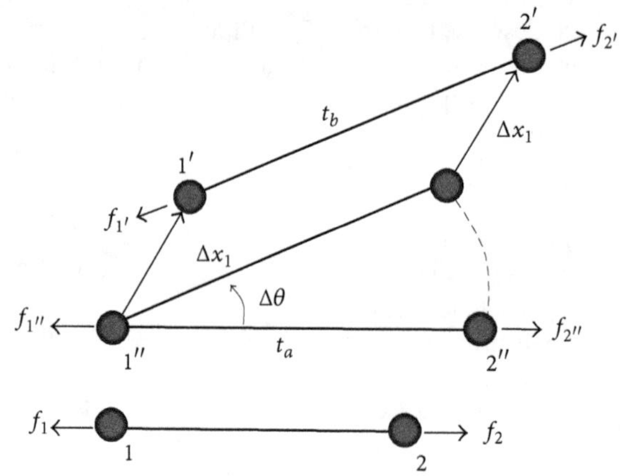

FIGURE 4: Particle internal force.

internal forces of truss bars [22]. As the bar element, its deformation is only related to the variation of the bar length, and the incremental deformation of the bar can be obtained.

$$\Delta u_2^d = (l_{1'2'} - l_{12})\, e_{12}, \quad (4)$$

where l_{12} and $l_{1'2'}$ are the length of element 1-2 at time t_a and t_b, respectively, and e_{12} is directional vector of element at time t_a.

The internal force (axial force) of bar element can be expressed as

$$f_{2''} = f + \Delta f = \left[\sigma A + \frac{EA}{l_{12}}\left(l_{1'2'} - l_{12}\right)\right] e_{12}, \quad (5)$$

where f and σ are axial force and axial stress of bar element 1-2 at time t_a, respectively; Δf is the incremental axial force of bar element 1-2 at time t_b; E and A are Young's modulus and the section area of bar element 1-2, respectively.

According to the static equilibrium condition

$$f_{1''} = -f_{2''}. \quad (6)$$

When the axial force at the virtual position $1''$-$2''$ has been obtained, let the bar element is imposed a rotation $\Delta\theta$ and a translation Δx_1 to return to its original position (Figure 4). In the process, only the direction of element axial force is changed, and the internal force of the bar element can be determined from

$$f_{2'} = -f_{1'} = \left[\sigma A + \frac{EA}{l_{12}}\left(l_{1'2'} - l_{12}\right)\right] e_{1'2'}, \quad (7)$$

where $f_{1'}$ and $f_{2'}$ are the internal forces of the particles $1'$ and $2'$, respectively.

Finally, the internal force of the particle can be obtained by the summation of all axial forces of bar elements connected to it.

2.4. Solution of Particles Motion Equation. The implicit time integration solution of motion equations requires iterations, so the calculation process is complex and there may be no convergence problem. In the FPM, explicit time integration is adopted to solve the motion equation particles. The advantages of explicit time integration lie in having a concise formula and requiring no iterations.

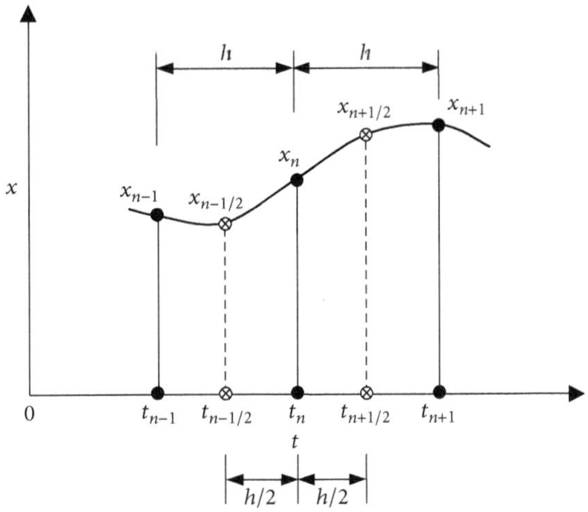

FIGURE 5: Central difference method.

According to the central difference method (Figure 5), the velocity of the particle dx_n/dt and the acceleration of the particle d^2x_n/dt^2 can be expressed as

$$\begin{aligned}\frac{dx_n}{dt} &= \frac{1}{2h}\left(x_{n+1} - x_{n-1}\right),\\ \frac{d^2x_n}{dt^2} &= \frac{1}{2\left((1/2)h\right)}\left(\frac{d}{dt}x_{n+1/2} - \frac{d}{dt}x_{n-1/2}\right)\\ &= \frac{1}{h^2}\left(x_{n+1} - 2x_n + x_{n-1}\right).\end{aligned} \quad (8)$$

Substituting (9) into (1), the difference calculation formula of particles under earthquake ground motions can be obtained as follows:

$$x_{n+1} = 2x_n - x_{n-1} + \frac{F^{\text{ext}} + F^{\text{int}} + F^{\text{damp}} + F^{\text{eaq}}}{m}h^2. \quad (9)$$

Thus, difference calculation can be started from $n = 1$ in a step-by-step manner using (9), and the corresponding displacement of particles of the structure can be solved in every time step; finally, structural behavior can be described.

2.5. Calculation Procedures. The calculation procedure of truss structures using the FPM is illustrated in the following steps.

Step 1. At the first time step, $t = 0$, the structure is dispersed into finite particles, the initial conditions of the structural calculation are determined, and the topological relationship between particles is defined, which is called as the connection of elements.

Step 2. When $n = 0$, calculate the one step displacement value x_{-1} before initial displacement of each particle according to (5).

Step 3. At any time step, $n \neq 0$, update the position of all particles, define special constraint points, and start the forward recursive calculation.

Step 4. Update internal and external forces of particles based on the new particle position.

Step 5. Define special constraint particles according to the boundary condition of the structure.

Step 6. Determine whether t_n is the finial time step, and compare t_n with total calculation time. If t_n is less than the total time, then repeat Steps 3–6. Otherwise, end the process.

Step 7. Conduct postprocess. Obtain the required results.

The calculation flow chart of the truss structure using the FPM is shown in Figure 6.

2.6. Failure Criterion and Failure Mode of Members. The collapse of structures is caused by the destruction of the material. It is necessary to define the failure criterion of material when simulating the collapse process of truss structures, the conditions must be confirmed when the material reaches the ultimate strength and members' damage. At the same time, the failure mode considering the destroyed behavior must be defined, and collapse analysis of truss structures can be conducted by defining the failure criterion and failure mode of members.

2.6.1. Failure Criterion. The main objective of this paper is to simulate the collapse process of transmission tower, which is a macroscopic process of spatial structural motion. However, the accurate fracture criterion of spatial steel structure is related to the research of fracture mechanics and structural material testing, which makes the problem too complicated. Thus, the spatial bar element in the FPM is simplified as the circular section without considering the emergence and development of cracks when material failure occurs. Therefore, the spatial bar element in steel structures is regarded as the ideal element that has no initial defects, and failure directly occurs when this failure reaches the stress-strain limit. There are three kinds of failure rule for 3D

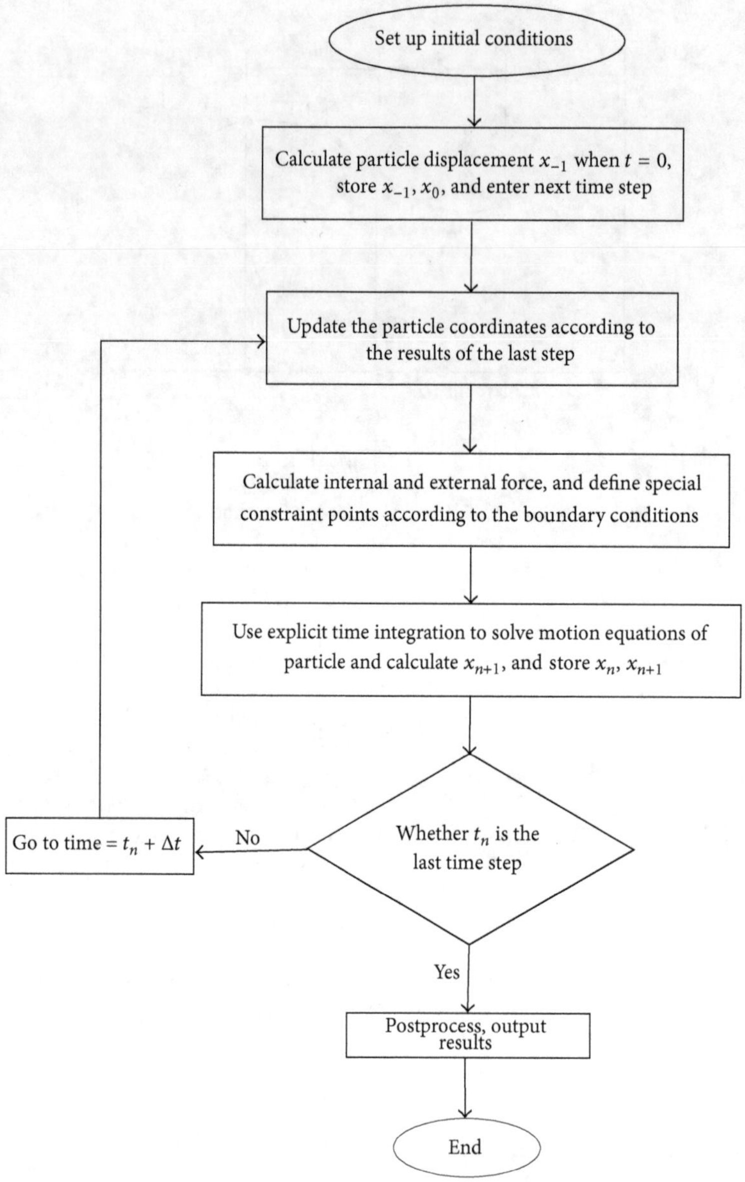

FIGURE 6: Calculation flow chart using the FPM.

truss element, that is, strength failure rule, ultimate strain failure rule, and compression member buckling and softening failure rule. Strength failure rule is suitable to the hard brittle member, ultimate strain failure rule applies to the member with smaller slenderness and good ductility, and compression member buckling and softening failure rule can account for several failure modes of the larger slenderness member [23]. In this paper, the ultimate strain failure will be adopted as shown in Figure 7. The relationship between stress and strain is linear until the member axial tensile or compressive stress reaches the yield stress value of the material σ_{crit}. Then, the material behavior is proposed to be plastic until failure happens. However, the exact critical strain value at the point at which failure happens is not quite clear. According to several experiments, the critical axial tensile value is suggested to be approximately three times the yield strain, $\varepsilon_{crit} = 3\varepsilon_y$ [24]. After failure happens, the fracture member may have one or two free ends. If the external force is not exerted, unloading will occur in the failure member. The unloading is elastic in nature with the same slope as the initial loading phase.

2.6.2. Failure Mode. Fracture behaviors of members involve a series of problems, such as the separation of particles, the redistribution of internal forces and masses, and the redefinition of topological relations of elements [25]. In order to simulate these behaviors in the analysis of the FPM, it is necessary to establish the fracture failure model. When the stress reaches the failure criterion of the material, the failure mode of members may be separated from the two ends of the particle simultaneously, and it is also possible to be separated from one side of the particle. Particles

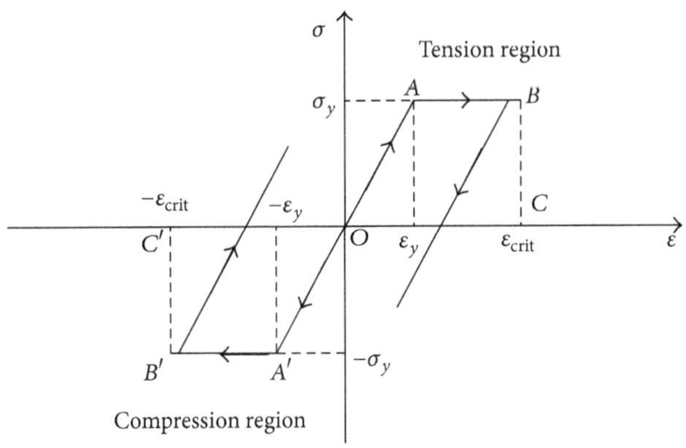

FIGURE 7: Ideal elastic-plastic stress-strain curve of steel.

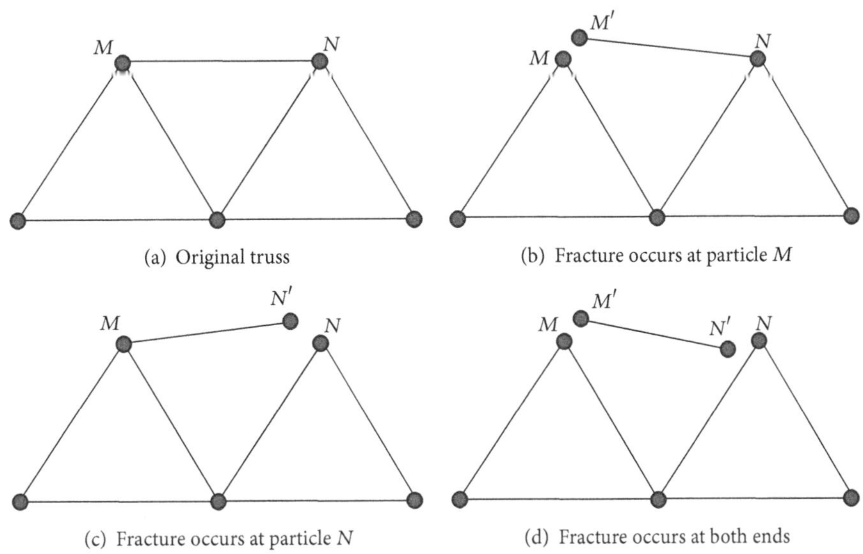

FIGURE 8: Failure modes of members of the truss.

maintain a dynamic balance under internal and external forces. Although axial forces at the both ends of the bar element are the same, the resultant forces of particles at the both sides of the bar element will not necessarily be the same. Thus, the resultant forces of particles at the both sides of the bar element are needed to be compared, and the fracture of members will occur at the end of greater resultant forces of particles. $|f_M|$ and $|f_N|$ are defined to resultant forces of particles at M end and N end of the bar element, respectively (Figure 8(a)). If $|f_M| > |f_N|$, the fracture of the member occurs at particle M, the new particle M' is generated and the element is connected to the new particle M' as shown in Figure 8(b). If $|f_M| < |f_N|$, the fracture of the member occurs at particle N, the new particle N' is generated, and the element is connected to the new particle N' as shown in Figure 8(c). If $|f_M| = |f_N|$, the fracture of the member happens at both ends of the member, then two new particles M' and N' are generated simultaneously, and the element is connected to the new particles M' and N'. Finally, the element is completely separated from the structure as shown in Figure 8(d).

2.7. Example Analysis. To verify the capabilities of the computation code, two examples are designed: Example 1: buckling analysis of 24-bar star dome; Example 2: elastic-plastic loading of three-bar plane truss.

Example 1 (buckling analysis of 24-bar star dome). This example is analyzed using FPM, and the results are compared with the nonlinear finite element method. The space truss element is adopted, and the top center node is subjected to the concentrated force F_P in the Z direction (as shown in Figure 9), elastic modulus $E = 3.03 \times 10^3$ MPa, cross-sectional area $A = 317\,\text{mm}^2$, and time step $\Delta t = 1 \times 10^{-3}$ s. Displacement control is used in the analysis, and the displacement step $\Delta d = 1 \times 10^{-4}$ mm.

The Z-direction displacement of node 1 and node 2 is tracked. The load-displacement curves are shown in Figures 10(a) and 10(b), respectively, and the results are in good agreement with the reference [26]. It is shown that the buckling analysis of the structure using the finite particle method is feasible.

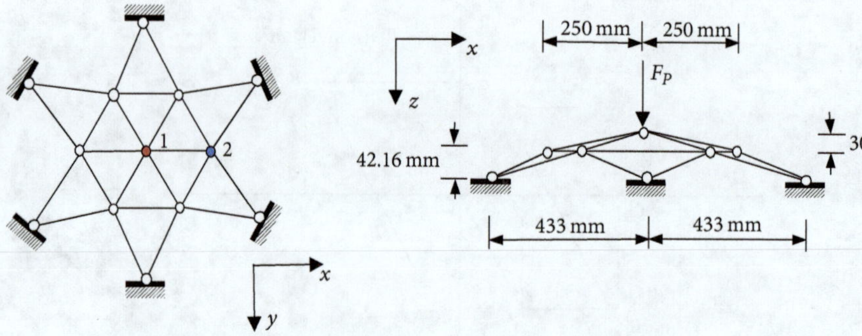

FIGURE 9: 24-bar star dome geometry and loading.

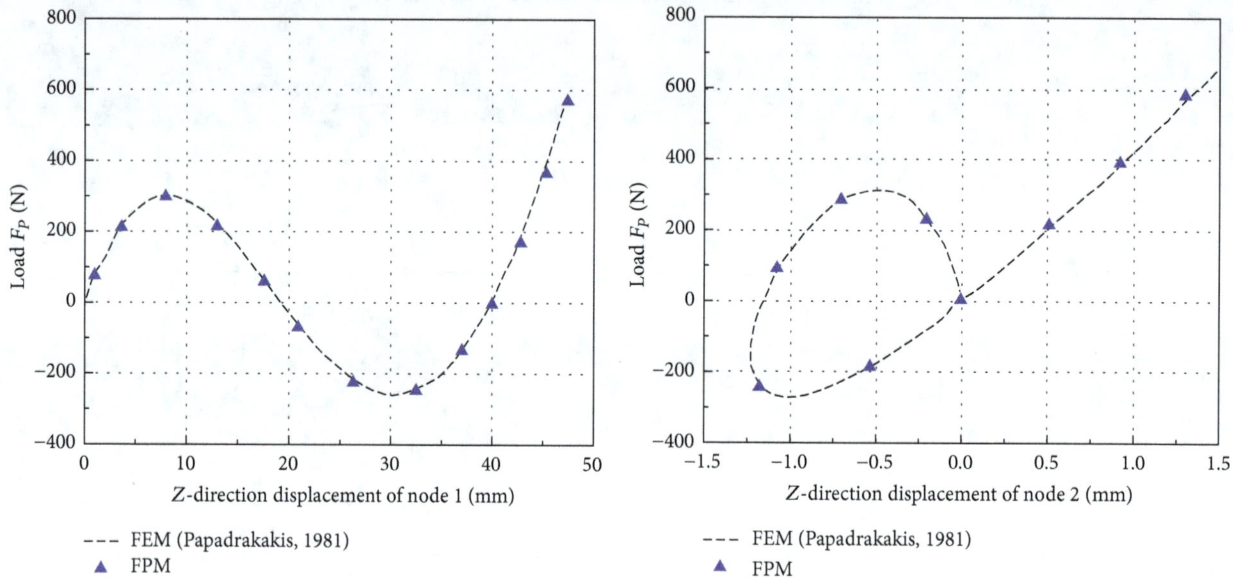

(a) Node 1

(b) Node 2

FIGURE 10: Load-displacement curves of node 1 and node 2 in the Z direction.

Example 2 (elastic-plastic loading of three-bar plane truss). The three-bar plane truss is subjected to the load F_P in the y direction (shown in Figure 11), the cross-sectional area of bars is A, elastic modulus at the elastic phase is E, yield strength is σ_y, the length of BD bar is l, and AD = CD.

Using the ideal elastic-plastic model, the theoretical solution of the load-displacement at point D is expressed as follows:

$$\text{When } F_P \le F_{Pe} \quad F_P = \frac{EA\left(1+2\cos^3\theta\right)\Delta}{l}, \tag{10}$$

$$\text{When } F_{Pe} < F_P \le F_{Py} \quad F_P = \sigma_y A + \frac{2EA\left(\cos^3\theta\right)\Delta}{l}, \tag{11}$$

$$\text{When } F_P > F_{Py} \quad F_P = F_{Py}, \tag{12}$$

where $F_{Pe} = \sigma_y A(1 + 2\cos^3\theta)$ and $F_{Py} = \sigma_y A(1 + 2\cos\theta)$.

The analysis is conducted using the finite particle method, elastic modulus value at the elastic phase is $E = 206$ GPA, yield strength value is $\sigma_y = 235$ MPa, the length of BC bar $l = 1$ m, density is $\rho = 7850\,\text{kg/m}^3$, $\theta = 45°$, and time step is $\Delta t = 1 \times 10^{-5}$ s. The truss structure is subjected to the slow load action, and the results are turned into the dimensionless curve and compared with the theoretical solution as shown in Figure 12.

As can be seen from Figure 12, when $\theta = 45°$ and $\Delta_y/\Delta_e = 2$, $F_{Py}/F_{Pe} = 1.41$. FPM analysis results are very close to the theoretical solution. However, the FPM result of the load-displacement curve is slightly higher than that of the theoretical solution when the bars go into the plastic stage. The reason is that the structure is continuously loaded in the actual analysis, and even if there is a small load increment, the great strain will be produced in the truss structure. But the theoretical solution does not consider the full plastic loading, and this small difference is considered to be reasonable. Simultaneously, the numerical examples show that the finite particle method can effectively simulate the elastic-plastic properties of truss structures.

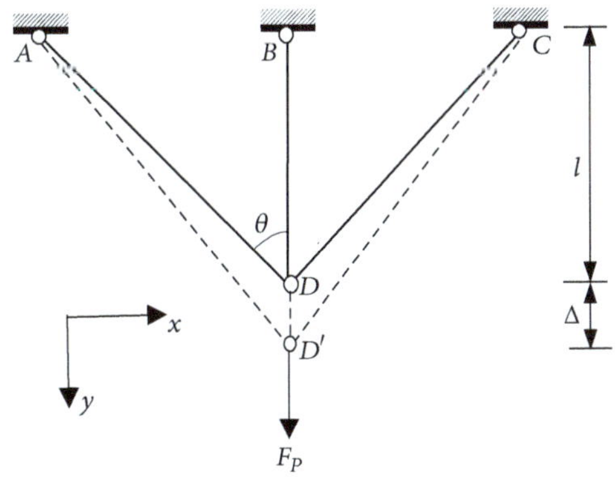

FIGURE 11: Three-bar plane truss.

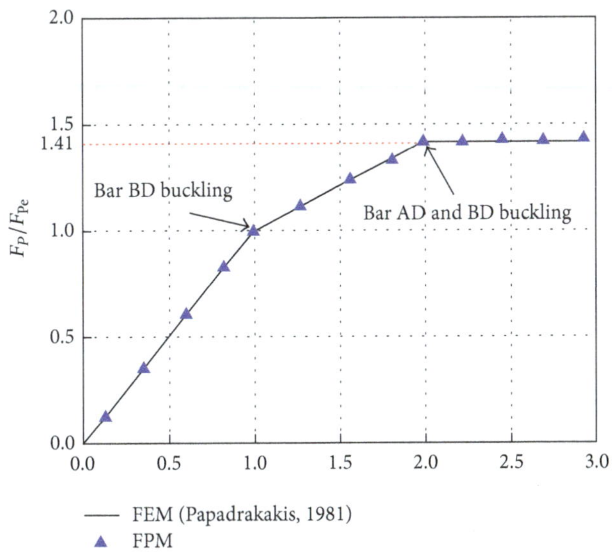

FIGURE 12: Load-displacement curve.

3. Establishment and Verification of Transmission Tower Model

3.1. Establishment of the Model of Transmission Steel Tower. A 110 kV double-circuit linear transmission tower structure is studied. The elevation drawing of transmission tower is shown in Figure 13. The total height of the transmission tower is 21 meters, the sections of tower columns are square, and materials are equilateral angle steel (Table 1). Software ANSYS and software MATLAB are used to establish the model of the transmission tower structure by the FEM and the FPM, respectively. Although ANSYS is widely used and is relatively mature finite element software, the study on finite particle method has just started by the MATLAB program. To verify the accuracy of the model of the FPM, the finite element model of the transmission tower is also established by ANSYS.

On one hand, the finite element model is established in ANSYS by adopting the element beam 188 in Figure 14, cross section of the elements is L-shaped equilateral steel angle, and the rigid joint of beam elements is assumed without considering eccentric connections. The finite element model has 522 elements and 172 nodes. On the other hand, a finite particle model of transmission tower is established by adopting the truss element in which only the axial force is considered. So, each particle has only three translation degrees of freedom in the directions of x, y, and z; thus, all joint connections that can be regarded as hinges and elements in the FPM were modeled by bar elements. During the modeling of the transmission tower, the node coordinates, connections, and section parameters of elements are derived directly from the ANSYS model and then imported into the compiled MATLAB program. When the initial parameters of the analysis are inputted, the finite particle model is established in MATLAB as shown in Figure 15. The number of elements and nodes in the FPM is the same as that of the FEM, that is, 522 elements and 172 nodes. In the FPM, the section area of the bar element is only considered and the shape of bar element is not considered.

3.2. Static Analysis and Verification of the Transmission Tower. The static analysis of the transmission tower is conducted using the FEM (in ANSYS) and FPM (in MATLAB), respectively. The calculation results were compared in order to verify the correctness of the finite particle model. Three kinds of static loads are applied, respectively. (1) The vertical load

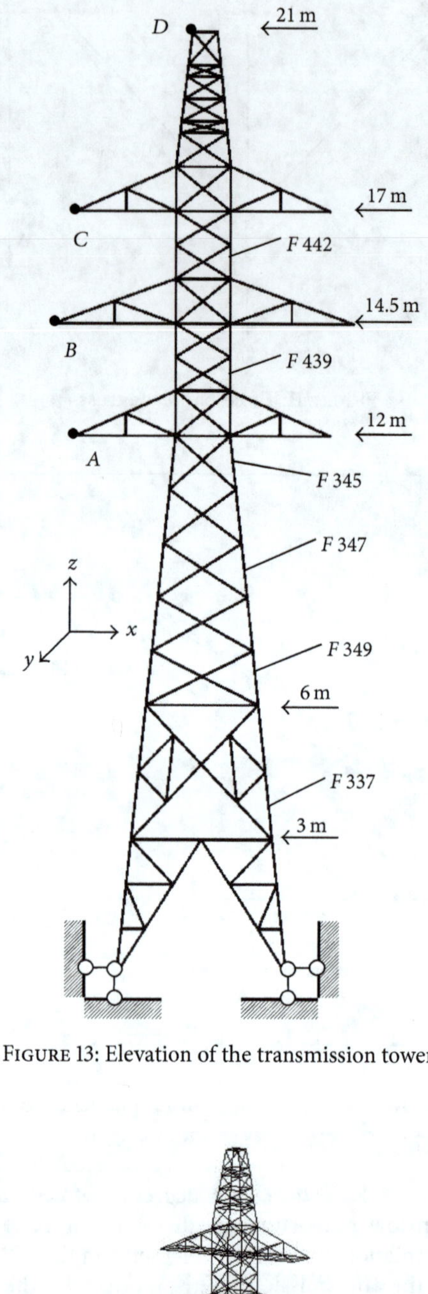

FIGURE 13: Elevation of the transmission tower.

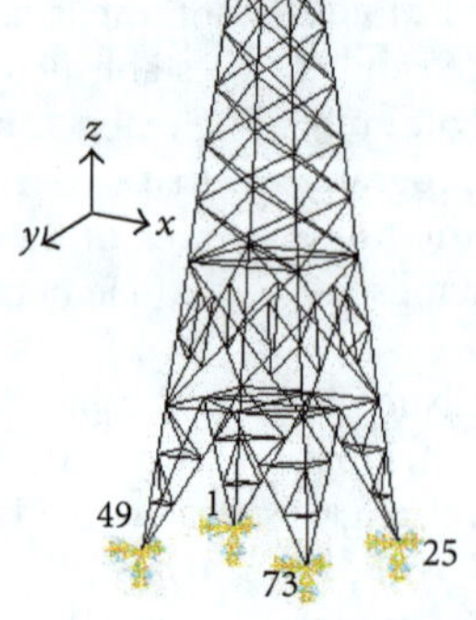

FIGURE 14: FEM model (ANSYS).

TABLE 1: Material parameters of transmission tower.

Number	Material type	Section area /m^2	Density /kg·m^{-3}	Elastic modulus /Pa	Member type
(1)	$L100 \times 10$	$1.93E-03$	$7.80E+03$	$2.06E+11$	Main material
(2)	$L50 \times 4$	$3.90E-04$	$7.80E+03$	$2.06E+11$	Auxiliary material
(3)	$L40 \times 3$	$2.36E-04$	$7.80E+03$	$2.06E+11$	Diagonal brace
(4)	$L90 \times 8$	$1.39E-03$	$7.80E+03$	$2.06E+11$	Main material
(5)	$L45 \times 4$	$3.49E-04$	$7.80E+03$	$2.06E+11$	Auxiliary material
(6)	$L63 \times 5$	$6.14E-04$	$7.80E+03$	$2.06E+11$	Main material
(7)	$L80 \times 7$	$1.09E-03$	$7.80E+03$	$2.06E+11$	Main material

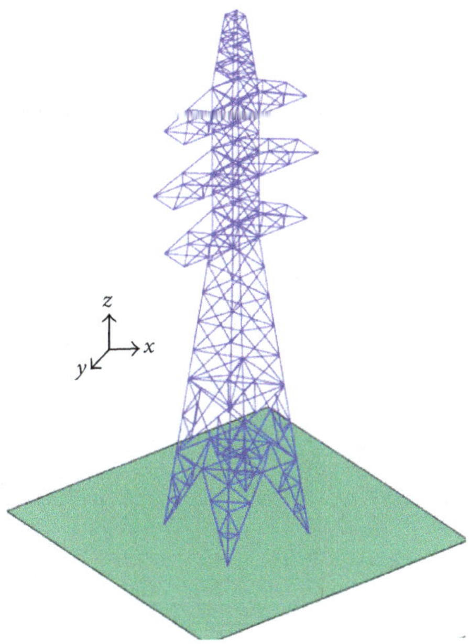

FIGURE 15: FPM model (MATLAB).

$F = -1000N$ is imposed on the four corners of the tower in the z-direction of the model. (2) The transverse load $F = 200N$ is applied at four corners of the tower in the x-direction of the model. (3) The gravity load is applied for the entire structure. The acceleration of gravity is $g = 9.8\,\mathrm{m/s^2}$. The analysis results are compared by using ANSYS and MATLAB when the structure is imposed on three kinds of static loads (Tables 2, 3, and 4). Table 2, Table 3, and Table 4 show the results, including the support reactions of the four supports (in Figure 11) and the corresponding displacements of four nodes A, B, C, and D at different heights (in Figure 10).

Tables 2, 3, and 4 show that the relative errors of support reactions are small under three kind of loads, all of them are no greater than 1%, and the relative deviation is only 0.06% under vertical load. Thus, this means that the finite particle method is very accurate in calculating the support reactions of the transmission tower. By comparison of the FEM and the FPM, relative deviations of displacement of the tower are less than 3% under vertical and transverse loads, respectively. It indicates that the lateral stiffness of the FPM model is very close to the FEM model. Because the global stiffness matrix is not assembled in the FPM, the dynamic characteristics of the tower can be obtained. In addition, the maximum relative deviation of displacements of the tower which is 4.11% under the gravity load caused by the distribution pattern of the mass in these two methods is different. In the FPM, the mass is concentrated on particles which are equivalent to the concentrated load at the node. However, in the FEM, mass is considered based on the distribution of the actual mass which is equivalent to the uniform loads on bar elements. In conclusion, the results of the static analysis of the finite particle method are consistent with that of the finite element method, and the accuracy of the model of the finite particle method is validated for the following application in seismic response analysis and the collapse simulation of the transmission tower.

3.3. Linear Seismic Response and Verification of the Transmission Steel Tower. The acceleration time history of El Centro earthquake ground motion record (NS) is selected to conduct

Table 2: Support reactions and vertical displacements of nodes under vertical loads.

Methods	Vertical support reactions/N				Vertical displacements/mm			
	Support point 1	Support point 25	Support point 49	Support point 73	Node A	Node B	Node C	Node D
FEM	967.8	999.3	999.3	1033.6	−0.0359	−0.0422	−0.0528	−0.0973
FPM	968.4	999.9	999.9	1031.8	−0.0367	−0.0427	−0.0540	−0.0990
Relative deviation	0.06%	0.06%	0.06%	−0.17%	2.17%	1.23%	2.24%	1.80%

TABLE 3: Support reactions and horizontal displacements of nodes under horizontal loads.

Methods	Transverse support reactions/N				Transverse displacements/mm			
	Support point 1	Support point 25	Support point 49	Support point 73	Node A	Node B	Node C	Node D
FEM	−198.8	−198.2	−201.6	−201.4	0.340	0.655	1.068	1.916
FPM	−197.2	−197.2	−202.8	−202.8	0.349	0.672	1.094	1.957
Relative deviation	−0.80%	−0.50%	0.60%	0.70%	2.65%	2.60%	2.34%	2.14%

TABLE 4: Support reactions and vertical displacements of nodes under gravity loads.

Methods	Vertical support reactions/N				Vertical displacements/mm			
	Support point 1	Support point 25	Support point 49	Support point 73	Node A	Node B	Node C	Node D
FEM	5637.4	5640.0	5635.9	5652.9	0.197	0.233	0.219	0.172
FPM	5651.7	5660.4	5656.2	5665.0	0.204	0.242	0.228	0.179
Relative deviation	0.25%	0.36%	0.36%	0.21%	3.55%	3.86%	4.11%	4.07%

the seismic response analysis. The peak ground acceleration (PGA) is adjusted to 1 m/s^2, and the time step of El Centro ground motion in the FEM is 0.02 s while the time step is 0.0001 s in the FPM by using linear interpolation method, and earthquake acceleration is inputted in the direction of transverse line direction of the transmission tower. Then the elastic seismic response analysis is carried out using FEM and FPM, respectively. The time history curves of the displacement of the particle at the top of the transmission tower by the two methods are shown in Figure 16.

Figure 16 shows that the results of the finite element method and the finite particle method to calculate the displacement response of the transmission tower under earthquake acceleration are very close and the maximum displacement is 5.28 mm in 4.96 s for the ANSYS analysis while that is 5.61 mm in 5.10 s for the MATLAB program. Thus, the relative error of displacement of particles is very small so that the time history of displacement is consistent.

Table 5 shows the maximum and minimal axial forces of some members of the transmission tower which are obtained by the FEM and the FPM. The axial forces calculated by the FEM are in good agreement with those by the FPM, and relative error is also within 5%. The seismic response analysis of the transmission tower is conducted by the FEM and the FPM using the same earthquake ground motion record, respectively. Results show that the time history of displacement maximum displacements and extreme value of axial forces are very close. These results also illustrated that the input method of the earthquake wave in the FPM is reliable. Therefore, the FPM can be applied to simulate the collapse analysis of the transmission tower by the MATLAB program.

4. Collapse Analysis of the Transmission Tower

4.1. Collapse Analysis under Different PGAs. When the collapse analysis of the transmission tower is carried out the effect of gravity is considered, the value of the virtual damping factor is set to 5, and the yield strength of material is 235 MPa. There are 522 elements in the finite particle model by MATLAB. According to "Code for Design of Steel Structures" [25], the stability factor of the member related to its slenderness ratio can be obtained. To simplify the analysis process, the stability factor φ of all members is set as 0.88. The first 10 seconds of acceleration time history of the El Centro earthquake ground motion record is selected. In order to study the collapse of the structure under different seismic PGA and determine the critical PGA of collapsed, the PGA value of the earthquake ground motion record is adjusted to 18 m/s^2, 20 m/s^2, 22 m/s^2, 24 m/s^2, and 26 m/s^2. Failure time of transmission tower structure and members can be obtained, such as the first member, 20% members of the structure, and the complete structure, and the results are listed in Table 6.

When PGA is 18 m/s^2, the structure does not destroy, and the elastic behavior occurs. The structure collapsed when PGA is greater than 18 m/s^2, and greater PGA values lead to earlier occurrences of collapse. According to the first damaged members, failure occurred at the 12 m height of the transmission tower when PGA is 20 m/s^2 and 22 m/s^2, respectively, whereas failure occurred at 6 m under 22 m/s^2 and 24 m/s^2. The damage location of the transmission tower will be different under different PGA values; namely, the destruction form is not the same. Because the space structure is very complex, no unified quantitative index to judge complete collapse of structure has been recorded. This paper uses the method in reference [22], which supposed that the structure completely collapsed when 20% of total members damaged. With the increase of the PGA, the time from first damage to complete collapse becomes much shorter. Furthermore, the time under the PGA for 22 and 24 m/s^2 is shorter than the time under 20 and 22 m/s^2, indicating a decrease of one order of magnitude. The greater PGA is, the quicker the structure collapses.

Failure modes under the PGA for 20, 22, 24, and 26 m/s^2 are shown in Figure 17. When PGA was 20 m/s^2, damage occurred mainly at heights of 6 and 12 m; when PG is 22 m/s^2,

TABLE 5: Maximum and minimal axial force of some members of tower.

Element number	F337	F349	F347	F345	F439	F442
Maximum (FEM)	−12816.4	−12665.6	−11463.9	−9838.9	−7166.8	−3151.7
Maximum (FPM)	−12556.6	−12263.6	−10982.1	−9481.6	−6946.7	−3133.3
Relative deviation	−2.03%	−3.17%	−4.20%	−3.63%	−3.07%	−0.59%
Minimum (FEM)	4222.9	4660.6	4658.9	4398.2	2901.7	842.0
Minimum (FPM)	4138.6	4471.3	4451.1	4292.2	2861.1	830.0
Relative deviation	−2.00%	−4.06%	−4.46%	−2.41%	−1.40%	−1.42%

TABLE 6: Failure time of the transmission tower subjected to different PGAs.

$PGA/m{\cdot}s^{-2}$	18	20	22	24	26
First damaged time/s	No damage	6.76	5.92	5.66	4.93
Element number	No damage	138	139	359	349
Damaged time for 20% of members/s	No damage	8.34	6.97	5.85	5.06
Time from first crack to collapse/s	No damage	1.57	1.05	0.18	0.13

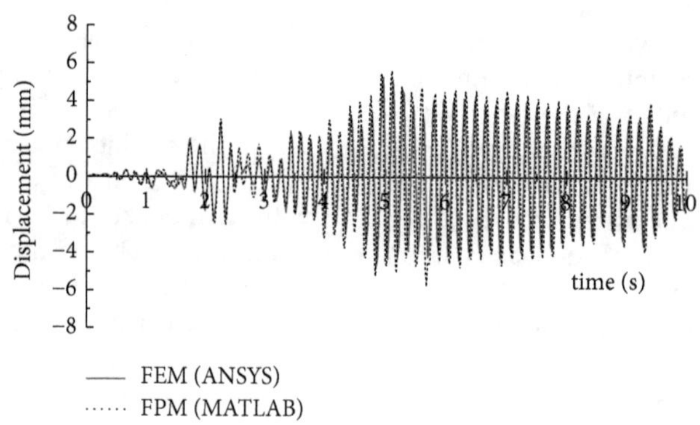

FIGURE 16: Time history curves of displacement of node D at the top of the transmission tower.

failure occurred at supports and the height of 6 and 12 m; when PGA reaches 24 m/s^2, failure occurs from the bottom to 14 m high; when PGA is 26 m/s^2, structural damage occurs at almost every height. Thus, when the PGA is large, the damage occurred early and the failure mode becomes massive.

4.2. Collapse Analysis under Unidirectional Earthquake Action. According to the discussion in the previous section, the critical PGA wherein the transmission tower started to collapse is between 18 and 20 m/s^2. With the increase of the PGA, collapse becomes severe and the duration of the process of collapse was abrupt. The PGA for 20 m/s^2 is selected to capture the details of the process of collapse more clearly. The process of transmission tower collapse is complex and involves a variety of nonlinear problems and fractures of materials as well as contact-impact problems. Contact-impact problems are a new research topic that is cumbersome to discuss; hence, this subject is not discussed. Only contacts between members and the ground are considered, which are simplified by using a simple spring model for simulation. Impacts, either among members or between members and structures, are not considered after fracture. The transmission tower starts to fail at $t = 6.76$ s and collapsed totally at $t = 8.34$ s, at which point the duration of the collapse is 1.57 s. The process of failure is simulated by using the FPM, and structural forms in every moment are captured. Elevation drawings of structural deformation are intercepted in several representative moments, as shown in Figure 18.

At $t = 6.9$ s, the fracture of elements firstly happen at the height of 12 m at $t = 7.0$ s; elements are subsequently destroyed at 6 m. Damage at these two heights is the main reason for the collapse of the transmission tower. At $t = 7.3$ s, destruction continues to worsen and elements of the transmission tower foot begin to damage. At $t = 7.6$ s, elements of the transmission tower at 17 m start to fail. Finally, the structural collapse processes continue at $t = 8.0$ s, and structure damage begins to worsen until the complete structural collapse appears due to gravity.

The transmission tower belongs to the statically indeterminate structures which possess many redundant constraints. Redundant constraints are considered to prevent the progressive collapse of the structure. However, for the transmission tower truss structure, the redistribution of internal forces

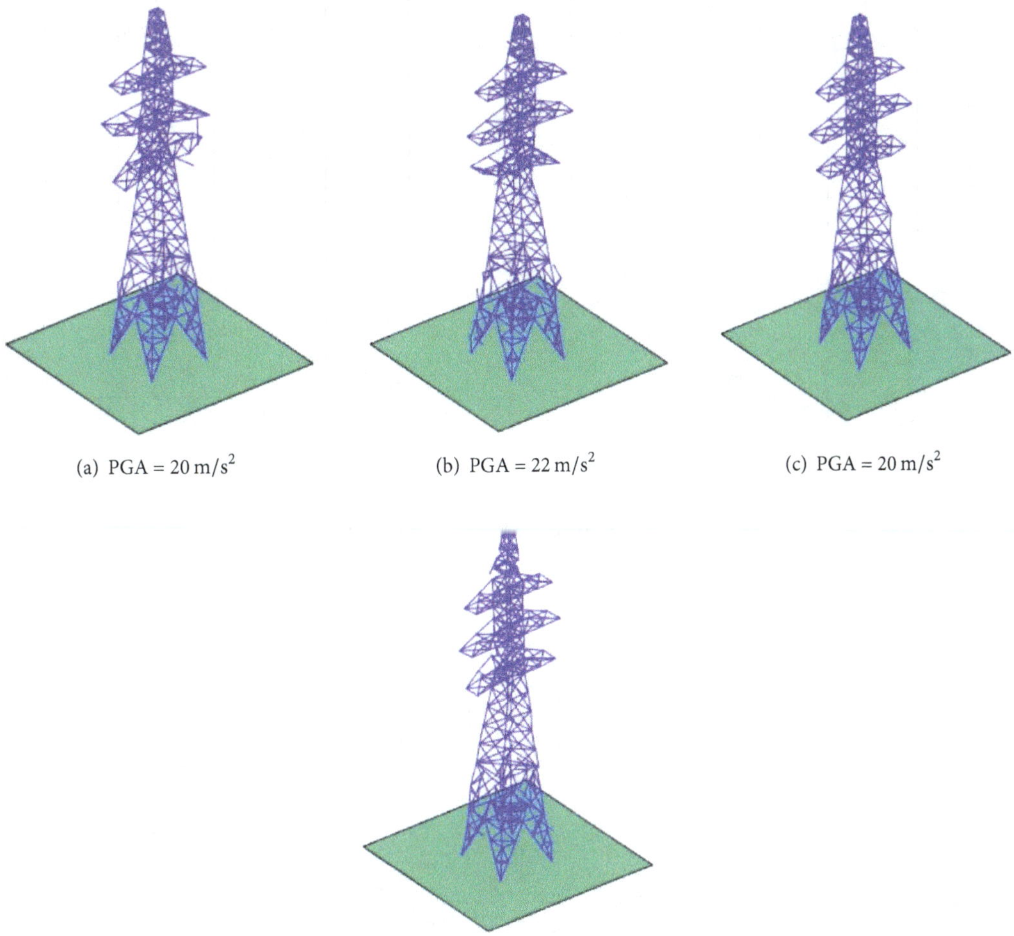

(a) PGA = 20 m/s^2 (b) PGA = 22 m/s^2 (c) PGA = 20 m/s^2

(d) PGA = 22 m/s^2

FIGURE 17: Damage drawing of the transmission tower under different PGAs.

will occur if a bar element is damaged. This redistribution will lead to the destruction of other elements, and the structure will present mechanism failure until it is eventually collapsing.

Some axial force responses of representative elements are extracted for a more intuitive understanding of the collapsing process of the transmission tower, as shown in Figure 19. For example, element 139 is the bar that is first cracked, in which the failure mode is buckling under compression and simultaneously breaks off at both ends of the bar element. Before the fracture, the axial force changed with the change of earthquake acceleration. At about $t = 6.7$ s, the element becomes a free element and the axial force gradually decreases due to damping energy dissipation. In addition, tensile failure first happens for element 137, and the fracture also happened at $t = 6.7$ s. Two cases are available to define the failure mode of the fracture: the fracture occurring at both ends simultaneously and a fracture occurring as just one end of the element. Element 137 belongs to the latter. Before damage, the axial force changed with the change of earthquake acceleration. Moreover, after damage, one end of the bar became free whereas another remains connected to the remaining structure. The axial forces irregularly change under earthquake ground motions.

4.3. Collapse Seismic Fragility Analysis. Seismic fragility in structural engineering is the relationship between ground motion intensity measure (IM) and structural damage. The probability of exceeding a damage state under a given ground motion can be quantitatively described by the fragility curve. The calculation method can be divided into parameter analysis method and nonparametric analysis method [27]. The difference between the two approaches is whether the probability distribution of engineering demand parameters (EDP) needs to be assumed in advance. Nonparametric analysis does not need to assume the probability distribution of EDP. By setting different IM levels, a large number of nonlinear time history analyses are conducted to calculate the failure probability of the structure in different damage states. When selecting EDP, the relevant performance parameters (such as force, displacement, velocity, acceleration, energy, and damage) are used to describe the

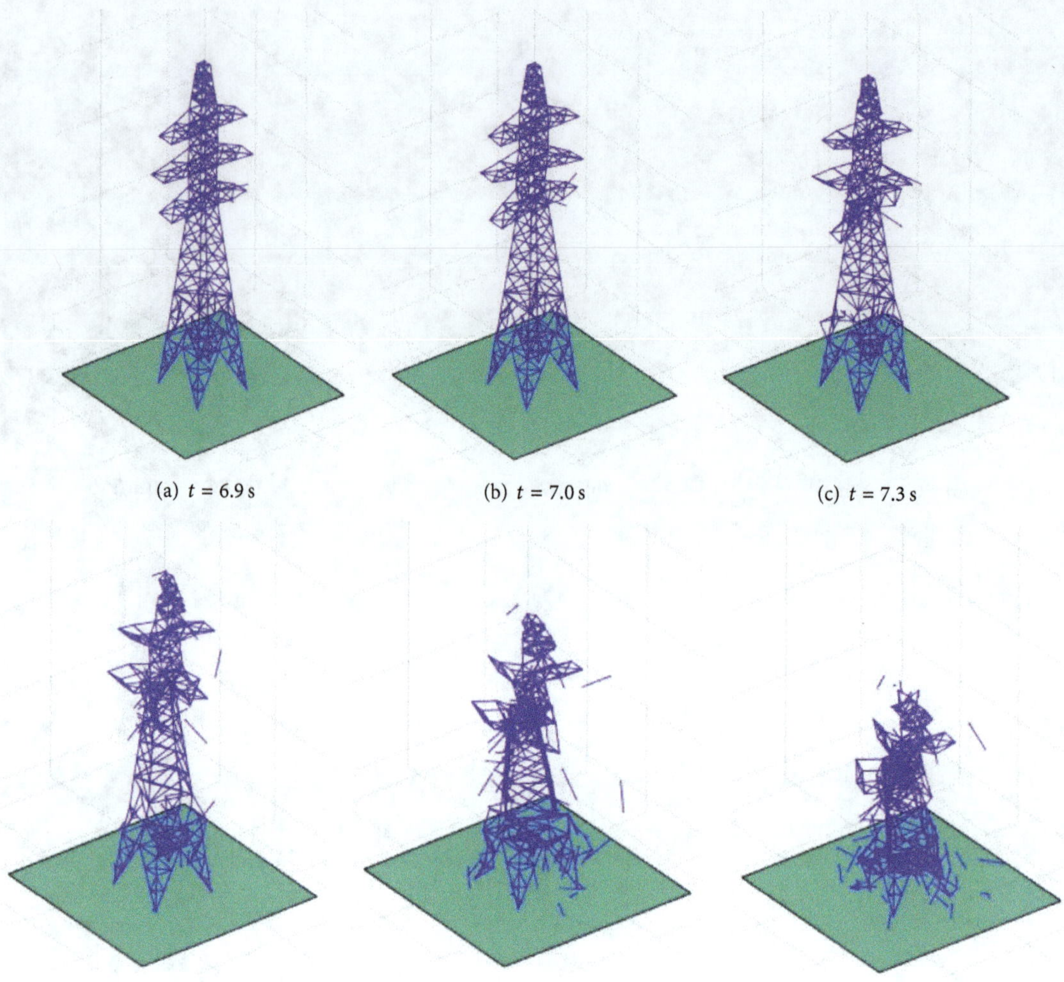

(a) $t = 6.9$ s (b) $t = 7.0$ s (c) $t = 7.3$ s (d) $t = 7.6$ s (e) $t = 8.0$ s (f) $t = 8.3$ s

FIGURE 18: Failure process of the transmission tower at different time.

strength degradation, stiffness degradation, damage state, and ductility of the structure under earthquake action. Displacement is easy to apply and quantitatively describe the ultimate state of the structure under earthquake action [28]. In this paper, the maximum horizontal top displacement of the transmission tower $u_{\max}$ is selected as EDP. Combined with Chinese code, the top displacement of the tower should not exceed $3h/1000$ (h is the tower height) to ensure that the transmission steel tower is in serviceability limit states. Top displacement should not exceed $h/80$ to ensure that the transmission steel tower does not collapse [29].

(1) Parameter Analysis Method. For the performance-based seismic evaluation of the transmission tower, the structural earthquake collapse fragility analysis is defined as the conditional probability of the collapse of the structure under given earthquake intensity measure. The mathematical expression is

$$P_f = P\left(\text{Collapse} \mid \text{IM} = \text{im}\right), \quad (13)$$

where P_f is structural collapse probability, IM is the ground motion intensity measure, and PGA is taken as IM in this paper.

It is assumed that the PGA value of the structure collapses from the lognormal distribution; the probability of the collapse of the structure under a given PGA can be expressed as follows:

$$P_f = \Phi\left(\frac{\ln \text{PGA} - \widehat{\mu}_{\ln \text{PGA}|\text{Collapse}}}{\widehat{\sigma}_{\ln \text{PGA}|\text{Collapse}}}\right), \quad (14)$$

where $\Phi(\cdot)$ is the standard normal distribution function and $\widehat{\mu}_{\ln \text{PGA}|\text{Collapse}}$ and $\widehat{\sigma}_{\ln \text{PGA}|\text{Collapse}}$ are the mean and standard deviation of $\ln \text{PGA}_i$ for structural collapse, which can be calculated as follows, respectively:

$$\widehat{\mu}_{\ln \text{PGA}|\text{Collapse}} = \ln\left(\frac{\widehat{\mu}_{\text{PGA}|\text{Collapse}}}{\sqrt{\delta^2 + 1}}\right),$$

$$\widehat{\sigma}_{\ln \text{PGA}|\text{Collapse}} = \sqrt{\ln\left(\delta^2 + 1\right)},$$

Table 7: Characteristics of the earthquake ground motion records.

Number	Earthquake	Year	Magnitude	Station	PGA (g)
(1)	Imperial valley	1951	5.6	EI Centro-9	0.0294
(2)	Hollister	1961	5.6	Hollister City Hall	0.1210
(3)	Imperial valley-06	1979	6.53	Parachute Test Site	0.1661
(4)	Imperial valley-06	1979	6.53	Plaster City	0.0501
(5)	Loma Prieta	1989	6.9	Capitola	0.4803
(6)	Loma Prieta	1989	6.93	Coyote Lake Dam (Downst)	0.1718
(7)	Northridge-01	1994	6.69	Arcadia-Arcadia Av	0.0951
(8)	Kobe, Japan	1995	6.90	Abeno	0.2219
(9)	Kobe, Japan	1995	6.90	Amagasaki	0.3011
(10)	Kobe, Japan	1995	6.90	Fukushima	0.0337
(11)	Chi-Chi, Taiwan	1999	7.62	CHY047	0.1825
(12)	Chi-Chi, Taiwan	1999	7.62	CHY088	0.1796

$$\delta = \frac{\widehat{\sigma}_{\mathrm{PGA|Collapse}}}{\widehat{\mu}_{\mathrm{PGA|Collapse}}},$$

$$\widehat{\sigma}_{\ln \mathrm{PGA|Collapse}} = \frac{1}{n-1}\sum_{i=1}^{n}\left(\mathrm{PGA}_i - \widehat{\mu}_{\mathrm{PGA|Collapse}}\right)^2. \tag{15}$$

(2) Nonparametric Analysis Method. A group of ground motion records are selected as the input of the structural seismic collapse analysis (the total number of ground motion records is N_{total}). If there are N_{collapse} ground motion records which lead to the collapse of the structure, then the collapse probability of the structure is $N_{\mathrm{collapse}}/N_{\mathrm{total}}$ under the IM. When the number of ground motion records is sufficient and representative, the collapse probability of the structure can be quantitatively analyzed. The process of collapse seismic fragility analysis of transmission tower based on IDA method is summarized as follows:

(1) The numerical model is established, which can accurately simulate earthquake collapse behavior.

(2) Select a group of ground motion records (the total number is N_{total}), which can reflect the ground motion characteristics of the site, and PGA is selected as IM.

(3) The elastic-plastic time history analysis of the structure is carried out under a certain IM, and the corresponding collapse number of ground motions N_{collapse} can be obtained; finally, the collapse probability of the structure under the IM is expressed as $N_{\mathrm{collapse}}/N_{\mathrm{total}}$.

(4) Monotonically increasing IM and repeating step (3), the collapse probability of the structure under different IMs can be calculated; finally, the structural collapse seismic fragility curve can be obtained.

(5) According to the structural collapse seismic fragility curve, the anticollapse safety and the corresponding reliability of the structure can be evaluated under different earthquake fortification levels.

(3) Collapse Seismic Fragility Analysis. In this paper, according to different earthquake sites and seismic intensity, 12 earthquake ground motion records are selected from the United States PEER strong earthquake record database (Table 7). Figure 20 shows the seismic acceleration response spectrum of record waves with the damping ratio is 2%. It can be seen that the response spectrum of the selected record waves is consistent with that of Chinese code [30]. Thus, 12 earthquake ground motion records are well representative.

The earthquake ground motion is input horizontally along the direction of the transmission tower. By adjusting the PGA amplitude of 12 seismic waves in Table 1, the PGA is adjusted from 0.1 g to 2.5 g and the acceleration increment is 0.1 g, and the PGA of each seismic wave is gradually increased until the transmission tower collapses (the top displacement of tower reaches to $h/80$); substituting the calculated values $\widehat{\mu}_{\ln \mathrm{PGA|Collapse}}$ and $\widehat{\sigma}_{\ln \mathrm{PGA|Collapse}}$ into (14), the collapse fragility curve of the transmission tower can be obtained.

It can be seen from Figure 21 that the seismic collapse probability of the transmission tower increases with the increase of PGA. And the parametric method agrees well with the nonparametric method. According to the Chinese seismic code [30] and seismic fortification standards for transmission towers, the PGA values are 0.11 g, 0.3 g, and 0.51 g under the action of minor earthquakes (exceeding probability is 50%), moderate earthquake (exceeding probability is 10%), and major earthquake (exceeding probability is 2%), respectively. When the collapse probability of the structure is 50%, the corresponding PGA value is 1.43 g using the nonparametric method, while the PGA value is 1.49 g by the parametric method. This value is about 3 times PGA value of major earthquake (the PGA value is 0.3 g). The result indicates that the transmission steel tower has better seismic safety performance and anticollapse ability.

5. Conclusions

This study presents the A general finite particle method framework for collapse simulation of the transmission tower subjected to earthquake ground motions. The FPM is used to establish motion equations of bar elements. Static and dynamic analyses of the transmission tower are conducted using software package MATLAB and ANSYS. Furthermore, failure criteria and the failure modes of structural elements

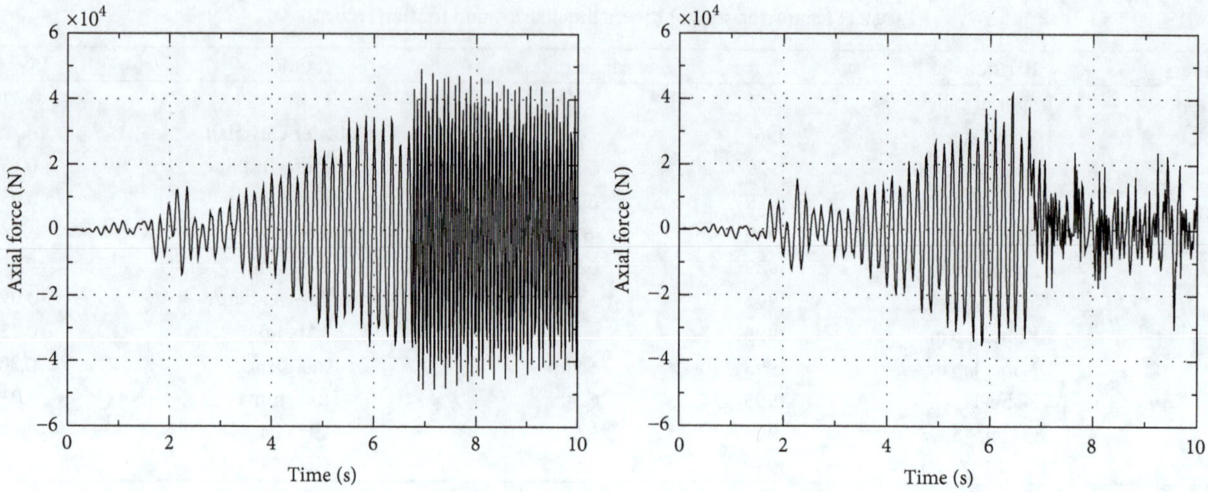

(a) Bar element 139 (b) Bar element 137

FIGURE 19: Axial force time history curves of bar elements 139 and 137.

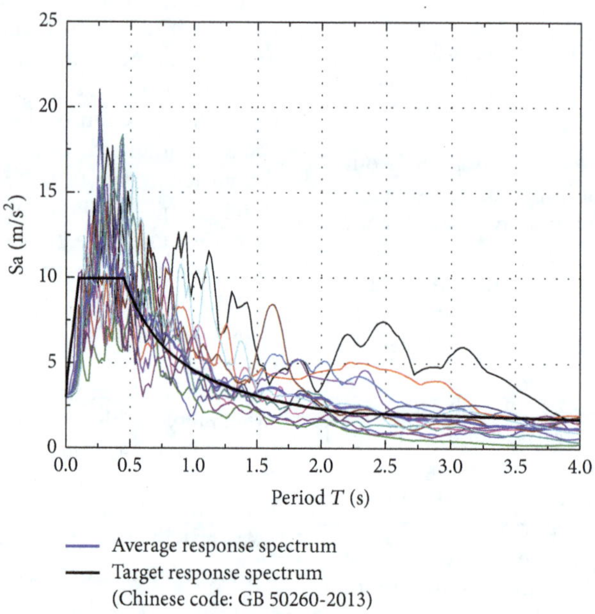

FIGURE 20: Seismic response spectrum.

are defined. Finally, the process of the transmission tower collapse is simulated. Based on the analysis the following conclusions can be drawn:

(i) The FPM does not need to generate the global stiffness matrix when calculating fracture of members and does not require iterative solution and special amendments in the calculation process; only the number of equations of motion for particles increases. Therefore, the FPM is suitable for solving strong nonlinear dynamic problems such as the seismic collapse analysis of transmission steel tower.

(ii) The model of the transmission tower is established using FEM and FPM, respectively. The correctness of the transmission tower model and the reliability of linear seismic response analysis are verified by comparing the analytical results of the FPM with those of the FEM.

(iii) The material nonlinearity, geometric nonlinearityn and fracture of the truss structure can be well considered in FPM. The collapse processes of the transmission tower are discussed under different PGAs, and the corresponding collapse modes are obtained. The collapse modes will provide a reference to the anticollapse design of the transmission tower subjected to earthquake ground motions.

(iv) The collapse seismic fragility analysis indicates that the parametric method agrees well with the nonparametric method, and seismic collapse probability of

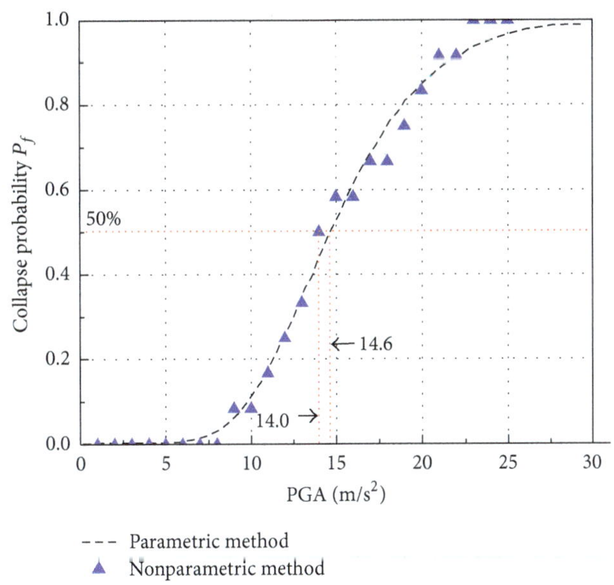

FIGURE 21: Collapse seismic fragility curves.

the transmission steel tower can meet the demand of Chinese seismic code.

The FPM is advantageous in the simulation of structural failure because particles are free to separate from one another. The collapse processes of the single transmission tower are simulated under one-directional ground motion. Further research on transmission tower-line coupling system and multisupport earthquake excitations can be taken into account in the future. And the reliability of collapse analysis should be verified through experiments or practical failure engineering.

Conflicts of Interest

The authors declare that there are no conflicts of interest regarding the publication of this paper.

Acknowledgments

The authors would like to acknowledge the joint financial support by National Natural Science Fund of China (51278213, 51378234) and the Innovation Foundation of the Fundamental Research Funds for the Central Universities (HUST: 2016YXMS093).

References

[1] F. J. Hall, T. W. Holmes, and P. Somers, "Northridge earthquake, January 17, 1994," Preliminary reconnaissance report, 1994, Preliminary reconnaissance report.

[2] Shinozuka, M. "The Hanshin-Awaji earthquake of January 17, 1995 performance of lifelines," Technical Report NCEER. US National Center for Earthquake Engineering Research (NCEER), 1995, 95.

[3] C. Scawthorn, "The Marmara, Turkey earthquake of August 17, 1999," Reconnaissance Report. Technical Report MCEER, US Multidisciplinary Center for Earthquake Engineering Research, 2000.

[4] Z. Zhang, B. Zhao, and W. Cao, "Investigation and preliminary analysis of damages on the power grid in the Wenchuan earthquake of M8.0," *Electric Power Technologic Economics*, vol. 20, no. 4, pp. 1–4, 2008.

[5] Y. Yu, G. Li, P. Li et al., "Investigation and analysis of electric equipment damage in Sichuan power grid caused by Wenchuan earthquake," *Power System Technology*, vol. 32, no. 11, pp. 1–6, 2008.

[6] D. Zhang, W. Zhao, and M. Liu, "Analysis on seismic disaster damage cases and their causes of electric power equipment in 5·12 Wenchuan earthquake," *Journal of Nanjing University of Technology (National Science Edition)*, vol. 31, no. 1, pp. 44–48, 2009.

[7] L. Kempner, S. Smith, and R. C. Stroud, "Transmission line dynamic/static structural testing," *Journal of the Structural Division*, vol. 107, no. 10, pp. 1895–1906, 1981.

[8] S. L. Chan and S. Kitipornchai, "Geometric nonlinear analysis of asymmetric thin-walled beam-columns," *Engineering Structures*, vol. 9, no. 4, pp. 243–254, 1987.

[9] S. Kitipornchai, K. Zhu, Y. Xiang, and F. G. A. Al-Bermani, "Single-equation yield surfaces for monosymmetric and asymmetric sections," *Engineering Structures*, vol. 13, no. 4, pp. 366–370, 1991.

[10] J. L. Meek and W. J. Lin, "Geometric and material nonlinear analysis of thin-walled beam-columns," *Journal of Structural Engineering (United States)*, vol. 116, no. 6, pp. 1473–1490, 1990.

[11] L. J. Meek and S. Loganathan, "Geometric and material nonlinear behavior of beam-columns," *Computers Structures*, vol. 34, no. 1, pp. 87–100, 1990.

[12] H. Yasui, H. Marukawa, Y. Momomura, and T. Ohkuma, "Analytical study on wind-induced vibration of power transmission towers," *Journal of Wind Engineering & Industrial Aerodynamics*, vol. 83, pp. 431–441, 1999.

[13] Q. Lin, W. Wang, D. Cao et al., "Progressive collapse analysis of high-voltage transmission tower under earthquake action," *Sichuan Building Science*, vol. 40, no. 3, pp. 157–161, 2014.

[14] Z. J. Zhang, G. H. Paulino, and W. Celes, "Extrinsic cohesive modelling of dynamic fracture and microbranching instability in brittle materials," *International Journal for Numerical Methods in Engineering*, vol. 72, no. 8, pp. 893–923, 2007.

[15] X. Lu, X. Lin, and L. Ye, "Simulation of structural collapse with coupled finite element-discrete element method," in *Computational Structural Engineering*, pp. 127–135, Springer, Netherlands, 2009.

[16] Y. Luo and C. Yang, "The finite particle method for solving geometric large deformation of planar solids," *Engineering Mechanics*, vol. 30, no. 4, pp. 260–268, 2013.

[17] W.-L. Jin and T. Fang, "Numerical simulation of failure behavior of reinforced concrete frame structures by discrete element method," *Gongcheng Lixue/Engineering Mechanics*, vol. 22, no. 4, pp. 67–73, 2005.

[18] F. Zhang and X. Lv, "Numerical simulation and analysis of different collapse patterns for RC frame structure," *Journal of Building Structures*, vol. 30, no. 5, pp. 119–125, 2009.

[19] E. C. Ting, C. Shih, and Y.-K. Wang, "Fundamentals of a vector form intrinsic finite element: Part I. Basic procedure and a plane frame element," *Journal of Mechanics*, vol. 20, no. 2, pp. 113–122, 2004.

[20] E. C. Ting, C. Shih, and Y.-K. Wang, "Fundamentals of a vector form intrinsic finite element: Part II. Plane solid elements," *Journal of Mechanics*, vol. 20, no. 2, pp. 123–132, 2004.

[21] W. J. Lewis, M. S. Jones, and K. R. Rushton, "Dynamic relaxation analysis of the non-linear static response of pretensioned cable roofs," *Computers & Structures*, vol. 18, no. 6, pp. 989–997, 1984.

[22] Y. Yu, G. H. Paulino, and Y. Luo, "Finite particle method for progressive failure simulation of truss structures," *Journal of Structural Engineering*, vol. 137, no. 10, pp. 1168–1181, 2011.

[23] C. D. Hill, G. E. Blandford, and S. T. Wang, "Post-buckling analysis of steel space trusses," *Journal of Structural Engineering (United States)*, vol. 115, no. 4, pp. 900–919, 1989.

[24] K. M. Lynn and D. Isobe, "Structural collapse analysis of framed structures under impact loads using ASI-Gauss finite element method," *International Journal of Impact Engineering*, vol. 34, no. 9, pp. 1500–1516, 2007.

[25] H.-D. Zheng, J. Fan, and X.-H. Long, "Analysis of the seismic collapse of a high-rise power transmission tower structure," *Journal of Constructional Steel Research*, vol. 134, pp. 180–193, 2017.

[26] M. Papadrakakis, "Post-buckling analysis of spatial structures by vector iteration methods," *Computers & Structures*, vol. 14, no. 5-6, pp. 393–402, 1981.

[27] W. J. Baker and C. A. Cornell, "Vector-valued ground motion intensity measures for probabilistic seismic demand analysis," PEER Report 10, 2006.

[28] M. Sasani and A. Der Kiureghian, "Seismic fragility of RC structural walls: displacement approach," *Journal of Structural Engineering*, vol. 127, no. 2, pp. 219–228, 2001.

[29] DL/T 5154-2012, *Technical code for design of tower and pole structures of overhead transmission line*, China Planning Press, Beijing, China, 2012.

[30] GB 50260-2013, *Code for Seismic Design of Power Installations*, China Planning Press, China, 2013.

11

Interaction of Submerged Breakwater by a Solitary Wave Using WC-SPH Method

Afshin Mansouri[1] **and Babak Aminnejad**[2]

[1] *Departmentof Civil Engineering, The Islamic Azad University of Roudehen Branch, Tehran, Iran*
[2] *Department of Civil Engineering, Faculty of Civil Engineering, The Islamic Azad University of Roudehen Branch, Tehran, Iran*

Correspondence should be addressed to Afshin Mansouri; afshinmansoori@gmail.com

Academic Editor: Agostino Bruzzone

Interaction of a solitary wave and submerged breakwater is studied in a meshless, Lagrangian approach. For this purpose, a two-dimensional smoothed particle hydrodynamics (SPH) code is developed. Furthermore, an extensive set of simulations is conducted. In the first step, the generated solitary wave is validated. Subsequently, the interaction of solitary wave and submerged breakwater is investigated thoroughly. Results of the interaction of solitary wave and a submerged breakwater are also shown to be in good agreement with published experimental studies. Afterwards, the effects of the inclination and length of breakwater as well as distance between two breakwaters are evaluated on damping ratio of breakwater.

1. Introduction

Due to development of ports and marine, protection of the seashores against the sea waves must be considered as an important issue in the field of marine structure. In this context, breakwaters have been used as appropriate tools in previous decades. A wide range of breakwaters has been introduced, so far. Submerged breakwater is one of them which has been utilized in many countries. Generally, some advantages of submerged breakwater can be mentioned as follows:

(1) simple structure,

(2) cost effectiveness,

(3) safety of marine transport vessels,

(4) short construction period,

(5) preserving beauty of the coast (touristy aspect),

(6) high strength of the structures against sea waves,

(7) preventing coastal erosion area with scattering of wave energy.

In addition, the performance of submerged breakwaters depends on geometry, material properties, and environmental characteristics of surrounding breakwater [1]. In the following paragraphs, several studies dealing with the submerged breakwaters are reviewed.

Effects of overtopping submerged obstacle on wave characteristics were investigated by Hara et al. [2]. They used a numerical model as well as a regression method to calculate the rate of the wave breaking. Hayakawa et al. [3] also studied flow field around a trapezoidal submerged breakwater numerically. They compared the obtained 2D and 3D numerical results with an experimental finding. Solitary wave interaction with submerged multibodies was considered by Wan Decheng and Guoxiong [4] theoretically. An experimental investigation of the wave transmission on submerged breakwaters was also considered by Armono and Hall [5]. An experimental comparison of wave field behind a submerged trapezoidal breakwater and double submerged trapezoidal breakwater was performed by Yoshida et al. [6]. In their study, the ratio between breakwaters and the wave height has been studied. Interaction of a solitary wave and permeable rectangular submerged breakwater had been evaluated by Huang et al. [7]. They measured the wave transmission as well as the amount of dragon the breakwater. Stamos et al. [8] performed an experimental study and considered performance of absorption and reflection of waves from hemicylindrical and rectangular submerged breakwaters. Johnson et al. [9] conducted an experimental and theoretical study of currents and waves around trapezoidal submerged breakwater. Muni-Reddy and Neelamani [10] presented a numerical and experimental investigation for measurement of wave breaking and force on a protected offshore base with trapezoidal breakwater. Chen et al. [11] experimentally studied transformation of waves with various profiles, between a trapezoidal submerged breakwater and sea wall. Christou et al. [12] presented numerical simulation of regular nonlinear waves interaction with rectangular submerged breakwaters for investigation of transmitted and reflected waves profile. Zaman et al. [13] investigated the amount of deformation of waves passing over the parabolic breakwater and the extent of waves reflected area. Losada et al. [14] investigated transmission, reflection, turbulence, and breakage of regular and irregular waves overtopping of rubble mound breakwaters. Cao et al. [15] numerically studied hydrodynamic properties of wave propagation which surrounded two impermeable trapezoidal submerged breakwaters.

In the area of breakwater computations, smoothed particle hydrodynamics (SPH) can also be implemented as a novel meshless numerical method that is being developed for the study of near shore waves and navy needs. In this context, several numerical models have been developed with SPH method for modeling free surface flows. Dalrymple et al. [16], Gotoh et al. [17], and Shao [18] have presented some numerical techniques in conjunction with SPH which have been widely used. Furthermore, Dalrymple and Rogers [19] showed that SPH method offers a variety of advantages for fluid modeling, especially those with a free surface. Yim et al. [20] investigated wave's interaction with porous rigid semisubmerged rectangular breakwater with RANS and SPH methods. Cherfils et al. [21] also considered flow surrounding a horizontal submerged plate as a breakwater near shores with SPH method. Didier and Neves [22] studied wave speed, wave height, and breakage of wave over sloped sea wall for a region of the Portuguese with SPH method.

Furthermore, to simulate water waves with SPH, many different approaches have been presented, so far. Shao [23] offered a coupled incompressible smoothed particle hydrodynamics (ISPH) porous medium method to simulate free surface profile. Liu et al. [24] optimized 3D SPH method for application of water wave modeling. Generally, interested reader can refer to some articles [25, 26] for finding review literature of SPH development.

Due to necessity of obtaining appropriate comparison between the different factors affecting the performance of submerged breakwater and lack of widespread study on submerged breakwaters with SPH method, a relatively extensive parametric study on trapezoidal submerged breakwaters is presented. For this purpose, various geometrical aspects of breakwater are considered and a comparative study is performed. Furthermore, it is being tried to introduce an appropriate combination of numerical techniques in SPH for accurate simulation of solitary wave-breakwaters interaction.

2. Governing Equations

In smooth particle hydrodynamics, fluid flow can be represented by N discrete particles. Each of these particles may be specified by an individual mass. Moreover, particles move according to the equations of Newtonian mechanics, including a pressure force. For each of these particles, the equations are based on Lagrangian form of the Navier-Stokes equations:

$$\frac{d\rho}{dt} = -\rho\nabla . v, \qquad \frac{dv}{dt} = -\frac{1}{\rho}\nabla p + g + \Pi, \tag{1}$$

where v, p, and ρ are the velocity, pressure, and density, respectively. g is the gravitational acceleration and Π refers to the diffusion terms.

3. SPH Formulation

Navier-Stokes equations solution in SPH method is by defining a kernel function $W(x - x', h)$ around every x' location of an SPH particle. The maximum of this function occurs at $x = x'$. The features of kernel function must be similar to Dirac Delta function. Therefore, behavior of kernel function is completely dependent on value of $r = x - x'$. The value h is the parameter which defines the "size" of influence area of kernel function and is called the smoothing length. The kernel is normalized to unity in the following way:

$$\int_{\Omega} W\left(x - x', h\right) dx' = 1, \tag{2}$$

and we also have

$$\lim_{h\to 0} W\left(x - x', h\right) = \delta\left(x - x'\right). \tag{3}$$

We should regard the SPH kernel as a defining region of influence of the SPH particle. It will turn out with only SPH particles which lie within each other's regions of influence and are able to interact with each other.

Furthermore, kernel function is implemented to approximate field variables at any point in a computational domain. For instance, an estimate value of a function $f(x)$ at the location x is given in a continuous form by an integral of the product of the function and a kernel function $W(x - x', h)$:

$$\langle f(x)\rangle = \int_{\Omega} f\left(x'\right) W\left(x - x', h\right) dx', \tag{4}$$

where the angle brackets $\langle\rangle$ denote a kernel approximation. In addition to the abovementioned characteristics, kernel function must satisfy several other features such as positivity condition [27]. Moreover, (4) can be also approximated by a summation as follows:

$$\langle f(x)\rangle \cong \sum_{j} \frac{m_j}{\rho_j} f\left(x_j\right) W\left(\left|x - x'\right|, h\right), \tag{5}$$

where m_j/ρ_j is the volume associated with the particle j. This equation can be used if the function $f(x)$ is only known at N discrete points.

The kernel function description can be finalized by defining a specific kernel function. Numerous possibilities exist. A large number of kernel function types are discussed in literature, ranging from polynomial to Gaussian. The most common is the B-spline kernel that was proposed by Monaghan and Lattanzio [28]:

$$w(r,h) = \alpha_D \begin{cases} 1 - \frac{3}{2}q^2 + \frac{3}{4}q^3 & 0 \le q \le 1 \\ \frac{1}{4}(2-q)^3 & 1 \le q \le 2 \\ 0 & q \ge 2 \end{cases} \tag{6}$$

which is considered in the present paper and $q = r/h$ and α_D is equal to $10/\pi h^2$ in two dimensional cases.

4. SPH Implementation on Governing Equations

Now, based on the SPH formulation, mentioned in previous section, governing equations will be discretized. In this context, the continuity equation for particle i can be considered in the SPH formulation as

$$\frac{d\rho_i}{dt} = -\sum_{j=1}^{N} m_j \left(v_j - v_j\right) \cdot \nabla W_{ij}, \tag{7}$$

where the sum extends over all neighboring particles and W is the smoothing kernel evaluated at the distance between particles i and j.

The SPH formulation of the momentum equations based on artificial viscosity can also be obtained through the particle approximation procedure as in

$$\frac{dv_i}{dt} = -\sum_{j=1}^{N} m_j \left(\frac{P_i}{\rho_i^2} + \frac{P_j}{P_j^2} + \Pi_{ij}\right) \nabla W_{ij} + g, \tag{8}$$

where Π_{ij} is the viscosity term. To introduce viscous dissipation, an artificial viscosity concept which has been developed using Monaghan [29] is implemented. This term is symmetric, conserves momentum, and introduces a shear viscosity into the momentum equation. The SPH form of the momentum equation is written as

$$\Pi_{ab} = \begin{cases} \frac{-\alpha \overline{C_{ab}} \mu_{ab}}{\overline{\rho_{ab}}} V_{ab} \cdot r_{ab} < 0, \\ 0, \text{ for other values,} \end{cases} \tag{9}$$

where

$$\mu_{ab} = \frac{h v_{ab} \cdot r_{ab}}{r_{ab}^2 + \eta^2}, \tag{10}$$

$$\overline{\rho}_{ab} = \frac{1}{2}\left(\rho_a + \rho_b\right), \qquad \overline{C}_{ab} = \frac{1}{2}\left(C_a + C_b\right); \qquad \eta^2 = 0.01h^2. \tag{11}$$

α is also a constant. Monaghan [29] suggests that a value of $\alpha = 0.01$ can be used for most computations.

5. Numerical Details

5.1. XSPH Technique. To modify the particle's velocity, XSPH technique is used and particle i is moved based on this technique. In XSPH scheme, it is formulated as follows:

$$\frac{dr_i}{dt} = v_i + \varepsilon \sum_j \frac{m_j}{\overline{\rho}_{ij}} v_{ij} W_{ij}. \tag{12}$$

Velocity of particle i is recalculated in such manner that the velocity of particle i and the average velocity of the closest neighborhood particles interacting with particle i are taken into account.

5.2. Time Stepping Procedure. In the present paper, predictor-corrector time stepping algorithm is implemented. In the predictor-corrector formulation, variation of particles properties (such as velocity) is estimated as [17]. Consider

$$v_i^{n+1/2} = v_i^n + \frac{\Delta t}{2} F_i^{n+1/2}. \tag{13}$$

And by implementing forces at half step, the values of velocity can be calculated at the end of any step which is formulated as follows [17]:

$$v_i^{n+1} = 2v_i^{n+1/2} - v_i^n. \tag{14}$$

5.3. Density Filter. In weakly compressible SPH formulation, large pressure oscillation has been observed. These pressure inaccuracies can be modified using some procedures. In addition to pressure modification, the free surface profile will be also improved. Therefore, MLS (moving least square) density filter which is a first order density correction is used to reproduce the linear variation of the density field:

$$\overline{\rho}_a = \sum_b \rho_b W_{ab}^{\mathrm{MLS}} \frac{m_b}{\rho_b} = \sum_b m_b W_{ab}^{\mathrm{MLS}}. \tag{15}$$

The corrected kernel is also evaluated as follows:

$$W_{ab}^{\mathrm{MLS}} = W_{ab}^{\mathrm{MLS}}\left(r_a\right) = \beta\left(r_a\right) \cdot \left(r_a - r_b\right) W_{ab}. \tag{16}$$

More details can be found in many references in the literature [30, 31].

5.4. Boundary Conditions. Two rows of stationary water particles at the boundary (Figure 1) remain fixed or move according to an externally imposed function for, say, wave maker boundaries. When a fluid particle approaches the boundary, the density and hence pressure exerted by the boundary particles increase. This generates the necessary repulsive force on the water particles. A major deficiency of this method is the generation of numerical boundary layers. When water particles get close to the boundary particles, they tend to stick to the boundary; thus they generate a boundary layer which is not physically observed.

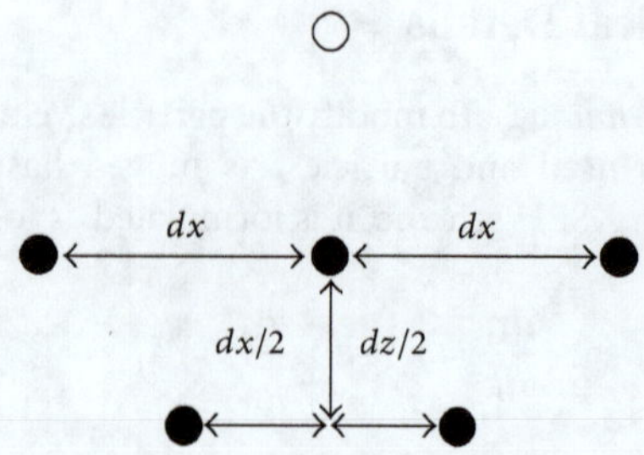

FIGURE 1: Interaction between fluid particles and boundary particles.

6. Solitary Wave Generation

Solitary wave is generated by Prescribe Piston Wave Maker [5]. Value of paddle displacement is applied in the following equation [22]:

$$X_p(t) = \frac{H}{k}\left[\tanh(X(t)) + \tanh\left(\frac{k}{d}\lambda\right)\right];$$
$$X(t) = \frac{k}{d}\left(ct - X_p(t) - \lambda\right), \tag{17}$$

where H is the wave height, $k = \sqrt{(3H/4d)}$ is wave number, and $c = \sqrt{(g(d+H))}$ is speed. Wave speed can also be obtained from the following equation [22]:

$$u_p(t) = \frac{cH}{d}\frac{1}{\cosh^2 X(t) + (H/d)}. \tag{18}$$

6.1. Validation of Generated Solitary Wave. According to Maiti and Sen [32] simulation which is a benchmark test, the length and flat length of the tank, water depth, and slope of the right hand side of the tank were, respectively, considered to be 10 m, 9.7 m, 0.3 m, and 45 degree (Figure 2). The motion of wave paddle is calculated for $H/d = 0.1$. Therefore, wave height and λ are 0.03 and 6.882.

The result of solitary wave generation is illustrated in Figure 3. Simulation is conducted from 0 till 10 sec. It is observed that the wave height is equal to 0.03 m and obtained free surface profile is relatively accurate for further studies. However, some inconsistencies are visible at second 4. In addition, average error of SPH simulation (point to point) is equal to 1.89 percent.

7. Results and Discussion

In this section, various aspects of solitary wave-breakwater interaction will be studied. At first step, free surface related to a specific breakwater under interaction with a solitary wave is computed using SPH to ensure that our numerical setup is suitable for investigation of more complicated problems. Afterward, various geometrical characteristics of breakwaters are also discussed. For more detailed studies, two breakwaters are considered and effects of distance between these breakwaters will be evaluated.

7.1. Validation. For further evaluation of SPH accuracy, the free surface profile, generated in solitary wave-breakwater

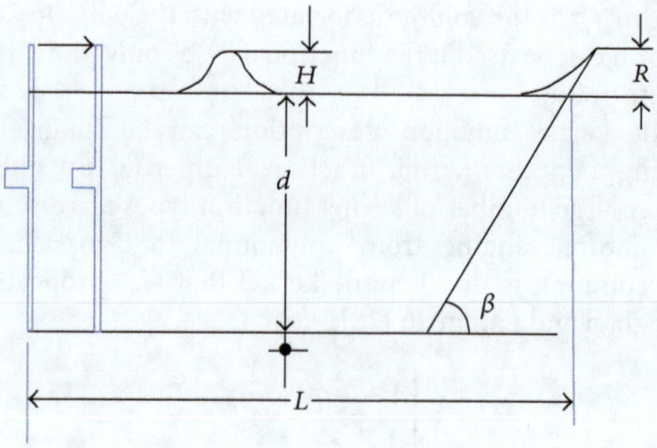

FIGURE 2: Computational domain.

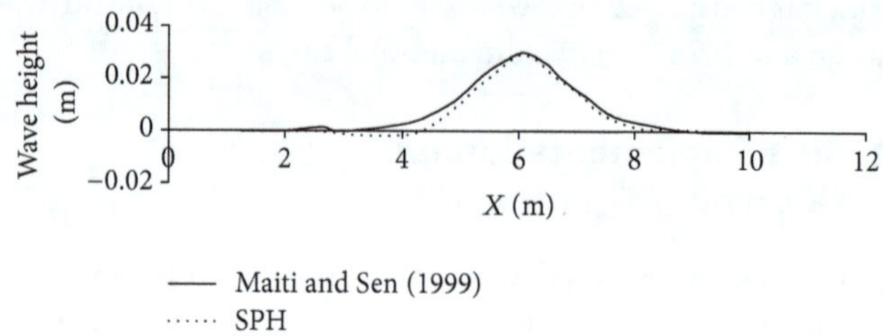

FIGURE 3: Comparison of generated solitary wave with Sagnita's results.

TABLE 1: Wave flume, solitary wave, and breakwater features.

Value	Characteristics
0.3 m	Wave height (H)
1.0 m	Water depth (d)
45 m	(L_0)
70 m	(L_{overal})
0.8 m	Height of breakwater (h)
0.4 m	Breakwater's beam
1 : 2	Slope of breakwater's wall

interaction, is compared with experimental work of Grilli et al. [33]. To perform this comparison, a computational domain is defined (Table 1). Furthermore, the characteristics of wave flume, solitary wave, and breakwater are presented. Results of this simulation are illustrated at Figure 4. Features of the presented simulation are also reported in Table 2. Free surface profiles in Figure 4(a) through 4(d) are at 10, 11.81, 13, and 13.92 sec, respectively. RMS of obtained solutions which are calculated relative to experimental findings seem to be reasonable. Therefore, it can be declared that our numerical setup can model solitary wave-breakwater interaction with satisfying precision.

7.2. Investigation of Breakwater Inclination. A subject studied here is considering the effects of breakwater's slope on H/H' ratio. For this purpose, computational domain as well as solitary wave is considered to be the same as previous section. Slope of breakwater, at aft part (β_2), which is shown in Figure 5 is varied systematically. All considered angles are

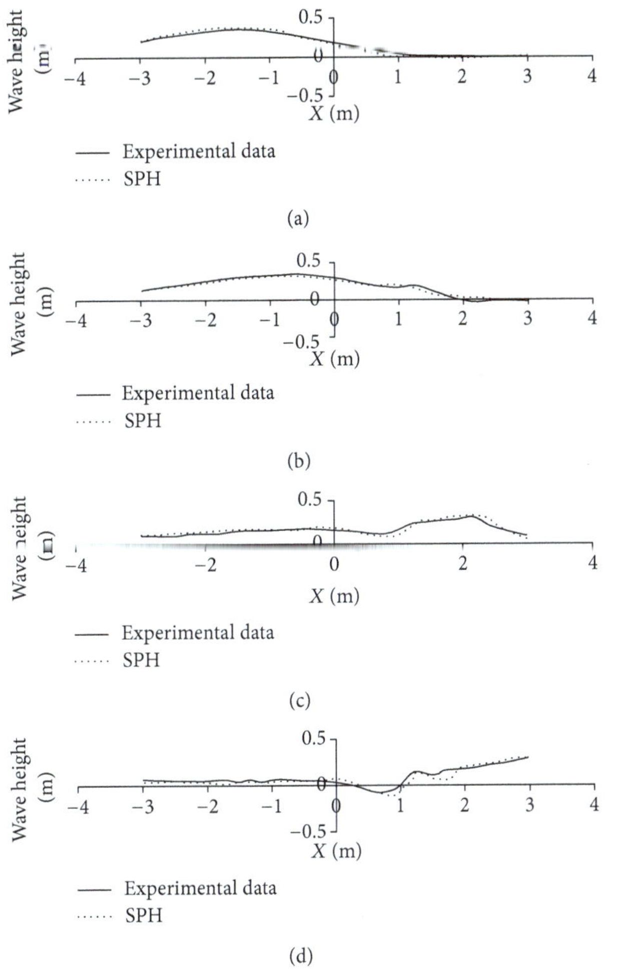

FIGURE 4: Comparison of SPH's free surface profile with experimental data [33].

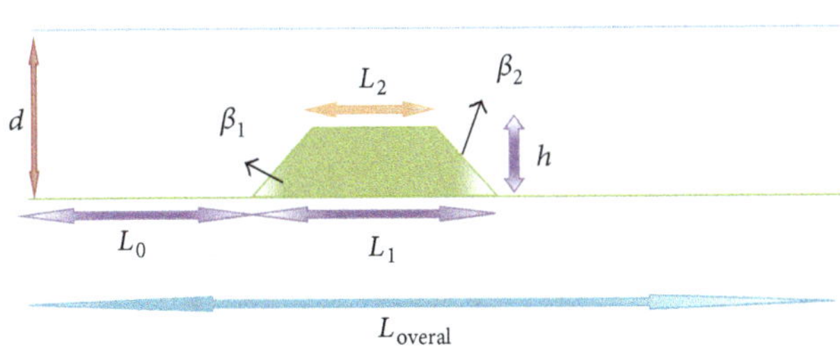

FIGURE 5: Computational domain.

TABLE 2: Computational error.

Time (sec)	Figure	RMS
10	Figure 4(a)	0.036
11.81	Figure 4(b)	0.034
13	Figure 4(c)	0.049
13.92	Figure 4(d)	0.053

presented in Table 3. Furthermore, geometrical characteristics of breakwater are also proposed in this table.

TABLE 3: Breakwater's characteristics.

β_2	β_1	h	L_1
0	$\tan^{-1}(0.5)$	0.4	2
15	$\tan^{-1}(0.5)$	0.4	2
45	$\tan^{-1}(0.5)$	0.4	2
60	$\tan^{-1}(0.5)$	0.4	2
75	$\tan^{-1}(0.5)$	0.4	2
90	$\tan^{-1}(0.5)$	0.4	2

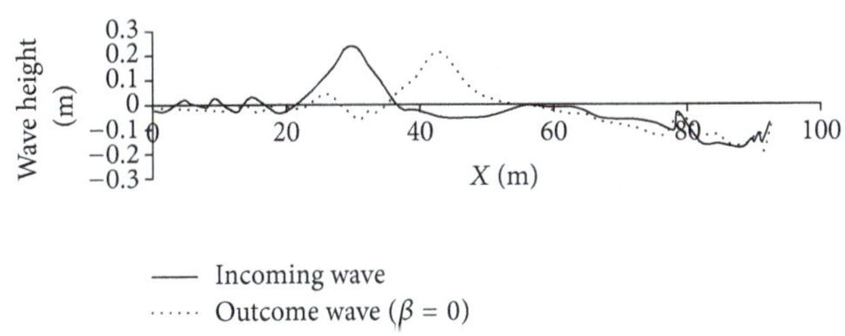

FIGURE 6: Free surface profile for $\beta_2 = 0$ degree.

Free surface profile related to each considered condition is illustrated in Figures 6, 7, 8, 9, 10, and 11. In each case, obtained free surface profile is compared with generated solitary wave without considering breakwater effect. Exact evaluation of breakwater's slope on the free surface profile is presented in Table 4. It is obvious that wherever $\beta_2 = 90$, maximum wave reduction can be achieved. Correspondingly, when β_2 decreased to zero, minimum breakwater effect occurs. Generally, Figure 12 also shows a representative frame of SPH solution.

7.3. Investigation of Breakwater Length. To perform further parametric studies, the effects of breakwater's length on free surface can also be evaluated. In this context, previous computational characteristics are kept and only the length ratio L_1/L_2 is changed systematically. Considered situations are presented in Table 5. Figures 13 and 16 show the obtained results at various length ratios. In all cases, breakwater generates a hollow in free surface at breakwater position which leads to a reduction in wave crest. Generally, based on Table 6, it is observed that L_1/L_2 ratio decreased, so breakwater performance will be improved. In addition, it can be concluded that smaller length ratio leads to more reduction of wave amplitude. To complete this section, a view of SPH solution is also presented in Figure 17.

7.4. Investigation of Distance between Two Submerged Breakwaters. In this part, influences of distance variation between two breakwaters are evaluated. Distance between two breakwaters (X) is considered to be less than wavelength at the beginning of simulations. Consequently, this distance will increase until the gap becomes larger than the wavelength. In addition, wave height, water depth, and length of wave flume are considered to be $H = 0.3$ m, 1.2 m, and 70 m, successively. Breakwater is located 30 meters from the left. Geometrical characteristics of breakwaters are also presented

TABLE 4: Effect of breakwater slope on H/H' ratio.

Values	$\beta_2 = 0$	$\beta_2 = 15$	$\beta_2 = 45$	$\beta_2 = 60$	$\beta_2 = 75$	$\beta_2 = 90$
H/H'	1.078973	1.185936	1.297549	1.32081	1.470258	1.503094

TABLE 5: Geometrical characteristics.

L_1/L_2	β_1	β_2	L_2	L_1
9	$\tan^{-1}(0.5)$	$\tan^{-1}(0.5)$	0.2	1.8
6.33	$\tan^{-1}(0.5)$	$\tan^{-1}(0.5)$	0.3	1.9
4.2	$\tan^{-1}(0.5)$	$\tan^{-1}(0.5)$	0.5	2.1
3.67	$\tan^{-1}(0.5)$	$\tan^{-1}(0.5)$	0.6	2.2

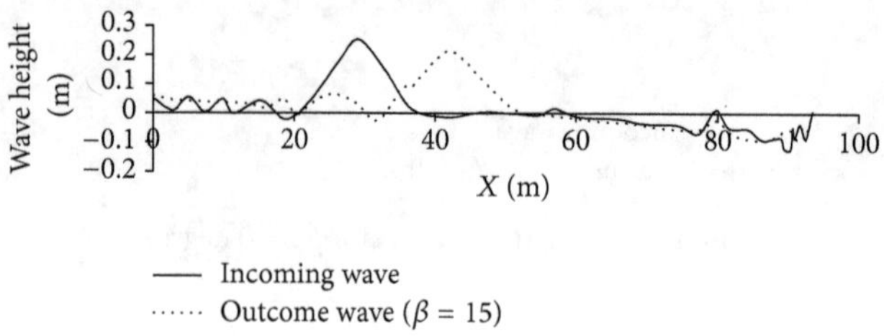

FIGURE 7: Free surface profile for $\beta_2 = 15$ degree.

in Table 7. Figure 18 shows the gap effects between two considered breakwaters. It is observed that when wave reaches the first breakwater, the wave height is equal to 0.25 m. However, second breakwater decreases the wave height in correspondence with considered distance between two breakwaters. In this context, two parameters, H (wave height above first breakwater) and H' (wave height above second breakwater), are introduced. Breakwater influence coefficient is also defined as H/H'. Precise values of defined parameter are presented in Table 8. It is obviously clear that, for $X = 0.3$ m, maximum wave reduction is achieved. Furthermore, it can be concluded that optimum gap between breakwaters is affected by wave length. In reality, when gap between breakwaters is smaller than wave length, effects of second breakwater will be decreased. Additionally, maximum effectiveness of breakwater is observed whenever the distance is equal to wave length. A representation of our SPH solution is also depicted in Figure 19.

8. Results

The present paper focuses on the breakwater performance under a solitary wave. Furthermore, we have tried to implement meshless and Lagrangian approach of smoothed particle hydrodynamics method to study considered physics. In this context, a two-dimensional SPH code is developed. To consider diffusion terms, an artificial viscosity model is included. Particles movement is also calculated using XSPH variant and predictor-corrector scheme which is investigated as time stepping algorithm. To reinitialize density field, MLS density filter is utilized.

To start computations, generated solitary wave is verified by comparing previous studies. Afterwards, a breakwater

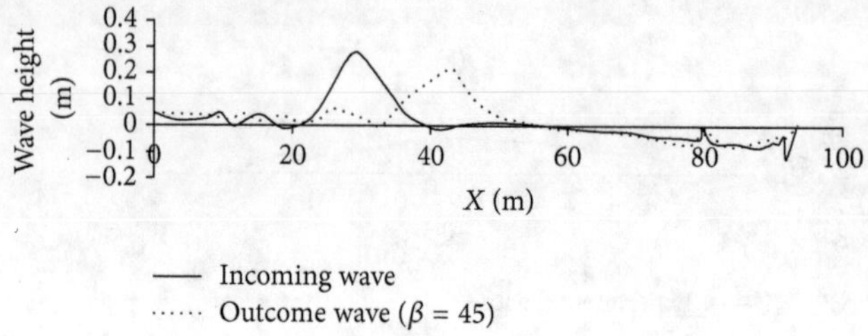

FIGURE 8: Free surface profile for $\beta_2 = 45$ degree.

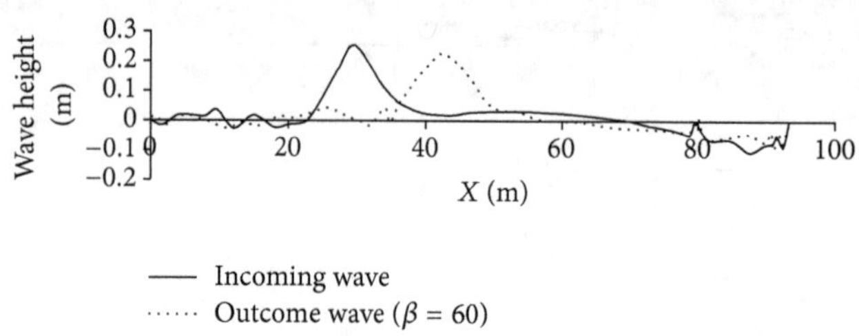

FIGURE 9: Free surface profile for $\beta_2 = 60$ degree.

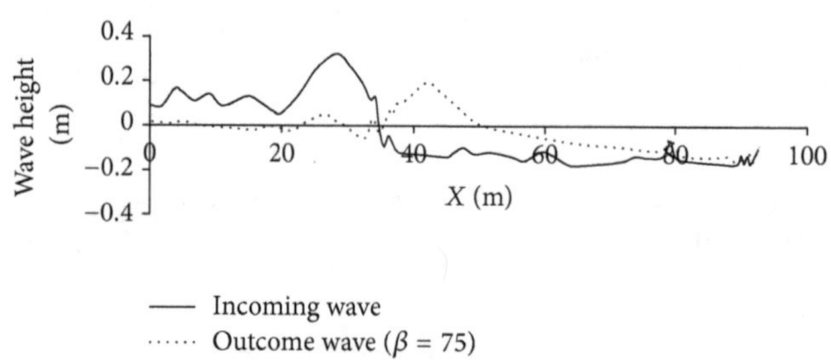

FIGURE 10: Free surface profile for $\beta_2 = 75$ degree.

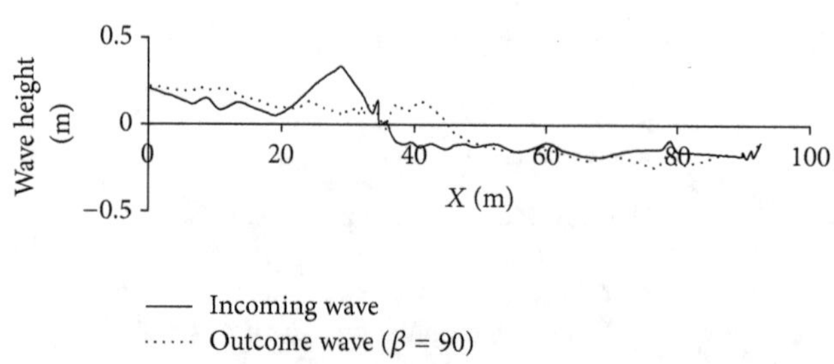

FIGURE 11: Free surface profile for $\beta_2 = 90$ degree.

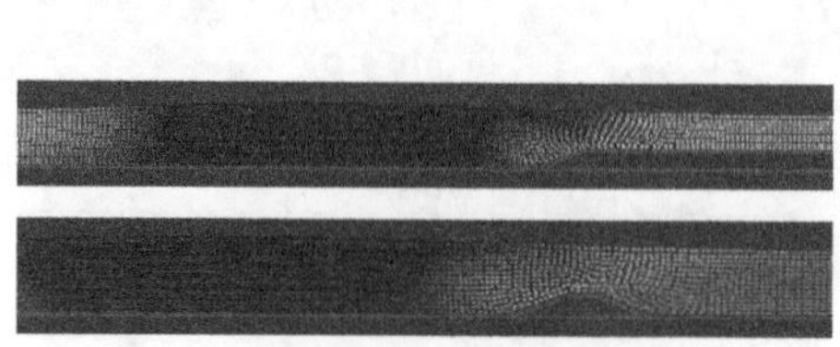

FIGURE 12: A representative frame of SPH solution.

TABLE 6: Exact evaluation of length ratio effect on free surface reduction.

Values	$L_1/L_2 = 9$	$L_1/L_2 = 6.33$	$L_1/L_2 = 4.2$	$L_1/L_2 = 3.67$
H/H'	1.094116	1.105263	1.111112	1.222223

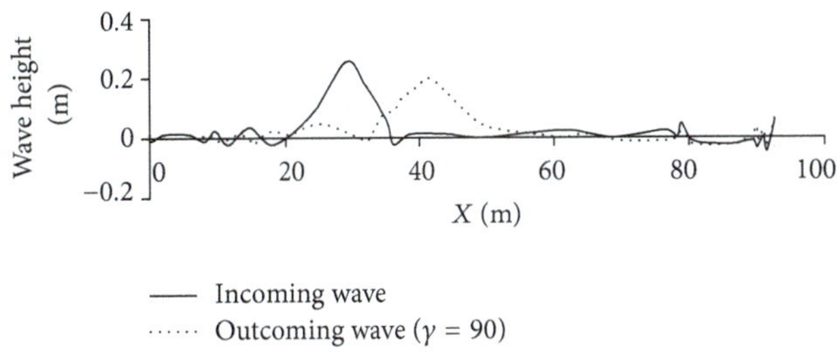

FIGURE 13: Free surface profile with ratio 9 in length.

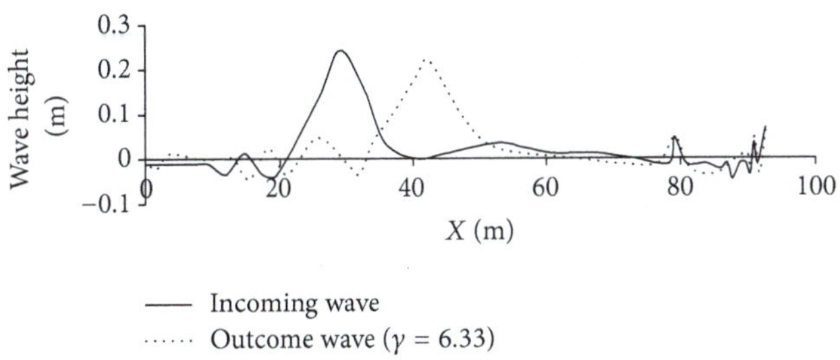

FIGURE 14: Free surface profile at length ratio 6.33.

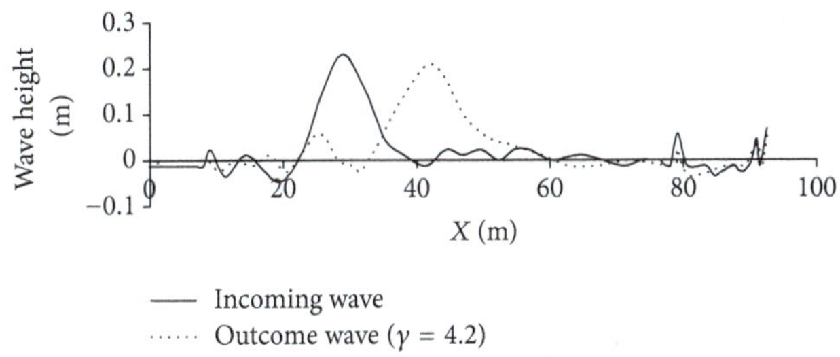

FIGURE 15: Free surface profile at length ratio 4.2.

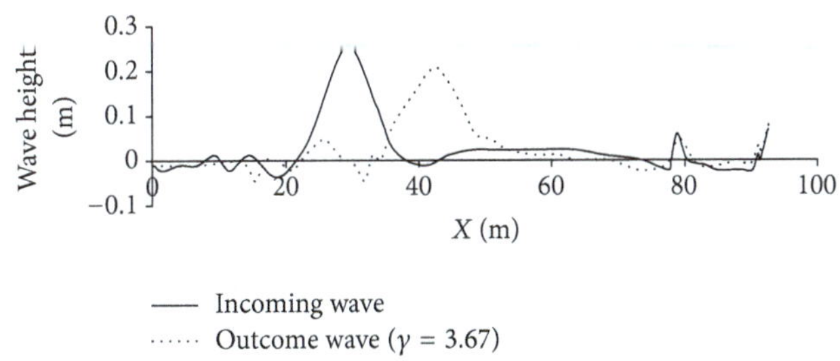

FIGURE 16: Free surface profile at length ratio 3.67.

TABLE 7: Breakwater characteristics.

(X)	Left side slope	Right side slope	Breakwater height	Lower length
0.2	$\tan^{-1}(0.5)$	$\tan^{-1}(0.5)$	0.4	2
0.3	$\tan^{-1}(0.5)$	$\tan^{-1}(0.5)$	0.4	2
0.4	$\tan^{-1}(0.5)$	$\tan^{-1}(0.5)$	0.4	2
0.5	$\tan^{-1}(0.5)$	$\tan^{-1}(0.5)$	0.4	2

TABLE 8: Effects of breakwaters distance.

Damping ratio	$X = 0.2$	$X = 0.3$	$X = 0.4$	$X = 0.5$
H/H'	1.247285	1.42718	1.3381	1.337985

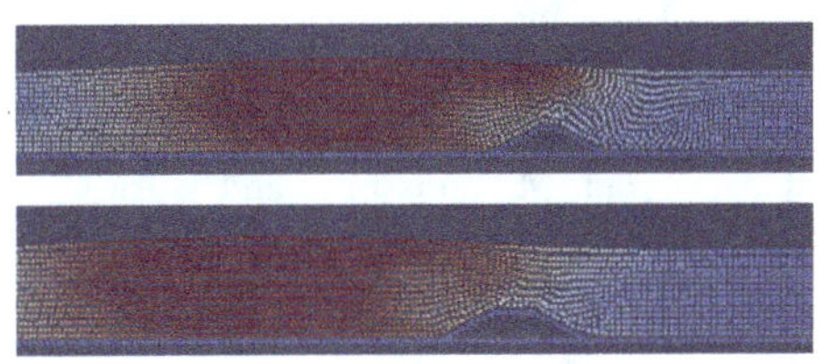

FIGURE 17: A view of SPH solution.

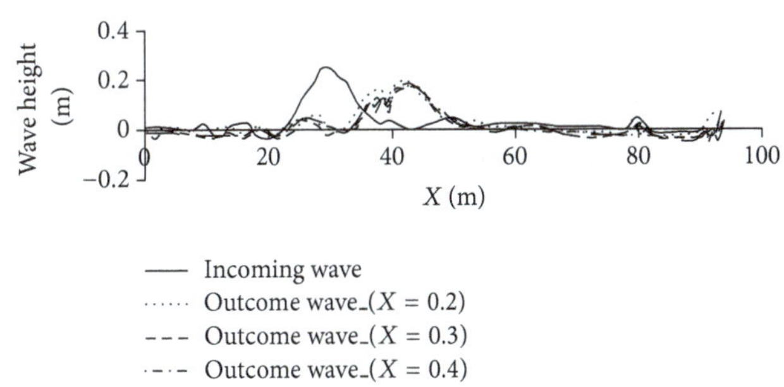

FIGURE 18: Simulated free surface profiles at different distances between breakwaters.

is also included in computational domain. To validate our SPH solutions, obtained free surface profile is compared with existing experimental results and it is observed that our SPH solutions have a reasonable accuracy. In the second phase, various geometrical aspects of breakwater are also studied.

Generally, effects of implementing two breakwaters, inclination of breakwater, and breakwater's length on free surface depression are evaluated. It is found that when gap between breakwaters is smaller than wave length, effects of second breakwater will be decreased. Maximum effectiveness of breakwater also occurred whenever the distance is equal to wave length. In addition, when the inclination of breakwater increased, free surface deformation will be modified. Furthermore, it is observed that length ratio of breakwater can effectively influence free surface profile (Figures 14 and 15).

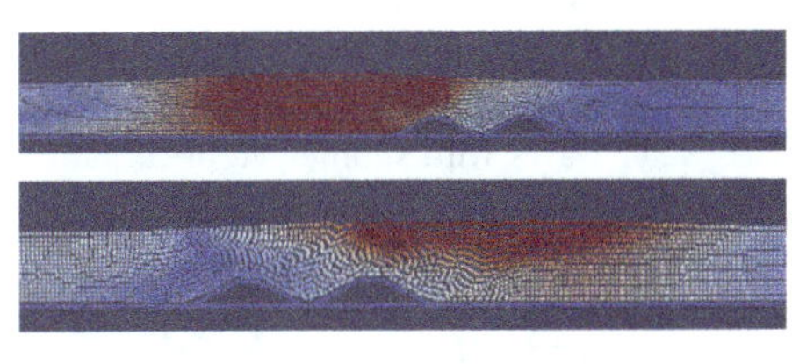

FIGURE 19: Particle representation of SPH solution.

To perform more precise computation, extending SPH solutions to three-dimensional case will be in attendance. Various simple breakwater geometries will also be introduced in near future.

Conflict of Interests

The authors declare that there is no conflict of interests regarding the publication of this paper.

References

[1] S. Tokura and T. Ida, "Simulation of wave dissipation mechanism on submerged structure using fluid-structure coupling capability in LS-DYNA," in *Proceedings of the 5th European LS-DYNA Conference*, 2005.

[2] M. Hara, T. Yasuda, and Y. Sakakibara, "Characteristics of a solitary wave breaking caused by a submerged obstacle," *Coastal Engineering*, vol. 1, no. 23, 1992.

[3] N. Hayakawa, T. Hosoyamada, S. Yoshida, and G. Tsujimoto, "Numerical simulation of wave fields around the submerged breakwater with SOLA-SURF method," in *Proceedings of the 26th International Conference on Coastal Engineering (ICCE '98)*, pp. 843–852, June 1998.

[4] W. Decheng and W. Guoxiong, "Numerical simulation of a solitary wave interaction with submerged multi-bodies," *Acta Mechanica Sinica*, vol. 14, no. 4, pp. 304–305, 1998.

[5] H. D. Armono and K. R. Hall, "Wave transmission on submerged breakwaters made of hollow hemispherical shape artificial reefs," in *Canadian Coastal Conference*, pp. 313–322, June 2003.

[6] A. Yoshida, S. Yan, M. Yamashiro, and I. Irie, "Wave Field Behind a Double-Submerged Breakwater," in *Proceedings of the 12th International Offshore and Polar Engineering Conference*, pp. 785–790, May 2002.

[7] C.-J. Huang, H.-H. Chang, and H.-H. Hwung, "Structural permeability effects on the interaction of a solitary wave and a submerged breakwater," *Coastal Engineering*, vol. 49, no. 1-2, pp. 1–24, 2003.

[8] D. G. Stamos, M. R. Hajj, and D. P. Telionis, "Performance of hemi-cylindrical and rectangular submerged breakwaters," *Ocean Engineering*, vol. 30, no. 6, pp. 813–828, 2003.

[9] H. K. Johnson, T. V. Karambas, I. Avgeris, B. Zanuttigh, D. Gonzalez-Marco, and I. Caceres, "Modelling of waves and currents around submerged breakwaters," *Coastal Engineering*, vol. 52, no. 10-11, pp. 949–969, 2005.

[10] M. G. Muni-Reddy and S. Neelamani, "Wave interaction with caisson defenced by an offshore low-crested breakwater," *Journal of Coastal Research*, no. 39, pp. 1767–1770, 2006.

[11] H.-B. Chen, C.-P. Tsai, and C.-C. Jeng, "Wave transformation between submerged breakwater and seawall," *Journal of Coastal Research*, no. 50, pp. 1069–1074, 2007.

[12] M. Christou, C. Swan, and O. T. Gudmestad, "The interaction of surface water waves with submerged breakwaters," *Coastal Engineering*, vol. 55, no. 12, pp. 945–958, 2008.

[13] M. H. Zaman, H. Togashi, and R. E. Baddour, "Deformation of monochromatic water wave trains propagating over a submerged obstacle in the presence of uniform currents," *Ocean Engineering*, vol. 35, no. 8-9, pp. 823–833, 2008.

[14] I. J. Losada, J. L. Lara, R. Guanche, and J. M. Gonzalez-Ondina, "Numerical analysis of wave overtopping of rubble mound breakwaters," *Coastal Engineering*, vol. 55, no. 1, pp. 47–62, 2008.

[15] Y.-G. Cao, C.-B. Jiang, and Y.-C. Bai, "Numerical study on flow structure near two impermeable trapezoid submerged breakwaters on slop bottoms," *Journal of Hydrodynamics*, vol. 22, no. 5, pp. 190–196, 2010.

[16] R. A. Dalrymple, O. Knio, D. T. Cox, M. Gesteira, and S. Zou, "Using a lagrangian particle method for deck overtopping," in *Proceedings of the 4th International Symposium Waves*, pp. 1082–1091, ASCE, September 2001.

[17] H. Gotoh, S. Shao, and T. Memita, "SPH-LES model for numerical investigation of wave interaction with partially immersed breakwater," *Coastal Engineering Journal*, vol. 46, no. 1, pp. 39–63, 2004.

[18] S. Shao, "Incrompressible SPH simulation of wave breaking and overtopping with turbulence modelling," *International Journal for Numerical Methods in Fluids*, vol. 50, no. 5, pp. 597–621, 2006.

[19] R. A. Dalrymple and B. D. Rogers, "Numerical modeling of water waves with the SPH method," *Coastal Engineering*, vol. 53, no. 2-3, pp. 141–147, 2006.

[20] S. C. Yim, D. Yuk, A. Panizzo, M. di Risio, and P. L.-F. Liu, "Numerical simulations of wave generation by a vertical plunger using RANS and SPH models," *Journal of Waterway, Port, Coastal and Ocean Engineering*, vol. 134, no. 3, pp. 143–159, 2008.

[21] J. M. Cherfils, L. Blonce, G. Pinon, and E. Rivoalen, "Simulation of water wave—coastal structure interactions by smoothed particle hydrodynamics," in *Proceedings of the 8th International Conference on Hydrodynamics*, Nantes, France, September 2008.

[22] E. Didier and M. G. Neves, "Wave over topping of a typical coastal structure of the portuguese coast using A SPH model," *Journal of Coastal Research*, vol. 56, pp. 496–500, 2009.

[23] S. Shao, "Incompressible SPH flow model for wave interactions with porous media," *Coastal Engineering*, vol. 57, no. 3, pp. 304–316, 2010.

[24] C. Liu, J. Zhang, and Y. Sun, "The optimization of SPH method and its application in simulation of water wave," in *Proceedings of the IEEE 7th International Conference on Natural Computation (ICNC '11)*, pp. 2327–2331, July 2011.

[25] P. W. Cleary, M. Prakash, J. Ha, N. Stokes, and C. Scott, "Smooth particle hydrodynamics: status and future potential," *Progress in Computational Fluid Dynamics*, vol. 7, no. 2-4, pp. 70–90, 2007.

[26] M. B. Liu and G. R. Liu, "Smoothed particle hydrodynamics (SPH): an overview and recent developments," *Archives of Computational Methods in Engineering*, vol. 17, no. 1, pp. 25–76, 2010.

[27] G. R. Liu and M. B. Liu, *Smoothed Particle Hydrodynamics—A Meshfree Particle Methods* , World Scientific Publishing, 2007.

[28] J. J. Monaghan and J. C. Lattanzio, "A refined particle method for astrophysical problems," *Astronomy and Astrophysics*, vol. 149, pp. 135–143, 1985.

[29] J. J. Monaghan, "Smoothed particle hydrodynamics," *Annual Review of Astronomy and Astrophysics*, vol. 30, no. 1, pp. 543–574, 1992.

[30] A. Colagrossi and M. Landrini, "Numerical simulation of interfacial flows by smoothed particle hydrodynamics," *Journal of Computational Physics*, vol. 191, no. 2, pp. 448–475, 2003.

[31] A. Crespo, *Application of smoothed particle hydrodynamics model SPHysics to free surface hydrodynamics [Doctoral thesis]*, 2008.

[32] S. Maiti and D. Sen, "Computation of solitary waves during propagation and runup on a slope," *Ocean Engineering*, vol. 26, no. 11, pp. 1063–1083, 1999.

[33] S. T. Grilli, M. A. Losada, and F. Martin, "Characteristics of solitary wave breaking induced by breakwaters," *Journal of Waterway, Port, Coastal & Ocean Engineering*, vol. 120, no. 1, pp. 74–92, 1994.

12

On Bending of Bernoulli-Euler Nanobeams for Nonlocal Composite Materials

Luciano Feo and Rosa Penna

Department of Civil Engineering, University of Salerno, Via Giovanni Paolo II, 132, 84084 Fisciano, Italy

Correspondence should be addressed to Luciano Feo; l.feo@unisa.it

Academic Editor: Theodoros C. Rousakis

Evaluation of size effects in functionally graded elastic nanobeams is carried out by making recourse to the nonlocal continuum mechanics. The Bernoulli-Euler kinematic assumption and the Eringen nonlocal constitutive law are assumed in the formulation of the elastic equilibrium problem. An innovative methodology, characterized by a lowering in the order of governing differential equation, is adopted in the present manuscript in order to solve the boundary value problem of a nanobeam under flexure. Unlike standard treatments, a second-order differential equation of nonlocal equilibrium elastic is integrated in terms of transverse displacements and equilibrated bending moments. Benchmark examples are developed, thus providing the nonlocality effect in nanocantilever and clampled-simply supported nanobeams for selected values of the Eringen scale parameter.

1. Introduction

Analysis of nanodevices is a subject of special interest in the current literature. Particular attention is given to the static behavior of beam-like components of nanoelectromechanical systems (NEMS). Nonlocal constitutive behaviors are adequate in order to evaluate the size phenomenon in nanostructures; see, for example, [1–9]. Investigations on random elastic structures have been carried out in [10–13]. Many research efforts have been devoted to theoretical and computational advances about specific structural models [14–19]. Recent variational formulations of nonlocal continua have been developed in [20–22]. Noteworthy theoretical results on functionally graded nanobeams have been contributed in [23–25]. Nevertheless, exact solutions are not always available so that finite element strategies are needful; see, for example, [26]. Micromechanical approaches are broadly used in order to analyze the effective behavior of composite structures [27].

Innovative applications of engineering interest are proposed in [28–30]. Numerical and experimental methodologies for composite structures are developed in [31, 32]. Effective applications of tensionless models concerning crack propagation are reported in [33–37]. A skillful analysis of equilibrium configurations of hyperelastic cylindrical bodies and compressible cubes is carried out in [38].

The present paper deals with one-dimensional nanostructure by making recourse to the tools of nonlocal continuum mechanics. Small-scale effects exhibited by functionally graded nanobeams under flexure are analyzed in Section 2.

2. Nonlocal Elasticity

In local linear elasticity for isotropic materials, stress and strain at a point $\mathbf{x}$ of a Cauchy continuum are functionally related by the following classical law:

$$\boldsymbol{\sigma}_{ij}(\mathbf{x}) = 2\mu\mathbf{e}_{ij}(\mathbf{x}) + \lambda\mathbf{e}_{rr}(\mathbf{x})\,\delta_{ij}, \tag{1}$$

with μ and λ Lamé constants.

Such a constitutive behavior is not adequate to evaluate size effects in nanostructures. An effective law able to capture scale phenomena was developed by Eringen in [39] who defined the following nonlocal integral operator:

$$\mathbf{t}_{ij}(\mathbf{x}) = \int_V K\left(\left|\mathbf{x}' - \mathbf{x}\right|, \tau\right)\boldsymbol{\sigma}_{ij}\left(\mathbf{x}'\right) dV, \tag{2}$$

where

(1) $\mathbf{t}_{ij}$ is the nonlocal stress,

(2) $\boldsymbol{\sigma}_{ij}$ is the macroscopic stress given by (5),

(3) K is the influence function,

(4) $\tau = e_0 a/l$ a dimensionless nonlocal parameter defined in terms of the material constant e_0 and of the internal and external characteristic lengths a and l, respectively.

In agreement with the Eringen proposal in choosing the following influence function $1-c^2\nabla^2$, the nonlocal elastic law (2) rewrites as

$$\left(1-(e_0 a)^2\nabla^2\right)\mathbf{t}_{ij} = \boldsymbol{\sigma}_{ij}, \tag{3}$$

where ∇^2 denotes the Laplace operator. The differential form adopted for bending of nanobeams, analogous to (3), is provided by

$$\overline{\sigma} - c^2\frac{d^2\overline{\sigma}}{dx^2} = \sigma, \tag{4}$$

where $\overline{\sigma}$ is the nonlocal normal stress and σ is the macroscopic normal stress on cross sections. Note that the stress σ is expressed in terms of elastic axial strains by

$$\sigma = E\varepsilon, \tag{5}$$

with E Young modulus.

3. Bending of Nonlocal Nanobeams

Let us consider a bent nanobeam of length L, with Young modulus E functionally graded in the cross section Ω and uniform along the beam axis x. The cross-sectional elastic centre and the principal axes of elastic inertia, associated with the scalar field E, are, respectively, denoted by $\mathbf{G}$ and by the pair (y, z).

The nanobeam is assumed to be subjected in the plane (x, y) to the following loading conditions:

q_t, distributed load per unit length in the interval $[0, L]$,

$\mathscr{F}_t$, concentrated forces at the end cross sections $\{0, L\}$,

$\mathscr{M}$, concentrated couples at the end cross sections $\{0, L\}$.

The bending stiffness is defined by

$$k_b := \int_\Omega E y^2 dA. \tag{6}$$

Differential and boundary conditions of equilibrium are expressed by

$$\begin{aligned} \frac{d^2 M_b}{dx^2} &= q_t, \quad \text{in } [0, L], \\ \frac{dM_b}{dx} &= \pm\mathscr{F}_t, \quad \text{at } \{0, L\}, \\ M_b &= \mp\mathscr{M}, \quad \text{at } \{0, L\}, \end{aligned} \tag{7}$$

where M_b is the bending moment.

The bending curvature, corresponding to the transverse displacement v, is given by

$$\chi(x) = \frac{d^2 v}{dx^2}(x). \tag{8}$$

The differential equation of nonlocal elastic equilibrium of a nanobeam under flexure is formulated as follows. Let us preliminarily multiply (4) by the coordinate y along the bending axis and integrate on the cross section Ω:

$$\int_\Omega \overline{\sigma} y\, dA - (e_0 a)^2 \int_\Omega \frac{d^2\overline{\sigma}}{dx^2} y\, dA = \int_\Omega E\varepsilon y\, dA, \tag{9}$$

with the axial dilation provided by the known formula $\varepsilon(x) = -\chi(x) y$.

Enforcing (8) and $(7)_1$ and imposing the static equivalence condition

$$M_b = -\int_\Omega \sigma y\, dA, \tag{10}$$

we obtain the relation

$$M_b - (e_0 a)^2 q_t = k_b \frac{d^2 v}{dx^2}. \tag{11}$$

This equation can be interpreted as decomposition formula of the bending curvature $\chi_b = d^2v/dx^2$ into elastic χ_{EL} and inelastic χ_{IN} parts

$$\chi_b = \chi_{\mathrm{EL}} + \chi_{\mathrm{IN}}, \tag{12}$$

with

$$\begin{aligned} \chi_{\mathrm{EL}} &= \frac{M_b}{k_b}, \\ \chi_{\mathrm{IN}} &= -\frac{(e_0 a)^2}{k_b} q_t. \end{aligned} \tag{13}$$

Accordingly, the scale effect exhibited by bending moments and displacements of a FG nonlocal nanobeam can be evaluated by solving a corresponding linearly elastic beam subjected to the bending curvature distortion χ_{IN} $(13)_2$.

4. Examples

The solution methodology of the nonlocal elastic equilibrium problem of a nanobeam enlightened in the previous section is here adopted in order to assess small-scale effects in nanocantilever and clamped-simply supported nanobeams under a uniformly distributed load q_t. The nonlocality effect on the transverse displacement is thus due to the uniform bending curvature distortion formulated in $(13)_2$. Graphical evidences of the elastic displacements are provided in Figures 1 and 2, in terms of the following dimensionless parameters $\xi = x/L$ and $v^*(\xi) = (100 k_b)/(q_t L^4) v(\xi)$, for selected values of the nonlocal parameter $\tau := e_0 a/L$. Details of the calculations and some comments are reported below.

4.1. Cantilever Nanobeam. The bending moment is given by

$$M_b(x) = \frac{q_t}{2}x^2 - q_t L x + \frac{q_t L^2}{2}. \tag{14}$$

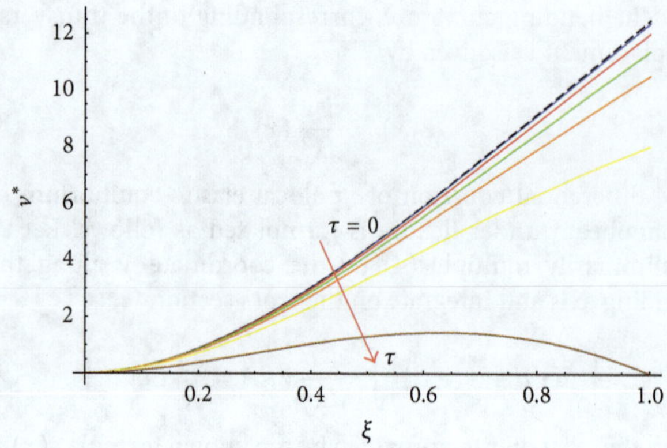

FIGURE 1: Dimensionless transverse displacements v^* for the nanocantilever versus ξ for $\tau \in \{0, 0.05, 0.1, 0.15, 0.2, 0.3, 0.5\}$.

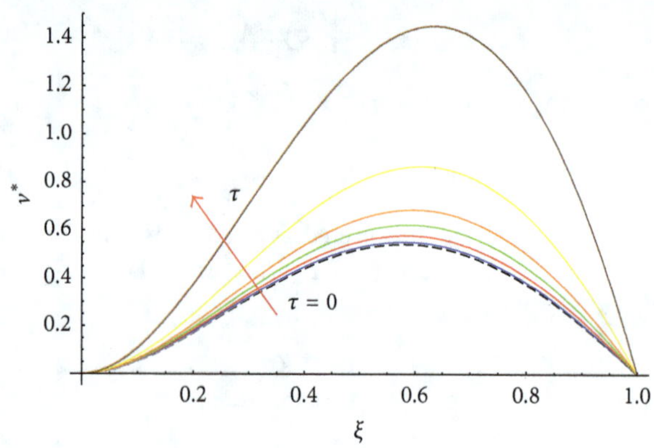

FIGURE 2: Dimensionless transverse displacement v^* for the clamped-simply supported nanobeam versus ξ for $\tau \in \{0, 0.05, 0.1, 0.15, 0.2, 0.3, 0.5\}$.

The l.h.s. of (11) is hence known, so that the differential condition of nonlocal elastic equilibrium to be integrated writes explicitly as

$$k_b \frac{d^2 v}{dx^2}(x) = M_b(x) - \tau^2 L^2 q_t. \tag{15}$$

The general integral of (15) takes the form

$$v(x) = ax + b + \overline{v}(x), \tag{16}$$

where

$$\overline{v}(x) = -\frac{q_t}{24k_b}x^4 + \frac{q_t L}{6k_b}x^3 - \frac{q_t L^2}{4}x^2 - \frac{\tau^2 L^2 q_t}{2k_b}x^2 \tag{17}$$

is a particular solution of (15). The evaluation of the integration constants a and b is carried out by prescribing the boundary conditions

$$v(0) = 0, \qquad \frac{dv}{dx}(0) = 0. \tag{18}$$

The transversal deflection follows by a direct computation

$$v(x) = \frac{q_t}{24k_b}x^4 - \frac{q_t L}{6k_b}x^3 + \frac{q_t L^2}{4k_b}x^2 - \frac{\tau^2 q_t L^2}{2k_b}x^2. \tag{19}$$

The maximum displacement is given by

$$v_m = v(L) = \left(\frac{1}{8} - \frac{\tau^2}{2}\right)\frac{q_t L^4}{k_b}. \tag{20}$$

Nanocantilever becomes stiffer with increasing the nonlocal parameter τ. Indeed, according to the analysis proposed in Section 3, the sign of the prescribed distortion χ_{IN}, describing the nonlocality effect, is opposite to the one of the elastic curvature χ_{EL}. In particular, the displacement of the free end vanishes if $\tau = 0.5$ (see Figure 1) according to (20).

It is worth noting that, with the structure being statically determinate, the bending moment (14) is not affected by the scale phenomenon.

4.2. Clamped-Simply Supported Nanobeam. Equilibrated bending moments are provided by the relation

$$M_b(x) = a_1(x - L) + \frac{q_t L^2}{2} - \frac{q_t}{2}x^2, \tag{21}$$

with $a_1 \in \mathscr{R}$ being an arbitrary constant. The differential condition of nonlocal elastic equilibrium (11) to be integrated takes thus the form

$$-k_b \frac{d^2 v}{dx^2}(x) = a_1(x - L) + \frac{q_t L^2}{2} - \frac{q_t}{2}x^2 + \tau^2 L^2 q_t. \tag{22}$$

Equation (22) is solved by imposing the following kinematic boundary conditions:

$$v(0) = 0, \qquad \frac{dv}{dx}(0) = 0, \qquad v(L) = 0. \tag{23}$$

A direct computation gives the transversal displacement field

$$v(x) = \frac{q_t}{24k_b}x^4 - \frac{5q_t L}{48k_b}x^3 + \frac{q_t L^2}{16k_b}x^2 + \frac{\tau^2 q_t L}{4k_b}x^2(L - x) \tag{24}$$

and the bending moment

$$M_b(x) = \frac{q_t}{2}x^2 - \frac{q_t L}{2}\left(\frac{5}{4} + 3\tau^2\right)x + \frac{q_t L^2}{2}\left(3\tau^2 + \frac{1}{4}\right). \tag{25}$$

The maximum displacement is given by

$$v_m = v(\overline{x}), \tag{26}$$

with $\overline{x} = 3L((3/2)(5/24 + \tau^2/2) - \sqrt{(9/16)\tau^4 + 11/768 + (13/96)\tau^2})$ solution of the equation (the known local maximum point $\overline{x} \simeq 0.578L$ is recovered by setting $\tau = 0$) $(dv/dx)(x) = 0$.

In agreement with the equivalence method exposed in Section 3, (24) and (25) provide the deflection and

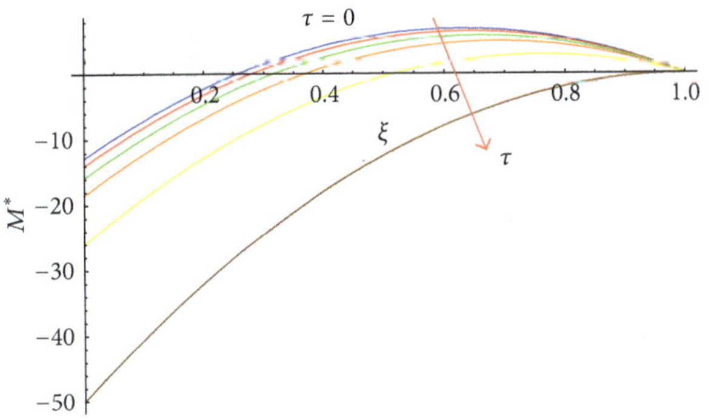

FIGURE 3: Dimensionless bending moment M^* versus dimensionless abscissa ξ of a clamped-simply supported nanobeam.

bending moment of a corresponding local nanobeam under the transversal load distribution q_t and the distortion χ_{IN} equivalent to the nonlocality effect.

A plot of the dimensionless bending moment $M^* = -(100/q_t L^2)M$ versus the dimensionless parameter $\xi = x/L$ is given in Figure 3 for increasing values of the nonlocal parameter τ.

5. Conclusion

The Eringen nonlocal law has been used in order to assess size effects in nanobeams formulated according to the Bernoulli-Euler kinematics. The treatment extends to functionally graded materials the analysis carried out in [24] under the special assumption of elastically homogeneous nanobeams. Transverse deflections of cantilever and clamped-simply supported nanobeams have been established for different values of the nonlocal parameter. Such analytical solutions could be conveniently adopted by other scholars as simple reference examples for numerical evaluations in nonlocal composite mechanics.

Competing Interests

The authors declare that they have no competing interests.

References

[1] R. de Borst and H.-B. Muehlhaus, "Gradient-dependent plasticity: formulation and algorithmic aspects," *International Journal for Numerical Methods in Engineering*, vol. 35, no. 3, pp. 521–539, 1992.

[2] E. C. Aifantis, "Gradient deformation models at nano, micro, and macro scales," *Journal of Engineering Materials and Technology*, vol. 121, pp. 189–202, 1999.

[3] B. Arash and Q. Wang, "A review on the application of nonlocal elastic models in modeling of carbon nanotubes and graphenes," *Computational Materials Science*, vol. 51, no. 1, pp. 303–313, 2012.

[4] R. Rafiee and R. M. Moghadam, "On the modeling of carbon nanotubes: a critical review," *Composites Part B: Engineering*, vol. 56, pp. 435–449, 2014.

[5] J. N. Reddy, "Nonlocal theories for bending, buckling and vibration of beams," *International Journal of Engineering Science*, vol. 45, no. 2–8, pp. 288–307, 2007.

[6] Ö. Civalek and Ç. Demir, "Bending analysis of microtubules using nonlocal Euler-Bernoulli beam theory," *Applied Mathematical Modelling*, vol. 35, no. 5, pp. 2053–2067, 2011.

[7] H.-T. Thai and T. P. Vo, "A nonlocal sinusoidal shear deformation beam theory with application to bending, buckling, and vibration of nanobeams," *International Journal of Engineering Science*, vol. 54, pp. 58–66, 2012.

[8] Q. Wang and K. M. Liew, "Application of nonlocal continuum mechanics to static analysis of micro- and nano-structures," *Physics Letters A*, vol. 363, no. 3, pp. 236–242, 2007.

[9] R. Barretta, L. Feo, R. Luciano, and F. Marotti de Sciarra, "Application of an enhanced version of the Eringen differential model to nanotechnology," *Composites Part B: Engineering*, vol. 96, pp. 274–280, 2016.

[10] R. Luciano and J. R. Willis, "Bounds on non-local effective relations for random composites loaded by configuration-dependent body force," *Journal of the Mechanics and Physics of Solids*, vol. 48, no. 9, pp. 1827–1849, 2000.

[11] R. Luciano and J. R. Willis, "Non-local constitutive response of a random laminate subjected to configuration-dependent body force," *Journal of the Mechanics and Physics of Solids*, vol. 49, no. 2, pp. 431–444, 2001.

[12] R. Luciano and J. R. Willis, "Boundary-layer corrections for stress and strain fields in randomly heterogeneous materials," *Journal of the Mechanics and Physics of Solids*, vol. 51, no. 6, pp. 1075–1088, 2003.

[13] R. Luciano and J. R. Willis, "Hashin-Shtrikman based FE analysis of the elastic behaviour of finite random composite bodies," *International Journal of Fracture*, vol. 137, no. 1–4, pp. 261–273, 2006.

[14] A. Caporale, R. Luciano, and L. Rosati, "Limit analysis of masonry arches with externally bonded FRP reinforcements," *Computer Methods in Applied Mechanics and Engineering*, vol. 196, no. 1–3, pp. 247–260, 2006.

[15] F. Marmo and L. Rosati, "Analytical integration of elasto-plastic uniaxial constitutive laws over arbitrary sections," *International Journal for Numerical Methods in Engineering*, vol. 91, no. 9, pp. 990–1022, 2012.

[16] F. Marmo and L. Rosati, "The fiber-free approach in the evaluation of the tangent stiffness matrix for elastoplastic uniaxial constitutive laws," *International Journal for Numerical Methods in Engineering*, vol. 94, no. 9, pp. 868–894, 2013.

[17] R. Barretta, "Analogies between Kirchhoff plates and Saint-Venant beams under flexure," *Acta Mechanica*, vol. 225, no. 7, pp. 2075–2083, 2014.

[18] L. Rosati and F. Marmo, "Closed-form expressions of the thermo-mechanical fields induced by a uniform heat source acting over an isotropic half-space," *International Journal of Heat and Mass Transfer*, vol. 75, pp. 272–283, 2014.

[19] F. Marmo and L. Rosati, "A general approach to the solution of boussinesq's problem for polynomial pressures acting over polygonal domains," *Journal of Elasticity*, vol. 122, no. 1, pp. 75–112, 2016.

[20] F. Marotti de Sciarra and M. Salerno, "On thermodynamic functions in thermoelasticity without energy dissipation," *European Journal of Mechanics—A/Solids*, vol. 46, pp. 84–95, 2014.

[21] F. Marotti de Sciarra, "Finite element modelling of nonlocal beams," *Physica E*, vol. 59, pp. 144–149, 2014.

[22] F. Marotti de Sciarra and R. Barretta, "A new nonlocal bending model for Euler-Bernoulli nanobeams," *Mechanics Research Communications*, vol. 62, no. 1, pp. 25–30, 2014.

[23] R. Barretta, R. Luciano, and F. Marotti de Sciarra, "A fully gradient model for Euler-Bernoulli nanobeams," *Mathematical Problems in Engineering*, vol. 2015, Article ID 495095, 8 pages, 2015.

[24] R. Barretta and F. Marotti de Sciarra, "Analogies between nonlocal and local Bernoulli–Euler nanobeams," *Archive of Applied Mechanics*, vol. 85, no. 1, pp. 89–99, 2015.

[25] R. Barretta, L. Feo, R. Luciano, F. Marotti de Sciarra, and R. Penna, "Functionally graded Timoshenko nanobeams: a novel nonlocal gradient formulation," *Composites Part B: Engineering*, vol. 100, pp. 208–219, 2016.

[26] F. Greco and R. Luciano, "A theoretical and numerical stability analysis for composite micro-structures by using homogenization theory," *Composites Part B: Engineering*, vol. 42, no. 3, pp. 382–401, 2011.

[27] A. Caporale, L. Feo, D. Hui, and R. Luciano, "Debonding of FRP in multi-span masonry arch structures via limit analysis," *Composite Structures*, vol. 108, no. 1, pp. 856–865, 2014.

[28] V. A. Salomoni, C. Majorana, G. M. Giannuzzi, and A. Miliozzi, "Thermal-fluid flow within innovative heat storage concrete systems for solar power plants," *International Journal of Numerical Methods for Heat & Fluid Flow*, vol. 18, pp. 969–999, 2008.

[29] V. A. Salomoni, G. Mazzucco, C. Pellegrino, and C. E. Majorana, "Three-dimensional modelling of bond behaviour between concrete and FRP reinforcement," *Engineering Computations*, vol. 28, no. 1, pp. 5–29, 2011.

[30] V. A. Salomoni, C. E. Majorana, B. Pomaro, G. Xotta, and F. Gramegna, "Macroscale and mesoscale analysis of concrete as a multiphase material for biological shields against nuclear radiation," *International Journal for Numerical and Analytical Methods in Geomechanics*, vol. 38, no. 5, pp. 518–535, 2014.

[31] G. Dinelli, G. Belz, C. E. Majorana, and B. A. Schrefler, "Experimental investigation on the use of fly ash for lightweight precast structural elements," *Materials and Structures*, vol. 29, no. 194, pp. 632–638, 1996.

[32] G. Xotta, G. Mazzucco, V. A. Salomoni, C. E. Majorana, and K. J. Willam, "Composite behavior of concrete materials under high temperatures," *International Journal of Solids and Structures*, vol. 64, pp. 86–99, 2015.

[33] A. M. Tarantino, "Nonlinear fracture mechanics for an elastic Bell material," *The Quarterly Journal of Mechanics & Applied Mathematics*, vol. 50, part 3, pp. 435–456, 1997.

[34] A. M. Tarantino, "The singular equilibrium field at the notch-tip of a compressible material in finite elastostatics," *Journal of Applied Mathematics and Physics*, vol. 48, no. 3, pp. 370–388, 1997.

[35] A. M. Tarantino, "On extreme thinning at the notch-tip of a neo-Hookean sheet," *The Quarterly Journal of Mechanics & Applied Mathematics*, vol. 51, part 2, pp. 179–190, 1998.

[36] A. M. Tarantino, "On the finite motions generated by a mode I propagating crack," *Journal of Elasticity*, vol. 57, no. 2, pp. 85–103, 1999.

[37] A. M. Tarantino, "Crack propagation in finite elastodynamics," *Mathematics and Mechanics of Solids*, vol. 10, no. 6, pp. 577–601, 2005.

[38] A. M. Tarantino, "Homogeneous equilibrium configurations of a hyperelastic compressible cube under equitriaxial dead-load tractions," *Journal of Elasticity*, vol. 92, no. 3, pp. 227–254, 2008.

[39] A. C. Eringen, "On differential equations of nonlocal elasticity and solutions of screw dislocation and surface waves," *Journal of Applied Physics*, vol. 54, no. 9, pp. 4703–4710, 1983.

13

Initiation of Failure for Masonry Subject to In-Plane Loads through Micromechanics

V. P. Berardi

Department of Civil Engineering, University of Salerno, Via Ponte don Melillo, 84084 Fisciano, Italy

Correspondence should be addressed to V. P. Berardi; berardi@unisa.it

Academic Editor: Theodoros C. Rousakis

A micromechanical procedure is used in order to evaluate the initiation of damage and failure of masonry with in-plane loads. Masonry material is viewed as a composite with periodic microstructure and, therefore, a unit cell with suitable boundary conditions is assumed as a representative volume element of the masonry. The finite element method is used to determine the average stress on the unit cell corresponding to a given average strain prescribed on the unit cell. Finally, critical curves representing the initiation of damage and failure in both clay brick masonry and adobe masonry are provided.

1. Introduction

Materials used in civil engineering such as cement concrete [1–4] and masonry are subject to deterioration; moreover, constructions built in the past need to carry increasing seismic horizontal as well as vertical loads, leading to the need for strengthening interventions. Fibre reinforced plastics (FRP), polymeric nets embedded in the plaster, and other more classical materials and techniques are effectively used to strengthen existing structures [5–8]. More complex techniques require the insertion of seismic isolation bearings made of rubber, usually modelled as a hyperelastic material [9–16]. New materials are being tested for civil engineering applications; recently, carbon nanotubes (CNTs) are being used in various research works in order to improve the mechanical properties of cement mortar (usable also in strengthening existing structures); therefore many authors are investigating the behaviour of structures at nanoscale with nonlocal material models [17–21].

Numerical methods for determining the structural response of unreinforced and reinforced masonry constructions, such as arches, wall, and dome, are available in the literature [22–27]. It is important to have available methods for estimating not only the structural response but also the material behaviour. To this end, micromechanics and homogenization methods have been applied to the masonry material. In this work, a micromechanical procedure is used in order to evaluate the initiation of damage and failure of masonry panels with in-plane loads following the approach developed in [28–30].

2. Micromechanical Analysis and Material Model

The masonry material is composed of two constituents: the brick and the mortar. These constituents have different mechanical properties and the resulting masonry material can be viewed as a composite so that classical techniques based on micromechanics and homogenization are used in order to determine the local and overall response of the masonry material. The overall mechanical properties of the composite are derived from the geometric and mechanical properties of the constituents, the microstructure of the composite, and so forth. Such information is contained in a reference volume element (RVE), which is statistically representative of the masonry under consideration. The following analyses are performed on the representative volume element. Instead of modelling the masonry as a random composite material (see, e.g., [31–36]), it is assumed that the constituents are arranged in a periodic way [37–45] and a unit cell is adopted as the representative volume element. The

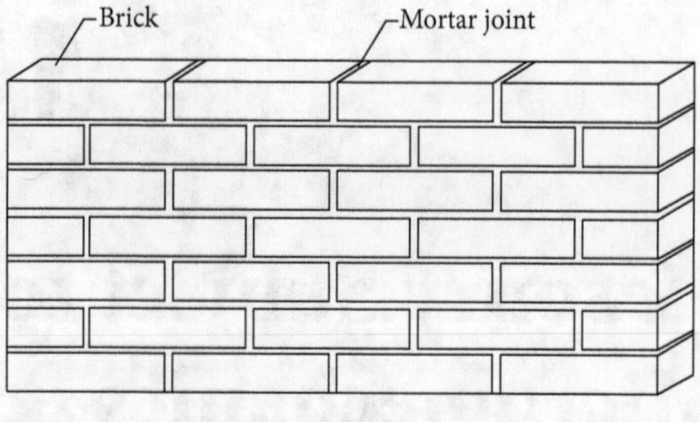

FIGURE 1: A masonry wall.

masonry considered in this work is shown in Figure 1, where a wall is depicted. The aim is to determine the strength of the masonry when it is subjected to in-plane loads.

The wall in Figure 1 is a three-dimensional structure whose three-dimensional unit cell is shown in Figure 2(a), where $2a_1$, $2a_2$, and $2a_3$ are the dimensions of the unit cell along the x_1-, x_2-, and x_3-axes of the coordinate system with origin in the centre of the unit cell.

In micromechanical and homogenization analyses of periodic microstructure, the three-dimensional unit cell is subjected to the following periodic boundary conditions:

$$\begin{aligned} &u_i(a_1, x_2, x_3) - u_i(-a_1, x_2, x_3) = 2E_{i1}a_1 \\ &\qquad \forall x_2 \in [-a_2, a_2], \ \forall x_3 \in [-a_3, a_3], \\ &u_i(x_1, a_2, x_3) - u_i(x_1, -a_2, x_3) = 2E_{i2}a_2 \\ &\qquad \forall x_1 \in [-a_1, a_1], \ \forall x_3 \in [-a_3, a_3], \\ &u_i(x_1, x_2, a_3) - u_i(x_1, x_2, -a_3) = 2E_{i3}a_3 \\ &\qquad \forall x_1 \in [-a_1, a_1], \ \forall x_2 \in [-a_2, a_2], \end{aligned} \tag{1}$$

where u_i is the displacement along the x_i-axis and E_{ij} are the components of the average strain prescribed to the unit cell. Together with (1), the following constraint must be imposed in order to avoid the rigid-body translations:

$$\mathbf{u}(\mathbf{x}^*) = \mathbf{0}, \tag{2}$$

where $\mathbf{x}^*$ is the position vector of the centre of the unit cell. Since the in-plane behaviour must be investigated, the plane unit cell shown in Figure 2(b) can be used as representative volume element for a two-dimensional (2D) analysis. In the homogenization analysis, the boundary conditions prescribed to the plane unit cell are

$$\begin{aligned} &u_i(a_1, x_2) - u_i(-a_1, x_2) = 2E_{i1}a_1 \quad \forall x_2 \in [-a_2, a_2], \\ &u_i(x_1, a_2) - u_i(x_1, -a_2) = 2E_{i2}a_2 \quad \forall x_1 \in [-a_1, a_1]. \end{aligned} \tag{3}$$

The average stress in the unit cell is denoted by $\boldsymbol{\Sigma}$, whereas the local stress in a point of the constituent of the unit cell is the microstress $\boldsymbol{\sigma}$ and, for simplicity, will be called stress in the following. The constituents (brick and mortar) are considered linear elastic before their damage and failure. The bond between the constituents is considered perfect. It is assumed that the plane unit cell and the constituents are subject to a plane stress state. Therefore, the principal stress perpendicular to the plane unit cell is equal to zero whereas the principal stresses parallel to the plane of the unit cell can be different from zero. The failure in a point of a constituent occurs when one of the following three failure criteria is satisfied.

First Criterion. If the two principal stresses parallel to the plane of the unit cell are nonnegative the failure occurs when the maximum principal stress in a point is equal to the tensile strength of the constituent.

Second Criterion. If the two principal stresses parallel to the plane of the unit cell are nonpositive the failure occurs when the following equation is satisfied:

$$CJ_2 + (1 - C) I_1 + CI_2 = 1, \tag{4}$$

where $C = 1.6$,

$$\begin{aligned} J_2 &= \frac{1}{f_c^2} (\sigma_2 - \sigma_3)^2, \\ I_1 &= \frac{1}{f_c} (\sigma_2 + \sigma_3), \\ I_2 &= \frac{\sigma_2 \sigma_3}{f_c^2}, \end{aligned} \tag{5}$$

where σ_2 and σ_3 are the two nonpositive principal stresses parallel to the plane of the unit cell and f_c is the compressive strength of the material of the constituent.

Third Criterion. When a principal stress σ_1 parallel to the plane of the unit cell is nonnegative and the other principal stress σ_3 parallel to the plane of the unit cell is nonpositive, the failure occurs when the following equation is satisfied:

$$\frac{\sigma_1}{f_t} + \frac{\sigma_3}{f_c} = 1, \tag{6}$$

where f_t is the tensile strength of the material of the constituent.

Next, the average strain $\mathbf{E}$ (which is a tensor with components E_{ij}) is imposed on the plane unit cell by means of the conditions (2) and (3). Then, the average stress $\boldsymbol{\Sigma}$ corresponding to the prescribed average strain $\mathbf{E}$ is evaluated and the critical curves of the masonry are determined with a finite element homogenization technique. The critical curves may be plotted in the plane defined by Σ_{11}- and Σ_{22}-axes or in the plane defined by Σ_{11}- and Σ_{12}-axes. There are two kinds of critical curves: a generic critical curve may be related to the brick or to the mortar. The generic point of the critical curve related to a constituent represents the average stresses corresponding to the initiation of failure of that constituent. Specifically, the initiation of failure of a constituent is evaluated by prescribing a linear load path $\mathbf{E} = \lambda \mathbf{E}^0$, where $\mathbf{E}^0$ is a constant tensor, which does not vary in the loading process, and the scalar λ is an increasing loading positive parameter. Considering both constituents linear

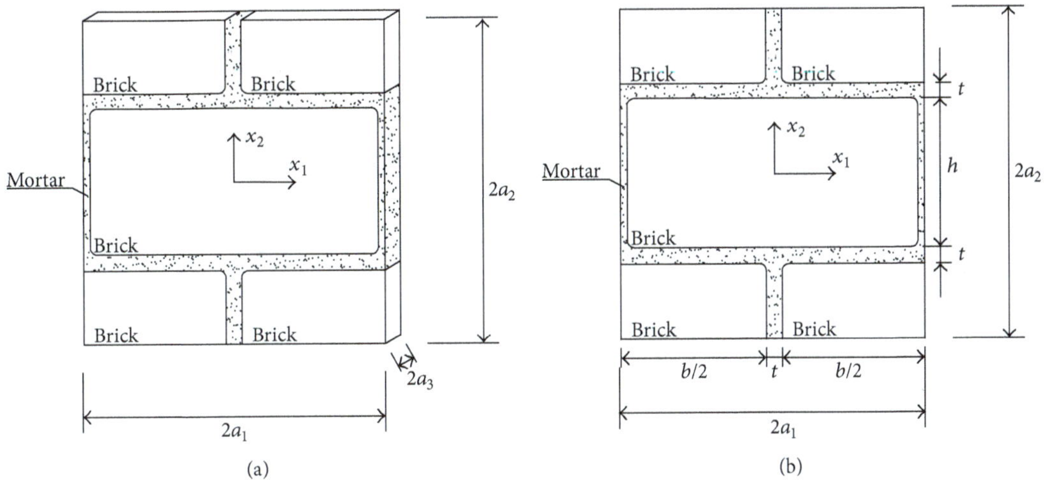

FIGURE 2: (a) Three-dimensional unit cell; (b) plane unit cell.

elastic, there exists a value $\lambda^{\mathrm{cr},i}$ such that the microstresses in a constituent i (i = brick or mortar) of the unit cell subject to the average strain $\mathbf{E} = \lambda^{\mathrm{cr},i}\mathbf{E}^0$ satisfy some of the above-mentioned criteria in the same constituent i, whereas failure criteria are not satisfied in the same constituent i when the unit cell is subject to $\mathbf{E} = \lambda\mathbf{E}^0$ for $0 \leq \lambda < \lambda^{\mathrm{cr},i}$. The average stress of the unit cell subject to the average strain $\mathbf{E} = \lambda^{\mathrm{cr},i}\mathbf{E}^0$ is denoted by $\mathbf{\Sigma}^{\mathrm{cr},i}$. In the plane defined by Σ_{11}- and Σ_{22}-axes, the components $(\Sigma_{11}^{\mathrm{cr},i}, \Sigma_{22}^{\mathrm{cr},i})$ of $\mathbf{\Sigma}^{\mathrm{cr},i}$ define a closed curve which represents the masonry critical curve related to the initiation of failure of the constituent i. An analogous critical curve can be plotted in the plane defined by Σ_{11}- and Σ_{12}-axes.

3. Numerical Examples

If the constituents have the same mechanical properties and the bond between the brick and the mortar is perfect, the masonry behaves like a homogeneous material and the homogenization procedure is not required. Critical cases occur when the mechanical properties of one constituent are significantly different from the mechanical properties of the other constituents. For example, in adobe masonry (AM), bricks and mortar have similar elastic properties, whereas in some kinds of clay brick masonry (CBM) Young's modulus of the brick may be significantly different from Young's modulus of the mortar. The masonry critical curves are very sensitive to the ratio E_b/E_m, where E_b and E_m are Young's moduli of the brick and the mortar, respectively. This is shown in Figure 3, where the masonry critical curve related to the initiation of failure of the constituent i is denoted by "b" for i = brick and by "m" for i = mortar in the legend of the figure.

The critical curves in Figure 3 are obtained by prescribing the following average strain:

$$\mathbf{E} = \begin{bmatrix} E_{11} & E_{12} \\ E_{21} & E_{22} \end{bmatrix} = \lambda \begin{bmatrix} \cos\varphi & 0 \\ 0 & \sin\varphi \end{bmatrix} \quad \text{for } \varphi \in [0, 2\pi]. \quad (7)$$

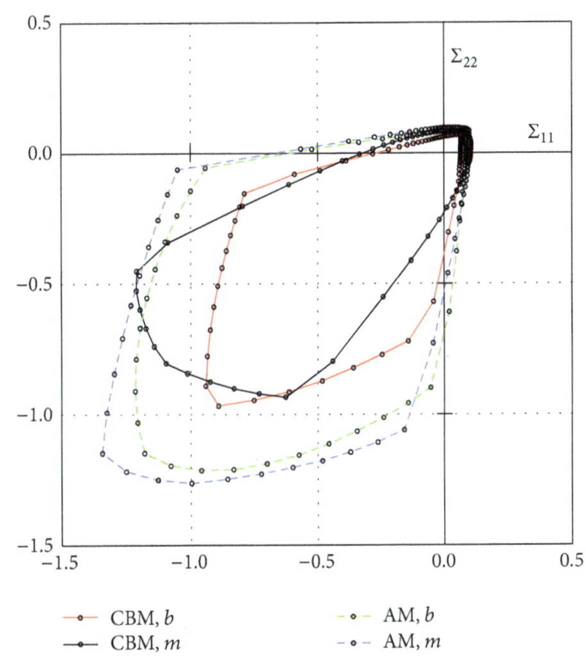

FIGURE 3: CBM and AM critical curves, in the Σ_{11}-Σ_{22} plane, related to the initiation of failure of brick and mortar.

In this section, the mechanical properties of the constituents of CBM are $\nu_b = 0.23$, $\nu_m = 0.15$, $E_b = 10E_m = 10000$ MPa, $f_{cb} = 10f_{tb} = 1$ MPa, $f_{cm} = f_{cb}$, and $f_{tm} = f_{tb}$; the mechanical properties of the constituents of AM are $\nu_b = \nu_m = 0.15$, $E_b = 2E_m = 100$ MPa, $f_{cb} = 10f_{tb} = 1$ MPa, $f_{cm} = f_{cb}$, and $f_{tm} = f_{tb}$. In this work, ν_b and ν_m are Poisson's ratios of the brick and the mortar, respectively, f_{tb} and f_{tm} are the tensile strengths of the brick and the mortar, and f_{cb} and f_{cm} are the compressive strengths of the brick and the mortar. Both CBM and AM have the following geometric properties: b = 400 mm (width of the bricks), h = 100 mm (height of the bricks), t = 10 mm (thickness of mortar joints), and

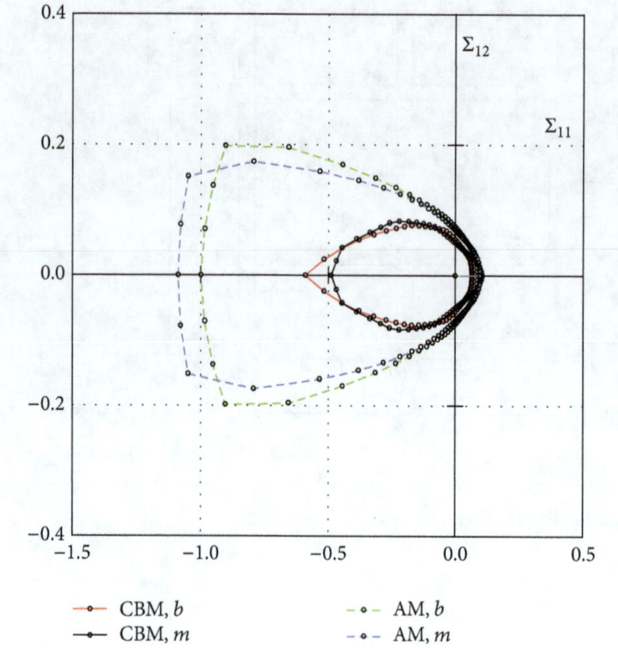

FIGURE 4: CBM and AM critical curves, in the Σ_{11}-Σ_{12} plane, related to the initiation of failure of brick and mortar.

$\rho = h/40$ (each brick corner was approximated as arc of circle with radius ρ).

In Figure 3, two different values of the ratio $c = E_b/E_m$ are considered: $c = 10$ for the CBM critical curves and $c = 2$ for the AM critical curves. Greater values of c determine greater concentration of the local stress and, as a consequence, smaller critical areas (the critical area is the area bounded by the critical curve related to the initiation of failure of a constituent). A similar behaviour is observed in Figure 4, where the critical curves are obtained by prescribing the following average strain:

$$\mathbf{E} = \begin{bmatrix} E_{11} & E_{12} \\ E_{21} & E_{22} \end{bmatrix} = \lambda \begin{bmatrix} \cos\varphi & \sin\varphi \\ \sin\varphi & 0 \end{bmatrix} \quad \text{for } \varphi \in [0, 2\pi]. \quad (8)$$

The contrast c between the masonry constituents also influences the deformed shape of the unit cell and the distribution of local stress in the unit cell, as shown in Figures 5 and 6, where the local stress σ_{12} on the deformed unit cell subject to the average strain (8) with $\varphi = \pi/2$ and $\lambda = 1$ is plotted for CBM ($c = 10$) and AM ($c = 2$), respectively.

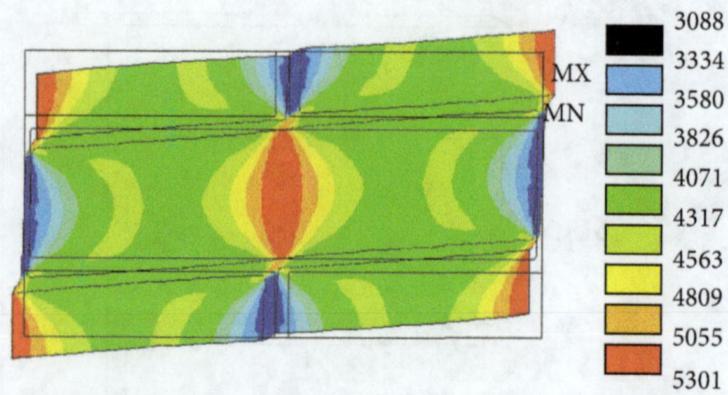

FIGURE 5: Local stress σ_{12} on the deformed CBM unit cell subject to a pure shear average strain $E_{12} = E_{21} = 1$.

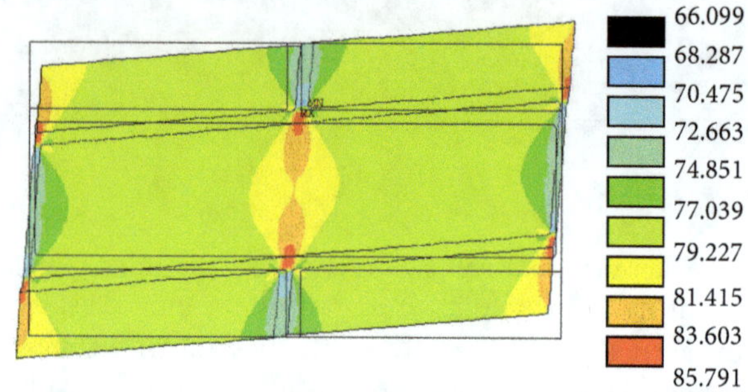

FIGURE 6: Local stress σ_{12} on the deformed AM unit cell subject to a pure shear average strain $E_{12} = E_{21} = 1$.

4. Conclusions

In this work, masonry is viewed as a composite constituted of two components (namely, brick and mortar) so that classical techniques based on micromechanics and homogenization are used in order to determine the local and overall response of the masonry material. The critical curves of the masonry are determined with a finite element homogenization technique. The generic point of the critical curve related to a constituent represents the average stresses corresponding to the initiation of failure of that constituent. The numerical analyses show that the critical curves are sensitive to the contrast $c = E_b/E_m$ between brick and mortar: greater contrast involves earlier initiation of damage in the constituents.

Competing Interests

There are no competing interests related to this paper.

References

[1] A. Caporale, L. Feo, and R. Luciano, "Damage mechanics of cement concrete modeled as a four-phase composite," *Composites Part B: Engineering*, vol. 65, pp. 124–130, 2014.

[2] A. Caporale and R. Luciano, "A micromechanical four-phase model to predict the compressive failure surface of cement concrete," *Frattura ed Integrita Strutturale*, vol. 8, no. 29, pp. 19–27, 2014.

[3] G. Dinelli, G. Belz, C. E. Majorana, and B. A. Schrefler, "Experimental investigation on the use of fly ash for lightweight precast structural elements," *Materials and Structures*, vol. 29, no. 10, pp. 632–638, 1996.

[4] G. Xotta, G. Mazzucco, V. A. Salomoni, C. E. Majorana, and K. J. Willam, "Composite behavior of concrete materials under high temperatures," *International Journal of Solids and Structures*, vol. 64, pp. 86–99, 2015.

[5] A. D'Ambrisi, F. Focacci, and R. Luciano, "Experimental investigation on flexural behavior of timber beams repaired with CFRP plates," *Composite Structures*, vol. 108, no. 1, pp. 720–728, 2014.

[6] A. D'Ambrisi, F. Focacci, R. Luciano, V. Alecci, and M. De Stefano, "Carbon-FRCM materials for structural upgrade of

masonry arch road bridges," *Composites Part B: Engineering*, vol. 75, no. 3373, pp. 355–366, 2015.

[7] A. D'Ambrisi, M. Mezzi, and A. Caporale, "Experimental investigation on polymeric net-RCM reinforced masonry panels," *Composite Structures*, vol. 105, pp. 207–215, 2013.

[8] V. Salomoni, G. Mazzucco, C. Pellegrino, and C. Majorana, "Three-dimensional modelling of bond behaviour between concrete and FRP reinforcement," *Engineering Computations*, vol. 28, no. 1, pp. 5–29, 2011.

[9] A. M. Tarantino, "The singular equilibrium field at the notch-tip of a compressible material in finite elastostatics," *Zeitschrift für Angewandte Mathematik und Physik*, vol. 48, no. 3, pp. 370–388, 1997.

[10] A. M. Tarantino, "On extreme thinning at the notch tip of a neo-Hookean sheet," *The Quarterly Journal of Mechanics and Applied Mathematics*, vol. 51, no. 2, pp. 179–190, 1998.

[11] A. M. Tarantino, "On the finite motions generated by a mode I propagating crack," *Journal of Elasticity. The Physical and Mathematical Science of Solids*, vol. 57, no. 2, pp. 85–103 (2000), 1999.

[12] A. M. Tarantino, "Asymmetric equilibrium configurations of symmetrically loaded isotropic square membranes," *Journal of Elasticity*, vol. 69, no. 1–3, pp. 73–97, 2002.

[13] A. M. Tarantino, "Asymmetric equilibrium configurations of hyperelastic cylindrical bodies under symmetric dead loads," *Quarterly of Applied Mathematics*, vol. 64, no. 4, pp. 605–615, 2006.

[14] A. M. Tarantino and A. Nobili, "Constitutive branching analysis of cylindrical bodies under in-plane equibiaxial dead-load tractions," *International Journal of Non-Linear Mechanics*, vol. 41, no. 8, pp. 958–968, 2006.

[15] A. M. Tarantino, "Homogeneous equilibrium configurations of a hyperelastic compressible cube under equitriaxial dead-load tractions," *Journal of Elasticity. The Physical and Mathematical Science of Solids*, vol. 92, no. 3, pp. 227–254, 2008.

[16] L. Lanzoni and A. M. Tarantino, "Damaged hyperelastic membranes," *International Journal of Non-Linear Mechanics*, vol. 60, pp. 9–22, 2014.

[17] A. Apuzzo, R. Barretta, and R. Luciano, "Some analytical solutions of functionally graded Kirchhoff plates," *Composites Part B: Engineering*, vol. 68, pp. 266–269, 2015.

[18] R. Barretta, L. Feo, and R. Luciano, "Torsion of functionally graded nonlocal viscoelastic circular nanobeams," *Composites Part B: Engineering*, vol. 72, pp. 217–222, 2015.

[19] R. Barretta, L. Feo, R. Luciano, and F. Marotti de Sciarra, "A gradient Eringen model for functionally graded nanorods," *Composite Structures*, vol. 131, pp. 1124–1131, 2015.

[20] R. Barretta, L. Feo, R. Luciano, and F. Marotti de Sciarra, "Variational formulations for functionally graded nonlocal Bernoulli-Euler nanobeams," *Composite Structures*, vol. 129, pp. 80–89, 2015.

[21] R. Barretta, R. Luciano, and F. Marotti De Sciarra, "A fully gradient model for Euler-Bernoulli nanobeams," *Mathematical Problems in Engineering*, vol. 2015, Article ID 495095, 8 pages, 2015.

[22] A. Caporale, R. Luciano, and L. Rosati, "Limit analysis of masonry arches with externally bonded FRP reinforcements," *Computer Methods in Applied Mechanics and Engineering*, vol. 196, no. 1–3, pp. 247–260, 2006.

[23] A. Caporale, L. Feo, and R. Luciano, "Limit analysis of FRP strengthened masonry arches via nonlinear and linear programming," *Composites Part B: Engineering*, vol. 43, no. 2, pp. 439–446, 2012.

[24] A. Caporale and R. Luciano, "Limit analysis of masonry arches with finite compressive strength and externally bonded reinforcement," *Composites Part B: Engineering*, vol. 43, no. 8, pp. 3131–3145, 2012.

[25] A. Caporale, L. Feo, R. Luciano, and R. Penna, "Numerical collapse load of multi-span masonry arch structures with FRP reinforcement," *Composites Part B: Engineering*, vol. 54, no. 1, pp. 71–84, 2013.

[26] A. D'Ambrisi, F. Focacci, and A. Caporale, "Strengthening of masonry-unreinforced concrete railway bridges with PBO-FRCM materials," *Composite Structures*, vol. 102, pp. 193–204, 2013.

[27] A. Caporale, L. Feo, D. Hui, and R. Luciano, "Debonding of FRP in multi-span masonry arch structures via limit analysis," *Composite Structures*, vol. 108, no. 1, pp. 856–865, 2014.

[28] A. Caporale, F. Parisi, D. Asprone, R. Luciano, and A. Prota, "Critical surfaces for adobe masonry: micromechanical approach," *Composites Part B: Engineering*, vol. 56, pp. 790–796, 2014.

[29] A. Caporale, F. Parisi, D. Asprone, R. Luciano, and A. Prota, "Micromechanical analysis of adobe masonry as two-component composite: influence of bond and loading schemes," *Composite Structures*, vol. 112, no. 1, pp. 254–263, 2014.

[30] A. Caporale, F. Parisi, D. Asprone, R. Luciano, and A. Prota, "Comparative micromechanical assessment of adobe and clay brick masonry assemblages based on experimental data sets," *Composite Structures*, vol. 120, pp. 208–220, 2015.

[31] R. Luciano and J. R. Willis, "Bounds on non-local effective relations for random composites loaded by configuration-dependent body force," *Journal of the Mechanics and Physics of Solids*, vol. 48, no. 9, pp. 1827–1849, 2000.

[32] R. Luciano and J. R. Willis, "Non-local effective relations for fibre-reinforced composites loaded by configuration-dependent body forces," *Journal of the Mechanics and Physics of Solids*, vol. 49, no. 11, pp. 2705–2717, 2001.

[33] R. Luciano and J. R. Willis, "Non-local constitutive response of a random laminate subjected to configuration-dependent body force," *Journal of the Mechanics and Physics of Solids*, vol. 49, no. 2, pp. 431–444, 2001.

[34] R. Luciano and J. R. Willis, "Boundary-layer corrections for stress and strain fields in randomly heterogeneous materials," *Journal of the Mechanics and Physics of Solids*, vol. 51, no. 6, pp. 1075–1088, 2003.

[35] R. Luciano and J. R. Willis, "FE analysis of stress and strain fields in finite random composite bodies," *Journal of the Mechanics and Physics of Solids*, vol. 53, no. 7, pp. 1505–1522, 2005.

[36] R. Luciano and J. R. Willis, "Hashin-Shtrikman based FE analysis of the elastic behaviour of finite random composite bodies," *International Journal of Fracture*, vol. 137, no. 1–4, pp. 261–273, 2006.

[37] A. Caporale, R. Luciano, and E. Sacco, "Micromechanical analysis of interfacial debonding in unidirectional fiber-reinforced composites," *Computers and Structures*, vol. 84, no. 31-32, pp. 2200–2211, 2006.

[38] A. Caporale and R. Luciano, "Micromechanical analysis of periodic composites by prescribing the average stress," *Annals of Solid and Structural Mechanics*, vol. 1, no. 3, pp. 117–137, 2010.

[39] A. Caporale, R. Luciano, and R. Penna, "Fourier series expansion in non-orthogonal coordinate system for the homogenization of linear viscoelastic periodic composites," *Composites Part B: Engineering*, vol. 54, no. 1, pp. 241–245, 2013.

[40] A. Caporale, L. Feo, and R. Luciano, "Eigenstrain and Fourier series for evaluation of elastic local fields and effective properties of periodic composites," *Composites Part B: Engineering*, vol. 81, pp. 251–258, 2015.

[41] F. Greco and R. Luciano, "A theoretical and numerical stability analysis for composite micro-structures by using homogenization theory," *Composites Part B: Engineering*, vol. 42, no. 3, pp. 382–401, 2011.

[42] G. Cricrì and R. Luciano, "Homogenised properties of composite materials in large deformations," *Composite Structures*, vol. 103, pp. 9–17, 2013.

[43] D. Bruno, F. Greco, R. Luciano, and P. Nevone Blasi, "Nonlinear homogenized properties of defected composite materials," *Computers and Structures*, vol. 134, pp. 102–111, 2014.

[44] L. Feo, F. Greco, L. Leonetti, and R. Luciano, "Mixed-mode fracture in lightweight aggregate concrete by using a moving mesh approach within a multiscale framework," *Composite Structures*, vol. 123, pp. 88–97, 2015.

[45] F. Greco, L. Leonetti, and R. Luciano, "A multiscale model for the numerical simulation of the anchor bolt pull-out test in lightweight aggregate concrete," *Construction and Building Materials*, vol. 95, no. 6986, pp. 860–874, 2015.

14

Scenario Based Municipal Wastewater Estimation: Development and Application of a Dynamic Simulation Model

Yong Zhang, Tingsheng Zhao, Aijiao Zhou, Zhengzhu Zhang, and Wen Liu

School of Civil Engineering & Mechanics, Huazhong University of Science and Technology, No. 1037, Luoyu Road, Hongshan District, Wuhan 430074, China

Correspondence should be addressed to Wen Liu; 379047775@qq.com

Academic Editor: Luis Carlos Rabelo

This paper develops causal loop diagrams and a system dynamics model for estimation of wastewater quantity changes as a function of future socioeconomic development and the municipal water environment of the city under the influence of several key factors. Using Wuhan (a city with population more than 10 million in China) as a case study, the variability of Wuhan's wastewater quantity and water environment is modeled under different development patterns by year 2030. Nine future scenarios are designed by assigning different values to those key factors, including GDP growth rate, water consumption of annual ten thousand GDP, and wastewater treatment fee. The results show that (1) GDP growth leads to an increase in municipal wastewater quantity, but an increase in wastewater treatment fee can be in favor of reducing urban water pollution, and (2) the impact of per ten thousand yuan GDP water consumption on the amount of municipal wastewater is larger in the near future, while the impact of GDP growth rate is much larger in the long term. The dynamic model has proven to be reliable for simulating the municipal wastewater changes, and it could help decision makers to make the scientific and reasonable decisions.

1. Introduction

Over the years, a number of methods and models have been developed to explore the links between economy and water demand or wastewater quantity of a city, such as (1) the Environmental Kuznets Curve (EKC) method [1, 2]; (2) Input-Output Model [3–6]; (3) Econometric Methods [7, 8] and Time Series [9, 10]; and (4) Artificial Intelligence Methods [11–13]. However, in these methods, the correlations among different system/process factors and their combined effects on the amount of wastewater are not considered systematically; they are not easy to explain the feedback mechanism(s) among these factors. The system dynamics (SD) method could have great advantages for simulating the future scenarios with simultaneous consideration of the correlations among different system/process factors and their combined effects [14].

System dynamic (SD), originated by Jay Forrester of the Massachusetts Institute of Technology in 1960s, was primarily used in the management of industry and enterprises. *Industrial Dynamics* which was published in 1961 is regarded as the classical literary work about SD. Since being originated in 1960s, SD has been applied to more and more research fields. In recent years, SD also has been used in the research of water resource, water environment, and urban water supply. Xi and Poh [15] researched the sustainable water resources management in Singapore by using SD. Davies and Simonovic [16] discussed the global sustainable water resources and developed an integrated SD model that involves social, economic, and environmental subsystem. Ahmad and Prashar [14] applied SD model to evaluate the effectiveness of the municipal water conservation policies. For water environment, Guo et al. [17] utilized a SD model to plan the environmental management system of the basin and tested four planning alternatives that was desired for sustainable for regional development. Zhu et al. [18] investigated the influence of the total nitrogen (TN) and total phosphorus (TP) concentrations in socioeconomic water on the Normalized Difference Vegetation Index (NDVI) and the feedback mechanism among them, which could help to promote the water quality and the ecological functions in Baiyangdian Lake, of China. For water supply, Sahin et al. [19] researched the effect of water price in dealing with the water scarcity when suffering temporary drought and SD modeling could help the planners to formulate reasonable price regime and scheme water supply infrastructure. Zarghami and Akbariyeh [20] established the SD model for the complex urban water system of Tabriz, which comprehensively considered the potential water supply resources as potential sources of demand for water resources and corresponding management measures. Dawadi and Ahmad [2] concerned the influence of climate change, population growth, water price, and water management policies on the water resources in Las Vegas, and SD model that integrated the abovementioned factors was developed to simulate changes in the urban water supply and demand in a long term in future under implementation of the policies. Compared with the forecast of water consumption, wastewater forecasting has its own characteristics. For example, the irrigation water demand is huge, but it actually seeps into underground or flows into rivers through surface runoff, and due to the imperfectness of wastewater network, collected agricultural wastewater is finite; and deterioration of water environment pushes the government to make more stringent standards to restrict the emission of wastewater.

Urban wastewater that contains large amount of organics (including nitrogen and phosphorus), inorganics, and varieties of bacteria and viruses is mainly composed of domestic wastewater and industrial wastewater. In the aspect of wastewater treatment technologies, biological treatment processes, such as activated sludge and biofilm, are widely used for the good biodegradability of urban wastewater. According to the requirements of water quality, physicochemical methods, such as coagulation, precipitation, and membrane filtration, can be combined with biological technologies to further improve treatment effect. In the generation of wastewater, information exchanges and material communications among different factors (such as population and socioeconomic growth) can affect the amount of municipal wastewater. And previous studies neglected the correlations among different factors and their combined influences on the amount of wastewater. On the whole, both domestic water consumption forecast and industrial water consumption forecast by SD approach have been reported. But there are a few studies taking into account the dynamic changes of municipal wastewater and its influence on basin environment and regional economic development.

In light of the aforementioned analysis, a SD method is proposed in this study to research the dynamic changes of municipal wastewater and municipal water environment under different policies for socioeconomic development. The objectives of this work are to (1) develop a SD model for analyzing long-term fluctuations in the amount of municipal wastewater under the influence of socioeconomic factors and the dynamic relationship among them and (2) through case studies verify and calibrate the SD model developed and the further use it to simulate changes in municipal wastewater quantity under different patterns of economic development.

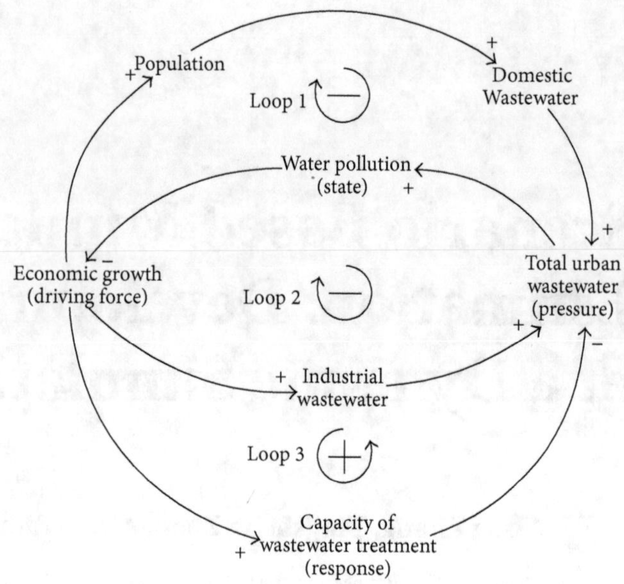

FIGURE 1: The causal loop diagram of the model.

2. Method

2.1. Development of SD Model

2.1.1. System Boundary. This study mainly explores the fluctuation of municipal wastewater under the different development patterns and investigates the complex relationship among economic development, municipal wastewater, and water pollution system. In this model, municipal wastewater is forecasted though the economy and population level, and water pollution is caused by the discharge of untreated wastewater, and the quantity of untreated wastewater depends on the capacity of wastewater collection and disposal of the city. The above three parts are the system boundaries and form the system of "economic development-municipal wastewater-water pollution," which is the main structure of the model. Besides, the time bound of the SD model is from the year of 2000 to 2030, and the time step is 1 year.

2.1.2. Causal Loop. The system boundary is briefly described in the previous section, and numerous factors have influence on the system of "economic development-municipal wastewater-water pollution." The main factors affecting the behavior of the system are to be identified based on the logic thinking of "Pressure-State-Response." A causal loop diagram is established to describe the dynamic behavior of the entire model. According to the basic principle of SD, three feedback loops are proposed in the causal loop diagram, among which two are negative and one is positive. As shown in Figure 1, loop 1 is a negative feedback loop. This loop starts from the economic development, which attracts more people into the city; the increase of population inevitably leads to the increase demand of domestic water and the increase of domestic wastewater which is a major part of the total municipal wastewater, thus increasing the total amount of wastewater. As a result, the pollution of

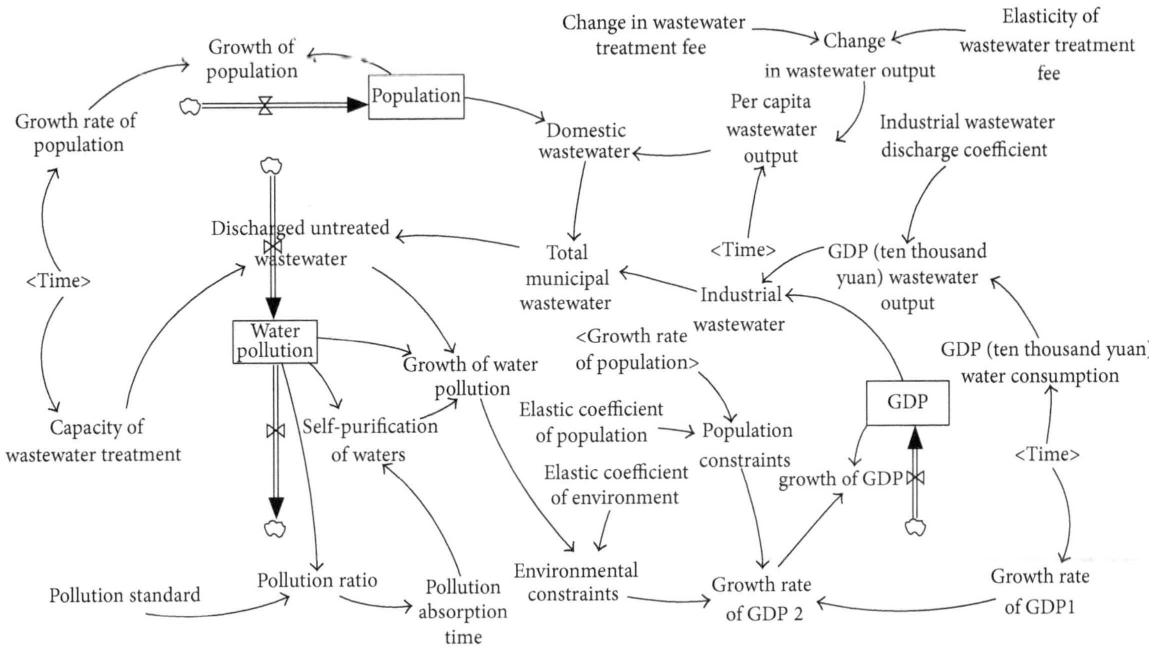

FIGURE 2: The stock-flow diagram of the model.

natural water body is exacerbated. And in reverse, water pollution can indirectly hinder the economic development. Therefore, the factor *economic growth* can affect itself through the negative feedback chain. Loop 2 is a negative feedback loop as well. In terms of economic growth element, the increase of population enlarges the labor resources that accelerate the economic growth; thus the industrial water use increases as economic growth speeding up. Finally, the industrial wastewater increases. As it is in loop 1, the variables affect itself negatively by the feedback mechanism. Loop 3 is a positive feedback loop, which is about capacity of municipal wastewater treatment. The economic development promotes the development of industries related to wastewater and the investment of wastewater treatment increases then. Consequently, the capacity of wastewater treatment can be largely improved. The wastewater discharged directly into water bodies can be reduced; namely, the water pollution diminishes. It can be seen that the abovementioned three feedback loops eventually contain the feedback of the environmental pollution on economic development. This is the first attempt to take into account the retarding effect of water pollution on the development of regional economy which is usually neglected by other researches using SD model.

The model structure is visually described through the causal loop diagram that preliminarily reflects the dynamic relationships of the wastewater forecast and management system. Though the causal loop diagram cannot represent the whole system and it is only a rough description of the system structure, it is essential for building a formal model (the stock-flow diagram).

2.1.3. Stock-Flow Diagram. Those cited in the causal loop diagram are the primary variables that control the behavior changes of the system, which are not the whole of the wastewater forecast and management system, and the nature of different variables cannot be distinguished through the causal loops diagram. Therefore, more variables and flow need to be added to complete the SD model; thus it is necessary to draw the stock-flow diagram which highlights the role of the stock and the flow rate in the entire system. Stock, the most important element in system dynamics, reflects a process of accumulation. Flow, reflecting the input and output rate of the stock, is the variable that affects the value of the stock directly through changing the rate of accumulation process. In this study, the modeling software is Vensim. The stock-flow diagram is shown as Figure 2.

Some necessary explanation and assumptions for the SD model established in this study are as follows:

(a) Municipal wastewater treatment capacity is one of the key factors that determine whether water pollution can be effectively controlled. Though the wastewater treatment capacity is largely influenced by the investment of wastewater treatment industry and economic development, it is regarded as a variable that only changes over time and can be controlled by the policy makers.

(b) Wastewater is divided into two categories depending on its source: domestic wastewater and industrial wastewater. Although the irrigation water demand is huge, most of agricultural wastewater, especially no-point wastewater, actually seeps into underground or flows into rivers through surface runoff. Due to the imperfectness of wastewater network, most agricultural wastewater could not be collected into wastewater treatment plants. The real amount of agricultural wastewater that gets into wastewater network is finite. In this model, agricultural wastewater is excluded.

(c) In cities, separate sewers drainage system are constructed in overwhelming majority of new districts. Rainwater is collected separately, and most of the collected rainwater is discharged directly into natural water bodies, while the wastewater is delivered to the wastewater treatment plant. But the combined sewers still exist in some old districts, which can be divided into two types: intercepting combined sewers and direct-discharging combined sewers. Due to water pollution caused by combined sewers, more and more combined sewers are reformed to implement rainfall-wastewater separation in old communities nowadays. Therefore, the amount of rain does not appear as a variable in the SD model.

(d) In this study, there are three state variables in SD model: *GDP* (Gross Domestic Product) standing for the economic development; *population* referring to the resident population in Wuhan; *water pollution* equivalent to the amount of discharged wastewater. Besides, there are four rate variables: *growth of GDP* which controls the changes of *GDP*; *growth of population* which changes the population size; *discharged wastewater* that is untreated changes the pollution degree of waters; and *self-purification of waters*. In addition, there are many auxiliary variables whose values are calculated according to other variables in the system. And there are four constants: *per capita water use quota, discharge coefficient of domestic wastewater, discharge coefficient of industrial wastewater*, and *pollution standard*. According to the objective of this study, the model can be divided into three parts: wastewater forecast, water pollution, and pollution control. They are interrelated and interacted and jointly determine the dynamic behavior of the system.

2.1.4. Major Equations

(1) Population. Population is one of the main driving factors of municipal water consumption. When per capita water use quota is constant, the population growth directly results in the growth of municipal water consumption, inevitably leading to the growth of wastewater. The factors affecting the size of population include the birth rate, death rate, immigration, and emigration, but only the net rate of population growth is considered in this study. The data comes from *Wuhan Statistical Yearbook*. The initial time set by the model is the year of 2000, in which the population of Wuhan was 8,0481,000, that is, the initial value of *population*. In the time bounds of the model, the annual value of *population* is determined by following equations:

$$\text{Pop}(t) = \text{Pop}(t_0) + \int_{t_0}^{t} \text{growth}(t)\,dt, \tag{1}$$

where Pop(t) is the population at time "t", growth(t) is the population growth at time "t", and t is the simulation step, in one year.

(2) GDP. Economic development is equivalent to economic growth, which is illustrated by Gross Domestic Product (GDP) in most literature. Neoclassical growth theory suggests that GDP growth is the result of the capital accumulation, the increase of labor, and the technological change acting in a long term. But the new theory of economic growth regards the labor as labor capital investment, including not only the absolute number of the labor force, but also the education level and production skills, and so forth. They are the production factors of during the economic growth. Certainly, they do not represent all production factors which also include the land and entrepreneurs. Regional economy cannot grow without the aggregation of production factors, and the aggregation of those factors brings the industry aggregation, thus promoting regional economic prosperity. Some scholars have clearly stated that the economic development is equivalent to the aggregation. However, more and more researches pay attention to regional economic development under the constraints of economic resources and environmental in recent years. In other words, resources and environment are also the important production factor that shall be taken into account when using the production function. In this paper, the influence of population and negative impact of environmental pollution on the economic growth are considered in the SD model for the first time. Population means labor resources, which is the most important production factor. Although whether the influence of population on the economic development is favorable, there is no consensus in different literatures. But on the whole, the economy of China is developing rapidly while China's population growth rate is declining, and the reason is that China has a large size of population and enjoys a huge demographic dividend.

The resource abundance of a region and environmental quality can limit the economic growth of the region. Total Factor Productivity (TFP) [22] considers resources and environments certainly lower than traditional productivity excluding resources and environment. In that way, economic growth slows down. Additionally, Nordhaus and Weitzman [23] first proposed the notion of *growth drag* and explained it as the reduction degree of per capita output growth rate with the resource constraints comparing with that without the resource constraints. Bruvoll et al. [24] clearly put forward the notion of *environmental drag* and considered that environment pollution could reduce the economic output of a region and the well-being of consumers. A dynamic Computable General Equilibrium (CGE) model was developed in their study to measure the damage extent of environmental constraints to Norwegian residents' welfare. Chinese scholars have also studied the "blocking effect" of environmental pollution to the regional economic development. Zhang et al. [25] illustrated the long-term dynamic relationships between the industry wastewater and industrial solid waste and the economic development by a VAR model in Guanzhong, Shanxi. It was found that the environmental pollution has obvious restraining effect on the economic development of Guanzhong. Li [26] has also measured the "blocking effect" on Chinese economic growth under the environmental constraints with a mean value of 0.44%; namely, Chinese economic growth declines 0.44% per year in the presence of resource and environmental constraints. The constraint of resources and environment is an objective reality, but neither resource nor environment was considered as a factor of production in traditional statistical methods of economic growth. Therefore, this study suggests that the environmental pollution pays negative impacts on the economic growth. Coordinated development of economy and environment is the essence of sustainable development; SD model of this article takes negatively effect of the environmental deterioration to the economic development into consideration, which directly reflects in the dragon GDP growth. The value of GDP is calculated by the following equations:

$$\begin{aligned} \mathrm{GDP}(t) &= \mathrm{GDP}(t_0) + \int_{t_0}^{t} \mathrm{Growth}(t)\,dt, \\ \mathrm{Growth}(t) &= \mathrm{GDP}(t) \times \mathrm{rate}_2(t), \\ \mathrm{rate}_2(t) &= \mathrm{rate}_1(t) \times (1+e_1)(1-e_2), \end{aligned} \tag{2}$$

where GDP(t) is GDP at time "t", Growth(t) is GDP growth at time "t", t is the simulation step, in one year, $\mathrm{rate}_1(t)$ is the growth rate without constraints, $\mathrm{rate}_2(t)$ is the growth rate under constraints, e_1 is the influence coefficient of population to economy, and e_2 is the influence coefficient of environment to economy.

(3) Water Pollution. For the pollution degree of natural water bodies in the study region, only the influence of untreated municipal wastewater is considered according to the foregoing assumptions. The receiving water of that region is regarded as a whole and the pollution degree depends on the stock of wastewater in the waters that is determined by the amount of untreated wastewater and self-purification of waters. Specifically, the removal of pollutant depends on the stock of *water pollution* and *pollution absorption time (pat)* that represent the time spent in entirely removing a certain amount of contamination. So the amount of water self-purification for each unit of time is the ratio of *water pollution* and *pollution absorption time.* In this study, *pat* is a nonlinear function of the *pollution ratio (polr).* Lookup function in Vensim software is used to build an equation to describe the nonlinear relationship. *Pollution ratio,* represented by the *water pollution* and *pollution standard (pols),* is the characterization of the pollution levels. The main equations in this part are as follows:

$$\begin{aligned} \mathrm{Pol}(t) &= \mathrm{Pol}(t_0) + \int_{t_0}^{t} [\mathrm{dis}(t) - \mathrm{pur}(t)]\,dt, \\ \mathrm{pur}(t) &= \frac{\mathrm{Pol}(t)}{\mathrm{pat}(t)}, \\ \mathrm{polr}(t) &= \frac{\mathrm{Pol}(t)}{\mathrm{pols}}, \end{aligned} \tag{3}$$

where Pol(t) is the value of water pollution at time "t", dis(t) is the amount of untreated wastewater at time "t", pur(t) is the water self-purification at time "t", and t is the simulation step, in one year.

(4) Wastewater Treatment Fee. The increase of water price can generally reduce the water demand [27]. Furthermore, the price leverage can ease the contradiction between water supply and demand. Wastewater treatment fee generally consists of two parts, namely, the government subsidy and the actual payment by polluters. The range of wastewater treatment fee that the residents in Chinese capital city actually pay is from 0.3 to 1.4 yuan/m^3, which accounts for 14% to 46% of living water price. Industrial wastewater must be treated to meet the water quality discharge standard before discharged into the urban wastewater pipe network, and its treatment fee is the same as that of domestic wastewater. The wastewater treatment fee in Wuhan was 0.8 yuan/m^3 before August 1, 2015, thereafter 1.1 yuan/m^3, rising 37.5%. Municipal wastewater treatment fee has not received adequate attentions because it is included in the actual water charges. To increase the wastewater treatment fee can reduce the discharge of municipal wastewater, thereby cutting down the burden of municipal wastewater treatment system which is caused by the growing amount of wastewater. And what should be particular of concern is how much the rise of wastewater treatment fees can lead to the wastewater reduction. The double logarithmic analysis model [28] is commonly utilized to decide the price elasticity of water demand water:

$$\ln w = \alpha \ln p + \gamma \ln y + \eta, \tag{4}$$

where w is water demand, p is water price, y is household income, and η is errors.

The impact of water price and household income on water demand is considered in this model and α and γ are the price elasticity and income elasticity. In this paper, the model above is simplified to get the equation for the quantitative relationship between the increase of wastewater treatment fee and the changes of wastewater:

$$c = \left(\frac{p_2}{p_1}\right)^{\alpha}, \tag{5}$$

where c is the coefficient of wastewater variation, α is the wastewater price elasticity, p_1 is the current wastewater treatment fee, and p_2 is the adjusted wastewater treatment fee.

2.2. Model Validation. According to Barlas [29], the ultimate objective of system dynamics model validation is to test the validity of the structure of the model which can be classified as two types, namely, direct structure test and structure-oriented behavior test. To test direct structure validity of the model, we have checked that whether the model includes all important variables that are to be investigated. The SD model developed in this study is to simulate the changes of municipal waste water under the socioeconomic development and the scenarios after some related policy implemented. Simultaneously, to find out whether the feedback relationship is illogical, the correctness of the relationships between variables must be checked. Eventually, it is found that the SD model for wastewater forecast and management not only meets the modeling purpose, but also guarantees the completeness and accuracy of the relationship between the variables. In the model with numerous variables and complex causal loops, units of these variables and their quantitative relationships in the equations are fallible. Unit check can ensure the consistency of variables in the causal chain. When Vensim warns an error in checking units, it is necessary to reassign unit for the variable that has wrong unit. If the error still exists, all the other variables related to that variable should be checked to assure they are defined correctly.

After completing the SD model and model validation, a case study in Wuhan is carried out. In the next section, the study region is introduced and the variables are assigned; then the simulation for status quo is conducted in order to calibrate the SD model established previously. In the end, the scenarios of different future development patterns are designed for the scenario analysis.

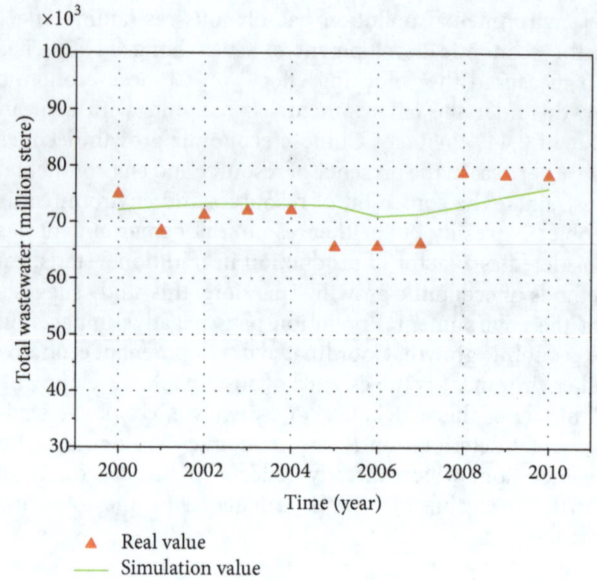

FIGURE 3: The comparison of the real value and the simulation value.

3. Case Study

3.1. Study Region. Wuhan, the capital of Hubei province, is located in central China. Wuhan has a vast land of 8,494.41 km^2 and a population of 102,200,000 [30], which is subject to a typical north subtropical monsoon climate with four distinctive seasons, abundant rainfall, and heat, with annual temperature of 15.8°C–17.5°C. Wuhan is rich in water resources and has a total water area of 2,217.6 km^2 accounting for 26.1% of the city's land [31]. However, the adequate water resources and relatively cheap price lead to low efficiency of water utilization and low wastewater reuse rate which result in a huge amount of wastewater output. This is contrary to the resource conservation policy advocated by the government, and a more serious problem is that wastewater output increases as a result of the low utilization rate of water resource. Currently, 12 wastewater treatment plants with a total capacity of 1,885,000 tons per day have been built in the main city of Wuhan so far. The amount of wastewater changes drastically under different policies, which would bring enormous pressures on the collection, treatment, planning, and management of municipal wastewater.

3.2. Values of Variables. There are three level variables in the SD model, namely, population, GDP, and water pollution, whose initial values need to be set. The initial value of population and GDP is 8,048,100 and 1.20648 billion which is based on the statistical data of 2000, the starting year of this model. The values of population growth rate, GDP growth rate, and water consumption of annual ten thousand GDP are extracted from historical statistics from 2000 to 2012. The data is from the *Wuhan Statistical Yearbook* [32] during the period from 2000 to 2012. The initial value of pollution is assumed to be 1 in 2000. In this study, the population elasticity to GDP, e_1, is set as 0.2; namely, each 10% population growth results in 2% GDP growth. The coefficient environmental drag to GDP growth, e_2, is set as 0.12 [25]. The elasticity of wastewater treatment fee for wastewater output α is set as 0.1. The initial values of the variables and the main parameters in the SD model are listed in Table 1.

3.3. Model Calibration. The simulation results of status quo simulation were compared with the historical statistical data to evaluate the reliability of the SD model. Based on the guidelines of "the process of estimating the model parameters (structure) to obtain a match between observed and simulated structures and behaviors [as] a stringent test of a hypothesis linking structure to behavior" [33], a sensitivity analysis was conducted and the wastewater system in Wuhan during 2000–2010 was simulated with the model. The mean absolute percentage errors (MAPE) between the simulated data and the historical data are shown in Table 2. The average of MAPE is 5.25% for the amount of treated wastewater. Figure 3 shows the comparison between the simulated data and real values of the amount of treated wastewater. As can be seen, the trend of simulated values coincides with the real values during the period of 2000–2010 during which the statistical data are available. These results indicate that the model can effectively simulate the real municipal wastewater system, and thus it is reliable.

3.4. Simulation Results

3.4.1. Simulation Results of Status Quo. Status quo simulation results of total wastewater have been shown in last section, and Figure 4 shows the comparison of total wastewater (TW) and wastewater treatment capacity (WTC). The maximum difference between those two variables appears in the period of 2000 to 2002 and decreases yearly. This gap flattens after the wastewater treatment capacity increasing sharply by the year of 2007. In fact, the government of Wuhan had been aware of the seriousness of pollution and constructed a number of wastewater treatment facilities making use of

TABLE 1: Details of the parameter values used in the SD model.

Parameter type		Value	Source	Note
Initial value	GDP	120.648 billion	*Wuhan Statistical Yearbook*	The data of 2000
	Population	8,048,100	*Wuhan Statistical Yearbook*	
	Water pollution	1	Assumption	
Constant	Change in wastewater treatment fee	37.5%	The government of Wuhan	Increase (compared to the data of 2013)
	Industrial wastewater discharge coefficient	0.75	Code for design of outdoor wastewater engineering (GB 50014, 2006)	
	Pollution standard	1	Assumption	
	Elastic coefficient of population	0.2	Kelley and Schmidt, 1995 [21]	
	Elastic coefficient of environment	0.12	Zhang et al., 2009	
	Elasticity of wastewater treatment fee	−0.1	Ruijs et al., 2008	Combination of assumption and the reference
Lookup function	Growth rate of population		*Wuhan Statistical Yearbook*	Time series from 2000 to 2010
	GDP (ten thousand yuan) water consumption		*Wuhan Statistical Yearbook*	
	Growth rate of GDP1		*Wuhan Statistical Yearbook*	
	Capacity of wastewater treatment		Wuhan Water Authority	
	Per capita wastewater output	If (time ≤ 2014,65 m^3), then it is determined by other variables	Code for design of outdoor wastewater engineering (GB 50014, 2006)	
	Pollution absorption time		World model	

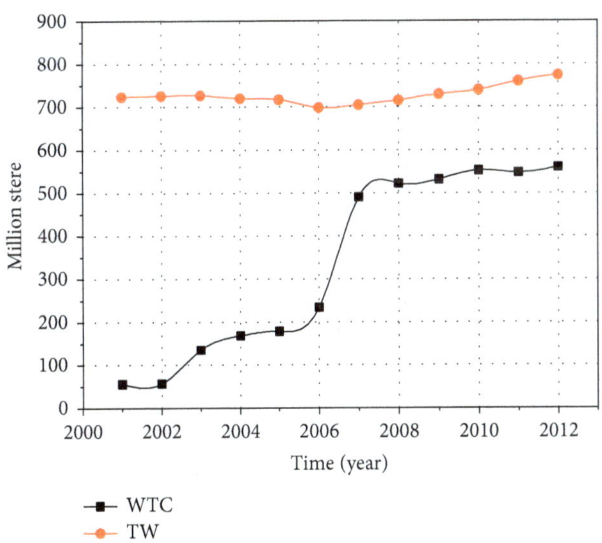

FIGURE 4: The comparison of wastewater treatment capacity and total wastewater.

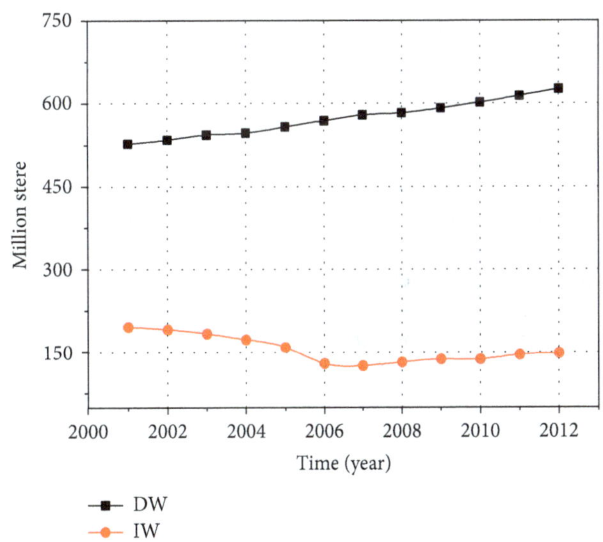

FIGURE 5: The component of total wastewater: domestic wastewater (DW) & industrial wastewater (IW).

Asian Development Bank (ADB) loans and other funds from 2001. So the gap between the wastewater treatment capacity and the amount of wastewater was becoming smaller. But there is still a gap of 200 million tons by the year of 2012. As a result, a large amount of untreated wastewater was discharged into the receiving waters, which has an extremely negative impact on the urban water environment.

Figure 5 shows the trend of domestic wastewater and industrial wastewater; we can see that domestic wastewater rises slowly and makes up a major part of municipal wastewater, accounting for more than 60% of the total wastewater.

TABLE 2: The comparison of the real value and the simulation value.

Year	The real value	The simulation value	MAPE (percent)
2000	74971.03	72183.4	3.72
2001	68456	72681.8	6.17
2002	71189.72	73146	2.75
2003	72026.27	73193.8	1.62
2004	72144.77	73116.1	1.35
2005	65665.27	72926.5	11.06
2006	65746.34	71075	8.10
2007	66283.2	71488.1	7.85
2008	78858.76	72973.3	7.46
2009	78435.06	74689.8	4.77
2010	78376.66	76127	2.87
		MAPE	5.25

TABLE 3: The design of scenarios.

Scenarios	Growth rate of GDP (%)	Water consumption of annual ten thousand GDP (m^3)	Increase of wastewater treatment fee (%)
1	10	35	37.5
2	10	25	37.5
3	8	35	37.5
4	8	35	50
5	8	35	100
6	8	30	37.5
7	8	25	37.5
8	6	35	37.5
9	6	25	37.5

Also, it can be seen that there is a downward trend for industrial effluent over time.

3.4.2. Scenario Design. Scenario analysis can also be called policy simulation, which treats some major factors as the control variables; their values are to be adjusted to form different simulation schemes. This simulation tests the policies in the period of 2015 to 2030. The goal of resource-saving and environment-friendly society is to obtain rapid economic growth in the case of low resource consumption and less waste generation. The economic growth is characterized as the growth of GDP, and resource conservation in this article mainly refers to reducing the industrial water consumption. Thus three main factors are chosen as scenario variables for the scenario analysis, namely, GDP growth rate, water consumption of annual ten thousand GDP, and wastewater treatment fee. The different assignments of each variable constitute different future scenarios, which are listed in Table 3.

Scenarios 1 and 2. These two scenarios represent the future scenario of municipal wastewater under rapid economic development; that is, the average annual growth rate of GDP is 10% in Wuhan during the period of 2015–2030. For the water consumption of annual ten thousand GDP, two levels are set in these two scenarios. Namely, the water consumption of annual ten thousand GDP is set as 35 m^3 (current level) and 25 m^3 by the year of 2030. The increase of wastewater treatment fee is 37.5% (compared to the data of 2014) which was adjusted by the government in 2015.

Scenarios 3~7. The future municipal wastewater is simulated under the moderate pace of economic development, and the average annual growth rate of GDP is set as 8% which is the current GDP growth rate of Wuhan. The impact of resource conservation and wastewater treatment fee on municipal wastewater are investigated in these scenarios. In scenarios 3~5, only the increase of wastewater treatment fee is changed, while the input value of water consumption of annual ten thousand GDP is set as 35 m^3, the current level. Three different levels are set for the increase of wastewater treatment fee; one is 37.5%; the second is 50% which is higher than the current; the last is the highest, 100%. In scenario 6 and scenario 7, the industrial water consumption of annual ten thousand GDP is set as 30 m^3 and 25 m^3, respectively, while the increase of wastewater treatment fee is set as the current value, 37.5%.

Scenarios 8 and 9. Scenarios 8 and 9 are the analysis of future municipal wastewater under the economic growth of a low speed while the values of other variables are the same as scenarios 1 and 2. The average annual growth rate of GDP drops to 6%. Scenario 8 is the modeling of the response to policies changes under the development pattern of low

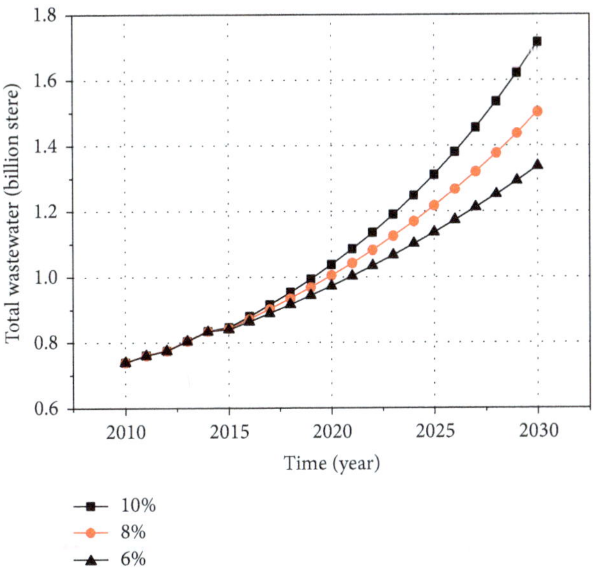

FIGURE 6: Total wastewater under different GDP growth rate.

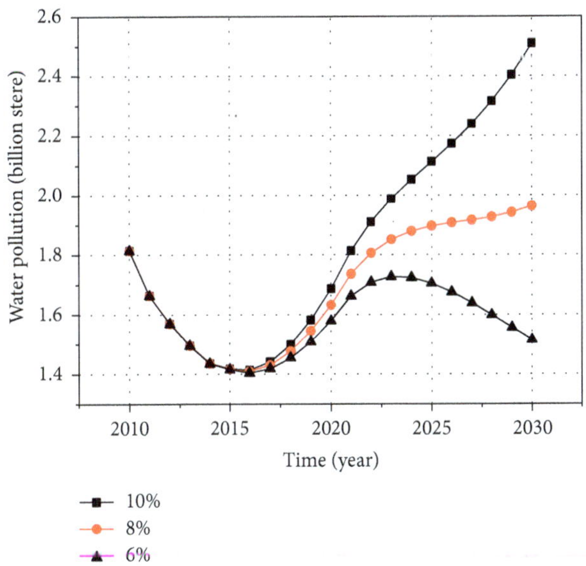

FIGURE 7: Water pollution under different GDP growth rate.

growth and high consumption. On the contrary, scenario 9 is to simulate the dynamic change of the entire system under the scenario of a low growth and low consumption.

3.4.3. Simulation Results of Future Scenarios

(1) GDP Growth Rate. Wuhan is a central developing city and the GDP growth is always regarded as one of the most important indicators of social development. However, rapid growth of GDP is pursued while environment issues are ignored, even at the expense of the environment. In this section, the influence of GDP growth to the wastewater amount and water environment is analyzed. Figure 6 shows the change of the amount of municipal wastewater under different GDP growth rate (10%, 8%, and 6%) while the other factors are constant. Obviously, there is a positive correlation between the GDP growth rate and the amount of municipal wastewater. The growth of GDP leads to the growth of municipal wastewater, and greater GDP growth leads to faster growth in wastewater. On the basis of 10%, 8%, and 6% GDP growth rate, the total amount of wastewater in 2030 increases by 50.6%, 43.8%, and 37.2%, respectively, compared to that in 2015. Compared to the current GDP growth rate 8%, when GDP growth rate increases to 10%, the total amount of wastewater by 2030 increases by 14.3%, and when it decreases to 6% that would decrease by 10.8%.

As shown in Figure 7, the urban water pollution is analyzed in the situation that the GDP growth rate changes, while maintaining the current capacity of municipal wastewater treatment unchanged. During the years 2010 and 2020, the turning point appears at time point of 2017 in these three curves; urban water pollution shows a rising trend after first cut, presenting U-shape. It demonstrates that the current municipal wastewater treatment capacity will alleviate the environmental pollution to some extent. But in the long term (2020–2030), when GDP is at the high growth level of 10%, water pollution (the cumulative amount of untreated wastewater) shows a growing trend; when GDP growth rate keeps at current 8%, water pollution gradually becomes stable after 2025; when GDP growth rate decreases to the low level of 6%, water pollution appears as a turning point in 2023, showing a trend of rising first and then falling and finally keeping to fall. By 2030, if GDP growth rate increases from the current level to 10%, water pollution would increase by 27.8%; if GDP growth rate decreases from the current level to 6%, water pollution would decrease by 22.8%. With the increase of wastewater output, when GDP growth rate of Wuhan keeps at the current level or higher than the current level, urban water pollution can be aggravated to some extent, which indicates that current municipal wastewater treatment capacity cannot meet the requirements of environment-friendly society and needs to be expanded.

(2) Water Consumption of Annual Ten Thousand Yuan GDP. GDP (ten thousand yuan) water consumption is another important factor affecting the amount of wastewater output, and it is related to the way of economic growth, the industrial structure of the region, the level of production equipment, and natural conditions.

Figure 8 shows that municipal wastewater changes are caused by different magnitude of water consumption of annual ten thousand GDP (35 m^3, 30 m^3, and 25 m^3). In the case of more extensive use of water resources, output of wastewater increases significantly. Compared to the current level of high water consumption, by 2030, when the GDP growth rate is 8%, the increase of the wastewater treatment fee is 37.5%, at the situation of resource saving (then thousand yuan GDP water consumption is 30 m^3 and 25 m^3), and the wastewater output of the whole city decreases by 6.0% and 12.0%, which is a great relief to the wastewater treatment system.

When other variables (GDP growth rate, wastewater treatment fee) are all kept at the current level, the total amount of water pollution can be reduced significantly as

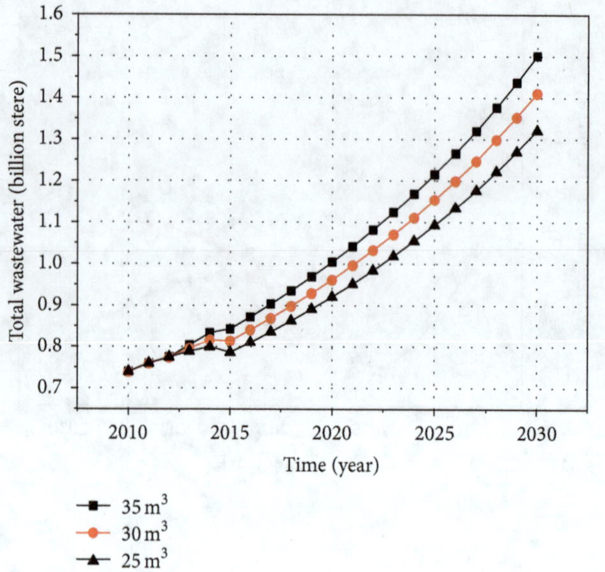

FIGURE 8: Total wastewater under different per ten thousand yuan GDP water consumption.

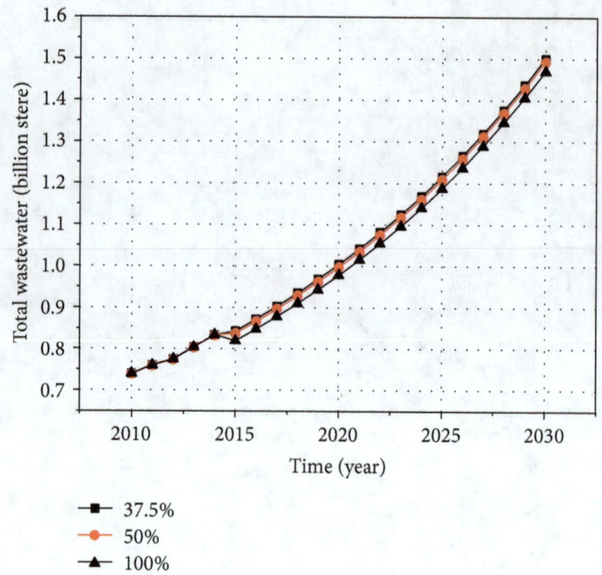

FIGURE 10: Total wastewater under different wastewater treatment fee.

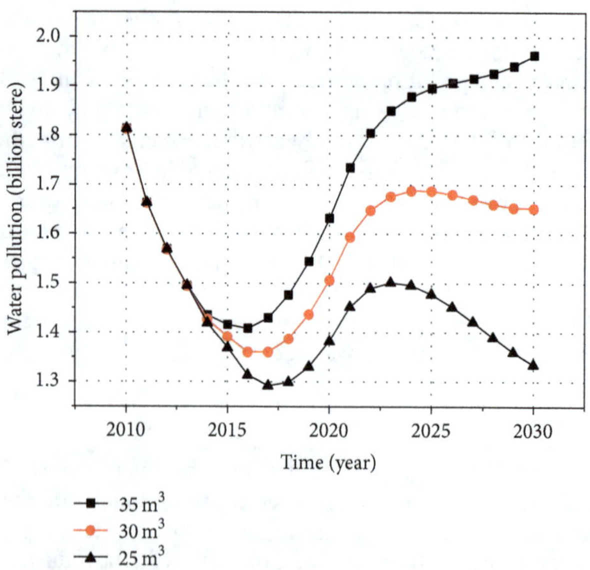

FIGURE 9: Water pollution under different per ten thousand yuan GDP water consumption.

soon as the implementation of the policy of water consumption reduction. Moreover, the more the water consumption is reduced, the more obvious the water pollution is. By 2030, due to the reduction of GDP water consumption (ten thousand yuan), decreasing from 35 m^3 to 30 m^3 and 25 m^3,respectively, the total amount of water pollution reduces by 15.8% and 32.0%, respectively. As seen in Figure 9, during the years 2010 and 2020, the three curves show a trend of decrease first and then increase. The difference is that the greater the GDP water consumption (ten thousand yuan) is, the earlier the turning point appears. This is because the water pollution is a status variable in the model and its accumulation depends on the input and output of status variables. The increase of water consumption leads to the increase of input, and the output depends on the self-purification capacity of water bodies and it has little changes to some extent and can be seen as a fixed value, so the status variables can grow rapidly. In the long term (2020–2030), high water consumption (35 m^3) leads to a rapid rise in water pollution, and the rise speed becomes slow and trends to stable after 2025; the rise speed under the medium water consumption (30 m^3) is lower than that under high water consumption, and the speed begins to slow down after 2024; under the situation of low water consumption (25 m^3), after the year 2020, water pollution experiences 3 years of rising and then begins to drop quickly. It can be seen that reducing the water consumption per unit output value is important way to reduce wastewater output and water pollution and it is a major measure of clean production process for enterprises. In the process of wastewater treatment, both source control and end treatment should be paid attention to.

(3) Wastewater Treatment Fee. With the continuous expansion of the water industry marketization in China, the pricing of the wastewater treatment fee is getting more and more attention. Reasonable pricing can bring a certain income for wastewater treatment enterprise to ensure its normal operation and also can indirectly encourage the implementation of water-saving measures. In this section, the influence of wastewater treatment fee on the future amount of wastewater and the situation of water pollution in the city in this case is analyzed. As shown in Figure 10, when the increase of wastewater treatment fee is added from 37.5% to 50% and 100%, respectively, the total amount of wastewater decreases to some extent (by 0.5% and 2.1%, resp.) during the period of policy simulation; the greater the increase is, the more obvious the wastewater amount decreases. For water pollution, the increase of wastewater treatment fee can also bring significant benefits. Similar to other two control factors,

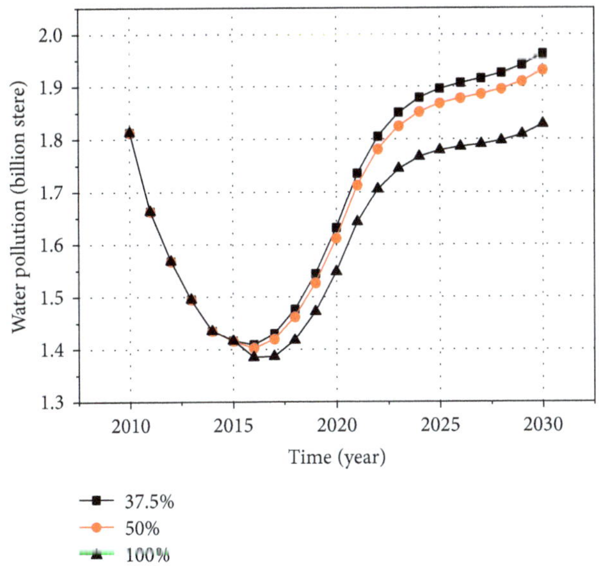

FIGURE 11: Water pollution under different wastewater treatment fee.

the influence curve of wastewater treatment fee on water pollution shows a trend of decrease first and then increase. The control role of the wastewater treatment fee to water pollution is relatively weak; the water pollution under three situations of different wastewater treatment fees shows a trend of decrease during the years 2020 and 2030 as shown in Figure 11. In general, the increases of wastewater treatment fee to some extent can restraint the aggravation of water pollution although its influence is limited.

Meanwhile, make longitudinal comparison to the three control factors selected in this article; changing such three factors can reduce the total amount of wastewater and restraint water pollution to some extent, and the influence of the wastewater treatment fee is the smallest.

(4) Different Development Modes. Finally, the future development modes of Wuhan are summarized as high-growth, high water consumption (H-H; scenario 1); high-growth, low water consumption (H-L; scenario 2); low growth, high water consumption (L-H; scenario 8), and low-growth, low water consumption (L-L; scenario 9). Under the situation of H-H, the total amount of wastewater in the future gets the fastest growth and it increases by 50.6% in 2030 when comparing with that in 2015; under the situation of L-L, it gets the slowest growth and it increases by 34.9% when comparing with that in 2015; the growth rate of wastewater during the years 2015 and 2023 under the situation of H-L is lower than that of L-H, but during the years 2023 and 2030, it will be higher than that of L-H. By 2030, the total amount of wastewater in Wuhan under different situations is sorted by size as follows: H-H > H-L > L-H > L-L. Comparing the situations of H-L and L-H, it can be seen in Figure 12 that the influence of GDP water consumption (then thousand yuan) to the total amount of wastewater is greater than that of GDP growth rate in the near future (2015–2023), and the influence of GDP growth rate is more obvious in the long term (2023–2030).

As shown in Figure 13, urban wastewater under different development model presents great differences. Under the

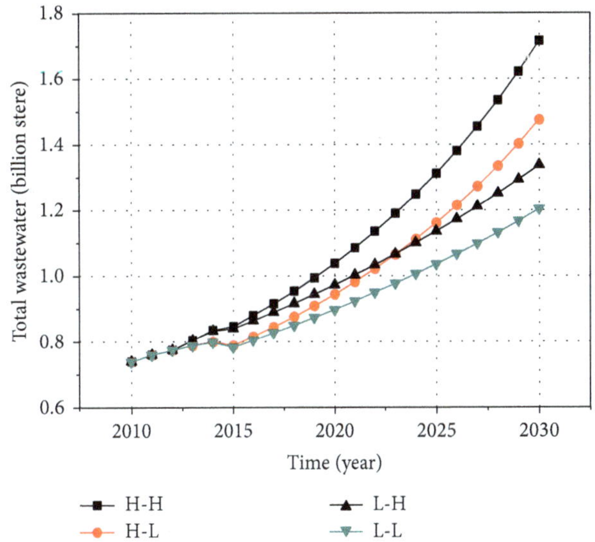

FIGURE 12: Total wastewater under different development patterns.

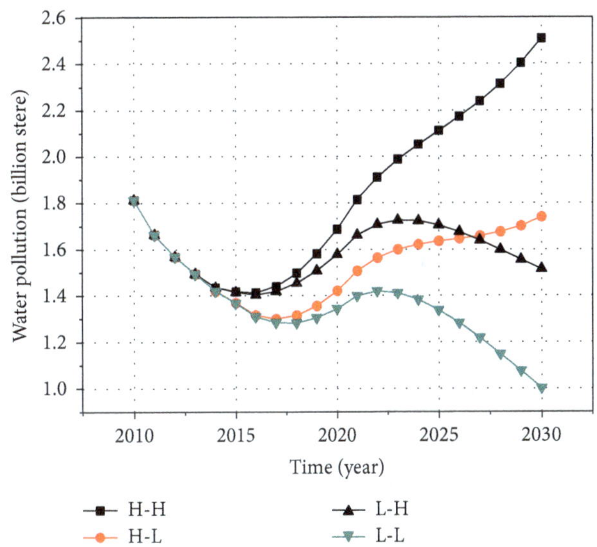

FIGURE 13: Water pollution under different development patterns.

development model of low economic growth, water pollution experiences a process of "decrease-increase-decrease" in the simulation period (2015–2030) and it presents a trend of decrease in the long term; but under two development models of high economic growth, water pollution presents a process of "decrease-increase" and finally presents a trend of increase and becomes stable. Thus, it can be seen that the wastewater treatment capacity of Wuhan would not meet current wastewater treatment needs and water pollution would continue to be aggravated if under the situation of high economic development. The two curves of H-L and L-H are analyzed and the intersection point appears in the year 2027. Before 2027, the water pollution under the development model of L-H is more serious than that of H-L. After 2027, the water pollution presents a trend of rising under the development model of H-L and dropping under the development model of L-H. In the long term, the influence of economic development to water pollution is greater.

4. Conclusion

The current work systematically investigated the socioeconomic, municipal wastewater, and urban water environment by the system dynamics for the first time, which provided a new strategy for the management of municipal wastewater system. In addition, it also offers the overlook of dynamic change of the whole system and the theoretical foundation to the management to make the right decision. Based on the discussion above, it can be concluded as follows:

(1) The factors affecting the sustainable development of the municipal wastewater system and their relationship and feedback mechanism were investigated. Based on the results, a SD model for the sustainable municipal wastewater system was developed from the socioeconomic, municipal wastewater, and urban water environment to simulate the future municipal wastewater and to explore the key factors that affect the municipal wastewater system.

(2) The reliability of the SD model was demonstrated by case study. The SD model was able to effectively simulate the real system, even the system with insufficient historical data, by the combination of qualitative and quantitative analyses. The dynamics simulation of municipal wastewater is an important part of urban water pollution control and management. It is significant for reducing pollution, ensuring the health of urban water environment, improving the life quality of the urban residents, and achieving the sustainable development.

(3) The case study also indicated that the municipal wastewater increased with the growth of GDP and increasing the wastewater treatment fee could significantly reduce the emission of wastewater. The municipal wastewater and water pollution increase most rapidly in the scenario of rapid economic growth and extensive resource utilization, and thus it cannot blindly pursue economic growth while ignoring resource conservation.

(4) The SD model was used to evaluate the sustainable development of the municipal wastewater using GDP growth rate, water consumption of annual ten thousand GDP, and wastewater treatment fee as the control variables. The results indicated that, compared with the GDP growth rate, the water consumption per ten thousand GDP affected the municipal wastewater more significantly in the short term. However, the influence of GDP growth rate was more obvious in the long term. Therefore, the effective implementation of water conservation measures is an efficient way to reduce the emission of wastewater and to eventually realize the environment-friendly wastewater system.

Competing Interests

The authors declare that there are no competing interests regarding the publication of this paper.

Acknowledgments

The project was financed by the National Natural Science Foundation of China (Grant no. 71390524). The authors would like to thank Professor Minhui Yu in School of Automation, Huazhong University of Science and Technology, for his guidance in developing the SD model, and Wuhan Urban Drainage Development Co., Ltd., for its basic data.

References

[1] T. Panayotou, *Empirical Tests and Policy Analysis of Environmental Degradation at Different Stages of Economic Development*, vol. 4 of *Ilo Working Papers*, International Labour Office, 1993.

[2] S. Dawadi and S. Ahmad, "Evaluating the impact of demand-side management on water resources under changing climatic conditions and increasing population," *Journal of Environmental Management*, vol. 114, pp. 261–275, 2013.

[3] I. Cazcarro, R. Duarte, and J. Sánchez-Chóliz, "Economic growth and the evolution of water consumption in Spain: a structural decomposition analysis," *Ecological Economics*, vol. 96, pp. 51–61, 2013.

[4] R. Duarte, J. Sánchez-Chóliz, and J. Bielsa, "Water use in the Spanish economy: an input–output approach," *Ecological Economics*, vol. 43, no. 1, pp. 71–85, 2002.

[5] T. Okadera, M. Watanabe, and K. Q. Xu, "Analysis of water demand and water pollutant discharge using a regional input-output table: an application to the City of Chongqing, upstream of the Three Gorges Dam in China," *Ecological Economics*, vol. 58, no. 2, pp. 221–237, 2006.

[6] W. Wang, L. Gao, P. Liu, and A. Hailu, "Relationships between regional economic sectors and water use in a water-scarce area in China: a quantitative analysis," *Journal of Hydrology*, vol. 515, pp. 180–190, 2014.

[7] M. S. Babel, A. D. Gupta, and P. Pradhan, "A multivariate econometric approach for domestic water demand modeling: an application to Kathmandu, Nepal," *Water Resources Management*, vol. 21, no. 3, pp. 573–589, 2007.

[8] Y. Bai, P. Wang, C. Li, J. Xie, and Y. Wang, "A multi-scale relevance vector regression approach for daily urban water demand forecasting," *Journal of Hydrology*, vol. 517, pp. 236–245, 2014.

[9] S. Alvisi, M. Franchini, and A. Marinelli, "A short-term, pattern-based model for water-demand forecasting," *Journal of Hydroinformatics*, vol. 9, no. 1, pp. 39–50, 2007.

[10] Y. Cai, "Forecasting for quantile self-exciting threshold autoregressive time series models," *Biometrika*, vol. 97, no. 1, pp. 199–208, 2010.

[11] G. J. Bowden, G. C. Dandy, and H. R. Maier, "Input determination for neural network models in water resources applications. Part 1—background and methodology," *Journal of Hydrology*, vol. 301, no. 1–4, pp. 75–92, 2005.

[12] C. Bennett, R. A. Stewart, and C. D. Beal, "ANN-based residential water end-use demand forecasting model," *Expert Systems with Applications*, vol. 40, no. 4, pp. 1014–1023, 2013.

[13] M. Firat, M. E. Turan, and M. A. Yurdusev, "Comparative analysis of neural network techniques for predicting water consumption time series," *Journal of Hydrology*, vol. 384, no. 1-2, pp. 46–51, 2010.

[14] S. Ahmad and D. Prashar, "Evaluating municipal water conservation policies using a dynamic simulation model," *Water Resources Management*, vol. 24, no. 13, pp. 3371–3395, 2010.

[15] X. Xi and K. L. Poh, "Using system dynamics for sustainable water resources management in Singapore," in *11th Annual Conference on Systems Engineering Research, CSER 2013*, pp. 157–166, usa, March 2013.

[16] E. G. R. Davies and S. P. Simonovic, "Global water resources modeling with an integrated model of the social-economic-environmental system," *Advances in Water Resources*, vol. 34, no. 6, pp. 684–700, 2011.

[17] H. C. Guo, L. Liu, G. H. Huang, G. A. Fuller, R. Zou, and Y. Y. Yin, "A system dynamics approach for regional environmental planning and management: A Study for The Lake Erhai Basin," *Journal of Environmental Management*, vol. 61, no. 1, pp. 93–111, 2001.

[18] J. Zhu, X. Wang, L. Zhang, H. Cheng, and Z. Yang, "System dynamics modeling of the influence of the TN/TP concentrations in socioeconomic water on NDVI in shallow lakes," *Ecological Engineering*, vol. 76, pp. 27–35, 2015.

[19] O. Sahin, R. A. Stewart, and M. G. Porter, "Water security through scarcity pricing and reverse osmosis: a system dynamics approach," *Journal of Cleaner Production*, vol. 88, pp. 160–171, 2015.

[20] M. Zarghami and S. Akbariyeh, "System dynamics modeling for complex urban water systems: application to the city of Tabriz, Iran," *Resources, Conservation and Recycling*, vol. 60, pp. 99–106, 2012.

[21] A. C. Kelley and R. M. Schmidt, "Aggregate population and economic growth correlations: the role of the components of demographic changes," *Demography*, vol. 32, no. 4, pp. 543–555, 1995.

[22] R. M. Solow, "A contribution to the theory of economic growth," *The Quarterly Journal of Economics*, vol. 70, no. 1, pp. 65–94, 1956.

[23] W. D. Nordhaus and M. L. Weitzman, "Lethal model 2: the limits to growth revisited," *Brookings Papers on Economic Activity*, vol. 23, no. 2, pp. 1–59, 1969.

[24] A. Bruvoll, S. Glomsrød, and H. Vennemo, "Environmental drag: evidence from norway," *Ecological Economics*, vol. 30, no. 2, pp. 235–249, 1999.

[25] Q. Zhang, H. F. Xue, H. Z. Dong, and R. Liu, "Empirical research on the relationship between the regional economic development and environmental pollution," *China. Statistics &Decision*, no. 7, pp. 77–79, 2009.

[26] G. Li, "The analysis of the blocking effect to economic growth under the constraints of resource and environment," *China. Chin. Econ.*, no. 4, pp. 28–33, 2008.

[27] A. Ruijs, A. Zimmermann, and M. van den Berg, "Demand and distributional effects of water pricing policies," *Ecological Economics*, vol. 66, no. 2-3, pp. 506–516, 2008.

[28] S. M. Olmstead, W. M. Hanemann, and R. N. Stavins, "Water demand under alternative price structures," *Journal of Environmental Economics and Management*, vol. 54, no. 2, pp. 181–198, 2007.

[29] Y. Barlas, "Formal aspects of model validity and validation in system dynamics," *System Dynamics Review*, vol. 12, no. 3, pp. 183–210, 1996.

[30] Wuhan Government. The population.

[31] Wuhan Government, The natural resource.

[32] Wuhan Bureau of Statistics. Statistical Yearbook.

[33] R. Oliva, "Model calibration as a testing strategy for system dynamics models," *European Journal of Operational Research*, vol. 151, no. 3, pp. 552–568, 2003.

15

A Review on Radiation Damage in Concrete for Nuclear Facilities: From Experiments to Modeling

Beatrice Pomaro

Department of Civil, Environmental and Architectural Engineering, University of Padova, via F. Marzolo 9, 35131 Padova, Italy

Correspondence should be addressed to Beatrice Pomaro; beatrice.pomaro@dicea.unipd.it

Academic Editor: Anna Pandolfi

Concrete is a relatively cheap material and easy to be cast into variously shaped structures. Its good shielding properties against neutrons and gamma-rays, due to its intrinsic water content and relatively high-density, respectively, make it the most widely used material for radiation shielding also. Concrete is so chosen as biological barrier in nuclear reactors and other nuclear facilities where neutron sources are hosted. Theoretical formulas are available in nuclear engineering manuals for the optimum thickness of shielding for radioprotection purposes; however they are restricted to one-dimensional problems; besides the basic empirical constants do not consider radiation damage effects, while its long-term performance is crucial for the safe operation of such facilities. To understand the behaviour of concrete properties, it is necessary to examine concrete strength and stiffness, water behavior, volume change of cement paste, and aggregate under irradiated conditions. Radiation damage process is not well understood yet and there is not a unified approach to the practical and predictive assessment of irradiated concrete, which combines both physics and structural mechanics issues. This paper provides a collection of the most distinguished contributions on this topic in the past 50 years. At present a remarkable renewed interest in the subject is shown.

1. Introduction

Despite the increasing interest in renewable energy, nuclear power still has an important role in the energy supply worldwide. Almost 12% of world's electricity is generated by nuclear power plants; nuclear energy is used by 30 countries in the world, for total 439 units, 64 new ones being under design. Europe depends on nuclear power for more than one-quarter of its electricity; in fact almost 30% of electricity demand is satisfied this way; 40% comes from fossil fuel and the remaining from renewables. 131 nuclear power reactors were operating in 2014 in Europe and nuclear plant construction is currently underway in three EU member states: Finland, France, and Slovakia [1].

Understanding the conditions that lead to concrete deterioration is object of a new attention worldwide in matter of nuclear plants. In USA many plants have already had a life extension from 40 to 60 years and serious consideration is now being given to a further extension to 80 years. In fact according to the Atomic Energy Act of 1954 and the Nuclear Regulatory Commission regulations the operating licenses for commercial power reactors are issued for 40 years and can be renewed for additional 20 years with no limit to the number of renewals.

Japan has now entered the period of safety management of nuclear plants; 20 nuclear power plants in Japan have undergone an examination called *Plant Life Management* by the Japanese government to determine their ability to carry on continuous operation. So in this country work is in progress to improve a quantitative evaluation method based on the time-dependent performance of structures and their components in relation to the deterioration mechanism, which could serve as a maintenance program to extend the operational period of up to 20 years in the near future.

Most of the EU's operational nuclear plants were built in the 1970s and 1980s and were designed to last for around 40 years; therefore they are due to retire by the 2020s unless they get extensions [2]. Some of the oldest reactors are in Belgium, two of which just reached age 40, and 10 extra years has been recently approved for them until 2025 [3]; France is the EU's nuclear leader, its 58 reactors producing nearly three-quarters of the country's electricity; this country also is considering whether to extend their lifespan to 50 years in 2018 or 2019 [1].

Nuclear reactors, however, are not the only structural type undergoing nuclear radiation damage which must address durability considerations.

For instance, radioisotopes are extensively used in medicine; in over 10,000 hospitals worldwide radiotherapy is the way in which some medical conditions are treated, using radiation to weaken or destroy particular targeted cells; these facilities as well require the design of bunker-type rooms, able to sustain photon and neutron radiation.

This paper is aimed at picturing out state of the art of the most relevant contributions from the late 1950s up to now on the topic of radiation damage of concrete employed for nuclear shielding, from the phenomenological evidences to the modeling aspects.

Many contemporary authors agree that the technical information necessary for determining an appropriate threshold value of nuclear radiation in assessing the soundness of concrete structures under irradiation conditions is poor, especially in regard to concrete durability evaluation. Consequently, further experimental tests are needed. On the other hand, from the first studies some advancements on the interaction between concrete and neutrons or gamma-rays have been made, which would now need implementation in predictive models of concrete deterioration and validation against representative laboratory tests to evaluate the related rate effects.

The understanding of radiation effects on concrete was largely based on Hilsdorf and coauthors' curves compiled in 1978 [5] up to a few years ago, when the need to develop comprehensive safety evaluation systems for aging nuclear power plants has become a compelling issue.

Fujiwara et al. [6] have recently conducted an irradiation test on concrete specimens simulating the exercise scenario of a boiling water reactor in Japan (maximum temperature 65°C; maximum neutron fluence 12×10^{18} n/cm^2 ($E > 0.1$ MeV)) by far consistently with a 60-year operation. With this work they confirmed that, within the range of radiation doses adopted in the study, radiation exposure did not significantly affect the basic long-term material characteristics; on the other hand the tendency pictured in [5] in terms of the decrease of concrete compressive strength in response to rise in neutron fluence was not detected, due to some critical observations supported also by [7–9].

In fact, despite the big effort in compiling a first data base of experimental results on irradiated concrete, the applicability of the reference values by Hilsdorf and coauthors to the soundness evaluation of concrete in light-water reactors is questioned in [8], basically in reason of some inconsistency of the experimental conditions with the real irradiation conditions (namely, Portland cement for the concrete mix; irradiation from fast neutron spectrum ($E > 0.1$ MeV); temperature field less than 65°C); in particular the experimental database in [5] is said to show wide scattering due to different choices on measured properties (bending strength instead of compressive strength or bending tension strength instead of splitting tension strength), binder type of concrete, specimen size (even dimensions of 8–15 mm), temperature of concrete (100°C or higher up to 550°C), and neutron spectrum (slow or fast neutrons or not available spectrum).

The work by Field et al. [10] very recently expanded the data collection provided in [5] with those coming from later experimental campaigns on neutron-irradiated concrete and mortars; it provides a complete up to date database to start from to investigate the key variables associated with radiation-induced degradation in concrete structures. Effects on concrete in terms of compressive and tensile strength, modulus of elasticity, weight loss, and volumetric change are here plotted and classified under the applied laboratory conditions. It is here pointed out for a wider data collection that the effects of some important parameters such as the neutron fluence magnitude at which concrete deterioration becomes significant, the neutron fluence energy cutoff associated with the threshold fluence level, and the effect of high concrete temperature during irradiation, to mention a few, cannot be readily separated out and clearly evaluated from the existing data.

2. Phenomenological Aspects on Irradiated Concrete

Radiation on shielding materials is known to affect them deep in their chemical structure. In metals radiation leads to the displacements of atoms from their equilibrium lattice sites, causing lattice defects, which are responsible for an increase of hardness but also embrittlement, and so loss of ductility [11–13]. In polymers the formation of additional cross-links is due to the surplus energy brought by radiation [14]. In geomaterials, like concrete, nuclear radiation leads to the break of atomic bonds, which is supposed to explain the decay in the mechanical properties envisaged in exposed concrete [15, 16] as it will be illustrated in the following.

The first comprehensive collection of published experimental data on the effects of nuclear radiation on the properties of concrete dates back to Hilsdorf et al. [5] and Kaplan [17], with the recent observations on the first work already mentioned in the previous section. It stands out that up to integrated neutron fluence of the order of 10^{19} n/cm^2 the effects of the irradiation are relatively small, while higher fluence may have detrimental effects on concrete compressive and tensile strength and modulus of elasticity.

In particular, distinguishing between fast or slow neutrons, Gray [18] found that for fast neutron fluence between 7×10^{18} and 3×10^{19} n/cm^2 the modulus of irradiated concrete was between 10% and 20% less than that of nonirradiated unheated concrete. Alexander [19] reported similar reductions in values of elastic modulus for slow neutron fluence of about 2×10^{19} n/cm^2. Negligible changes in mechanical properties and weight/volume were found for fast neutrons with $E > 1$ MeV at a fluence of 3×10^{18} n/cm^2 by Stoces et al. [20].

There is also evidence that concrete hardens under irradiation [21, 22]: from these studies the interference of penetrating ionizing radiation with the process of setting of cement paste has been found in the form of increased compressive and tensile strength, especially at the very beginning of the hardening period.

Almost in all the cases in which samples were tested to investigate the effects of neutron radiation, the concrete samples were also exposed to secondary gamma radiation, produced by nuclear reactions. Not many tests are in literature on the effects of gamma radiation alone without the simultaneous exposure to neutron radiation. Alexander [19] reports that for gamma radiation doses of about 10^{10} rad there is no reduction in the compressive strength of concrete, when compared with the strength of companion specimens which have been neither irradiated nor heated; there is, however, evidence of reductions between 25% and 60% in compressive strength for doses exceeding 10^{11} rad [23]. In this last test campaign the specimens were immersed in demineralized water in order to shield them against neutrons and study the effect of gamma radiation alone, but demineralized water was found to contribute to deteriorating concrete and after several years of exposure the surface of the irradiated samples was partially destroyed.

Soo and Milian [24] found that a loss in the compressive strength could occur at gamma doses that are much less than the threshold dose of 2×10^8 Gy (2×10^{10} rad). They postulated that the loss of strength could be connected with the radiolysis of the water of hydration in the cement as well as pore water. A loss of hydrogen and oxygen radiolytic species during irradiation would decrease the level of cement hydration and thus the strength of the cement.

The experimental campaign in [25] on concrete under gamma irradiation allowed concluding that the interaction with the shielding material leads to lowering both its strength and its porosity. The mechanism is explained to happen as a series of chemical reactions within the material, starting from the radiolysis of water and ending with the formation of calcite ($CaCO_3$); crystallites of calcite grow into pores, decreasing their size and destroying the tobermorite gel, a calcium silicate hydrate mineral responsible for concrete strength, by their crystallization pressure.

Another experimental study was conducted by Vodák et al. [26] to investigate further on the porosity variation in cement paste under gamma irradiation and under natural carbonation.

The radiolytic process seems to be confirmed by more recent studies [27], taking advantage of X-ray diffraction and scanning electron microscopy methods. The absorption of gamma radiation is found to induce the amorphization of cement hydrates and finally cause their decomposition. Moreover, bubbles were envisaged after irradiation exposure, which must be the effect of separation of the chemically bonded water, and numerous cracks in the cement matrix were also observed.

The presence of calcium is pointed out as a major cause for the progression of radiolysis; in fact the withdrawal of calcium under gamma-rays is responsible for the formation of $CaO_2{\cdot}8H_2O$, a highly insoluble phase which forms at the expense of portlandite and ettringite, leading to their elimination in case of very high radiation doses [28].

The basic mechanisms controlling the radiolysis in cementitious matrices are reviewed in the specific context of gamma irradiation in [29]. Two main outcomes result from water radiolysis: (i) buildup of internal gas overpressure (mainly hydrogen and oxygen) that may lead to cracking and/or explosive gas mixture when hydrogen gas is mixed with atmospheric oxygen in fraction of 4% and (ii) corrosion of steel bars in the long-term due to the attack by oxidizing products at the steel-concrete interface [30].

Thermal expansion coefficient, thermal conductivity, and shielding properties appear to be little affected by radiation. As for the coefficient of thermal expansion, Hilsdorf et al. [5] and Granata and Montagnini [21] indicate that for neutron fluence less than 5×10^{19} n/cm^2 there is no significant difference between the coefficient of thermal expansion of neutron-irradiated concrete samples and the coefficient of nonirradiated samples subjected to high temperature. Also, radiation energy is converted to heat when absorbed by a shield. The heat generated by irradiation can reach quite high temperatures, depending on the shielding material used and the configuration of the shielding structure. Hilsdorf et al. [5] found that temperatures even of the order of 250°C developed in some of the investigations he summarized, which is high enough to generate considerable damage in shielding materials.

An extensive study on the effect of elevated temperature exposure on properties and shielding effectiveness is reported in [31].

As for creep experiments under irradiation, limited reference data exist: [18, 32] for neutron and gamma irradiation, respectively. In particular the internal gas pressures resulting from water hydrolysis under gamma irradiation seem to affect concrete shrinkage and creep in some measure.

2.1. Physical Interpretation of the Experimental Results. Because the shielding properties of concrete are found to be driven mainly by the selection of the aggregates in the concrete mix; the choice must be taken in reason of the different attenuation mechanisms characterizing the specific particles to be addressed: gamma-rays or neutrons [17, 33–35].

Gamma-rays are attenuated by interactions at the level of the electrons of the atoms composing the shielding medium; in particular they show having a negligible effect on the solid materials composed of ionic and metallic bonds, but they can break anisotropic chemical bonds such as covalent bonds. Typically, siliceous materials are subject to gamma-rays degradation because the Si-O bond is a covalent bond. At this purpose concrete to be used as a gamma-ray shielding material should be designed to maximize the density of electrons within it. Practically, high-density iron ore aggregates are chosen at this purpose. This is the case of barytes, ferrophosphorus, magnetite, and hematite.

Instead, neutrons interact with nuclei of atoms leading to change in the lattice spacing within the material during these collision events. Therefore, neutrons have a more significant effect on dense and well-crystallized materials than on randomly structured materials with high porosity. In concrete, aggregates (coarse and fine aggregates) are in a crystallized phase, while cement paste is an amorphous phase; therefore neutron radiation causes more distortion and damage to the internal structure of aggregates than to that of cement paste. Moreover neutron attenuation is more effective when fast neutrons are made to collide with lightweight atoms (typically hydrogen) so hydrous aggregates are to be preferred for neutron shielding; hydrous aggregates are in fact capable of increasing the fixed-water content of concrete or retaining their water of crystallization at high temperatures. This is the case of serpentine, limonite, goethite, and bauxite.

In some cases the enrichment of minerals with certain elements, such as boron, is considered to enhance the thermal neutron attenuation of concrete and cut down secondary gamma-rays production. Boron-containing aggregates include colemanite, borocalcite, and ferroboron.

The specific effect of radiation-induced damage of aggregates is of primary interest in recent research, in view of the durability assessment of old nuclear structures.

Kelly et al. [36] and Elleuch et al. [37] emphasize the predominant role of the aggregate in the development of neutron irradiation-induced swelling and the potential creation or aggravation of damage in the surrounding paste.

Vanelstraete and Laermans [38] have shown that fast neutrons cause displacement cascades in quartz resulting in disordered regions of the crystal. For sufficiently high doses, damage regions overlap, reducing long-range ordering and resulting in amorphization of the SiO_2 phase. The loss of ordering is observed as a reduction of density and increase in swelling of the quartz phase. Complete amorphization is supposed to be reached at a neutron dose greater than 2×10^{20} n/cm^2.

The effect of nuclear irradiation, bot neutrons, and gamma-rays on quartz was studied extensively by [39]. Crystalline quartz or α-quartz, with specific gravity of about 2.65 is found to convert to distorted amorphous quartz with specific gravity of 2.27 under a fluence of 10^{20} n/cm^2 for fast neutrons with an energy level >0.1 MeV and under a dose of 10^{12} Gy for beta- and gamma-rays.

Maruyama et al. [8] confirm that an expansion of siliceous aggregate takes place due to neutron collisions. Atom alignments are deformed and a part of the energy imparted by neutron collision remains as a kind of strain energy causing permanent distortion. Also, denser siliceous aggregates have a larger risk of expansion due to neutron irradiation.

Because the atomic structure of some aggregates can be converted from crystalline structure to distorted amorphous structure with an increase in volume and a decrease in weight, the loss of concrete mechanical properties is conceived to be definitely correlated with radiation-induced aggregate swelling in [10].

The antagonist effect of neutrons and gamma-rays in radiation damage must be considered: the crystalline lattice structure present in the aggregates is much more affected by neutrons than the more vitreous lattice structure of the cement paste, which is much more attacked by gamma-rays.

At this purpose, many authors [5, 17, 18, 40, 41] state that different types of aggregates lead to concrete with different resistance against neutron radiation. Concrete made with flint aggregate shows considerably larger volume changes than concrete made with limestone aggregates; the phenomenon can be explained, once more, by the weakness of covalent bonds in quartz aggregates with respect to the ionic bonds in calcareous aggregates.

On the other hand, the cement paste undergoes shrinkage due to (i) the radiolysis process under gamma radiation, as illustrated above, and (ii) the evaporation of pore water under radiation heat.

So the mismatch in the volumetric change of concrete components (expansion in aggregates and shrinkage of the mortar) may cause damage at the interface between the two phases.

The uneven shrinkage properties of aggregates and cement paste may lead to loss of concrete compressive strength. While the overall volume growth of the composite compromises the tensile strength, this last is shown to decay on average of 62% and 47%, respectively, for flint and limestone aggregates in the range of 2×10^{19}–4×10^{19} n/cm^2 [17]. According to [21] a neutron fluence of less than 10^{19} n/cm^2 does not lead to a volume increase of the irradiated samples; rather in this range the volume change of irradiated samples is the expected shrinkage due to temperature exposure of the specimens.

The conversion of crystalline quartz into distorted quartz has a twofold detrimental consequence: (i) microcracking due to differential volume change in the composite and (ii) higher reactivity to certain aggressive chemicals, for example, calcium hydroxide responsible in concrete of alkali-silica reactions (ASR) [42, 43]. Both of these effects are detrimental to the long-term performance of irradiated concrete.

2.2. Radiation-Induced Effects on ASR. The alkali-silica reaction evolves in concrete as follows: OH^- ions present in the alkaline solution in the micropores of concrete react with SiO_2 in aggregates to perform the scission of the Si-O bonds and the subsequent expansion of the aggregates by hydration of SiO_2. The consumption of OH^- ions due to hydrolysis leads to the dissolution of Ca^{2+} ions into the solution. The Ca^{2+} ions then react with hydrated SiO_2 gels (ASR gel) to generate calcium silicate. Rigid calcium silicate shells typically form on the surfaces of the aggregates as the reaction by-product is generated. The alkaline solution can penetrate into the aggregates through the calcium silicate shells and dissolve SiO_2 groups. Since the rigid shells prevent the deformation of the aggregates, the expansion pressure generated by the penetration of the solution is accumulated in the aggregates under the confining pressure of the silicate shells, thus leading to cracks and the final expansion of the aggregates.

ASR may be accelerated both by lattice defects in SiO_2 minerals from neutrons irradiation and by the preexistence of cracks in the aggregates.

The experiments by Ichikawa and Koizumi [39], Ichikawa and Kimura [44], and Ichikawa and Miura [45] prove that nuclear radiation significantly increases the reactivity of silica-rich aggregates to alkalis; the decrease of the resistance to nuclear radiation with increasing the content of SiO_2 in aggregates strongly indicates that the deterioration is due to the acceleration of ASR in concrete, according to the authors.

In particular the work in [39] shows that nuclear irradiation may change the ASR expansion potential of aggregates, which is driven by the stiffness of the ASR gel and contributes to the free expansion capability of the aggregate in determining the extent of damage due to ASR. In fact when the ASR gel is soft and its stiffness is low, it is able to permeate into the surrounding porous cement paste; therefore the swelling pressure is not high and the cracking potential is low. In that case a large amount of ASR gel formation may not create severe damage and cracking in concrete. On the other hand, when the ASR gel is stiff, a small amount of ASR can generate significant damage [35]. The stiffness of ASR gel depends mainly on its chemical composition, such as the ratio of Na_2O to SiO_2. Struble and Diamond [46] conducted a study to investigate the swelling pressure of ASR in terms of ASR chemical composition.

Though not specifically addressing ASR, the work by Vodák et al. [25] indicates that irradiation generates a succession of chemical reactions, leading to a decrease in the size of pore space and hence inhibiting concrete to absorb some of the ASR gel produced prior to expansion.

Saouma and Hariri-Ardebili [47] critically reviewed existing available literature on ASR and they conclude that radiation effects on concrete degradation are minimal for the first 40-year operation of a nuclear power plant; however they agree with [6] that a structural life extension to 60–100 years may prove problematic, though the data to fully support this concern are insufficient. The work is contextualized to Seabrook site, the first reported nuclear plant in the USA known to possibly suffer from ASR, which saw in 2010 its operating license extended from 2030 to 2050. The work is aimed at developing an aging management program for old nuclear plants; it expresses the concern of overreliance on surface crack observation and structural component testing and at the same time it encourages the development of reliable Finite Element Method (FEM) based simulations to address the long-term assessment of such structures.

Graves et al. [48] offer a comprehensive evaluation of potential aging-related degradation modes for light-water reactor materials and components. Here ASR is pointed out as one of the high-ranked causes of damage that can potentially affect in the long-term the concrete containment of nuclear plants, together with creep of the posttensioned system and irradiation of concrete itself, which, as shown, may accelerate ASR.

3. Particle Transport Simulations for Nuclear Radiation Problems

The description of all the possible interactions between a radiation particle and the absorber medium is the topic of the so-called *transport theory*, a special branch of statistical mechanics.

Accurate calculations of particle transport are needed, nowadays more than in the past, for many reasons.

(i) The necessity to develop theoretical models of radiation transport also for particle energies above 20 MeV: attenuation coefficients, such as neutron cross sections, need to be evaluated for such energies, while complete libraries are available for a representative set of nuclides up to 20 MeV [49]. Neutrons are known to be easily transported over many energy decades, from the hundreds of MeV down to the meV range; therefore the knowledge of the cross section over the whole energy range is a concern.

Many new facilities, working at high energy, have been recently built or are planned, in order to perform experiments in this sense, to calibrate and validate predictive theoretical models.

(ii) The need to accurately describe the proton- and neutron-induced interaction mechanisms: a number of new applications in the last fifteen years are devoted to the production of intense neutron fluxes by protons impinging on a thick target of a heavy element (mercury, lead, uranium, etc.) for the purposes of point (i), thus producing a cascade of nuclear reactions, known as *spallation reactions*, which involve protons, neutrons, nuclear fragments (alpha-particles, tritium, and deuterium), and secondary gamma-rays.

(iii) The assistance in the design of accelerator-driven systems, which are conceived as nowadays solution to the problem of the transmutation, a possible mechanism for reducing the volume and hazard of the radioactive waste (spent fuel) in nuclear power reactors: accelerator-driven systems consist of a reactor coupled with a high-intensity proton accelerator impinging on a high-density target. The spallation reactions taking place in the target result in high neutron fluxes here and in the surrounding core, which requires accurate evaluation, for shielding design purposes and for the radiological protection assessment (e.g., thermomechanical assessment and study of the activation of components and materials).

(iv) Similarly, the assistance in the design of other nuclear facilities, such as new-generation reactors (generation IV): the concept with these next-generation reactors is to use different coolants (sodium-, gas-, or liquid-metal cooling system) and to consider recycling of the fuel, in order to both reach higher power (improved efficiency) and minimize the waste production. For these systems a radiation damage assessment of the materials subject to the intense neutron fluxes (fuel cladding, vessel, and cooling system) is mandatory for fulfilling safety requirements [50].

(v) The scientific knowledge of the biological effects of radiation, to study particle transport for biomedical purposes: Monte Carlo simulations of neutron transport in human tissues have been conducted in recent years, due to the difficulty in carrying out experiments, in order to determine the tissue-specific weighting factors at different energies, which are useful for the assessment of the effective dose of exposed subjects.

Two approaches are possible to quantify the main variable in transport theory, the radiation flux density, throughout a certain domain: the deterministic approach given by the solution of the Boltzmann transport equation and the statistical approach given by Monte Carlo simulations.

As for the first method, an exact solution is possible only in few cases [51, 52], but not for complex geometries; consequently much effort has been done in the development of approximations to the transport equation, both numerically and analytically. In this second case the diffusion theory and the two-group theory are worth mentioning.

Numerically speaking, the solution of the Boltzmann equation requires (i) energy discretization of the neutron spectrum into discrete energy intervals based on a multigroup formalism; (ii) angular discretization of the angular direction of the radiation flux; (iii) space discretization or mesh, according to FEM formulation.

As regards the second methods, some software programs [53–57] are under continuous upgrade by their developers to solve radiation transport problems via the Monte Carlo technique.

3.1. Modeling Approaches for Irradiated Concrete. When addressing the specific issues involving irradiated concrete, especially with regard to its durability, ad hoc numerical methods must be defined to catch all the complex coupled mechanisms illustrated above, which might be used in conjunction with radiation transport calculations.

In the following some recent distinguishing models of radiation-related issues concerning concrete are reviewed.

Radiation-induced volumetric expansion is identified as a predominant source of cracking in the cement paste of pressurized water reactors in [58]. Based on an extensive literature review on postirradiation experimental data, the authors here developed a 2D micromechanical formulation accounting for the radiation-induced swelling of aggregates, which uses upscaling techniques to address the macroscopic scale starting from the material scale (aggregates and cement paste), that is, homogenization theory applied to random media. Mechanical properties of concrete vary at varying neutron fluence; no gamma-rays effects are considered nor thermohydraulic mechanisms in concrete.

In a companion study Giorla et al. [59] proposed a 2D FEM model of concrete at the mesoscale, where aggregates are supposed to be elastic and subject to thermal and radiation-induced swelling while the cement matrix is viscoelastic and it undergoes damage, drying shrinkage, and thermal expansion. Therefore this model embodies thermohydraulic aspects, though they are not coupled with the mechanical field because actually thermal expansion, radiation-induced volumetric expansion, and shrinkage result in imposed deformations. In particular here radiation-induced volumetric expansion is a predominant cause of the development and propagation of damage around the aggregate and surroundings.

In [60] a combined use of a thermohydromechanical FEM model with a Monte Carlo code is suggested to analyse radiation damage on a real nuclear facility under its design operational life. Damage is here made to depend on the neutron fluence based on Hilsdorf and coauthors' data [5], with all the limitations discussed in Section 1; the model does not account for a specific radiation-induced volumetric expansion but it allows catching the thermal aspect of the phenomenon related to the energy deposition due to radiation attenuation together with its implications on the hydraulic field.

A chemomechanical approach to model aggregate dissolution and product precipitation in a microstructural reaction-transport environment has been used in [61] to simulate near-surface degradation of Portland cement paste in contact with a sodium sulfate solution. With this tool microstructural changes due to new solids in a confined-growth condition are responsible for localized stress and damage in a coupled way.

So far, then, in both the theoretical and the numerical field no fully coupled radiothermohydromechanical models have been accomplished yet to simulate the deterioration mechanism of irradiated concrete, though an approach in this sense has been proposed in [7], where the mismatch in the volume changes between cement paste and aggregate is pointed out as the main cause of cracking. For the sake of completeness, the proposed model nowadays could take also advantage of the recent advances on microscopic modeling in the field of fracture mechanics and crack propagation [62–66]. Recent theoretical studies and benchmark examples of engineering interest at the nanoscale can be found in [67–69], also in the context of thermoelasticity and viscoelasticity [70].

3.2. Modeling Approaches for ASR. Though not always addressing radiation-induced effects specifically, a considerable interest in modeling ASR for concrete durability investigations is found in literature. The effort in the past years up to now was spent on the development of both: macroscopic models and micro-/mesoscopic models.

The first are aimed at investigating the overall structural behavior of entire concrete elements in their real context, while the latter look at the phenomenon at a material scale, often coupling the mechanical field with the chemical reactions involved in the ASR gel production.

The work by Ulm et al. [71] belongs to the first kind of approach. They developed a chemothermomechanical model in the framework of Biot's theory, where the concrete is conceived as a two-phase material including the expansive gel and the homogenized concrete skeleton. Here the volumetric expansion of the gel is evaluated as a function of the reaction kinetics, which is influenced by temperature.

Saouma and Perotti [72] proposed a chemothermomechanical model based on two main assumptions: the volumetric expansion of the gel and its redistribution in relation of weights depending on the stress tensor.

In [73] a chemoelastic damage model, validated on the basis of the accelerated multiaxial experimental tests on small specimens, is formulated to simulate the mechanical effects of ASR in large dams.

As for the applications focusing on a lower scale, various analytical models based on empirical equations were developed to explain ASR at the mesoscale of concrete: Bazant and Steffens [74] proposed that the chemical reaction kinetics is related to the diffusion process of the reactants, leading to subsequent fracture in the characteristic unit cell of the concrete, modeled with one spherical glass particle; Multon et al. [75] developed a microscopic chemomechanical model based on damage theory in order to assess the decrease of stiffness of the mortar due to cracking caused by ASR and to quantify the related expansion. The diffusion and the fixation of alkalis are here assessed with the mass balance equation and by defining a threshold alkalis concentration, above which the formation of the gel is meant to start.

An analytical model, for which a numerical solution is provided, is studied in [76]. The formulation is based on the description of the transport and reaction of alkalis and calcium ions within a relative elementary volume, particularly taking into account the influence of the reactive aggregate size grading on ASR evolution.

Comby-Peyrot et al. [77] introduced a 3D mesoscopic FEM model of concrete as a biphasic medium with coarse aggregates randomly embedded in the cement matrix. Damage in the cement matrix is caused by the phenomenon of isotropic dilation of the reactive aggregates, specifically induced by ASR.

Dunant and Scrivener [78] developed a 2D mesoscale extended finite element (XFEM) model which defines the geometry of the gel swelling by updating the enrichment function. A specific damage parameter accounts for triggering of the fracture in the aggregates, due to growing gel pockets in them.

Alnaggar et al. [79] proposed a mesoscale formulation of ASR deterioration in concrete specimens via the Lattice Discrete Particle Model (LDPM), in a chemothermomechanical way, including nonuniform expansions, expansion transfer, and heterogeneous cracking.

In [80] 3D micro-CT scans of the microscale hardened cement paste are used to calibrate a coupled chemothermohydromechanical FEM model in which the gel is assumed to be produced in the micropores of the paste and it exerts uniform pressure on the neighborhood. The expansion coefficient of the gel at the microscale was obtained through a two-step homogenization approach, thus enabling correlating microscale damage and macroscale failure.

These references are not exhaustive; however; a comprehensive review of modeling of ASR in concrete is provided in [81, 82].

4. Conclusions

Concerns over aged nuclear power stations are mounting in several countries today. This is the primary reason for a renewed interest in radiation damage assessment of biological concrete shields in such facilities.

The behavior of irradiated concrete has reached up to date commonly accepted explanations. Particularly, cement paste and aggregates are known to behave differently under radiation exposure conditions: cement paste shrinks under drying conditions, the decrease in volume being only partially covered by the thermal expansion due to high temperatures arisen by associated radiation heat. Shrinkage of the cement paste is mostly explained by water hydrolysis and water evaporation occurring under gamma-ray irradiation. On the other hand, aggregates expand due to accumulation of defects in their crystal structures generated mainly by neutron collisions. This uneven volume expansion between cement matrix and aggregates is pointed out as the primary cause of damage in irradiated cementitious materials, to date, based on the available experimental studies. Much of the research related to this topic was conducted from the 1960s to the 1970s in support of the development of prestressed concrete reactor vessels for high temperature reactors and radioactive waste storage facilities.

The need for long-term predictions of concrete shielding of old reactors has enhanced the effort towards the modeling of the related deterioration mechanisms, both theoretically and numerically. The most relevant features which allow fully addressing the whole complex physical process are here summarized and the recent achievements in matter of irradiated concrete modeling are reviewed.

In this framework further research is recommended aimed at consolidating the new outstanding micro-/mesomechanical models in order to (i) fill the knowledge gaps from past experiments and (ii) possibly allow the integration of radiation-induced effects into the chemothermohydromechanical aspects of the problem in a coupled way.

Competing Interests

The author declares no competing interests.

References

[1] http://world-nuclear.org/info/Facts-and-Figures/World-Nuclear-Power-Reactors-and-Uranium-Requirements/.

[2] http://www.carbonbrief.org/the-trouble-with-europes-ageing-nuclear-power-plants.

[3] https://euobserver.com/beyond-brussels/129795.

[4] http://uk.reuters.com/article/us-europe-nuclear-power-insight-idUKKBN0GH05U20140818.

[5] H. K. Hilsdorf, J. Kropp, and H. J. Koch, "The effects of nuclear radiation on the mechanical properties of concrete," in *Proceedings of the Douglas McHenry International Symposium on Concrete and Concrete Structures*, ACI SP 55-10, pp. 223–251, American Concrete Institute, Mexico City, Mexico, 1978.

[6] K. Fujiwara, M. Ito, M. Sasanuma et al., "Experimental study of the effect of radiation expo-sure to concrete," in *Proceedings of the 20th International Conference on Structural Mechanics in Reactor Technology (SMIRT '09)*, paper 1891, SMIRT20-Division I, Espoo, Finland, 2009.

[7] I. Maruyama, O. Kontani, A. Ishizawa, M. Takizawa, and O. Sato, "Development of system for evaluating concrete strength deterioration due to radiation and resultant heat," IAEA-CN-194-093, Topic: 3-6 Non-metallic material Ageing Management, 2012.

[8] I. Maruyama, O. Kontani, S. Sawada, O. Sato, G. Igarashi, and M. Takizawa, "Evaluation of irradiation effects on concrete structure-background and preparation of neutron irradiation test," in *Proceedings of the ASME Power Conference (POWER '13)*, no. 98114, 9 pages, ASME, Boston, Mass, USA, August 2013.

[9] O. Kontani, S. Sawada, I. Maruyama, M. Takizawa, and O. Sato, "Evaluation of irradiation effects on concrete structure—gamma-ray irradiation tests on cement paste," in *Proceedings of the ASME 2013 Power Conference (POWER '13)*, no. 98099, ASME, Boston, Mass, USA, August 2013.

[10] K. G. Field, I. Remec, and Y. Le Pape, "Radiation effects in concrete for nuclear power plants—part I: quantification of radiation exposure and radiation effects," *Nuclear Engineering and Design*, vol. 282, pp. 126–143, 2015.

[11] D. S. Billington, "Radiation damage in reactor materials," in *Proceedings of the International Conference on the Peaceful Uses*

of Atomic Energy, vol. 7 of *Nuclear Chemistry and Effects of Irradiation*, pp. 121–132, United Nations, Geneva, Switzerland, 1955.

[12] F. E. Faris, "The effect of irradiation on structural materials," in *Proceedings of the International Conference on the Peaceful Uses of Atomic Energy*, vol. 7 of *Nuclear Chemistry and Effects of Irradiation*, pp. 484–489, United Nations, Geneva, Switzerland, August 1955.

[13] S. T. Konobeevsky, N. F. Pravdyuk, and V. I. Kutaitsev, "The effect of irradiation on the structure and properties of structural materials," in *Proceedings of the International Conference on the Peaceful Uses of Atomic Energy*, vol. 7 of *Nuclear Chemistry and Effects of Irradiation*, pp. 479–483, United Nations, Geneva, Switzerland, August 1955.

[14] A. Holmes-Siedle and L. Adams, *Handbook of Radiation Effects*, Oxford University Press, New York, NY, USA, 2002.

[15] B. T. Price, C. C. Horton, and K. T. Spinney, *Radiation Shielding*, Pergamon Press, London, UK, 1957.

[16] R. G. Jaeger, *Engineering Compendium on Radiation Shielding*, vol. 2, Springer, Berlin, Germany, 1970.

[17] M. F. Kaplan, *Concrete Radiation Shielding: Nuclear Physics, Concrete Properties, Design, and Construction*, John Wiley & Sons, New York, NY, USA, 1989.

[18] B. S. Gray, "The effect of reactor radiation on cements and concrete," in *Proceedings of the Conference on Prestressed Concrete Pressure Vessels*, pp. 17–39, Commission of European Communities, Luxembourg City, Luxembourg, 1972.

[19] S. C. Alexander, *Effects of Irradiation on Concrete: Final Results*, Atomic Energy Research Establishment, United Kingdom Atomic Energy Authority, Harwell, UK, 1963.

[20] B. Stoces, P. Otopal, V. Juricka, and J. Gabriel, "Effect of radiation on the mechanical properties of concrete," Tech. Rep. UJV-2390.R,T, Translated from the Czech, 1970.

[21] S. Granata and A. Montagnini, "Studies on behavior of concretes under irradiation," in *Concrete for Nuclear Reactors*, vol. 2 of *Special Publication SP-34*, pp. 1163–1172, American Concrete Institute, 1972.

[22] B. Nowakowski, "Influence of penetrating ionizing radiation on the curing of grouts and cement mortars," *Building Science*, vol. 7, no. 4, pp. 271–276, 1972.

[23] J. F. Sommers, "Gamma radiation damage of structural concrete immersed in water," *Health Physics*, vol. 16, no. 4, pp. 503–508, 1969.

[24] P. Soo and L. M. Milian, "The effect of gamma radiation on the strength of Portland cement mortars," *Journal of Materials Science Letters*, vol. 20, no. 14, pp. 1345–1348, 2001.

[25] F. Vodák, K. Trtík, V. Sopko, O. Kapičková, and P. Demo, "Effect of γ-irradiation on strength of concrete for nuclear-safety structures," *Cement and Concrete Research*, vol. 35, no. 7, pp. 1447–1451, 2005.

[26] F. Vodák, V. Vydra, K. Trtík, and O. Kapičková, "Effect of γ-irradiation on properties of hardened cement paste," *Materials and Structures*, vol. 44, no. 1, pp. 101–107, 2011.

[27] A. Lowińska-Kluge and P. Piszora, "Effect of gamma irradiation on cement composites observed with XRD and SEM methods in the range of radiation dose 0–1409 MGy," *Acta Physica Polonica A*, vol. 114, no. 2, pp. 399–411, 2008.

[28] P. Bouniol and A. Aspart, "Disappearance of oxygen in concrete under irradiation: the role of peroxides in radiolysis," *Cement and Concrete Research*, vol. 28, no. 11, pp. 1669–1681, 1998.

[29] P. Bouniol and E. Bjergbakke, "A comprehensive model to describe radiolytic processes in cement medium," *Journal of Nuclear Materials*, vol. 372, no. 1, pp. 1–15, 2008.

[30] P. Bouniol, B. Muzeau, and V. Dauvois, "Experimental evidence of the influence of iron on pore water radiolysis in cement-based materials," *Journal of Nuclear Materials*, vol. 437, no. 1–3, pp. 208–215, 2013.

[31] D. J. Naus, "A compilation of elevated temperature concrete material property data and information for use in assessments of nuclear power plant reinforced concrete structures," Tech. Rep. NUREG/CR-7031 ORNL/TM-2009/175, US Nuclear Regulatory Commission, Oak Ridge National Laboratory, 2010.

[32] D. McDowall, "The effects of gamma radiation on the creep properties of concrete," in *Proceedings of the Information Exchange Meeting on "Results of Concrete Irradiation Programs"*, EUR 4751 f-e, pp. 55–69, Commission des Communautés Européennes, Brussels, Belgium, 1971.

[33] T. Rockwell, *Reactor Shielding Design Manual*, McGraw-Hill, New York, NY, USA, 1956.

[34] A. N. Komarovskii, *Shielding Materials for Nuclear Reactors*, Pergamon Press, London, UK, 1961.

[35] K. William, Y. Xi, and D. Naus, "A review of the effects of radiation on microstructure and properties of concretes used in nuclear power plants," Tech. Rep. NUREG/CR-7171 ORNL/TM-2013/263, US Nuclear Regulatory Commission, Oak Ridge National Laboratory, 2013.

[36] B. Kelly, J. Brocklehurst, D. Mottershead, and S. McNearney, "The effects of reactor radiation on concrete," in *Proceedings of the 2nd Information Meeting on Prestressed Concrete and Reactor Pressure Vessels and Their Thermal Isolation*, EUR-4531, pp. 237–265, Commission des Communautés Européennes, Brussels, Belgium, 1969.

[37] M. F. Elleuch, F. Dubois, and J. Rappeneau, "Effects of neutron radiation on special concretes and their components," in *ACI Special Publication SP-34: Concrete for Nuclear Reactors*, Paper SP34-51, pp. 1071–1108, 1972.

[38] A. Vanelstraete and C. Laermans, "Tunneling states in neutron-irradiated quartz: measurements of the ultrasonic attenuation and velocity change," *Physical Review B*, vol. 42, no. 9, pp. 5842–5854, 1990.

[39] T. Ichikawa and H. Koizumi, "Possibility of radiation-induced degradation of concrete by alkali-silica reaction of aggregates," *Journal of Nuclear Science and Technology*, vol. 39, no. 8, pp. 880–884, 2002.

[40] D. Campbell-Allen, E. W. E. Low, and H. Roper, "An investigation on the effect of elevated temperatures on concrete for reactor vessels," *Nuclear Structural Engineering*, vol. 2, no. 4, pp. 382–388, 1965.

[41] D. J. Naus, "Primer on durability of nuclear power plant reinforced concrete structures: a review of pertinent factors," Tech. Rep. NUREG/CR-6927 ORNL/TM-2006/529, US Nuclear Regulatory Commission, Oak Ridge National Laboratory, 2007.

[42] A. F. Cohen, "Low-temperature thermal conductivity in neutron irradiated vitreous silica," *Journal of Applied Physics*, vol. 29, no. 3, pp. 591–593, 1958.

[43] L. W. Hobbs and M. R. Pascucci, "Radiolysis and defect structure in electron-irradiated α-quartz," *Journal de Physique Colloques*, vol. 41, no. 7, pp. 237–242, 1979.

[44] T. Ichikawa and T. Kimura, "Effect of nuclear radiation on alkali-silica reaction of concrete," *Journal of Nuclear Science and Technology*, vol. 44, no. 10, pp. 1281–1284, 2007.

[45] T. Ichikawa and M. Miura, "Modified model of alkali-silica reaction," *Cement and Concrete Research*, vol. 37, no. 9, pp. 1291–1297, 2007.

[46] L. Struble and S. Diamond, "Unstable swelling behaviour of alkali silica gels," *Cement and Concrete Research*, vol. 11, no. 4, pp. 611–617, 1981.

[47] V. E. Saouma and M. A. Hariri-Ardebili, "A proposed aging management program for alkali silica reactions in a nuclear power plant," *Nuclear Engineering and Design*, vol. 277, pp. 248–264, 2014.

[48] H. Graves, Y. Le Pape, D. Naus et al., "Expanded materials degradation assessment (EMDA), 4: aging of concrete," Tech. Rep. NUREG/CR-7153, ORNL/TM-2011/545, US Nuclear Regulatory Commission, Oak Ridge National Laboratory, 2013.

[49] P. Vaz, "Neutron transport simulation (selected topics)," *Radiation Physics and Chemistry*, vol. 78, no. 10, pp. 829–842, 2009.

[50] A. Alemberti, "Verso i reattori di IV generazione," in *Proceedings of the Workshop: IV Generation Nuclear Reactors: Which Issues for Metals?* AIM (Italian Association of Metallurgy) with Ansaldo Nucleare and IENI (National Council for Research, Institute for Energy and Interphases), Genova, Italy, February 2011, (Italian).

[51] J. R. Harrison, *Nuclear Reactor Shielding*, Temple Press, London, UK, 1958.

[52] J. K. Shultis and R. E. Faw, *Radiation Shielding*, Prentice Hall, New York, NY, USA, 1996.

[53] D. B. Pelowitz, Ed., "MCNPX user's manual, version 2.7.0", Tech. Rep. LA-CP-11-0438, Los Alamos National Laboratory, 2011.

[54] D. B. Pelowitz, J. W. Durkee, J. S. Elson et al., "MCNPX 2.7.0 extensions," Tech. Rep. LA-UR-11-02295, Los Alamos National Laboratory, 2011.

[55] S. Agostinelli, "Geant4—a simulation toolkit," *Nuclear Instruments and Methods in Physics Research Section A: Accelerators, Spectrometers, Detectors and Associated Equipment*, vol. 506, no. 3, pp. 250–303, 2003.

[56] J. Allison, K. Amako, J. Apostolakis et al., "Geant4 developments and applications," *IEEE Transactions on Nuclear Science*, vol. 53, no. 1, pp. 270–278, 2006.

[57] A. Ferrari, P. R. Sala, A. Fass, J. Ranft, and A. Fassò, "FLUKA: a multi-particle transport code," Tech. Rep. CERN-2005-10, INFN/TC_05/11, SLAC-R-773, 2011.

[58] Y. Le Pape, K. G. Field, and I. Remec, "Radiation effects in concrete for nuclear power plants, part II: perspective from micromechanical modeling," *Nuclear Engineering and Design*, vol. 282, pp. 144–157, 2015.

[59] A. Giorla, M. Vaitová, Y. Le Pape, and P. Štemberk, "Meso-scale modeling of irradiated concrete in test reactor," *Nuclear Engineering and Design*, vol. 295, pp. 59–73, 2015.

[60] B. Pomaro, V. A. Salomoni, F. Gramegna, G. Prete, and C. E. Majorana, "Radiation damage evaluation on concrete shielding for nuclear physics experiments," *Annals of Solid and Structural Mechanics*, vol. 2, no. 2–4, pp. 123–142, 2011.

[61] P. Feng, E. J. Garboczi, C. Miao, and J. W. Bullard, "Microstructural origins of cement paste degradation by external sulfate attack," *Construction and Building Materials*, vol. 96, pp. 391–403, 2015.

[62] A. M. Tarantino, "The singular equilibrium field at the notch-tip of a compressible material in finite elastostatics," *Journal of Applied Mathematics and Physics*, vol. 48, no. 3, pp. 370–388, 1997.

[63] A. M. Tarantino, "On extreme thinning at the notch tip of a neo-Hookean sheet," *The Quarterly Journal of Mechanics and Applied Mathematics*, vol. 51, no. 2, pp. 179–190, 1998.

[64] A. M. Tarantino, "On the finite motions generated by a mode I propagating crack," *Journal of Elasticity*, vol. 57, no. 2, pp. 85–103, 1999.

[65] A. M. Tarantino, "Crack propagation in finite elastodynamics," *Mathematics and Mechanics of Solids*, vol. 10, no. 6, pp. 577–601, 2005.

[66] A. Caporale, L. Feo, and R. Luciano, "Damage mechanics of cement concrete modeled as a four-phase composite," *Composites Part B: Engineering*, vol. 65, pp. 124–130, 2014.

[67] R. Barretta, L. Feo, R. Luciano, and F. Marotti de Sciarra, "An Eringen-like model for Timoshenko nanobeams," *Composite Structures*, vol. 139, pp. 104–110, 2016.

[68] R. Barretta, L. Feo, R. Luciano, and F. Marotti de Sciarra, "A gradient Eringen model for functionally graded nanorods," *Composite Structures*, vol. 131, pp. 1124–1131, 2015.

[69] R. Barretta, L. Feo, R. Luciano, and F. Marotti de Sciarra, "Variational formulations for functionally graded nonlocal Bernoulli-Euler nanobeams," *Composite Structures*, vol. 129, pp. 80–89, 2015.

[70] R. Barretta, L. Feo, and R. Luciano, "Torsion of functionally graded nonlocal viscoelastic circular nanobeams," *Composites Part B: Engineering*, vol. 72, pp. 217–222, 2015.

[71] F.-J. Ulm, O. Coussy, L. Kefei, and C. Larive, "Thermo-chemo-mechanics of ASR expansion in concrete structures," *Journal of Engineering Mechanics*, vol. 126, no. 3, pp. 233–242, 2000.

[72] V. Saouma and L. Perotti, "Constitutive model for alkali-aggregate reactions," *ACI Materials Journal*, vol. 103, no. 3, pp. 194–202, 2006.

[73] C. Comi, R. Fedele, and U. Perego, "A chemo-thermo-damage model for the analysis of concrete dams affected by alkali-silica reaction," *Mechanics of Materials*, vol. 41, no. 3, pp. 210–230, 2009.

[74] Z. P. Bazant and A. Steffens, "Mathematical model for kinetics of alkali-silica reaction in concrete," *Cement and Concrete Research*, vol. 30, no. 3, pp. 419–428, 2000.

[75] S. Multon, A. Sellier, and M. Cyr, "Chemo-mechanical modeling for prediction of alkali silica reaction (ASR) expansion," *Cement and Concrete Research*, vol. 39, no. 6, pp. 490–500, 2009.

[76] S. Poyet, A. Sellier, B. Capra et al., "Chemical modelling of alkali silica reaction: influence of the reactive aggregate size distribution," *Materials and Structures*, vol. 40, no. 2, pp. 229–239, 2007.

[77] I. Comby-Peyrot, F. Bernard, P.-O. Bouchard, F. Bay, and E. Garcia-Diaz, "Development and validation of a 3D computational tool to describe concrete behaviour at mesoscale. Application to the alkali-silica reaction," *Computational Materials Science*, vol. 46, no. 4, pp. 1163–1177, 2009.

[78] C. F. Dunant and K. L. Scrivener, "Micro-mechanical modelling of alkali-silica-reaction-induced degradation using the AMIE framework," *Cement and Concrete Research*, vol. 40, no. 4, pp. 517–525, 2010.

[79] M. Alnaggar, G. Cusatis, and G. Di Luzio, "A discrete model for alkali-silica-reaction in concrete," in *Proceedings of the 8th International Conference on Fracture Mechanics of Concrete and Concrete Structures (FraMCoS '12)*, J. G. M. Van Mier, G. Ruiz, C. Andrade, R. C. Yu, and X. X. Zhang, Eds., 2012.

[80] T. Wu, I. Temizer, and P. Wriggers, "Multiscale hydro-thermo-chemo-mechanical coupling: Application to alkali-silica reaction," *Computational Materials Science*, vol. 84, pp. 381–395, 2014.

[81] J. W. Pan, Y. T. Feng, J. T. Wang, Q. C. Sun, C. H. Zhang, and D. R. J. Owen, "Modeling of alkali-silica reaction in concrete: a review," *Frontiers of Structural and Civil Engineering*, vol. 6, no. 1, pp. 1–18, 2012.

[82] L. F. M. Sanchez, S. Multon, A. Sellier, M. Cyr, B. Fournier, and M. Jolin, "Comparative study of a chemo-mechanical modeling for alkali silica reaction (ASR) with experimental evidences," *Construction and Building Materials*, vol. 72, pp. 301–315, 2014.

16

Identification and Prioritization of "Black Spots" without Using Accident Information

Mahmoudreza Keymanesh,[1] Hasan Ziari,[2] Samira Roudini,[3] and Ali Nasrollahtabar Ahangar[3]

[1]*Faculty of Transport and Traffic Engineering, Payam Noor University of North Tehran, Tehran, Iran*
[2]*Faculty of Transport and Traffic Engineering, Iran University of Science & Technology, Tehran, Iran*
[3]*Payam Noor University of North Tehran, Tehran, Iran*

Correspondence should be addressed to Mahmoudreza Keymanesh; mrkeymanesh@pnu.ac.ir

Academic Editor: Zhiping Qiu

It is attempted to identify and prioritize the accident prone points (black spots) in "Iraanshahr-Sarbaaz-Chabahar" road located in Baluchistan, Iran, without no use of accident data but rather using Analytic Hierarchy Process (AHP), which is the enhanced procedure of road safety audit technique. First, by surveying the whole route, all factors that could influence accidents in this road were specified; then the route was divided into eight sections; this division was performed based on the uniformity and homogeny of each section in terms of geometric design and regional conditions. In each section, potentially hazardous locations were identified and some questionnaires were prepared, which were filled by 5 road traffic experts familiar with the route; then the collected data were analyzed by Analytical Hierarchy Process (AHP) using Expert Choice Software and the black spots were identified and prioritized. Finally, these black spots were compared with the black spots that had been obtained by traffic police based on accident data.

1. Introduction

Based on the latest estimation by World Health Organization [1], about 1.25 million people were killed due to road accident injuries, with another 20–50 millions of individuals being physically disabled as a result of road incidents. Road accidents are known as the 9th major cause of mortality among 15–29-year-old individuals all over the world, with more than 60% of the road casualties happening to 15–44-year-old individuals. While the number of registered vehicles during 2000–2013 has increased by 90%, road fatalities have been less significant. Interventions to improve road conditions have largely contributed into road safety and reduced road fatalities throughout the world. During 2010–2013, a number of 79 countries have succeeded to reduce their count of road casualties, indicating the feasibility to improve road conditions to save many people's lives.

Based on WHO statistics, in 2012, road fatalities are the third cause of death in Iran following ischemic heart disease and stroke; and since road casualties and associated physical disabilities impose a large economic cost on governments, road modification and improvement can be considered as highly justified. The locations along a road where highest number of accidents occur are called black spots. Identification and modification of these spots can contribute into reducing number of the accidents.

Analytic Hierarchy Process method has been extensively used in a large number of road safety researches and so many researches have been done in order to identify accident prone locations. For instance, Agarwal et al. [2] suggested a method for ranking black spots using AHP. Their method presents a four-step ranking approach for black spots wherein no accident information is used. Borsos et al. [3] compared the ranking of black spots in Italy and Hungary and have

FIGURE 1: The Iraanshahr-Sarbaaz-Chabahar road.

introduced the methods and performances that are employed in Italy and Hungary for identifying and prioritizing the black spots. Machado et al. [4] studied black spots for vulnerable users of the road including cyclists and pedestrians in Brazil and Italy. The paper addresses spots and areas defined by the severity of road accidents measured by fatalities and injuries in a developing country (Brazil) as well as a developed one (Italy). In order to characterize black spots, Chen [5] used GIS via spatial analysis. In his study, it was discussed how to implement geographical coordination technology in order to record accidents with geographical information attached, and the relation between black spots and traffic features is used analyzing factors affecting the accidents. Isen et al. [6] used GIS to identify and analyze black spots. Shabiraliyani et al. [7] investigated a black spot in Ghazi-Khan Valley in Pakistan. They aimed to study effects of associated risks with road traffic on society and the factors contributing to the occurrence of the accidents at the Ghazi-Khan Valley. In their study, it was shown that human mistakes are the largest cause of the road accidents and in more than 80% of accidents in Pakistan, drivers' mistakes ended up with serious injuries. Nguyen et al. [8] identified black spots based on saving road costs. Rouhi and Behzadi [9] prioritized four conventional methods, based on frequency, rate, severity, and rate-severity, and proposed a hybrid method using AHP. La Torre et al. [10] used crash modification factors in order to form an accident prediction model that could be applied to different road networks in Europe. Reshma and Sharif [11] presented a method in which the accident analysis includes ranking of some major black spots using ARC GIS 10 software package. This research studied some major accident spots in South Bangalore, Karnataka in India. Sadeghi et al. [12] proposed a new method for affecting environmental traffic and geometric features to identify black spots and for road safety performance evaluation Bao et al. [13] discovered a hierarchical fuzzy topsis as a support system for promising intelligent decision which can be used to enable taking the hierarchical structure of indicators into account and it was more suitable in combining layered road safety performances indicators into one total index. The main purpose's research of Randhawa et al. [14] is to excavate appropriate accident data to evaluate the validity of the OOS criteria as related to commercial vehicles by using a two-phase project that analyzed the commercial vehicle safety alliance's out of service criteria for vehicles. Barić et al. [15] used AHP model in order to rate road section design to enhance the quality of decision-making enterprises in transport infrastructure. Sufian et al. [16] presented a study on fault and deficiencies of Bangladesh transportation system by analyzing some major elements including pedestrians, none motorized vehicles, and heavy vehicles.

Identification and modification of black spots are necessary in order to have a safe road. However, a large number of methods proposed for identification of the black spots require accident data and since the information is time-consuming and casualty-intensive to acquire, it can be very beneficial to present a method which can identify the black spots within a short time, so as to reduce the fatalities and achieve safe roads.

2. Research Methodology

The methodology used in the present paper is a modified version of road safety audit technique. Road safety audit technique is a field study of an existing road or one under construction or a traffic project or any other project which is in relation with road users. In this technique, highly qualified auditors together with a team investigate possibility of the occurrence of road accidents and its safety performance. In the present research, Expert Choice Software was used to consider causes leading to the occurrence of an accident. This software allows auditors to pairwisely compare causes of accidents, so as to reduce human errors and save time. Moreover, this software makes it possible to perform surveys on a large number of experts and compare simultaneously numerous factors contributing into accidents. Also, if a cause of accident either rises or fades in future, one can conveniently apply corresponding changes into the software and be provided with the new set of results rapidly. The present research identifies and prioritizes black spots along the "Iraanshahr-Sarbaaz-Chabahar" road (Figure 1); this two-lane highway is an essential mountain road for the region connecting strategic port of Chabahar (Baluchistan, Iran) which is the only seaport in Iran with direct access to the Indian ocean. This highway is undertaken by investigating the road and environmental, traffic, and geometrical conditions. For this purpose, first, entire length of the road was checked for road imperfections and the factors which were expected

Table 1: Section 1: Iraanshahr-Polchamani.

Potential black spots	Cause of having potential
Shahrderaz	Lack of interchanges at residential areas and traffic of smugglers' vehicles

Table 2: Section 2: Polchamani-Kalaat.

Potential black spots	Cause of having potential
Kalaat	Lack of interchanges at residential areas and traffic of smugglers' vehicles

to contribute into accidents along the road; listed as (A) to (I), the factors were as follows:

(A) Inappropriate horizontal curves

(B) Inappropriate vertical curves

(C) Lack of traffic signs

(D) Lack of adequate lighting at night

(E) Inappropriate sight distance

(F) Lack of road shoulder

(G) Lack of guard rail and passage of animals across the road

(H) Traffic of smugglers' vehicles

(I) Lack of interchanges at residential areas

Following with the research, the "Iraanshahr-Sarbaaz-Chabahar" road was divided into 8 uniform and homogenous sections in terms of physical and road performance properties. The sections were as follows:

Section 1: Iraanshahr-Polchamani

Section 2: Polchamani-Kalaat

Section 3: Kalaat-Sarbaaz

Section 4: Sarbaaz-Pashaamag Mortaan

Section 5: Pashaamag Mortaan-Jakigor

Section 6: Jakigor-Dashtyari

Section 7: Dashtyari-Konaarak

Section 8: Konaarak-Chabahar

At each section, potential black spots along the section were identified based on the geometry of the road within that section, environmental factors, and also the 9 factors mentioned above. Unfortunately, due to high unemployment rate, many people go for gas oil and goods (cloths, food stuff, etc.) smuggling to be able to afford life expenses. Meanwhile, high speed of smuggling vehicles, particularly those used to transport gas oil, has resulted in disastrous accidents upon which a large number of casualties has been incurred. Potential black spots along Sections 1–8 are reported in Tables 1–8, respectively.

Table 3: Section 3: Kalaat-Sarbaaz.

Potential black spots	Cause of having potential
Sarbaaz	Lack of interchanges at residential areas, lack of adequate lighting at night, and lack of road shoulder

Table 4: Section 4: Sarbaaz-Pashaamag Mortaan.

Potential black spots	Cause of having potential
Dapkor	Lack of interchanges at residential areas and inappropriate horizontal curves
Bepaataan	Inappropriate horizontal curves and traffic of smugglers' vehicles

Table 5: Section 5: Pashaamag Mortaan-Jakigor.

Potential black spots	Cause of having potential
Garikan	Lack of interchanges at residential areas and inappropriate horizontal curves
Jangal	Lack of interchanges at residential areas and inappropriate horizontal curves
Firouzabad	Lack of interchanges at residential areas, inappropriate horizontal curves, and inappropriate vertical curves
Derakhshan	Lack of interchanges at residential areas and inappropriate horizontal curves
Jakigor	Inappropriate horizontal curves

Table 6: Section 6: Jakigor-Dashtyari.

Potential black spots	Cause of having potential
Dashtyari	Lack of interchanges at residential areas and lack of traffic signs

Table 7: Section 7: Dashtyari-Konaarak.

Potential black spots	Cause of having potential
Oraki	Lack of interchanges at residential areas and lack of traffic signs
Nobandian	Lack of interchanges at residential areas and lack of traffic signs
Negur	Lack of interchanges at residential areas

As can be observed from Figure 2, the inappropriate horizontal curves together with the traffic of smugglers' vehicles have made Bepaataan a potential black spot.

According to Figure 3, inappropriate horizontal and vertical curves along with lack of interchanges at residential area are among the causes contributing to car accidents across the locality.

FIGURE 2: Section 4: Bepaataan.

TABLE 8: Section 8: Konaarak-Chabahar.

Potential black spots	Cause of having potential
Konaarak	Lack of interchanges at residential areas
Tis-Kopan	Lack of interchanges at residential areas

FIGURE 3: Section 5: Firouzabad.

Based on Figure 4, lack of traffic signs and the grade byway intersection at Nobandian are among the causes contributing to car accidents across the locality.

According to Figure 5, Konaarak byway crossing serves as a local black spot due to lack of traffic signs and intersection.

For each section, two questionnaires were prepared and 5 experts who were familiar with the region were asked to score the effect of each factor contributing into accidents on 1–10-level scale. Questionnaire 1 was used to score each of the contributing factors into accidents, while Questionnaire 2 scored potential black spots based on each of the contributing factors into accidents.

The questionnaire results were averaged, with the contributing factors into accidents as well as potential black spots being subjected to pairwise comparisons; that is, they were

FIGURE 4: Section 7: Nobandian.

FIGURE 5: Section 8: Konaarak.

fed into a matrix of comparison in Expert Choice Software where they were analyzed based on numerical comparisons.

Table 9 that is presented by Saaty [17] was used to compare the criteria including inappropriate horizontal curves, inappropriate vertical curves, lack of traffic signs, lack of adequate lighting at night, inappropriate sight distance, lack of road shoulder, lack of guard rail and passage of animals across the road, traffic of smugglers' vehicles, and lack of interchanges at residential areas in a pairwise fashion and also to compare the alternatives including the Shahrderaz crossing, Kalaat byway crossing, Sarbaaz Bridge, Dapkor, Bepaataan, Garikan, Jangal, Firouzabad, Derakhshan, Jakigor, Dashtyari, Oraki, Nobandian, Negur, Konaarak byway crossing, and Tis-Kopan

TABLE 9: AHP preference scale.

Numeric value	Preference level
9	Extremely preferred
7	Very strongly preferred
5	Strongly preferred
3	Moderately preferred
1	Equally preferred
2, 4, 6, and 8	Intermediate values between the two adjacent judgments

byway crossing in a pairwise fashion; the table indicates the level of priority as measured numerically with a number ranging within 1 to 9. Undertaking the pairwise comparisons on the criteria and alternatives, we ended up with 8 and 9 matrixes, respectively. As examples, the resulting matrix from comparing, in a pairwise fashion, the criteria within the scope of Section 1 is demonstrated in Table 10, while Table 11 presents the corresponding pairwise comparison matrix to the inappropriate horizontal curve criterion.

3. Results and Analyses

The reported stats by highway police corresponds to 2015, and by 2016 when the present research has been performed, numerous instances of black spots were corrected or undergoing corrective actions. For example, although the inappropriate horizontal curves at three black spots (Garikan, Firouzabad, and Nobandian) were corrected even before this research was initiated, results of survey (using the designed questionnaire) on traffic experts and outputs of Expert Choice Software recognized the same three locations as black spots but now because of the lack of interchanges.

In Section 1, the most effective factor contributing into accidents was found to be lack of interchanges in residential areas. According to Table 1, there was only one potential black spot along the section with its cause being the lack of intersections in residential areas; as such, the potential black spot can be considered as an actual black spot.

According to the stats released by traffic police, in 2015, inappropriate horizontal curves have been the most significant cause of accidents along Sections 2 and 3; the curves were modified in early 2016. Accordingly, based on field investigations performed in the present research, in 2016, due to the factors mentioned in Tables 2 and 3, Kalaat and Sarbaaz have had the potential to become black spots. However, since the horizontal curves were corrected in early 2016, the inappropriate curves had no more served as a main cause of accident in the mentioned tables. Analysis of the questionnaires using Expert Choice Software shows that inappropriate horizontal curve represents the main cause of accidents at these points, and knowing that the curves are now corrected, the corresponding points may no more be black spots.

Establishment of a highway police station within Dashtyari and correction of inappropriate horizontal curves at two

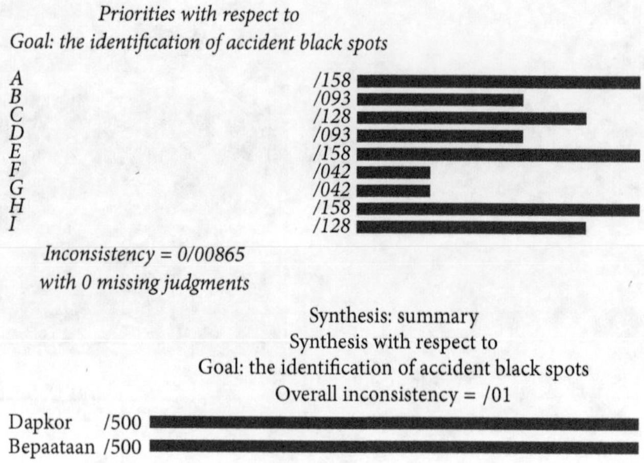

FIGURE 6: Section 4: Sarbaaz-Pashaamag Mortaan.

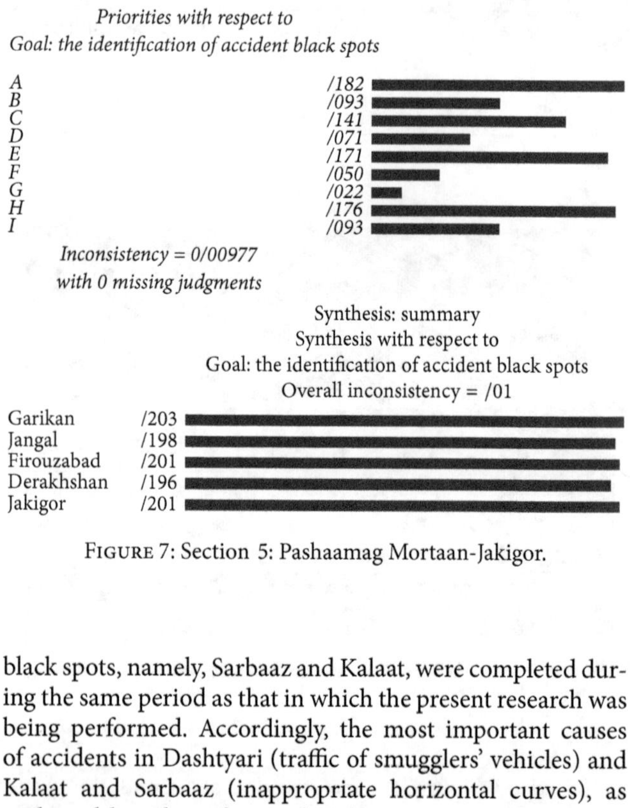

FIGURE 7: Section 5: Pashaamag Mortaan-Jakigor.

black spots, namely, Sarbaaz and Kalaat, were completed during the same period as that in which the present research was being performed. Accordingly, the most important causes of accidents in Dashtyari (traffic of smugglers' vehicles) and Kalaat and Sarbaaz (inappropriate horizontal curves), as evaluated based on the results of survey on traffic experts, were eliminated. That is to say, based on experts' opinions, these two points were not recognized as black spots. Other black spots are also undergoing corrective actions.

According to Figure 6, inappropriate horizontal curve and traffic of smugglers' vehicles were two important factors contributing, simultaneously, to the accidents occurring within Section 4. The potential black spots at Dapkor and Bepaataan were of the same priority as they suffered from both of the factors.

Inappropriate horizontal curve was the main contributing factor into accidents happening within Section 5. According to Table 5, all potential black spots within this section are actual black spots, and based on Figure 7, the spots

Table 10: Pairwise comparisons matrix on the criteria within Section 1.

		1	2	3	4	5	6	7	8	9
1	Inappropriate horizontal curves	1	1	$\frac{1}{4}$	$\frac{1}{7}$	$\frac{1}{2}$	$\frac{1}{3}$	$\frac{1}{2}$	$\frac{1}{4}$	$\mathbf{\frac{1}{8}}$
2	Inappropriate vertical curves		1	$\frac{1}{4}$	$\frac{1}{6}$	$\frac{1}{2}$	$\frac{1}{3}$	$\frac{1}{2}$	$\frac{1}{4}$	$\mathbf{\frac{1}{8}}$
3	Lack of traffic signs			1	$\frac{1}{2}$	2	1	2	1	$\mathbf{\frac{1}{2}}$
4	Lack of adequate lighting at night				1	3	2	3	2	**1**
5	Inappropriate sight distance					1	1	1	$\frac{1}{2}$	$\mathbf{\frac{1}{4}}$
6	Lack of road shoulder						1	1	1	$\mathbf{\frac{1}{3}}$
7	Lack of guard rail and passage of animals across the road							1	$\frac{1}{2}$	$\mathbf{\frac{1}{4}}$
8	Traffic of smugglers' vehicles								1	$\mathbf{\frac{1}{2}}$
9	Lack of interchanges at residential areas									**1**

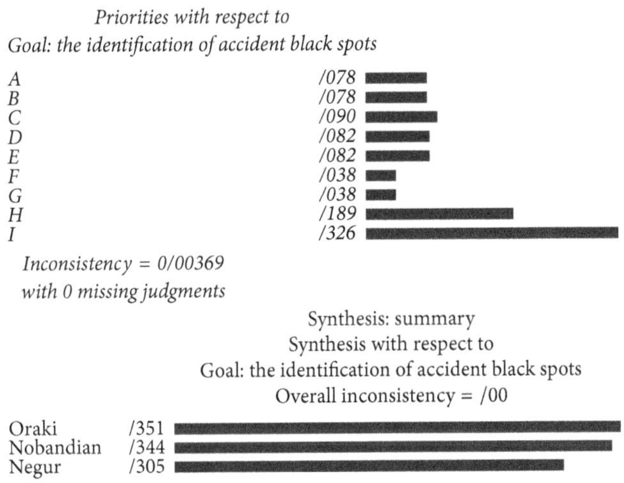

Figure 8: Section 7: Dashtyari-Konaarak.

were prioritized into the following order: Garikan, forest, Firouzabad, Derakhshan, and Jakigor.

In Section 6, the most significant factor contributing into accidents within this section was found to be smugglers' vehicles. However, based on Table 6, lack of interchanges was introduced as a potential factor contributing to the development of black spots within the section.

Lack of interchanges was found to be the most important cause of accidents within Section 7; and according to Table 7, all of the three locations mentioned were identified as actual black spots, with their prioritization (Figure 8) being as follows: Oraki, Nobandian, and Negur.

Lack of interchanges was an important cause of accidents within Section 8; as such, both potential black spots reported in Table 8 could be actual black spots as they had no interchanges. However, according to Figure 9, Konaarak was

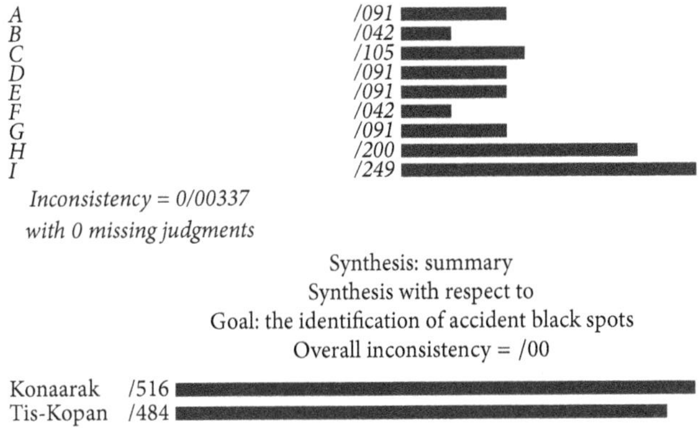

Figure 9: Section 8: Konaarak-Chabahar.

Table 11: Pairwise comparison matrix to the inappropriate horizontal curve criterion.

	1	2	3	4	5	6	7	8	9	10	11	12	13	14	15	16
	Shahrderaz	Kalaat	Sarbaaz	Dapkor	Bepaataan	Garikan	Jangal	Firouzabad	Derakhshan	Jakigor	Dashtyari	Oraki	Nobandian	Negur	Konaarak	Tis-Kopan
1	1	$\frac{1}{4}$	$\frac{1}{3}$	$\frac{1}{6}$	$\frac{1}{7}$	$\frac{1}{8}$	$\frac{1}{8}$	$\frac{1}{8}$	$\frac{1}{6}$	$\frac{1}{6}$	1	1	$\frac{1}{2}$	$\frac{1}{2}$	$\frac{1}{2}$	$\frac{1}{2}$
2		1	1	1	$\frac{1}{2}$	$\frac{1}{2}$	$\frac{1}{2}$	$\frac{1}{2}$	1	1	4	4	2	2	2	2
3			1	$\frac{1}{2}$	$\frac{1}{2}$	$\frac{1}{2}$	$\frac{1}{2}$	$\frac{1}{2}$	$\frac{1}{2}$	$\frac{1}{2}$	1	1	1	1	1	1
4				1	1	1	1	1	1	1	6	6	3	3	3	3
5					1	1	1	1	1	1	7	7	3	3	3	3
6						1	1	1	1	1	8	8	4	4	4	4
7							1	1	1	1	8	8	4	4	4	4
8								1	1	1	8	8	4	4	4	4
9									1	1	6	6	3	3	3	3
10										1	6	6	3	3	3	3
11											1	1	$\frac{1}{2}$	$\frac{1}{2}$	$\frac{1}{2}$	$\frac{1}{2}$
12												1	$\frac{1}{2}$	$\frac{1}{2}$	$\frac{1}{2}$	$\frac{1}{2}$
13													1	1	1	1
14														1	1	1
15															1	1
16																1

recognized as the top priority black spot followed by Tis-Kopan as second.

4. Conclusion

The three major causes of accident, respectively, are as follows:

(1) Lack of interchanges along the "Iraanshahr-Polchamani," "Dashtyari-Konaarak," and "Konaarak-Chabahar" roads.

(2) Inappropriate and nonstandard horizontal curves along "Polchamani-Kalaat," "Kalaat-Sarbaaz," "Sarbaaz-Pashaamag Mortaan," "Pashaamag Mortaan-Jakigor," and "Jakigor-Dashtyari."

(3) Traffic of smugglers' vehicles whose drivers are known to drive at high speed and minimum care along "Sarbaaz-Pashaamag Mortaan" and "Jakigor-Dashtyari" sections, imposing the largest contributions into road accidents at black spots along "Iraanshahr-Sarbaaz-Chabahar" road. The traffic of smugglers' vehicles shall be recognized as a cause of accident which is present every day with various severities along the day, so that serious interventions shall be undertaken to address this cause considering its large deal of risk at residential areas surrounding the road and intersections. The black spots recognized using the proposed method in the present research (with no use of accident information) are tabulated in Table 12, with the extracted black spots from accident information (as reported by traffic police) presented in Table 13.

Table 12: Identification of black spots without using accident information.

	Black spots
1	Shahrderaz
2	Dapkor
3	Bepaataan
4	Garikan
5	Jangal
6	Firouzabad
7	Derakhshan
8	Jakigor
9	Oraki
10	Nobandian
11	Negur
12	Konaarak
13	Tis-Kopan

Accident index is a number obtained for a given black spot based on the registry of accidents' information during the period under study; it indicates the significance of the black spot. The value of accident index can be calculated via

$$P = \frac{1000 \times I}{365 \times A \times T}, \tag{1}$$

where T is the length of the period during which accident information and data are collected (in years) and I is the accident intensity factor at accident point per T period, which can be calculated via

$$I = X + 3Y + 9Z, \tag{2}$$

where X is the number of accidents ending up with damage, Y is the number of accidents ending up with injuries, Z is the number of accidents ending up with fatality, at the accident point per T period, and A is the road performance factor which is considered to be 2, 4, 6, and 8 for byways, main roads (province-wide), main roads (national-wide), and expressways, respectively.

4.1. Comparing Black Spots Extracted with/without Accident Information. Comparing the extracted black spots according to Tables 12 and 13, police reports indicated the "Kalaat Crossing" as a black spot, while the proposed method failed to identify that. Further, the three crossings of "Shahrderaz," "Konaarak," and "Tis-Kopan" were recognized, by the proposed method in this research, as black spots, while they were excluded in the police reports.

4.2. Comparing Prioritization of Black Spots with/without Accident Information. Comparing the prioritization results of AHP with no use of accident information to those obtained directly from accident information indicates that the Dapkor and Bepaataan black spots at Section 4 were of the same priority in both methods. However, when not using accident information, the spots within Section 5 were prioritized as follows: Garikan, forest, Firouzabad, Derakhshan, and Jakigor, while the corresponding order of priority using accident information was Garikan, Firouzabad, forest, Derakhshan, and Jakigor. Further, in Section 7, the black spots were prioritized, using the proposed method, as follows: Oraki, Nobandian, and Negur, while traffic police statistics indicated the following order of priority: Nobandian, Oraki, and Negur.

5. Summary

Being based on geometrical plan of the road as well as a survey participated by several experts, identification of black spots using the proposed method in this research resulted in 10 of the reported black spots by traffic police, with the prioritization results being in good agreement with police reports. Therefore, this method can be used wherever accident information and statistics are not properly recorded due to either lack of required facilities or inadequate training of the registering agents.

TABLE 13: Identification of black spots using accident information reported by traffic police in 2015.

	Black spots	Code of region	Accident index
1	Kalaat	28034	54.8
2	Dapkor	28039	41.4
3	Bepaataan	28039	41.4
4	Garikan	28038	49.8
5	Jangal	28037	41.1
6	Firouzabad	28036	49.32
7	Derakhshan	28040	41.1
8	Jakigor	28041	41
9	Oraki	28051	41.1
10	Nobandian	28050	42.47
11	Negur	28050	20.55

Conflicts of Interest

The authors declare that there are no conflicts of interest regarding the publication of this paper.

References

[1] World Health Organization (2016), Specialized agency of the United Nations, United Nations Economic and Social Council (ECOSOC), Switzerland, 1948.

[2] P. K. Agarwal, P. K. Patil, and R. Mehar, "A methodology for ranking road safety hazardous locations using analytical hierarchy process," in *Proceedings of the 2nd Conference of Transportation Research Group of India (2nd CTRG)*, vol. 104, pp. 1030–1037, Agra, India, December 2013.

[3] A. Borsos, S. Cafiso, C. D'Agostino, and D. Miletics, "Comparison of Italian and Hungarian black spot ranking," in *Proceedings of the 6th Transport Research Arena*, Warsaw, Poland, April 2016.

[4] C. A. S. Machado, M. A. Giannotti, F. C. Neto, A. Tripodi, L. Persia, and J. A. Quintanilha, "Characterization of black spot zones for vulnerable road users in São Paulo (Brazil) and Rome (Italy)," *ISPRS International Journal of Geo-Information*, vol. 4, no. 2, pp. 858–882, 2015.

[5] H. Chen, "Black spot determination of traffic accident locations and its spatial association characteristic analysis based on GIS," *Journal of Geographic Information System*, vol. 4, no. 6, pp. 608–617, 2012.

[6] L. Isen, A. Shibu, and M. Saran, "Evaluation and treatment of accident black spots using Geographic Information System," *International Journal of Innovative Research in Science, Engineering and Technology*, vol. 2, no. 8, 2013.

[7] G. Shabiraliyani, M. Tariqaliyani, N. Hamad, and N. Iqbal, "A black spot on driving safety measures in Dera Ghazi Khan, Pakistan," *Journal of Applied Environmental and Biological Sciences*, vol. 5, no. 8, pp. 27–38, 2015.

[8] H. H. Nguyen, P. Taneerananon, and P. Luathep, "Approach to identifying black spots based on potential saving in accident costs," *Engineering Journal*, vol. 20, no. 2, pp. 109–122, 2016.

[9] A. Rouhi and G. A. Behzadi, "Prioritizing four methods and Presenting a combination method for determining black spots using Analytical Hierarchy Process(AHP)," *Journal of Science and Today's World*, vol. 5, no. 4, pp. 105–109, 2016.

[10] F. La Torre, L. Domenichini, M. Meocci et al., "Development of a transnational accident prediction model," *Transportation Research Procedia*, vol. 14, pp. 1772–1781, 2016.

[11] E. K. Reshma and S. U. Sharif, "Prioritization of accident black spots using GIS," *International Journal of Emerging Technology and Advanced Engineering*, vol. 2, no. 9, 2012.

[12] A. Sadeghi, E. Ayati, and M. P. Neghab, "Identification and prioritization of hazardous road locations by segmentation and data envelopment analysis approach," *Safety and Security in Traffic Preliminary Communication*, vol. 25, no. 2, pp. 127–136, 2013.

[13] Q. Bao, D. Ruan, Y. Shen, E. Hermans, and D. Janssens, "Improved hierarchical fuzzy TOPSIS for road safety performance evaluation," *Knowledge-Based Systems*, vol. 32, pp. 84–90, 2012.

[14] S. U. Randhawa, S. G. Miller, C. A. Bell, and P. E. Montagne, "A study of Commercial Vehicle Safety Alliance's out-of-service criteria," *Accident Analysis and Prevention*, vol. 30, no. 1, pp. 61–67, 1998.

[15] D. Barić, H. Pilko, and J. Strujić, "An analytic hierarchy process model to evaluate road section design," *Transport*, vol. 31, no. 3, pp. 312–321, 2016.

[16] A. A. Sufian, S. M. Khan, I. Ahmed, S. Islam, and N. Saha, "Safety analysis: observed deficiencies in existing transportation system of Bangladesh," *International Journal of Urban Planning and Transportation*, vol. 27, no. 1, 2016.

[17] T. L. Saaty, "Decision making with the analytic hierarchy process," *International Journal of Services Sciences*, vol. 1, no. 1, 2008.

17

Modelling of Wind Pressure Coefficients on C-Shaped Building Models

Monalisa Mallick,[1] **Abinash Mohanta**,[1] **Awadhesh Kumar**,[2] **and Vivek Raj**[3]

[1]*PhD Scholar, Department of Civil Engineering, National Institute of Technology Rourkela, Odisha, India*
[2]*Professor, Department of Civil Engineering, National Institute of Technology Rourkela, Odisha, India*
[3]*M Tech Scholar, Department of Civil Engineering, National Institute of Technology Rourkela, Odisha, India*

Correspondence should be addressed to Monalisa Mallick; monalisa.nitrkl@gmail.com

Academic Editor: Aiguo Song

Designs of buildings are changing with emerging demands of several aesthetical features and efficient design based on geometry. Development of new building materials and construction techniques have enabled us to build new buildings which are tall and unsymmetrical, but unfortunately such structures are more susceptible to wind loads. Thus it becomes necessary to estimate wind loads with higher degree of confidence. Although ample information regarding wind load on symmetrical and regular structure is available in various international codes, they lack the study of effect of wind forces on unsymmetrical structures. This paper presents experimental and numerical studies of the wind effect on commonly used C-shaped buildings with varying aspect ratio and its optimization caused by the alteration of angle of incidence. Furthermore, results obtained by numerical analysis have been validated with the experimental one. For this study, numerical analysis has been carried out using ANSYS Fluent with k-ε model of turbulence. Computational fluid dynamics (CFD) techniques is used to evaluate the surface pressure on various faces of the model for angle of attack of 0° to 180° at an interval of 30° in a subsonic open circuit wind tunnel. The results found by CFD technique are well compared with the experimental results which suggest the feasibility of using this technique of predicting wind pressures on building efficiently and accurately.

1. Introduction

Gradual increase of demand of tall and unsymmetrical buildings with efficient geometric design and planning needs futuristic visionary and scientific estimates of various kinds of forces. Wind force is one of them, which plays a crucial role in case of unsymmetrical buildings. Shortage of land and demand of good aesthetical view have forced us to construct nonconventional plan-shaped building. Although ample information regarding wind load on symmetrical structure (conventional plan-shaped structure) is available in various international codes, for example, IS: 875 (Part 3): 1987(code of practice for wind loads for buildings and structures) [1], no such direct reference is available for irregular plan-shaped structures. The effect is even more critical for unsymmetrical and tall structures due to dynamic response arising from vortex shedding and galloping. Among the loads which act on buildings, the wind load is quite difficult to predict. However, the evolution of computer programs for structural analysis has made the prediction of wind pressure relatively easy to design a building to suit any given set of applied loads.

Though now a days nonconventional plan shapes are very common building configurations, experimental as well as numerical data for such shapes for different wind directions are very limited. Various literatures provide the study of effect of wind pressure on tall and irregular shape of buildings. Paterson and Apelt [2] first presented the research on computation of wind flows over three-dimensional buildings by numerical analysis and found accurate results for some wind flows over simple three-dimensional buildings. Kwok [3] conducted the wind tunnel model test to analyse the effect of building shape on the wind-induced response of a tall building with a rectangular cross section. Afterwards, Stathopoulos and Zhou [4] presented the paper which shows the numerical prediction of turbulent wind pressures on the building surfaces with sharp corners of right angles for different wind incidence angle. Miyashita et al. [5] presented effects of corner cuts and openings in high-rise square buildings. Li et al. [6] have described the results obtained from the measurements of wind effects on two high-rise buildings with 70 storeys and 30 storeys, respectively. Gomes et al. [7] investigated pressure distributions on L-type and U-type building models under varying wind incidence angle. Chakraborty et al. [8] showed that pressure induced on a "+" plan-shaped tall building is widely different from that of a square plan-shaped tall building. Amin and Ahuja [9] presented the effects of side ratio on wind pressure distribution on rectangular buildings. Dagnew et al. [10] focused on numerical analysis of wind pressures on high-rise buildings by using the Commonwealth Advisory Aeronautical Council (CAARC) building model. The authors applied CFD techniques to predict wind load on CAARC irregular-shaped model. Chakraborty et al. [11] presented a study that showed irregular plan-shaped buildings are subjected to different pressure distributions, as compared to regular plan-shaped buildings after distribution. Yi and Li [12] conducted experiment in a tall building at Hong Kong as well as wind tunnel model test to explain force and pressure coefficients and high-frequency balance. The wind tunnel test result was seen to be fairly equivalent with a full model test. Li et al. [13] also conducted the wind tunnel test on L-plan-shaped tall building to quantify dynamic load across the wind. They proposed an empirical formulation by considering the side ratio and territory of the building to express wind load. Bhattacharyya and Dalui [14] investigated mean wind pressures on E-plan-shaped tall building and compared experimental results with numerical analysis using CFD. They compared their results by plotting pressure contours on all the surfaces using experimental as well as predicted values using CFD analysis. For better comparison of the results, the Fourier series expansion was obtained using the experimental data and compared with both the experimental data and CFD analysis. Li et al. [15] studied some recent efforts for credible computational fluid dynamics simulations in China. Chow et al. [16] investigated to simulate smoke filling in big halls by CFD. Mohanta and Patra [17] used CFD simulation for studying the flow field distribution in converging compound channel by the LES turbulence model.

In present study, numerical as well as experimental analysis has been carried out on irregular shape, i.e., C-shaped of the building for wind incidence angle of 0° to 180° at the interval of 30°. ANSYS FLUENT is used to numerically model the domain and study the wind flow, and experimental data were obtained by using wind tunnel. The main purpose of this paper is to assess the change in wind pressure on different faces of C-shaped model due to change in wind angle and aspect ratio (height) of the building by experimental analysis and numerical analysis, and then those results were validated. Validation of results is necessary because there is no such direct reference from where wind pressure on the irregular shape of the building could be calculated. So, it becomes necessary to compare the experimental and numerical data.

2. Experimental Setup

2.1. Details of Experiments and Models. Experiments are carried out in an open circuit subsonic wind tunnel in the Aerodynamics Laboratory of the Department of Aerospace Engineering, Indian Institute of Technology Kharagpur, India. The wind speed is kept constant at 12.9 m/s. The wind tunnel with a bottom surface made up of plywood, a test section is 1.83 m long, and cross-sectional dimensions of 0.61 m × 0.61 m. Models are placed within the boundary layer zone, centrally in the test section at a distance of 1.2 m from the beginning of the test section. To ascertain models within the boundary layer zone, wooden cubic blocks of 25 mm size, and clear spacing of 50 mm in all directions are fixed on a 4 mm thick plywood sheet as shown in Figure 1.

The experimental models are made of transparent Perspex sheet of 3 mm thickness. Details of the two C-shaped models, as C-1 and C-2 of varying height ratios configurations, are shown in Figures 2(a)–2(c) and given in Table 1. Figure 3 represents the photograph of the C-shaped building model placed in the test section.

The models are fitted with 90 to 120 numbers of pressure-tapping points in 4-5 rows and 3–5 columns on the surfaces. The pressure-tapping points are kept at less spacing near the wall boundaries to tap the sharp pressure variation due to flow separation and at larger spacing in the middle of the surfaces.

Free stream velocity during the experiment is measured with the help of the pitot tube. The models are fitted with 90 to 120 numbers of pressure-tapping points in 4-5 rows and 3–5 columns on the surfaces. The pressure-tapping points are kept at less spacing near the wall boundaries to tap the sharp pressure variation due to flow separation and at a larger spacing in the middle of the surfaces. The free ends of tubes are connected to Digital Sensor Array (DSA) to record the fluctuating wind pressure at the corresponding tapping points. All data are measured by a DSA scan valve corporation, model DSA 3217/16 pox, USA. This DSA device is set to give an average pressure of 5 s duration. At the same time, for greater accuracy, a pressure measurement at each tapping point is repeated for three times, and the mean of three pressure data are obtained. The pressure coefficient (C_P) for each tapping point is calculated by putting the pressure data in the following equation as follows:

$$C_P = \frac{P - P_\infty}{P_0 - P_\infty}, \tag{1}$$

where p = mean pressure data obtained experimentally, P_∞ = the static pressure in the reference tube, and P_0 = total pressure in the settling chamber.

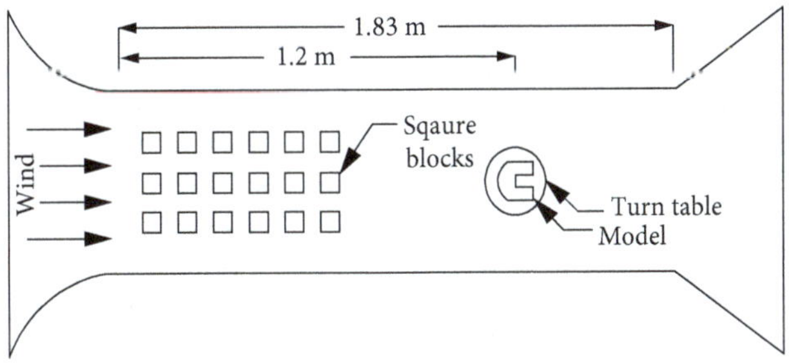

FIGURE 1: The schematic diagram of wind tunnel.

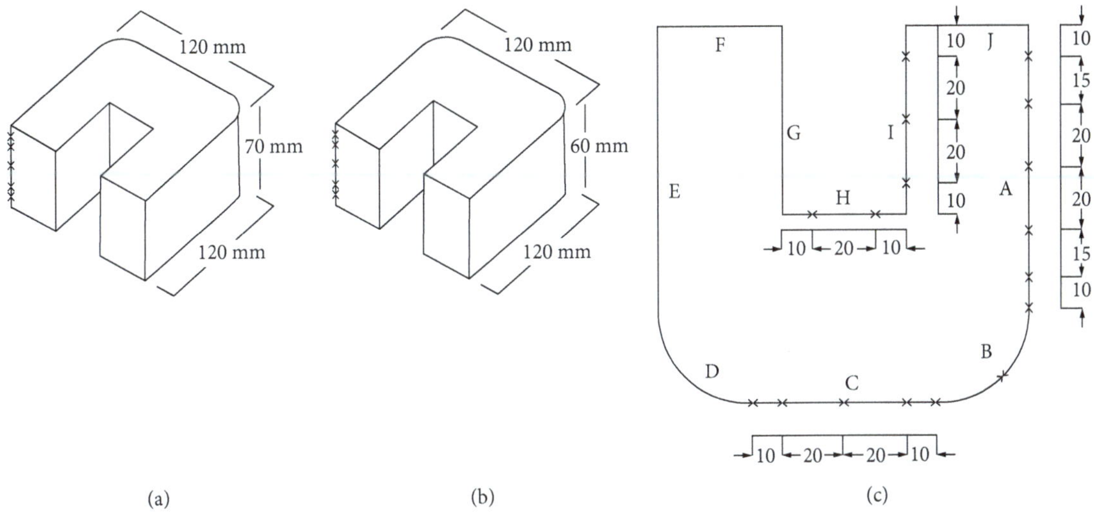

FIGURE 2: (a–c) Isometric view of C-1- and C-2-shaped building model and pressure-tapping locations along the periphery of models.

TABLE 1: Details of building models with different aspect ratios.

Sl. No. (1)	Overall depth, D (mm) (2)	Depth, d (mm) (3)	Overall breadth, B (mm) (4)	Breadth, b (mm) (5)	Height, H (mm) (6)	Plan area, (mm^2) (7)	Radius of curvature, R (mm) (8)
C-1	120	90	120	60	70	11613	30
C-2	120	90	120	60	60	11613	30

FIGURE 3: Photograph of C-shaped building model in the test section.

3. Numerical Modelling

Numerical simulations have been carried out in this study using ANSYS Fluent using the computational fluid dynamic (CFD) technique, based on the control volume method. The RNG $k-\varepsilon$ model is used to simulate the turbulence effects. The RNG model is developed using renormalization group (RNG) methods to solve the Navier–Stokes equations, to account for the effects of smaller scales of motion. The RNG approach is a mathematical technique that can be used to derive a turbulence model similar to the $k-\varepsilon$ model. Bhattacharyya and Dalui [18] studied wind force and pressure coefficients on the surfaces of an unsymmetrical E-shaped building using CFD ANSYS CFX software. They used two equations k-ε turbulence model for modelling their works which offers a good compromise between numerical effort and computational accuracy. Jena et. al [19] presented numerical studies on the interference effect on pentagon-shaped tall buildings and analysis performed with the realizable $k-\varepsilon$ model. Swaddiwudhipong and Khan [20] investigated wind-induced dynamic response of tall buildings using CFD. A comparative study between two widely used turbulence models, i.e., LES and k-ε Reynolds-average Navier–Stokes models have been made by Cheng et al. [21]. Villalpando et al. [22] studied the flow simulation of turbulence models around a wind turbine NACA 63–415 airfoil at various angles of attack. The RNG (renormalization group) model incorporates modifications over the standard k-ε [22] model, which improves its performance. Uchida [23] presented a new evaluation technique for wind turbine generators design using a CFD model-based unsteady flow simulation with changes of wind direction.

It is advantageous to use the $(k-\varepsilon)$ turbulence model as most commonly used in computational fluid dynamics (CFD). It simulates mean flow characteristics for turbulent flow conditions. It is a two-equation model which gives a general description of turbulence by means of two transport equations.

(i) The first transported variable determines the energy in the turbulence and is called turbulent kinetic energy (k).

(ii) The second transported variable is the turbulent dissipation (ε) which determines the rate of dissipation of the turbulent kinetic energy.

The $k-\varepsilon$ turbulence model is limited to the fully developed turbulent and nonseparated flows. It can have stability issues due to numerical stiffness. This model is still used widely despite its disadvantages. This performs poorly for complex flows involving severe pressure gradient, separation, and strong streamline curvature. The RNG produces lower turbulence levels and can underestimate the value of k. This produces a less viscous flow that creates more realistic flow features with the complex geometry.

3.1. Numerical Analysis. For numerical analysis, the models of C-shaped buildings of different aspect ratios have been created in ANSYS and then analysed by using the $k-\varepsilon$ model of turbulence. ANSYS (Fluent) uses the finite element method to solve the governing equations, i.e., the region of interest is divided into the finite number of cells (the mesh or grid).

Data considered in numerical analysis:

(i) Types of fluid: air

(ii) Density of air: 1.225 Kg/m^3

(iii) Viscosity of air: 1.789×10^{-5} kg·s/m

(iv) Turbulence model: $k-\varepsilon$ model

(v) Solver: pressure-based

3.2. Domain and Meshing. The domain size as shown in Figures 4(a) and 4(b) is considered as mentioned in Franke et al. [24]. The upstream side is considered as 5H from the face of the building, downstream side is taken as 15H from the face of the building, two side clearance of the domain is taken as 5H from the face of the building, and top clearance is also taken as 5H from the top surface of the building. Such large size of the domain helps in vortex generation at the leeward side of the building, and backflow of wind can also be prevented.

The finite volume discretization approach is used to discretize the whole domain so that separation of wind flow, upwash, and downwash mechanisms can happen similar to the experimental study. The discretization (meshing) of the C-shaped model is shown in Figure 5.

3.3. Boundary Conditions. For completing numerical modelling and simulation, there are several boundary conditions which are considered. The flow parameter at inlet, outlet wall, and surface are needed to be considered. At inlet, the velocity of flow is 12.9 m/s, and the same velocity is provided for experimental procedure so that the result could be compared. The flow velocity at inlet is along the positive X direction.

4. Results and Discussion

4.1. Wind Flow Pattern. Pressure variation on the building is directly influenced by wind flow pattern. Vortex generation and different types of mechanism such as separation of flow, upwind, and downwind happen due to dynamic behavior of wind flow. To investigate such mechanisms more accurately, wind flow patterns around the C-shaped building are being studied for different angle of incidences varying from 0° to 180° at an interval of 30° using the CFD technique. Flow pattern for different angles of incidences are shown in Figures 6(a)–6(e).

Wind is directly affecting Face C for wind incidence angle 0° to 180°, respectively, so the pressure distribution is symmetrical about vertical axis, and vortices generated at the wake region are also symmetrical for both the cases (Figure 6).

4.2. Variation of Pressure Coefficients. The study of pressure coefficient is necessary in every unconventional-shaped building which is designed under wind excitation. Chakraborty et al. [11] and Mukherjee et al. [25] analysed that the critical pressure coefficients may not be found perpendicular to wind incidence angles only. For this reason, a detailed study on wind pressure variation is required with respect to wind incidence angles.

4.3. Experimental Results. Pressure contour on different faces are plotted for every angle of incidence. Some pressure contours are shown in Figures 7 and 8, respectively, for all the faces. Also, mean pressure coefficients of all the faces are shown in Figures 9(a) and 10(a).

Figures 7 and 8 show the mean pressure coefficient contours on the faces of the model C-1 and C-2 at wind incidence angles of 0° to 180° at an interval of 30°. For wind

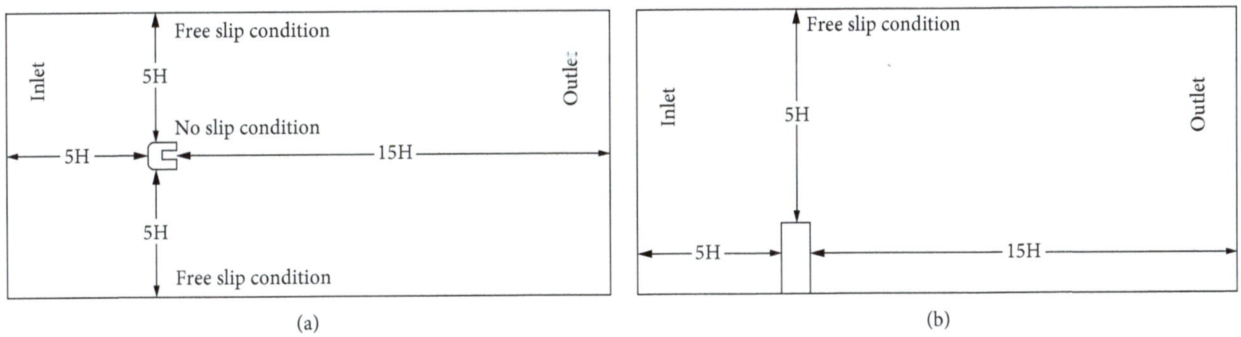

FIGURE 4: (a, b) Domain for the plan and elevation.

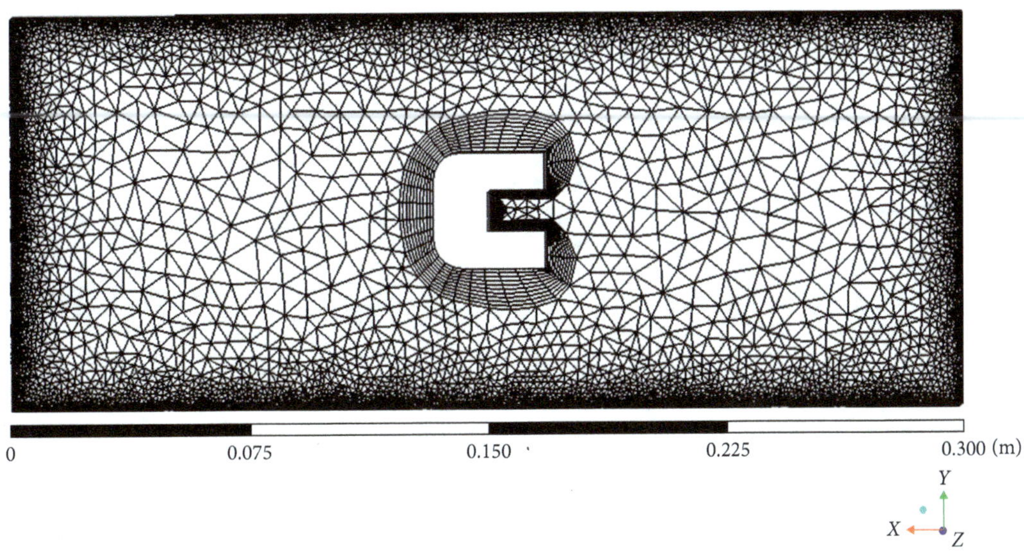

FIGURE 5: Meshing of C-shaped model.

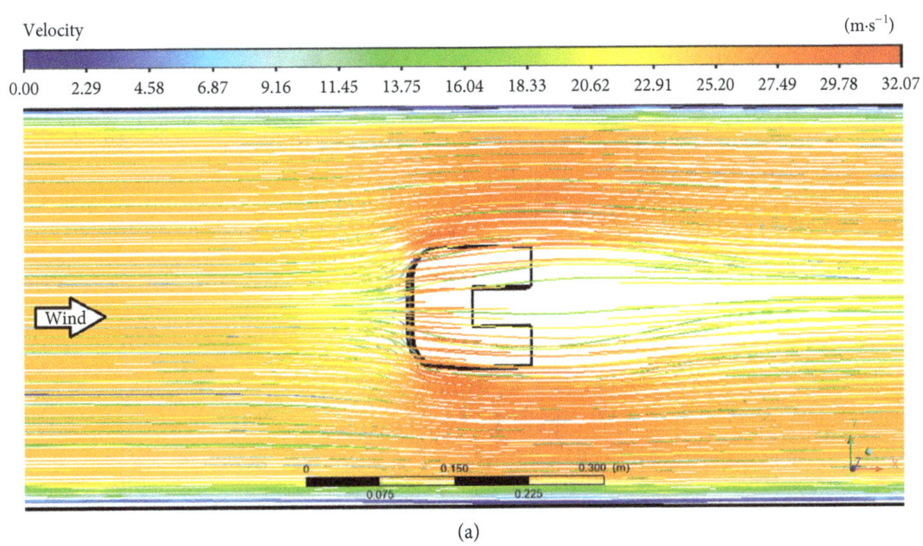

FIGURE 6: Continued.

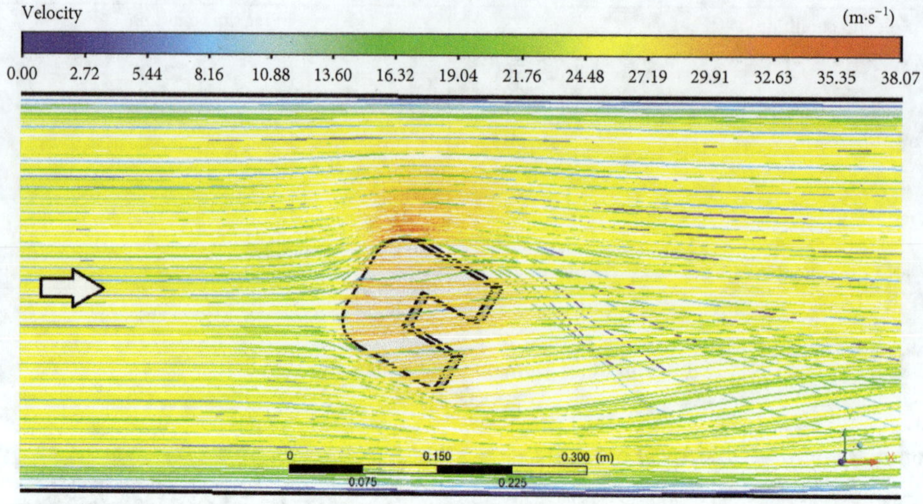

(b)

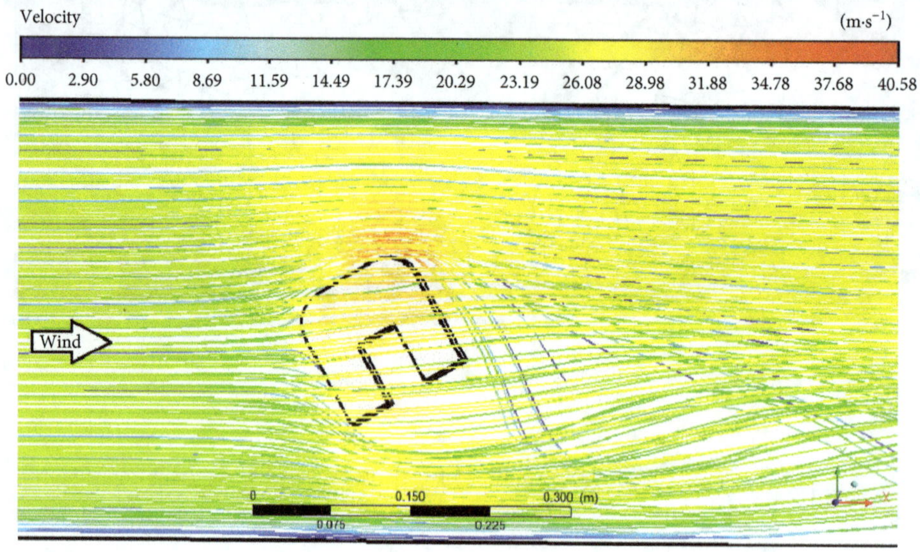

(c)

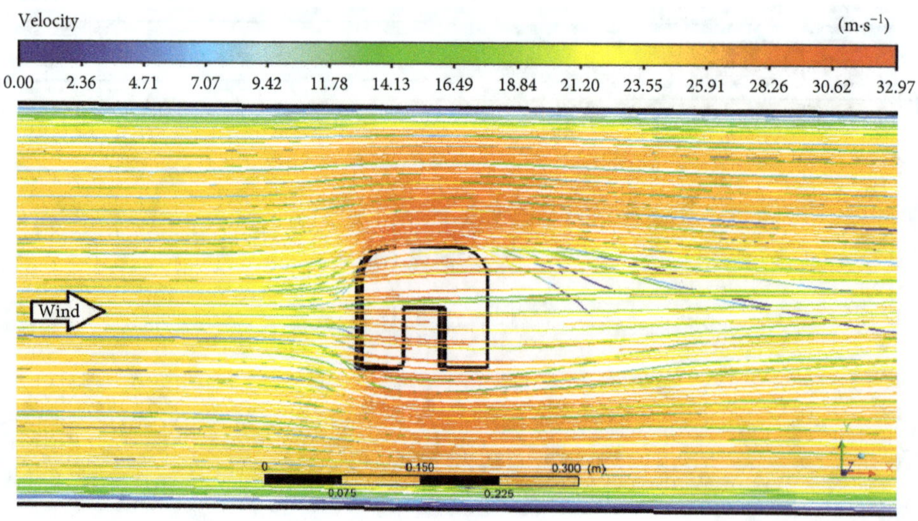

(d)

Figure 6: Continued.

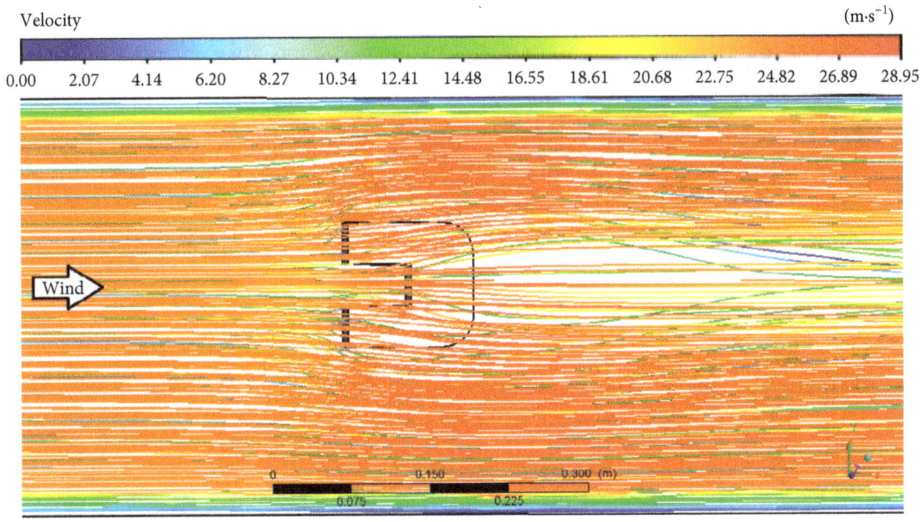

(e)

FIGURE 6: Wind flow pattern for different wind incidence angles. (a) 0°. (b) 30°. (c) 60°. (d) 90°. (e) 180°.

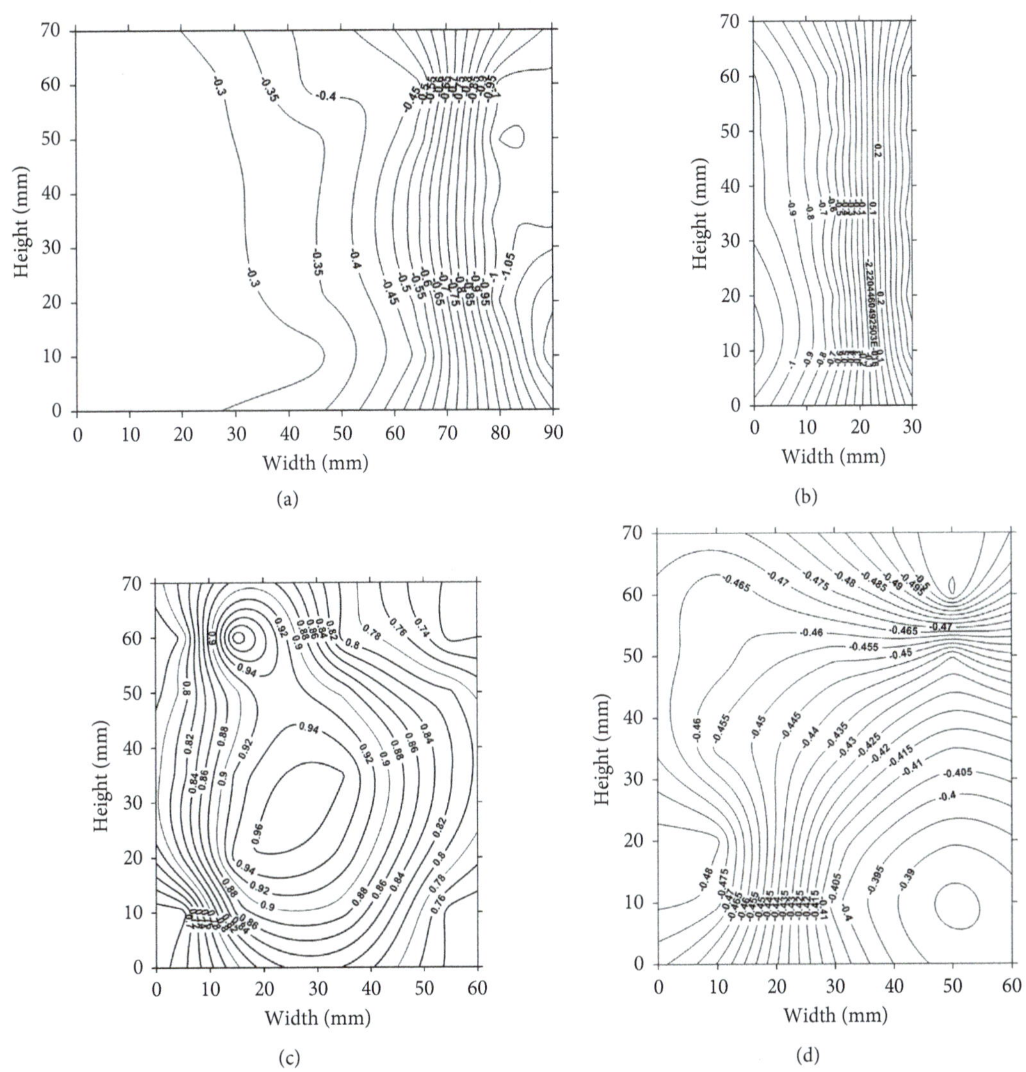

FIGURE 7: Continued.

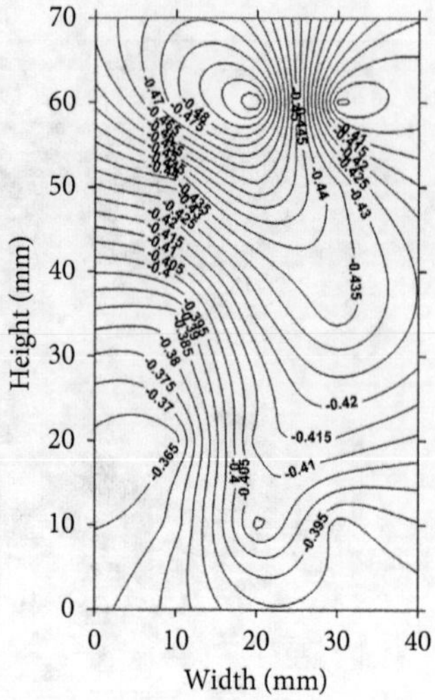

(e)

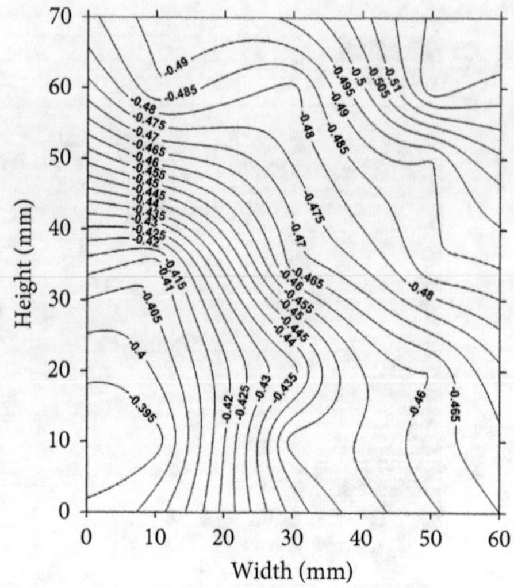

(f)

FIGURE 7: Pressure contour on different faces of the C-1 model at different wind angles. (a) Face A 90°. (b) Face B 45°. (c) Face C 0°. (d) Face G 90°. (e) Face H 180°. (f) Face I 90°.

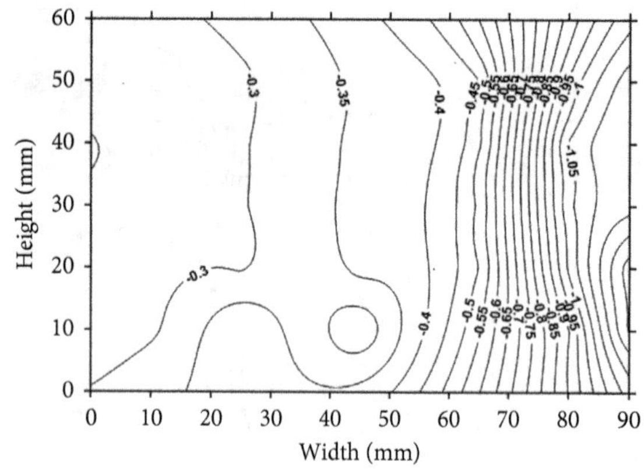

(a)

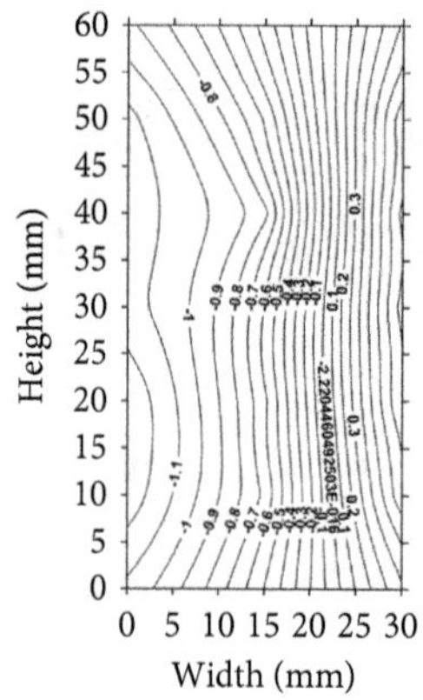

(b)

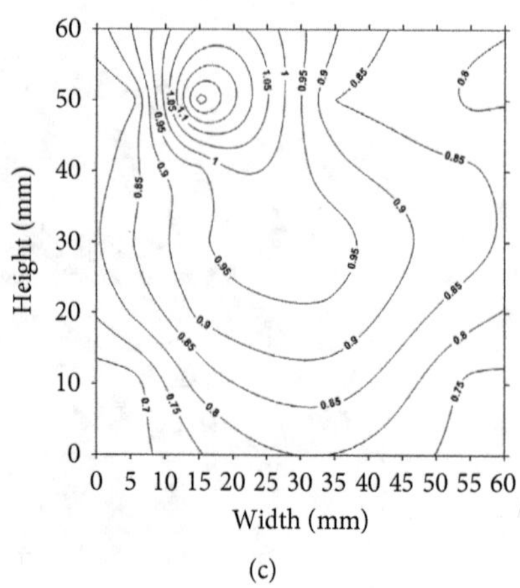

(c)

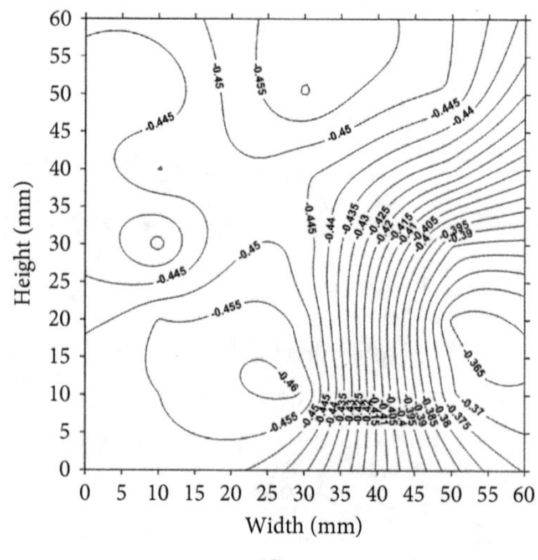

(d)

FIGURE 8: Continued.

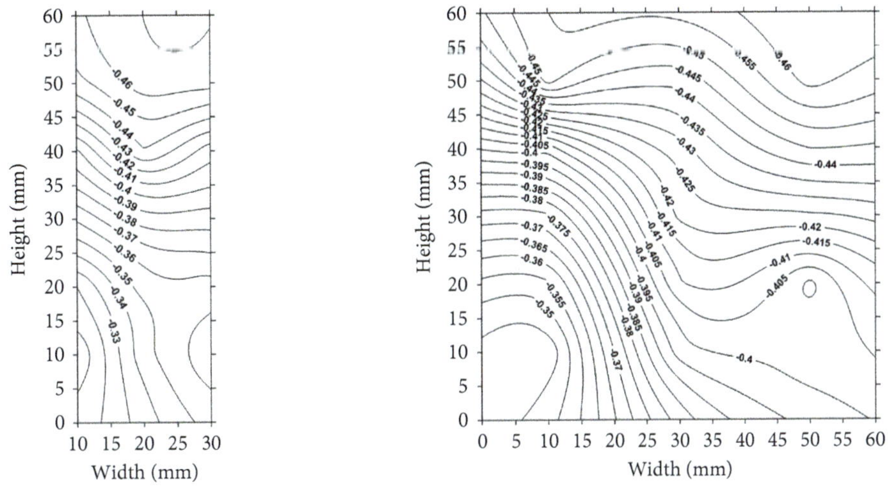

FIGURE 8: Pressure contour on different faces of the C-2 model at different wind angles. (a) Face A 90°. (b) Face B 45°. (c) Face C 0°. (d) Face G 90°. (e) Face H 180°. (f) Face I 90°.

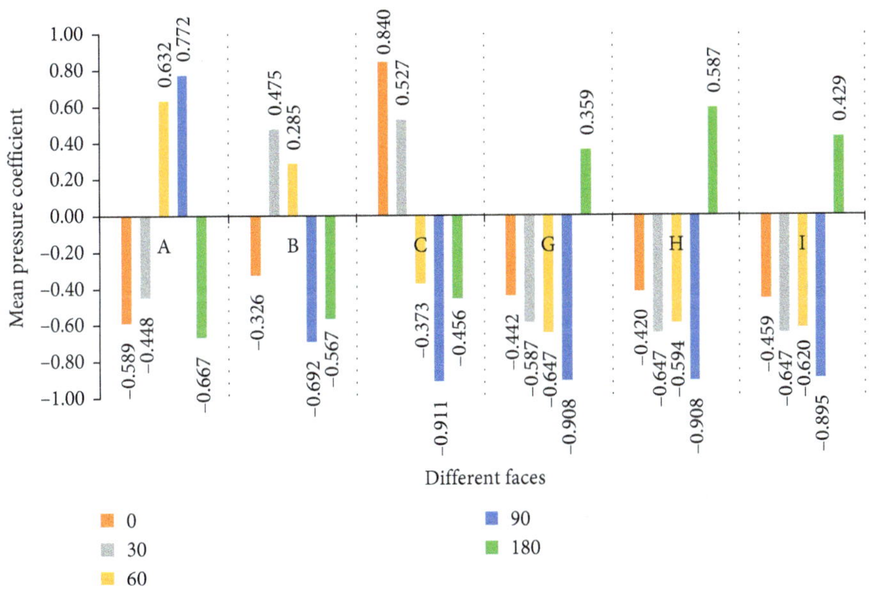

(a)

FIGURE 9: Continued.

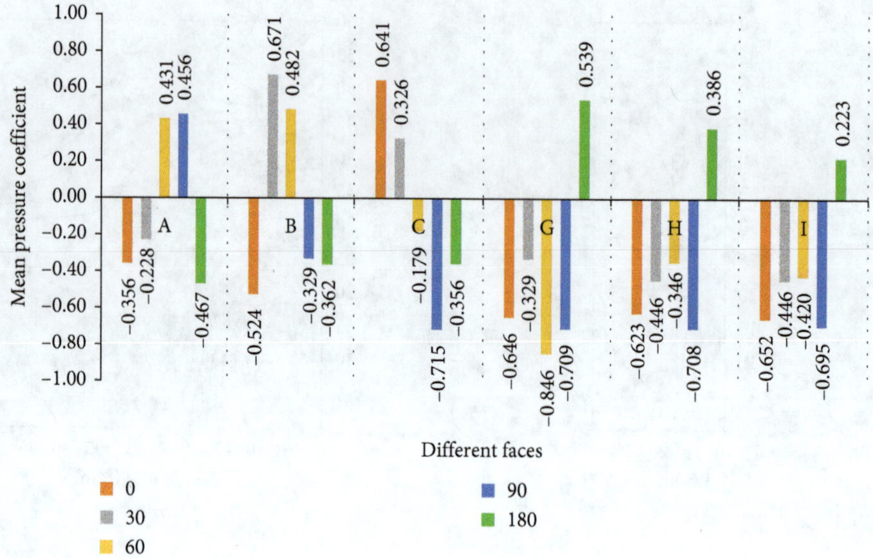

(b)

Figure 9: Mean pressure coefficients for all the faces of the C-1 model. (a) Experimental study. (b) Numerical study.

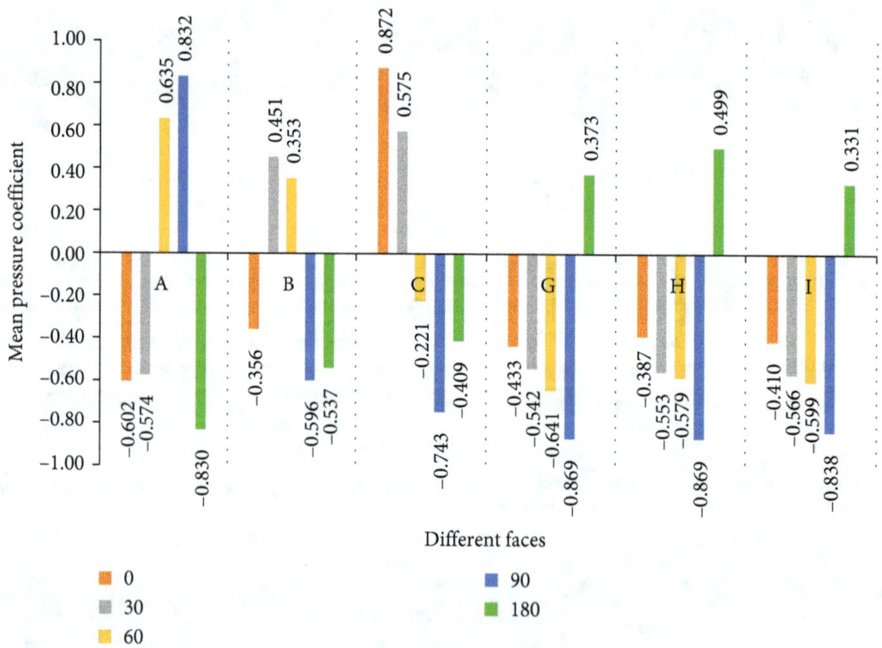

(a)

Figure 10: Continued.

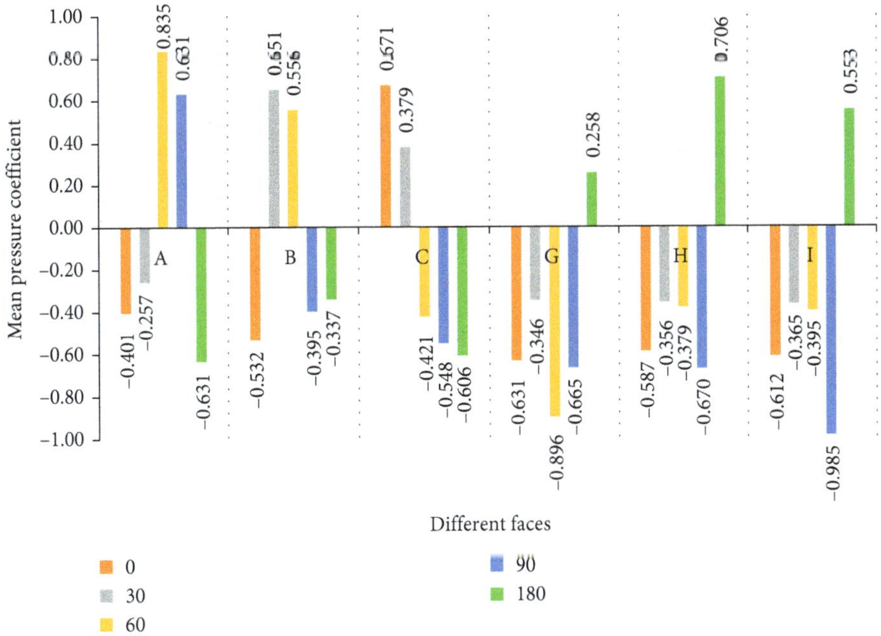

(b)

FIGURE 10: Mean pressure coefficients for all the faces of the C-2 model. (a) Experimental study. (b) Numerical study.

incidence angle 0°, Face A is parallel to the wind flow direction, and the front face of C is perpendicular to the direction of flow. Faces G, H, and I are inner windward faces. In Face A, maximum negative C_p occurs at the right side and minimum at the left side of the side face in Figure 7(a). Figure 7(b) indicates that the maximum pressure occurs in the left side and minimum in the right side of the curvature portion (Face B) due to flow separation of wind flow. In Figure 7(c), Face C is perpendicular to the wind flow, and the maximum positive pressure occurs at the central part of the face and minimum pressure at both edge sides due to curvature and edge effect as well. Inner windward Faces G and I have similar flow pattern (Figures 7(d) and 7(e)). Negative pressure decreases towards the edge. Similarly, the mean pressure coefficient contours of the C-2 shaped model are show in Figure 8.

4.4. Numerical Results. Pressure variation and mean pressure coefficients of all the faces of C-1 and C-2 models are also studied numerically in detail by CFD for different wind incidence angles. It can be seen clearly that numerically predicted pressure contours are converged well with the experimentally predicted results. But, very less variation in mean pressure coefficients are observed with respect to the experimental results (Figures 9 and 10).

4.5. Error Analysis. Error analysis is also performed by obtaining mean percentage error (ME), standard deviation (SD), coefficient of determination (R^2), mean absolute error (MAE), root mean square error (RMSE), and mean absolute percentage error (MAPE) [26, 27]. The percentage of mean error for prediction of C_p by various models and their standard deviation is shown in Figure 11. Tables 2 and 3 show the error analysis of the present experimental dataset with different faces of C-1 and C-2 models. All the predicted R^2 values nearly 0.99 which is very good for fitting other data and also the RMSE values are within the permissible limit for

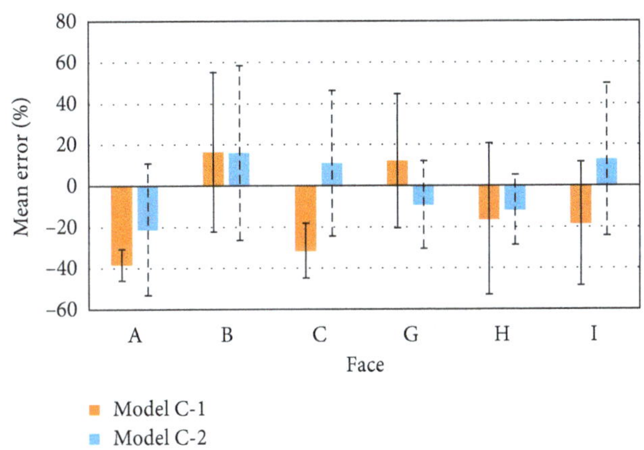

FIGURE 11: Percentage of mean error for different faces of the building.

TABLE 2: Error analysis of predicted C_p for model of C-1.

Face	R^2	MAE	RMSE	MAPE
A	0.99	0.234	0.238	−9.212
B	0.86	0.232	0.241	−7.812
C	0.99	0.178	0.182	−6.739
G	0.83	0.208	0.210	−18.488
H	0.87	0.211	0.212	−21.818
I	0.86	0.200	0.200	−15.976

TABLE 3: Error analysis of predicted C_p for model of C-2.

Face	R^2	MAE	RMSE	MAPE
A	0.95	0.223	0.228	−11.369
B	0.92	0.196	0.197	−3.718
C	0.95	0.198	0.198	−21.524
G	0.95	0.096	0.099	−6.717
H	0.96	0.094	0.128	−14.053
I	0.89	0.195	0.197	−13.872

all the faces. The better results regarding error analysis show the acceptability of the C-shaped model for the practical application.

5. Conclusions

The current study shows that pressure induced on the model building is significantly affected by the model geometry, configuration, aspect ratios, and angle of incidence. The significant outcomes of the present study are summarized as follows:

(i) Mean pressure coefficients are the main objective of this study. Maximum positive mean pressure coefficients occur at the front face of C, and maximum negative pressure occurs at the inner face of the model.

(ii) The study has been conducted with a C-shaped model experimentally as well as numerically using the wind tunnel test and CFD technique, respectively.

(iii) Pressure variation on the building is directly influenced by wind flow pattern.

(iv) Vortex generation and different types of mechanism such as separation of flow, upwash, and downwash are happened due to dynamic behavior of wind flow.

(v) The predicted values of the error analysis are measured by four accuracy measurement procedures such as R^2 value, mean absolute error (MAE), root mean square error (RMSE), and mean absolute percentage error (MAPE). It is observed that all the errors are within the permissible limit.

Conflicts of Interest

The authors declare that they have no conflicts of interest.

Acknowledgments

The authors express deep sense of gratefulness to Head of the Department of Aerospace Engineering, Indian Institute of Technology Kharagpur, India, for permitting and providing facilities to carry out the experiments. The authors are also thankful to Head of the Department of Civil Engineering, NIT Rourkela, and the National Institute of Technology Rourkela for the support.

References

[1] IS:875 (Part 3)–1987, *Code of Practice for Design Loads for Buildings and Structures*, Bureau of Indian Standards (BIS), Manak Bhawan, New Delhi, India, 2003.

[2] D. A. Paterson and C. J. Apelt, "Computation of wind flows over three-dimensional buildings," *Journal of Wind Engineering and Industrial Aerodynamics*, vol. 24, no. 3, pp. 193–213, 1986.

[3] K. C. S. Kwok, "Effect of building shape on wind-induced response of tall building," *Journal of Wind Engineering and Industrial Aerodynamics*, vol. 28, no. 1–3, pp. 381–390, 1988.

[4] T. Stathopoulos and Y. S. Zhou, "Numerical simulation of wind-induced pressures on buildings of various geometries," *Journal of Wind Engineering and Industrial Aerodynamics*, vol. 46-47, pp. 419–430, 1993.

[5] K. Miyashita, J. Katagiri, O. Nakamura et al., "Wind-induced response of high-rise buildings Effects of corner cuts or openings in square buildings," *Journal of Wind Engineering and Industrial Aerodynamics*, vol. 50, pp. 319–328, 1993.

[6] Q. S. Li, J. Q. Fang, A. P. Jeary, and C. K. Wong, "Full scale measurements of wind effects on tall buildings," *Journal of Wind Engineering and Industrial Aerodynamics*, vol. 74–76, pp. 741–750, 1998.

[7] M. G. Gomes, A. M. Rodrigues, and P. Mendes, "Experimental and numerical study of wind pressures on irregular-plan shapes," *Journal of Wind Engineering and Industrial Aerodynamics*, vol. 93, no. 10, pp. 741–756, 2005.

[8] S. Chakrabortya, S. K. Dalui, and A. K. Ahuja, "Experimental and numerical study of surface pressure on '+'plan shape tall building," *International Journal of Construction Materials and Structures*, vol. 1, pp. 45–48, 2013.

[9] J. A. Amin and A. K. Ahuja, "Effects of side ratio on wind-induced pressure distribution on rectangular buildings," *Journal of Structures*, vol. 2013, Article ID 176739, 12 pages, 2013.

[10] A. K. Dagnew, G. T. Bitsuamalk, and M. R., "Computational evaluation of wind pressures on tall buildings," in *Proceedings of 11th American conference on Wind Engineering*, San Juan, PR, USA, June 2009.

[11] S. Chakraborty, S. K. Dalui, and A. K. Ahuja, "Wind load on irregular plan shaped tall building–a case study," *Wind and Structures*, vol. 19, no. 1, pp. 59–73, 2014.

[12] J. Yi and Q. S. Li, "Wind tunnel and full-scale study of wind effects on a super-tall building," *Journal of Fluids and Structures*, vol. 58, pp. 236–253, 2015.

[13] Y. Li and Q. Li, "Across-wind dynamic loads on L-shaped tall buildings," *Wind Structures*, vol. 23, no. 5, pp. 385–403, 2016.

[14] B. Bhattacharyya and S. K. Dalui, "Investigation of mean wind pressures on 'E' plan shaped tall building," *Wind and Structures*, vol. 26, no. 2, pp. 99–114, 2018.

[15] L. Li, B. Wen, and L. Yihua, "Recent efforts for credible CFD simulations in China," *Modelling and Simulation in Engineering*, vol. 2011, Article ID 861272, 8 pages, 2011.

[16] W. K. Chow, C. L. Chow, and S. S. Li, "Simulating smoke filling in big halls by computational fluid dynamics," *Modelling and Simulation in Engineering*, vol. 2011, Article ID 781252, 16 pages, 2011.

[17] A. Mohanta and K. C. Patra, "LES modeling with experimental validation of a compound channel having converging floodplain," *Journal of The Institution of Engineers (India): Series A*, vol. 99, no. 3, pp. 519–537, 2018.

[18] B. Bhattacharyya and S. K. Dalui, "Along and across wind effects on irregular plan shaped tall building," in *Advances in Structural Engineering*, Springer, New Delhi, India, 2015.

[19] D. Jana, T. Bhaduri, and S. K. Dalui, "Numerical study of optimization of interference effect on pentagonal plan shaped tall building," *Asian Journal of Civil Engineering*, vol. 16, no. 8, pp. 1123–1153, 2015.

[20] S. Swaddiwudhipong and M. S. Khan, "Dynamic response of wind-excited building using CFD," *Journal of Sound and Vibration*, vol. 253, no. 4, pp. 735–754, 2002.

[21] Y. Cheng, F. S. Lien, E. Yee, and R. Sinclair, "A comparison of large eddy simulations with a standard k–ε Reynolds-averaged Navier–stokes model for the prediction of a fully developed turbulent flow over a matrix of cubes," *Journal of Wind Engineering and Industrial Aerodynamics*, vol. 91, no. 11, pp. 1301 1328, 2003.

[22] F. Villalpando, M. Reggio, and A. Ilinca, "Assessment of turbulence models for flow simulation around a wind turbine airfoil," *Modelling and Simulation in Engineering*, vol. 2011, Article ID 714146, 8 pages, 2011.

[23] T. Uchida, T. Maruyama, and Y. Ohya, "New evaluation technique for WTG design wind speed using a CFD-model-based unsteady flow simulation with wind direction changes," *Modelling and Simulation in Engineering*, vol. 2011, Article ID 941870, 6 pages, 2011.

[24] J. Franke, C. Hirsch, A. G. Jensen et al., "Recommendations on the use of CFD in Wind engineering," in *Proceedings of International conference on Urban Wind Engineering and Building Aerodynamics: COST C14: Impact of wind and storm on City life and Built Environment*, Rhode-Saint-Genèse, Belgium, May 2004.

[25] S. Mukherjee, S. Chakraborty, S. K. Dalui, and A. K. Ahuja, "Wind induced pressure on 'Y' plan shape tall building," *Wind and Structures*, vol. 19, no. 5, pp. 523–540, 2014.

[26] A. H. Gandomi, G. J. Yun, and A. H. Alavi, "An evolutionary approach for modeling of shear strength of RC deep beams," *Materials and Structures*, vol. 46, no. 12, pp. 2109–2119, 2013.

[27] A. Mohanta, K. Patra, and B. Sahoo, "Anticipate Manning's coefficient in meandering compound channels," *Hydrology*, vol. 5, no. 3, p. 47, 2018.

18

Development of Constitutive Model for Precast Prestressed Concrete Segmental Columns

M. Hafezolghorani Esfahani,[1] F. Hejazi,[1] R. Vaghei,[1] E. Nikbakht,[2] and D. C. J. Tze[1]

[1]*Department of Civil Engineering, University of Putra Malaysia, Serdang, Malaysia*
[2]*Department of Civil Engineering, Universiti Teknologi Petronas, Malaysia*

Correspondence should be addressed to F. Hejazi; farzad@fhejazi.com

Academic Editor: Anna Pandolfi

The interest of using precast segmental columns in construction of concrete bridges has significantly increased in recent years. One research area of concrete bridges is the application of Precast Prestressed Concrete Segmental (PPCS) Column in any structural analysis software or FE program code. Modeling a PPCS column, which consists of various materials with interaction between them, is complicated and time-consuming. This research attempts to formulate the stiffness matrix of PPCS columns in order to form the constitutive model in linear form to evaluate the response of the columns. A two-dimensional finite element model is presented in the finite element package ANSYS. Parametric studies are conducted by finite element models to verify the constitutive models for the PPCS column with a different number of concrete segments. Comparison between the constitutive model and the FE program results indicates that the constitutive model is accurate enough to predict the deformation of the PPCS columns.

1. Introduction

Construction of segmental bridge started in Europe in 1950. The first attempt of cast-in-place segmental concrete bridge was conducted across the Lahn River in Balduinstein, Germany, in 1950; however, the primary precast segmental concrete bridge was constructed in 1962, across the River Seine in France. Later, this construction method gained worldwide recognition. Bridge construction time, facilitating construction, and minimizing the traffic disruption are the main advantages of precast construction in contrast to cast-in-situ construction. Precast segmental concrete bridges are normally constructed in low seismic areas. Numerous precast segmental concrete columns and pier constructions have been carried out in the US in low seismic regions such as the states of Texas and California in the United States [1]. Studies of PPCS column can be categorized into two parts: firstly, with bonded and, secondly, with unbounded posttensioning systems.

Different researchers presented various aspects of prestressing with bonded tendons in their research [2, 3]. In bonded posttensioned systems, the lateral strength of columns could be enhanced by the bonding between prestressing strands and surrounding concrete; however, it might result in tendon yielding. To the best of our knowledge, there are just a few studies in the literature about unbounded tendons [4, 5]. Unbounded posttensioned steels might lead to reduced prestress loss during strong seismic excitations. The segmental column can be restored to its original configuration after earthquakes, which cannot be done for bonded structure. It also maintains the continuity between column segments and foundation. There are several components in precast segmental bridge columns, which are combined and associated with continuous posttensioning strands. The action of these types of bridges against seismic loading differs from conventional bridge columns; besides, segmental bridges act with rocking mechanism, which takes place while the segments opening is to be found, and, as a consequence, the resultant damage is much lower than that in the conventional monolithic bridge columns. The minor cracks and damage in this system make them more economical than the monolithic system due to the reparability of this system after severe earthquake loading. The PRESS program has been established (Precast Seismic Structural Systems) as an innovative alternative solution for precast connections by using prestressing strands and mild steel, in order to indicate appropriate ductility compared to conventional monolithic systems [6].

Prestressing strands have the capability of recentering against earthquake, which means they return the columns at the original place over the unloading stages. The mild steel reinforcements are able to dissipate the earthquake energy. The concept of a hybrid system has been introduced for the first time and applied for posttensioning in beam-column connections [7]. They claimed that the combination of prestressing strands and mild steel reinforcements can appropriately meet the design codes' requirements. Mild steel reinforcement and posttensioned tendons have an influential role in the ductility and strength of hybrid segmental systems due to their combination [8]. The performance of unbonded posttensioned prefabricated concrete segmental bridge column subjected to lateral earthquake loading has been analytically and experimentally investigated for four precast posttensioned columns with different thickness of steel jackets and the results indicated that these types of bridge columns have appropriate performance against seismic energy with negligible residual displacement. New criteria for functional and survival limits for posttensioned (PT) precast segmental bridge columns have been proposed [9]. At the functional level of earthquake, the objective is to keep the structure operational without occurrence of major damage while, at the survival limit, the objective is to prevent the structure from collapsing. They defined three criteria for the functionality of structure. First limit is yielding of the prestressing strands. The second criterion is the displacement which leads to 1% residual drift, and the third limit is 0.7 times the survival-level displacement. The surviving-level limit is the displacement at which the structure starts to collapse. Reduction of longitudinal reinforcements and increase in axial load result in minimal residual displacement [10]. Reduction by 86% in the residual displacement can be captured by replacing half of the rebar with prestressing strands and applying prestressing force that is equivalent to axial load due to dead load. They also investigated the dynamic analysis of the unbounded center strand columns. A combination ratio of energy dissipation of mild steel internal and/or external supplemental energy dissipation and self-centering posttensioning strands as one of the predominant design factors in hybrid posttensioned bridges has been proposed to achieve appropriate energy dissipation and self-centering capability with less damage and residual displacement against seismic loading [11, 12]. They used two rotational springs: one representing the behavior of prestressed tendons without the contribution of mild steel reinforcement, and the other representing only the mild steel reinforcement contribution. They investigated five specimens with different combination ratios. The experimental results proved the advantages of enhanced performance of the hybrid system in comparison with traditional monolithic solutions. The proposed hybrid system is shown to have minor flexural cracking, negligible residual displacement, and a stable hysteretic behavior up to high ductility level, so a combination ratio of prestressing strands and mild steel reinforcements in design is proposed. Some researchers have also investigated the behavior of hollow unbonded precast segmental bridge columns with rectangular box cross section, in which the prestressing strands were passed through the hollow ducts [12].

The actual simplicity at design and construction phase of high performance and low-cost hybrid bridge piers were studied by some researchers [13]. They also applied short lengths of unbonded mild steel reinforcements at the junction of footing-first segment to prevent premature yielding of longitudinal reinforcements. They indicated that a certain amount of unbonded length of mild reinforcements provides more energy dissipation and strength against earthquake loading. Furthermore, they claimed that the unbonded precast column could return to the undeformed position after earthquakes. Eight large-scale posttensioned precast columns were carried out to validate the detailed finite element model subjected to cyclic tests which was developed by ABAQUS platform, and the outcomes showed a good agreement [14]. A 3D finite element model of precast walls and connection was developed using finite element model [15]. A numerical analytical model with nonlinear factors of prestressing precast concrete bridge column systems, which consists of a segment model, prestressing tendon, and joint mode, was developed [16]. The detailed finite element model of prestressing bridge column was carried out with the structural analysis software "Open Sees" and, finally, the finite element outcomes met the experimental results. The monotonic behavior of precast segmental bridge columns has been investigated through three-dimensional finite element models [17, 18]. Later, they studied the accuracy of analytical results for predicting the damage when subjected to lateral loading. They also investigated the effects of posttensioning forces, the amount of segments, and aspect ratio (the relation between the heights of the column and the diameter of the column). Two types of numerical models for unbonded posttensioned (PT) precast concrete segmental bridge were presented by the computer program PISA [19]. A 3D nonlinear finite element model has been proposed for hybrid posttensioned precast segmental bridge columns to analyse the different prestressing strand levels subjected to nonlinear static and lateral seismic loading [20, 21].

Modeling a PPCS column with various materials and interaction has been studied by finite element model (FEM); however, the results show that this methodology is complicated and time-consuming. This paper presents the stiffness matrix of unbonded prestressed precast segmental (PPCS) column to form the constitutive model. Finite element formulations are derived in explicit form which is applicable in any structural analysis software or FEM program code. Finally, three finite element models of Precast Prestressed Concrete Segmental (PPCS) Column with respect to one, two, and three segments called RC1SS, RC2SS, and RC3SS are investigated.

2. Precast Prestressed Concrete Segmental Column

2.1. Description of the Case Study and Material Properties. Various experiments on different large-scale specimens have

Table 1: Material properties.

Concrete	Grade 40	$E_c = 30241$ MPa $\nu = 0.2$	$f_{cu} = 41.4$ MPa $f_t = 4.053$ MPa
Reinforcement	Grade 60	$E_s = 200000$ MPa	$f_{ys} = 410$ MPa
Prestressing steel	Grade 270	$E_p = 196500$ MPa	$f_{yp} = 1860$ MPa

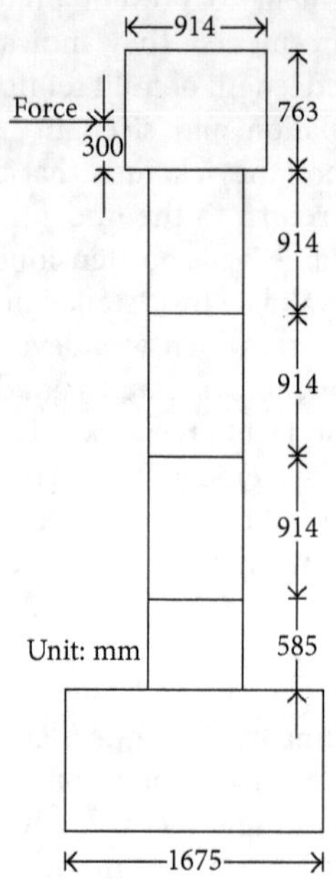

Figure 1: Geometry of the studied specimen [4].

been carried out in 2002 [4]. This study is chosen to develop the constitutive model. The geometry of the studied specimen is shown in Figure 1. Due to plane stress and plane strain analysis, two-dimensional view of the studied specimen has been chosen and equivalent section area is considered for 2D FE modeling and analysis.

The material properties of the specimens are presented in Table 1.

2.2. Constitutive Stress-Strain Relationships. Concrete is a quasi-brittle material which means concrete behaves differently against compression and tension. In general, concrete's tensile strength is only about 8% to 15% of the compressive strength. As shown in Figure 2(a), a stress-strain curve for confined concrete is presented in which f_c and f_{cu} are the peak compressive and ultimate strengths and ε_0 and ε_{cu} are the corresponding strains, respectively [22].

Idealized nonlinear prestressing steel with stress-strain model for 7-wire low-relaxation prestressing strand from ASTM A722 was proposed [4]. The curve, which is shown in Figure 2(b), can be derived by

$$\text{prestress steel limit of proportionality: } \varepsilon_{lp} = 0.0086,$$

$$\text{reduced ultimate prestress steel strain: } \varepsilon_{lp} = 0.0300,$$

$$\varepsilon_p \le 0.0086 : f_p = 28,500\varepsilon_p,$$
$$\varepsilon_p \ge 0.0086 : f_p = 270 - \frac{0.04}{\varepsilon_p - 0.007}. \quad (1)$$

Although the developed constitutive law is applicable for both linear and nonlinear analysis, linear analysis has been used in this paper to codify the finite element program due to its simplicity of application and verification.

3. Development of Constitutive Law

The proposed constitutive model for precast segmental columns comprises two stiffness matrixes for concrete and reinforcement. Each concrete segmental is modeled as two 4-node isoparametric elements as shown in Figure 3.

The shape function for isoparametric 4-node element is shown in Figure 4.

The displacement (u, v) is assumed as a bilinear function over the element and is given by the following [23]:

$$u = N_1u_1 + N_2u_2 + N_3u_3 + N_4u_4,$$
$$v = N_1v_1 + N_2v_2 + N_3v_3 + N_4v_4. \quad (2)$$

Element stiffness matrix is given as follows:

$$[k_c] = \int [B]^T [D] [B] \, dv = \iint_{-1}^{1} [B]^T [D] [B] \, t \, |J| \, d\xi \, d\eta, \quad (3)$$

where

$$dXdY = |J| \, d\xi d\eta,$$

$$D = \frac{E_c}{(1-\nu^2)} \begin{bmatrix} 1 & \nu & 0 \\ \nu & 1 & 0 \\ 0 & 0 & \frac{1}{2}(1-\nu) \end{bmatrix}$$

$$D \text{ matrix} \longrightarrow \text{plane Stress}, \quad (4)$$

$$D = \frac{E_c}{(1+\nu)(1-2\nu)} \begin{bmatrix} 1-\nu & \nu & 0 \\ \nu & 1-\nu & 0 \\ 0 & 0 & \frac{1}{2}-\nu \end{bmatrix}$$

$$D \text{ matrix} \longrightarrow \text{plane Strain}.$$

The matrix J is known in mathematics as the Jacobian matrix and can be numerically evaluated from

$$[J] = \begin{bmatrix} \frac{\partial x}{\partial \xi} & \frac{\partial y}{\partial \xi} \\ \frac{\partial x}{\partial \eta} & \frac{\partial y}{\partial \eta} \end{bmatrix} = \begin{bmatrix} J_{11} & J_{12} \\ J_{21} & J_{22} \end{bmatrix}. \quad (5)$$

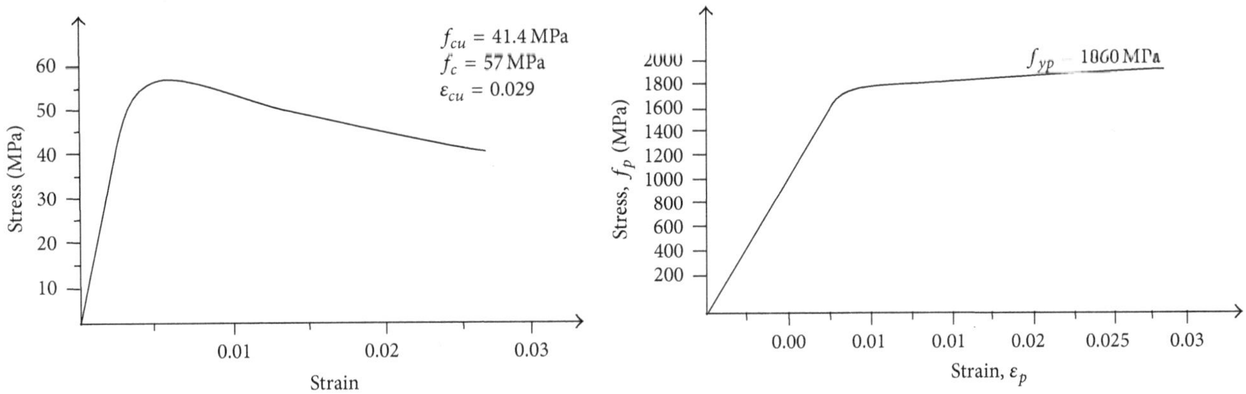

(a) Confined concrete

(b) Prestressing steel

FIGURE 2: Stress-strain curves for materials [4, 22].

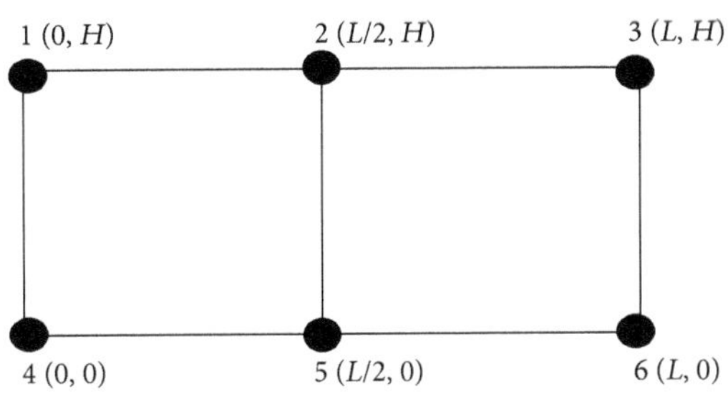

FIGURE 3: Modeling of concrete element.

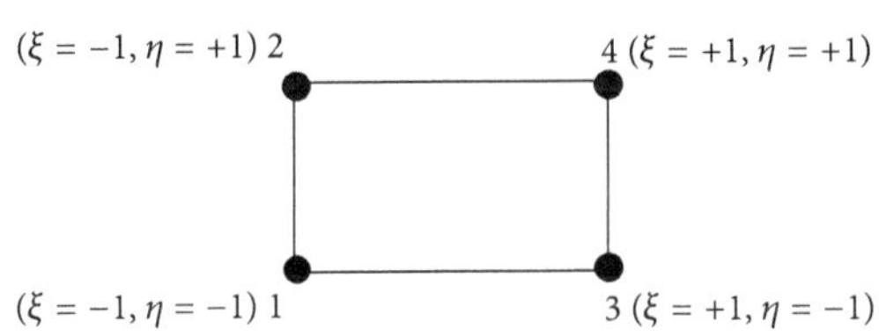

FIGURE 4: Shape function of isoparametric four nodes.

The force-displacement relation can be evaluated by

$$F = kd. \tag{6}$$

The strain-displacement relation is given by

$$\varepsilon = Bd. \tag{7}$$

The stress-strain relation can be obtained by

$$\sigma = D\varepsilon. \tag{8}$$

The discontinuous longitudinal reinforcement bar in the concrete segments and tendons is modeled as two (2) nodded bar elements. Each node of the truss is laterally restrained to the adjacent node of segmented column. The shape function of reinforcement bar is evaluated as shown in Figure 5.

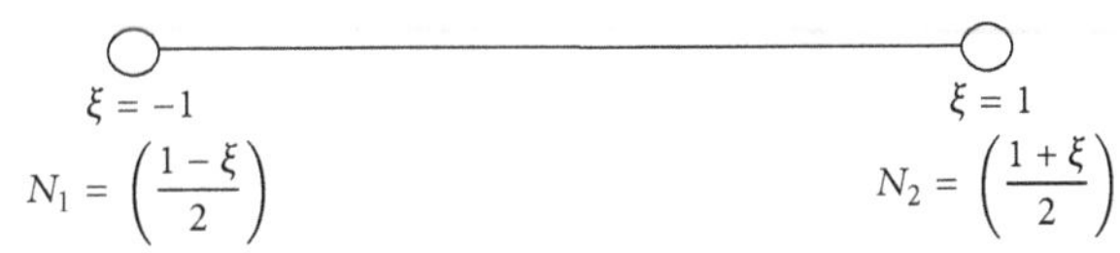

FIGURE 5: Shape function of bar element.

The stiffness matrix of reinforcement can be obtained by

$$\begin{aligned}[k_s] &= \int [B]^T [D] [B] \, dv = \int_l [B]^T [D] [B] A \, dx \\ &= \int_l [B]^T [D] [B] A \, |J| \, d\xi,\end{aligned} \tag{9}$$

where

$$dX = |J| \, d\xi. \tag{10}$$

It is known as

$$\begin{gathered}\frac{\partial N_i}{\partial x} = \left[\frac{\partial x}{\partial \xi}\right]^{-1} \times \frac{\partial N_i}{\partial \xi}, \\ J = \frac{\partial x}{\partial \xi}.\end{gathered} \tag{11}$$

Generally, the global stiffness of prestressed column in this study is formulated by assembling stiffness matrix for each concrete segment, prestressing tendon, longitudinal bar, and interface element.

4. Comparison of the Proposed Constitutive Model and FE Model

The contribution of tendons in precast concrete segments is investigated and new constitutive model is proposed and verified through modeling in FE software ANSYS.

Solution of many engineering problems is based on linear approximations and in this study the linear response is considered, although developed constitutive law is applicable for nonlinear analysis as well.

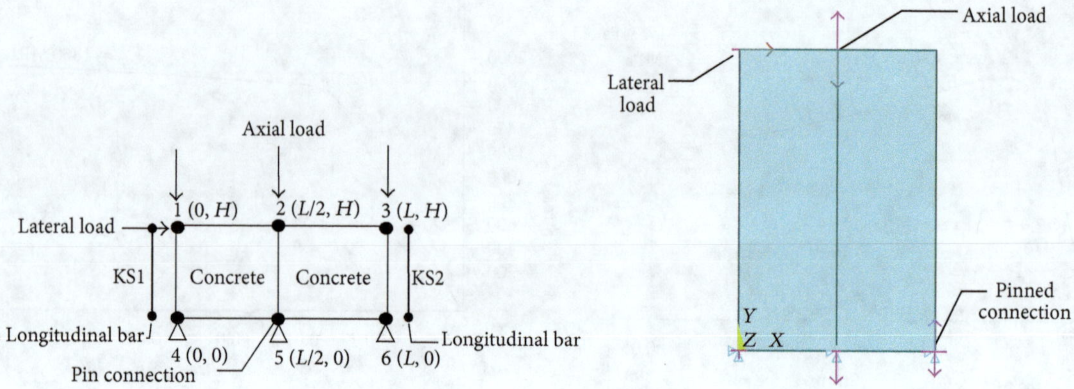

(a) Proposed constitutive model (b) FE model

FIGURE 6: One concrete segment model (RC1SS).

4.1. One Concrete Segment. Figure 6(a) shows the constitutive model of one concrete segment, where the longitudinal bars (KS1, KS2) are separated into sections equally and placed at both sides of the concrete segment; that is, KS1 is sharing node 1 and node 4 with the concrete segment in y direction while KS2 is sharing node 3 and node 6. The length of the longitudinal bar is the same as the height of single concrete segment. The lateral load is applied at node 1 of concrete segment with load step of 100 kN up to 500 kN and 500 kN is vertically and equally applied at top nodes (i.e., 1, 2, and 3). Figure 6(b) shows the finite element model for one concrete segment subjected to the lateral and axial loads.

The stiffness matrix for one concrete segment with longitudinal bar can be derived as shown in (12). The stiffness matrix variables A, B, C, a, b, c, d, e, f, and g are defined in Table 2.

A comparison between the constitutive model and FE model responses is carried out. As can be observed in Figure 7, it can be proven that the proposed constitutive model for one concrete segment has a good agreement with the FE model in plane stress analysis and plane strain analysis. As expected, displacement in plane strain analysis is about 80 mm which is four times higher than the displacement in plane stress analysis and both approaches show linear behavior:

$$K = C \times \begin{bmatrix} a & -c & -b & -d & 0 & 0 & b & d & -a & c & 0 & 0 \\ -c & e+g & d & f & 0 & 0 & -d & -f-g & c & -e & 0 & 0 \\ -b & d & 2a & 0 & -b & -d & -a & -c & 2b & 0 & -a & c \\ -d & f & 0 & 2e & d & f & -c & -e & 0 & -2b & c & -e \\ 0 & 0 & -b & d & a & c & 0 & 0 & -a & -c & b & -d \\ 0 & 0 & -d & f & c & e+g & 0 & 0 & -c & -e & d & -f-g \\ b & -d & -a & -c & 0 & 0 & a & c & -b & d & 0 & 0 \\ d & -f-g & -c & -e & 0 & 0 & c & e+g & -d & f & 0 & 0 \\ -a & c & 2b & 0 & -a & -c & -b & -d & 2a & 0 & -b & d \\ c & -e & 0 & -2b & -c & -e & d & f & 0 & 2e & -d & f \\ 0 & 0 & -a & c & b & d & 0 & 0 & -b & -d & a & -c \\ 0 & 0 & c & -e & -d & -f-g & 0 & 0 & d & f & -c & e+g \end{bmatrix}. \quad (12)$$

It can be seen from Table 3 that, in plane stress analysis, the maximum displacement by the proposed constitutive model is close to that obtained by FE model, with the former value a little bit more than the latter one. A maximum difference of 4.6% is observed between constitutive model and FE results. In plane strain analysis, the maximum displacement is 76.89 mm in the proposed constitutive model; however, the correspondence value in FE model is 73.94 mm. Table 3 shows about four percent difference in plane strain analysis between proposed constitutive model and FE results for one segmental concrete, whilst discrepancy percentage of aforementioned methods in plane stress analysis is increased to 4.6%.

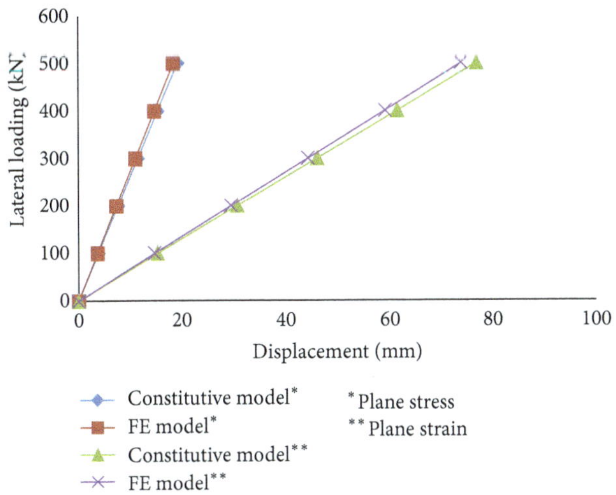

FIGURE 7: Lateral load-deflection of the proposed constitutive model and FE model for RC1SS.

4.2. Two Concrete Segments. Two concrete segments are attached together as shown in Figure 8(a) with prestressing tendon placed at the middle of the concrete segment, which connects node 2 and node 8 in order to connect the concrete segments together (RC2SS). The intention of designing longitudinal reinforcement discontinuously is to avoid the fracture of mild steel at the critical joint opening when a huge lateral load is applied. The developed constitutive model for RC2SS is examined under the following load condition as shown in Figure 8(b). The prestressing tendon role is to overcome weakness of concrete in tension by providing the clamping load between the concrete segment and support.

From (13), the stiffness matrix for the prestressing tendon can be derived as

$$
\begin{gathered}
\qquad\quad\; 8y \quad\;\; 2y \\
[K_t] = \frac{E_t A_t}{H} \begin{bmatrix} 1 & -1 \\ -1 & 1 \end{bmatrix}, \\
t = \frac{E_t A_t}{H} \times \frac{8LH\left(1-v^2\right)}{Et} \quad (\text{plane stress}), \\
t = \frac{E_t A_t}{H} \times \frac{16LH\left(1+v\right)B}{Et} \quad (\text{plane strain}).
\end{gathered} \tag{13}
$$

So by substituting these parameters in (14), the stiffness matrix can be derived as

$$
K_2 = C \times \begin{bmatrix}
a & -c & -b & 0 & 0 & b & d & -a & 0 & 0 & 0 & 0 & 0 & 0 & 0 & 0 & 0 & 0 \\
-c & e+g & d & 0 & 0 & -d & -f-g & c & -e & 0 & 0 & 0 & 0 & 0 & 0 & 0 & 0 & 0 \\
-b & d & 2a & 0 & -b & -d & -a & -c & 2b & 0 & -a & c & 0 & 0 & 0 & 0 & 0 & 0 \\
-d & f & 0 & 2e+t & d & f & -c & -e & 0 & -2f & c & -e & 0 & 0 & 0 & -t & 0 & 0 \\
0 & 0 & -b & d & a & c & 0 & 0 & -a & -c & b & -d & 0 & 0 & 0 & 0 & 0 & 0 \\
0 & 0 & -d & f & c & e+g & 0 & 0 & -c & -e & d & -f-g & 0 & 0 & 0 & 0 & 0 & 0 \\
b & -d & -a & -c & 0 & 0 & 2a & 0 & -2b & 0 & 0 & 0 & b & d & -a & c & 0 & 0 \\
d & -f-g & -c & -e & 0 & 0 & 0 & 2(e+g) & 0 & 2f & 0 & 0 & -d & -f-g & c & -e & 0 & 0 \\
-a & c & 2b & 0 & -a & -c & -2b & 0 & 4a & 0 & -2b & 0 & -a & -c & 2b & 0 & -a & c \\
c & -e & 0 & -2f & -c & -e & 0 & 2f & 0 & 4e & 0 & 2f & -c & -e & 0 & -2f & c & -e \\
0 & 0 & -a & c & b & d & 0 & 0 & -2b & 0 & 2a & c & 0 & 0 & -a & -c & b & -d \\
0 & 0 & c & -e & -d & -f-g & 0 & 0 & 0 & 2f & c & 2(e+g) & 0 & 0 & -c & e & d & -f-g \\
0 & 0 & 0 & 0 & 0 & 0 & b & -d & -a & -c & 0 & 0 & a & c & -b & d & 0 & 0 \\
0 & 0 & 0 & 0 & 0 & 0 & d & -f-g & -c & -e & 0 & 0 & c & e+g & -d & f & 0 & 0 \\
0 & 0 & 0 & 0 & 0 & 0 & -a & c & 2b & 0 & -a & -c & -b & -d & 2a & 0 & -b & d \\
0 & 0 & 0 & -t & 0 & 0 & c & -e & 0 & -2f & -c & -e & d & f & 0 & 2e+t & -d & f \\
0 & 0 & 0 & 0 & 0 & 0 & 0 & 0 & -a & c & b & d & 0 & 0 & -b & -d & a & -c \\
0 & 0 & 0 & 0 & 0 & 0 & 0 & 0 & c & -e & -d & -f-g & 0 & 0 & d & f & -c & e+g
\end{bmatrix}. \tag{14}
$$

Comparison between the constitutive and FE models for two concrete segments, RC2SS, is shown in Figure 9. From this figure, it can be seen that there is a good agreement between the proposed constitutive model results for RC2SS and the FE model results in plane stress and plane strain analysis. Consequently, it can be determined that the maximum displacement in plane strain analysis is about 200 mm, which is almost three times higher than the displacement in plane stress analysis. Linear behavior has been captured for both aforementioned approaches.

TABLE 2: Stiffness matrix variables.

Plane stress condition	Plane strain condition
$A = \frac{1}{2}(1-\nu)$	$B = (1-2\nu)$
$C = \frac{Et}{8LH(1-\nu^2)}$	$C = \frac{Et}{16LH(1+\nu)B}$
$a = 4H^2 + AL^2$	$a = 8H^2(1-\nu) + L^2(1-2\nu)$
$b = 4H^2 - AL^2$	$b = 8H^2(1-\nu) - L^2(1-2\nu)$
$c = 2HL(\nu + A)$	$c = 2HL$
$d = 2HL(\nu - A)$	$d = 2HL(4\nu - 1)$
$e = L^2 + 4H^2A$	$e = 2L^2(1-\nu) + 4H^2(1-2\nu)$
$f = L^2 - 4H^2A$	$f = 2L^2(1-\nu) - 4H^2(1-2\nu)$
$g = \frac{E_sA_s}{2H} \times \frac{8LH(1-\nu^2)}{Et}$	$g = \frac{E_sA_s}{2H} \times \frac{16LH(1+\nu)B}{Et}$

Table 4 represents the percentage difference between two types of analysis on two concrete segments. In plane stress analysis, maximum displacements of 74.46 mm for the proposed constitutive model and 73.94 mm in FE program are achieved subjected to applied load. Likewise, in plane strain analysis, applied forces produce 191.21 mm and 194.19 mm displacements by proposed constitutive model and FE model, respectively. Table 4 shows almost 0.65% difference between two approaches in plane stress analysis for different applied load; however, this percentage difference in plane strain analysis for two concrete segments is diminished from 6.73% to 1.53% owing to the applied load from 100 kN to 500 kN as a lateral load and 500 kN as a vertical load.

4.3. Three Concrete Segments. The PPCS column consisting of three concrete segments is shown in Figure 10(a). The constitutive model has 12 nodes where each node has two degrees of freedom and resulting stiffness matrix size is 24 × 24. The prestressing tendons are pulled between node 2 and node 11. There are discontinuous bars at both sides of the concrete segment. The stiffness matrix for three segmental precast columns is developed by assembling the stiffness matrixes and presented by (15). Based on the developed constitutive model, the PPCS column with three concrete segments is modeled by only 15 elements which reduce computation time and convergence issues of analysis considerably.

The analysis result of RC3SS is plotted in Figure 11 and tabulated in Table 5. It is clear from these plots that the constitutive model has smaller displacement than the FE model. As can be observed in Figure 11, there is a good agreement between the proposed constitutive model and the FE model in plane stress and plane strain analysis of RC3SS. Consequently, it can be found that displacement in plane stress analysis is about 200 mm and it is almost doubled in plane strain:

$$K = \frac{Et}{8LH(1-\nu^2)}$$

$$\begin{bmatrix}
2a & 0 & b & -d & -2b & 0 & -a & -c & 0 & 0 & 0 & 0 & b & d & -a & c & 0 & 0 & 0 & 0 & 0 & 0 & 0 & 0 \\
0 & 2(e+g) & d & -f-g & 0 & 2f & -c & -e & 0 & 0 & 0 & 0 & -d & -f-g & c & -e & 0 & 0 & 0 & 0 & 0 & 0 & 0 & 0 \\
b & d & a & -c & -a & c & -b & -d & 0 & 0 & 0 & 0 & 0 & 0 & 0 & 0 & 0 & 0 & 0 & 0 & 0 & 0 & 0 & 0 \\
-d & -f-g & -c & e+g & c & -e & d & f & 0 & 0 & 0 & 0 & 0 & 0 & 0 & 0 & 0 & 0 & 0 & 0 & 0 & 0 & 0 & 0 \\
-2b & 0 & -a & c & 4a & 0 & 2b & 0 & -2b & 0 & -a & -c & -a & -c & 2b & 0 & -a & c & 0 & 0 & 0 & 0 & 0 & 0 \\
0 & 2f & c & -e & 0 & 4a & 0 & -2f & 0 & 2f & -c & -e & -c & -e & 0 & -2f & c & -e & 0 & 0 & 0 & 0 & 0 & 0 \\
-a & -c & -b & d & 2b & 0 & 2a & 0 & -a & c & -b & -d & 0 & 0 & 0 & 0 & 0 & 0 & 0 & 0 & 0 & 0 & 0 & 0 \\
-c & -e & -d & f & 0 & -2f & 0 & 2e+t & c & -e & d & f & 0 & 0 & 0 & 0 & 0 & 0 & 0 & 0 & 0 & -t & 0 & 0 \\
0 & 0 & 0 & 0 & -2b & 0 & -a & c & 2a & 0 & b & d & 0 & 0 & -a & -c & b & -d & 0 & 0 & 0 & 0 & 0 & 0 \\
0 & 0 & 0 & 0 & 0 & 2f & c & -e & 0 & 2(e+g) & -d & -f-g & 0 & 0 & -c & -e & d & -f-g & 0 & 0 & 0 & 0 & 0 & 0 \\
0 & 0 & 0 & 0 & -a & -c & -b & d & b & -d & a & c & 0 & 0 & 0 & 0 & 0 & 0 & 0 & 0 & 0 & 0 & 0 & 0 \\
0 & 0 & 0 & 0 & -c & -e & -d & f & d & -f-g & c & e+g & 0 & 0 & 0 & 0 & 0 & 0 & 0 & 0 & 0 & 0 & 0 & 0 \\
b & -d & 0 & 0 & -a & -c & 0 & 0 & 0 & 0 & 0 & 0 & 2a & 0 & -2b & 0 & 0 & 0 & b & d & -a & c & 0 & 0 \\
d & -f-g & 0 & 0 & -c & -e & 0 & 0 & 0 & 0 & 0 & 0 & 0 & 2(e+g) & 0 & 2f & 0 & 0 & -d & -f-g & c & -e & 0 & 0 \\
-a & c & 0 & 0 & 2b & 0 & 0 & 0 & -a & -c & 0 & 0 & -2b & 0 & 4a & 0 & -2b & 0 & -a & -c & 2b & 0 & -a & c \\
c & -e & 0 & 0 & 0 & -2f & 0 & 0 & -c & -e & 0 & 0 & 0 & 2f & 0 & 4e & 0 & 2f & -c & -e & 0 & -2f & c & -e \\
0 & 0 & 0 & 0 & -a & c & 0 & 0 & b & d & 0 & 0 & 0 & 0 & -2b & 0 & 2a & 0 & 0 & 0 & -a & -c & b & -d \\
0 & 0 & 0 & 0 & c & -e & 0 & 0 & -d & -f-g & 0 & 0 & 0 & 0 & 0 & 2f & 0 & 2(e+g) & 0 & 0 & -c & -e & d & -f-g \\
0 & 0 & 0 & 0 & 0 & 0 & 0 & 0 & 0 & 0 & 0 & 0 & b & -d & -a & -c & 0 & 0 & a & c & -b & d & 0 & 0 \\
0 & 0 & 0 & 0 & 0 & 0 & 0 & 0 & 0 & 0 & 0 & 0 & d & -f-g & -c & -e & 0 & 0 & c & e+g & -d & f & 0 & 0 \\
0 & 0 & 0 & 0 & 0 & 0 & 0 & 0 & 0 & 0 & 0 & 0 & -a & c & 2b & 0 & -a & -c & -b & -d & 2a & 0 & -b & d \\
0 & 0 & 0 & 0 & 0 & 0 & 0 & -t & 0 & 0 & 0 & 0 & c & -e & 0 & -2f & -c & -e & d & f & 0 & 2e+t & -d & f \\
0 & 0 & 0 & 0 & 0 & 0 & 0 & 0 & 0 & 0 & 0 & 0 & 0 & 0 & -a & c & b & d & 0 & 0 & -b & -d & a & -c \\
0 & 0 & 0 & 0 & 0 & 0 & 0 & 0 & 0 & 0 & 0 & 0 & 0 & 0 & c & -e & -d & -f-g & 0 & 0 & d & f & -c & e+g
\end{bmatrix}. \quad (15)$$

TABLE 3: Displacement comparison of the constitutive model and FE model of RC1SS at node 1.

Lateral load (kN)	Constitutive model (plane stress) (mm)	FE program (plane stress) (mm)	Percentage difference (plane stress) (%)	Constitutive model (plane strain) (mm)	FE program (plane strain) (mm)	Percentage difference (plane strain) (%)
0	0	0	—	0	0	—
100	3.88	3.70	4.60	15.38	14.78	4
200	7.73	7.39	4.60	30.75	29.57	4
300	11.60	11.09	4.60	46.13	44.37	4
400	15.47	14.79	4.60	61.51	59.15	4
500	19.33	18.49	4.60	76.89	73.94	4

TABLE 4: Displacement comparison of the constitutive model and FE model of RC2SS at node 1.

Lateral load (kN)	Constitutive model (plane stress) (mm)	FE program (plane stress) (mm)	Percentage difference (plane stress) (%)	Constitutive model (plane strain) (mm)	FE program (plane strain) (mm)	Percentage difference (plane strain) (%)
0	0	0	—	0	0	—
100	14.82	14.93	0.68	37.38	40.08	6.73
200	29.73	29.62	0.39	75.83	78.95	3.94
300	44.64	44.35	0.66	114.29	117.60	2.8
400	59.55	59.13	0.71	152.75	156.02	2.09
500	74.46	73.94	0.7	191.21	194.19	1.53

TABLE 5: Comparison of constitutive model and FE model for displacement of RC3SS at node 1.

Lateral load (kN)	Constitutive model (plane stress) (mm)	FE program (plane stress) (mm)	Percentage difference (plane stress) (%)	Constitutive model (plane strain) (mm)	FE program (plane strain) (mm)	Percentage difference (plane strain) (%)
0	0	0	—	0	0	—
100	41.19	42.28	2.57	78.17	84.28	7.24
200	82.15	84.57	2.86	155.77	167.52	7
300	123.11	127.11	3.14	233.37	250.49	6.83
400	164.06	169.82	3.39	310.98	332.98	6.60
500	205.02	212.69	3.60	388.58	414.78	6.31

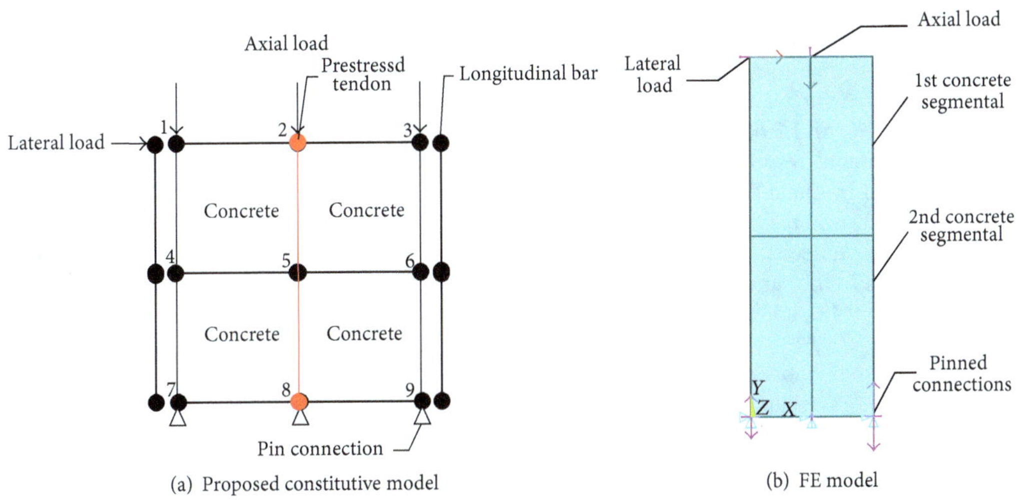

FIGURE 8: Two concrete segments' model (RC2SS).

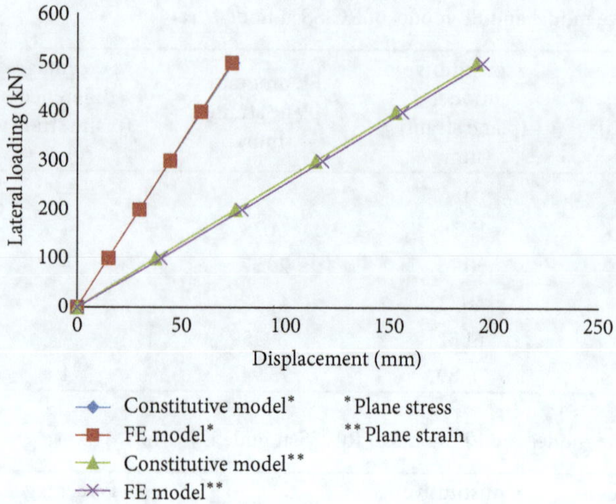

FIGURE 9: Lateral load-deflection of the proposed constitutive model and FE model for RC2SS.

A comparison between the constitutive model and FE model results is carried out under plane stress and plane strain analysis as shown in Table 5. The displacement increases clearly with applied load, for instance, in plane stress analysis; load applied is the same as in previous steps and results in 205.02 mm and 212.69 mm displacements in the proposed constitutive and FE models, respectively. Likewise, in plane strain analysis, applied force produces 388.58 mm and 414.78 mm displacements at node 1 in the proposed constitutive and FE models, respectively.

A maximum difference of 3.6% is observed between the results of the constitutive model and FE program for plane stress analysis of RC3SS model. Also, it can be seen that the constitutive model provides reasonable correlation with the FE program results under plane strain analysis for PPCS columns with three concrete segments.

Based on the outputs, it is clear from the plots that increasing the load increases the maximum displacement in plane stress or plane strain analysis. It can be determined that, in plane stress analysis with the proposed constitutive model, applying load incrementally up to 500 kN with 100 kN as each load step in a lateral direction and 500 kN as a vertical load results in almost 20 mm, 75 mm, and 200 mm as maximum displacements, which are close to the FEM results with about 4.6%, 0.7%, and 3.6% difference in one, two, and three concrete segments, respectively. Likewise, the proposed constitutive model in plane strain analysis shows about 77 mm, 191 mm, and 388 mm displacement in one, two, and three concrete segments subjected to imposed loading, while the FE program shows 74 mm and 194 mm to 414 mm as maximum displacements by 4%, 1.53%, and 6% difference, respectively.

Short-term stress losses may happen due to wobble and curvature frictions and anchorage slip. On the other hand, long-term stress losses include relaxation, elastic shortening, and losses due to creep, shrinkage, and superimposed loads.

Although in this study effects of losses are not considered, however, as a matter of fact, it can be easily incorporated into the stiffness matrix of prestress concrete particularly in elastic modulus of tendon and concrete individually as a degraded elastic modulus before forming global matrix (refer to (14) and (15)).

Short-term losses also could be considered in stiffness matrix of prestressed concrete element as a multiplayer of the yield stress of tendon which varies between 0.7 and 0.9. The effective strand strain after transfer is slightly overestimated because the model considered low-relaxation prestressing strand and does not take into account the long-term prestress losses due to concrete creep and shrinkage.

5. Conclusions

Modeling of PPCS column is not easy using available commercial software packages because all the parts should be modeled separately. Moreover, the interaction between different parts should be defined, which not only increases computation time but also requires experts.

In this study, an attempt was made to develop the constitutive model for PPCS columns in bridge structures. Moreover, FEM model for one, two, and three segments was developed in order to perform linear and nonlinear analysis for PPCS column subjected to static and dynamic loading.

A special FEM program was codified by using developed PPCS columns, and the accuracy of the developed constitutive model was evaluated by comparing analysis results with commercial software. Using this method would not only reduce modeling and computation time but also considerably facilitate the convergence problems due to diminishing meshing adaption issues by substituting meshing assemblage to one explicit matrix in which all characteristics of meshes are incorporated. The PPCS column is applicable as a supplementary subroutine in any structural analysis software or can be used to develop any FEM program code for analysis of PPCS columns.

Nomenclature

E: Young's modulus
E_c: Concrete Young's modulus
E_s: Steel Young's modulus
E_p: Prestressing steel Young's modulus
ν: Poisson's ratio
f_c: Stress of concrete in compression
f_t: Stress of concrete in tension
f_{ys}: Yield stress of reinforcing bar
f_{yp}: Yield stress of prestressing steel
f_{cu}: Ultimate compressive strength of concrete
k_c: Stiffness matrix of concrete
k_s: Stiffness matrix of reinforcing bar
k_t: Stiffness matrix of prestressing steel
B: Strain-displacement matrix
D: Constitutive matrix
N: Shape function
J: Jacobian matrix
ε_{lp}: Prestress steel limit
ε_p: Strain of prestressing steel
ξ: Natural coordinate.

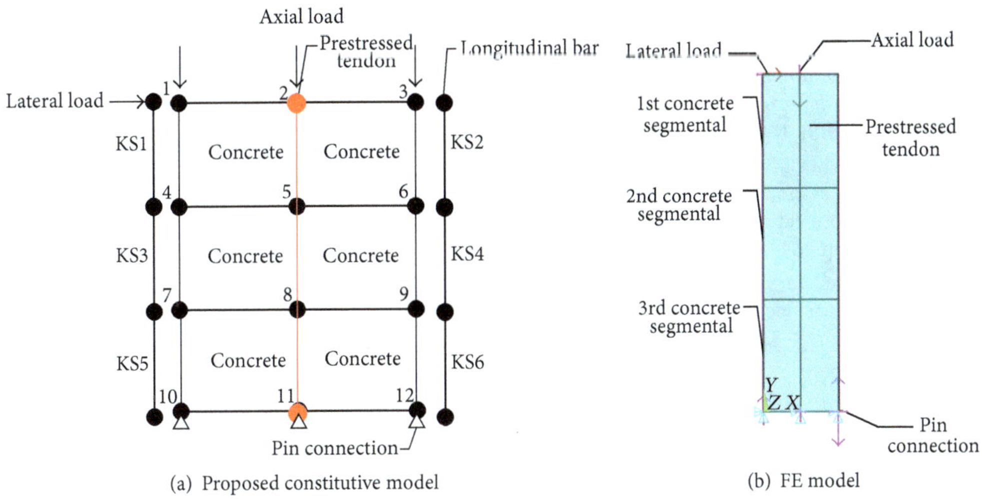

Figure 10: Three concrete segments' model (RC3SS).

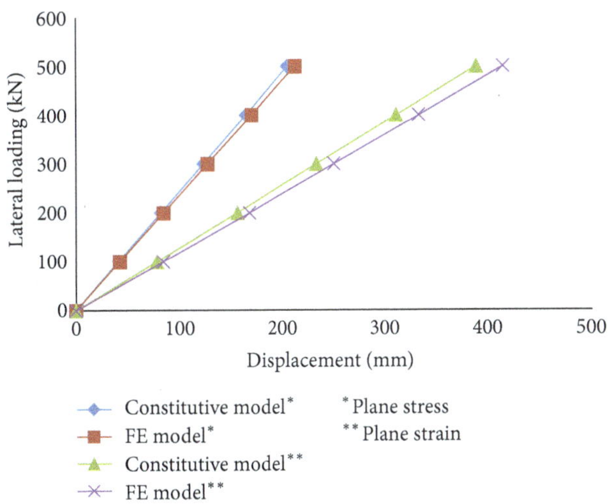

Figure 11: Lateral load-deflection of the proposed constitutive model and FE model for RC3SS.

Competing Interests

The authors declare that there are no competing interests regarding the publication of this paper.

Acknowledgments

This work received financial support from the Ministry of Higher Education of Malaysia under FRGS Research Projects no. 5524254 and no. 5524256 and was further supported by the University Putra Malaysia under Putra Grant no. 9415100. These supports are gratefully acknowledged.

References

[1] Y.-C. Ou, M. Chiewanichakorn, A. J. Aref, and G. C. Lee, "Seismic performance of segmental precast unbonded posttensioned concrete bridge columns," *Journal of Structural Engineering*, vol. 133, no. 11, pp. 1636–1647, 2007.

[2] K. Chang, C. H. Loh, H. S. Chiu et al., *Seismic Behavior of Precast Segmental Bridge Columns and Design Methodology for Applications in Taiwan*, Taiwan Area National Expressway Engineering Bureau, Taipei, Taiwan, 2002 (Chinese).

[3] T.-H. Kim, H.-M. Lee, Y.-J. Kim, and H. M. Shin, "Performance assessment of precast concrete segmental bridge columns with a shear resistant connecting structure," *Engineering Structures*, vol. 32, no. 5, pp. 1292–1303, 2010.

[4] J. T. Hewes and M. N. Priestley, *Seismic Design and Performance of Precast Concrete Segmental Bridge Columns*, University of California, Oakland, Calif, USA, 2002.

[5] S. L. Billington and J. K. Yoon, "Cyclic response of unbonded posttensioned precast columns with ductile fiber-reinforced concrete," *Journal of Bridge Engineering*, vol. 9, no. 4, pp. 353–363, 2004.

[6] M. J. Nigel Priestley, "Overview of PRESSS research program," *PCI Journal*, vol. 36, no. 4, pp. 50–57, 1991.

[7] W. C. Stone, G. S. Cheok, and J. F. Stanton, "Performance of hybrid moment-resisting precast beam-column concrete connections subjected to cyclic loading," *ACI Structural Journal*, vol. 92, no. 2, pp. 229–249, 1995.

[8] Y. Wang, Z. Y. Bu, and L. Hu, "Seismic behavior of precast segmental bridge columns with carbon fibre reinforcement as energy dissipation bars," *Applied Mechanics and Materials*, vol. 157, pp. 1148–1152, 2012.

[9] W.-P. Kwan and S. L. Billington, "Unbonded posttensioned concrete bridge piers. II. Seismic analyses," *Journal of Bridge Engineering*, vol. 8, no. 2, pp. 102–111, 2003.

[10] J. Sakai and S. A. Mahin, *Analytical Investigations of New Methods for Reducing Residual Displacements of Reinforced Concrete Bridge Columns*, Pacific Earthquake Engineering Research Center, University of California at Berkeley, Berkeley, Calif, USA, 2004.

[11] A. Palermo, S. Pampanin, and G. M. Calvi, "Concept and development of hybrid solutions for seismic resistant bridge systems," *Journal of Earthquake Engineering*, vol. 9, no. 6, pp. 899–921, 2005.

[12] Y.-C. Ou, M.-S. Tsai, K.-C. Chang, and G. C. Lee, "Cyclic behavior of precast segmental concrete bridge columns with high performance or conventional steel reinforcing bars as energy dissipation bars," *Earthquake Engineering & Structural Dynamics*, vol. 39, no. 11, pp. 1181–1198, 2010.

[13] A. Palermo, S. Pampanin, and D. Marriott, "Design, modeling, and experimental response of seismic resistant bridge piers with posttensioned dissipating connections," *Journal of Structural Engineering*, vol. 133, no. 11, pp. 1648–1661, 2007.

[14] M. A. Elgawady and A. Sha'Lan, "Seismic behavior of self-centering precast segmental bridge bents," *Journal of Bridge Engineering*, vol. 16, no. 3, pp. 328–339, 2011.

[15] R. Vaghei, F. Hejazi, H. Taheri, M. S. Jaafar, and A. A. Ali, "Evaluate performance of precast concrete wall to wall connection," *APCBEE Procedia*, vol. 9, pp. 285–290, 2014.

[16] Z. Wang, W. Song, Y. Wang, and H. Wei, "Numerical analytical model for seismic behavior of prestressing concrete bridge column systems," *Procedia Engineering*, vol. 14, pp. 2333–2340, 2011.

[17] H. Dawood, M. ElGawady, and J. Hewes, "Behavior of segmental precast posttensioned bridge piers under lateral loads," *Journal of Bridge Engineering*, vol. 17, no. 5, pp. 735–746, 2012.

[18] Z.-Y. Bu and Y.-C. Ou, "Simplified analytical pushover method for precast segmental concrete bridge columns," *Advances in Structural Engineering*, vol. 16, no. 5, pp. 805–822, 2013.

[19] C.-C. Chou, H.-J. Chang, and J. T. Hewes, "Two-plastic-hinge and two dimensional finite element models for post-tensioned precast concrete segmental bridge columns," *Engineering Structures*, vol. 46, pp. 205–217, 2013.

[20] E. Nikbakht, K. Rashid, F. Hejazi, and S. A. Osman, "A numerical study on seismic response of self-centring precast segmental columns at different post-tensioning forces," *Latin American Journal of Solids and Structures*, vol. 11, no. 5, pp. 864–883, 2013.

[21] E. Nikbakht, K. Rashid, F. Hejazi, and S. A. Osman, "Application of shape memory alloy bars in self-centring precast segmental columns as seismic resistance," *Structure and Infrastructure Engineering*, vol. 11, no. 3, pp. 297–309, 2015.

[22] J. B. Mander, M. J. Priestley, and R. Park, "Theoretical stress-strain model for confined concrete," *Journal of Structural Engineering*, vol. 114, no. 8, pp. 1804–1826, 1988.

[23] G. Strang and G. J. Fix, *An Analysis of the Finite Element Method*, Prentice-Hall, Englewood Cliffs, NJ, USA, 1973.

19

Numerical Modelling of Tailings Dam Thermal-Seepage Regime Considering Phase Transitions

Aniskin Nikolay Alekseevich[1,2] and Antonov Anton Sergeevich[1,3]

[1]*National Research Moscow State University of Civil Engineering (NRU MGSU), 26 Yaroslavskoye Shosse, Moscow 129337, Russia*
[2]*Hydrotechnical and Energy Construction Department, NRU MGSU, Moscow, Russia*
[3]*JSC "Institute Hydroproject", Moscow, Russia*

Correspondence should be addressed to Antonov Anton Sergeevich; antonov.an.s@yandex.ru

Academic Editor: Jean-Michel Bergheau

Statement of the Problem. The article describes the problem of combined thermal-seepage regime for earth dams and those operated in the permafrost conditions. This problem can be solved using the finite elements method based on the local variational formulation. *Results.* A thermal-seepage regime numerical model has been developed for the "dam-foundation" system in terms of the tailings dam. The effect of heat-and-mass transfer and liquid phase transition in soil interstices on the dam state is estimated. The study with subsequent consideration of these factors has been undertaken. *Conclusions.* The results of studying the temperature-filtration conditions of the structure based on the factors of heat-and-mass transfer and liquid phase transition have shown that the calculation results comply with the field data. Ignoring these factors or one of them distorts the real situation of the dam thermal-seepage conditions.

1. Introduction

Today, intensive development of the northern regions of Siberia and the Far East of Russia is associated with the exploitation of their rich natural resources. This development requires operation of already built hydraulic engineering structures of the power and mining industries, water supply facilities, and construction of new ones. Herewith, we need to deal with the problems of construction in the permafrost conditions [1-3]. Earth embankments and dams are the most frequent hydraulic engineering structures used in the permafrost conditions [1, 3]. Trouble-free operation of these structures depends on compliance with the required thermal-seepage regime of the dam-foundation system [2–4].

Any disturbance of this mode may cause emergency and even structure collapse [4–7]. According to experts, about 50% of 3rd and 4th category earth low-head dams built in the severe climatic conditions is damaged during the first three years of their operation due to thermal-seepage regime disturbance [4].

When designing the earth hydraulic engineering structures in the permafrost conditions, the extremely important task is the reliable determination of the thermal-seepage regime of the dam-foundation system. In most cases, the geotechnical setting in the construction area is nonhomogeneous geology and composed of frozen and thawed layers stratification. In order to determine the exact seepage pattern at the hydraulic engineering structure foundation, it is often necessary to solve the spatial three-dimensional problem [8–11]. In some cases, the seepage heating effect may result in negative consequences, such as increasing soil permeability and seepage losses in a reservoir, reduction of bearing capacity of frozen rocks, and seepage failure [1].

Therefore, for the facilities located in the area of permafrost occurrence, handling the problem of the combined thermal-seepage regime is actual based on the phenomena of heat-and-mass transfer and phase transitions, whose results are to be considered in the design.

Both in Russia and abroad there are numerous scientific studies devoted to the problems for calculations of seepage, thermal, combined thermal-seepage regimes of earth dams and their foundations [12–16]. In recent years, a mathematic modelling is widely used for these purposes [5, 13, 16–18]. However, most software systems including industrial ones handle the seepage and temperature problems with no interaction of two processes [8, 19–23]. Additionally, the effect of liquid phase transitions in various soils on thermal-seepage regime formation is not entirely clear. Thus, the need for further scientific studies in this sphere including calculation method improvement is quite obvious. The purpose of this work is to improve the method for calculating the thermal-seepage regime of the dam-foundation system based on heat-and-mass transfer and phase transitions in soil interstices.

2. Method

For the classical pattern, the temperature problem (Stephen) with no heat-and-mass transfer in the seepage flow can be solved by the differential equation [1]:

$$\frac{\partial t}{\partial \tau} = a_x \frac{\partial^2 t}{\partial x^2} + a_y \frac{\partial^2 t}{\partial y^2} + a_z \frac{\partial^2 t}{\partial z^2}, \quad (1)$$

where $t = f(x, y, z, \tau)$ is a temperature function; τ is a time; a_x, a_y, and a_z are temperature conductivity coefficients in x-, y-, and z-directions.

The set of equations describing dam thermal regime considering the temperature of seepage water is known from the heat-and-mass transfer theory:

$$\begin{aligned} &\frac{\partial t}{\partial \tau} = a_x \frac{\partial^2 t}{\partial x^2} + a_y \frac{\partial^2 t}{\partial y^2} + a_z \frac{\partial^2 t}{\partial z^2} + \beta_v (Q - t); \\ &\frac{\partial Q}{\partial \tau} + \left(v_x \frac{\partial Q}{\partial x} + v_y \frac{\partial Q}{\partial y} + v_z \frac{\partial Q}{\partial z} \right) = \alpha_v' (t - Q), \end{aligned} \quad (2)$$

where $t = f(x, y, z, \tau)$ is a foundation or dam material temperature; $Q(x, y, z, \tau)$ is a seepage water temperature; τ is a time; a_x, a_y, and α_z are conductivity coefficients of dam body or foundation material temperature in x-, y-, and z-directions; v_x, v_y, and v_z are seepage rates in x-, y-, and z-directions; β_v and α_v' are the coefficients that characterize the heat exchange between the material of the dam or foundation and filtering water in its pores. These coefficients are determined from the following dependencies:

$$\begin{aligned} \alpha_v' &= \frac{\alpha_v}{C_w \gamma_w}; \\ \beta_v &= \frac{\alpha_v}{C_s \gamma_s}, \end{aligned} \quad (3)$$

where α_v is a volume heat-exchange rate specifying heat exchange between the dam or foundation material and ambient water; C_w and C_s are volumetric specific heat of water and solid body; γ_w and γ_s are water and solid body density.

The solution of the set of equations (2) describes change of seepage water temperature after passing through the dam body or foundation material. Heat exchange in each space point is featured by a volumetric heat-exchange rate. The calculation does not use the coefficient of water temperature conductivity due to insignificant conductive heat transfer.

If the set of equations (2) is considered with regard to soil porosity, then the set changes to [5]

$$\begin{aligned} &\frac{\partial t}{\partial \tau} \\ &= a_x \frac{\partial^2 t}{\partial x^2} + a_y \frac{\partial^2 t}{\partial y^2} + a_z \frac{\partial^2 t}{\partial z^2} + \frac{\alpha_v}{C_s \gamma_s} \frac{(1 - m)}{m} (Q - t), \\ &\frac{\partial Q}{\partial \tau} + \left(v_x \frac{\partial Q}{\partial x} + v_y \frac{\partial Q}{\partial y} + v_z \frac{\partial Q}{\partial z} \right) \\ &= \frac{\alpha_v}{C_w \gamma_w} \frac{m}{1 - m} (t - Q), \end{aligned} \quad (4)$$

where m is a soil porosity.

If the problem contains a temperature of dam body material or foundation soil equal to a seepage water temperature, then the combined problem is the solution of Fourier-Kirchhoff equation [5]:

$$\begin{aligned} (1 + \delta) \frac{\partial t}{\partial \tau} &= \delta \left(a_x \frac{\partial^2 t}{\partial x^2} + a_y \frac{\partial^2 t}{\partial y^2} + a_z \frac{\partial^2 t}{\partial z^2} \right) \\ &+ \left(v_x \frac{\partial t}{\partial x} + v_y \frac{\partial t}{\partial y} + v_z \frac{\partial t}{\partial z} \right), \end{aligned} \quad (5)$$

where

$$\delta = \frac{C_s \gamma_s}{C_w \gamma_w} \frac{(1 - m)^2}{m^2}. \quad (6)$$

For soils with minimum porosity, (5) changes to

$$\begin{aligned} \frac{\partial t}{\partial \tau} &= a_x \frac{\partial^2 t}{\partial x^2} + a_y \frac{\partial^2 t}{\partial y^2} + a_z \frac{\partial^2 t}{\partial z^2} \\ &- \frac{C_s \gamma_s}{C_w \gamma_w} \left\{ v_x \frac{\partial t}{\partial x} + v_y \frac{\partial t}{\partial y} + v_z \frac{\partial t}{\partial z} \right\}. \end{aligned} \quad (7)$$

Search for solving the combined thermal-seepage problem comes to solving differential equations ((5), (7)) for the area with seepage liquid flowing and (1) with no seepage.

The differential equations solution comes to minimization of functional F [5]. The standard form of (7) based on boundary conditions and phase transition is given below:

$$\begin{aligned} F = &\iiint_V \left\{ \frac{1}{2} \left[a_x \left(\frac{\partial t}{\partial x} \right)^2 + a_y \left(\frac{\partial t}{\partial y} \right)^2 + a_z \left(\frac{\partial t}{\partial z} \right)^2 \right] \right. \\ &+ \left(\frac{\partial t}{\partial \tau} \right) t + \left(\vec{v}, \text{gradt} \right) t \pm \left(\frac{\gamma_{\text{ice}} n L}{\gamma_s c} \right. \\ &\left. \left. \cdot \frac{\partial W}{\partial \tau} t_{\text{phase}} \right) \right\} dx\, dy\, dz \\ &+ \iint_{\Omega_1} q t \, d\Omega + \iint_{\Omega_2} \frac{1}{2} \beta (t - t_{\text{air}})^2 \, d\Omega, \end{aligned} \quad (8)$$

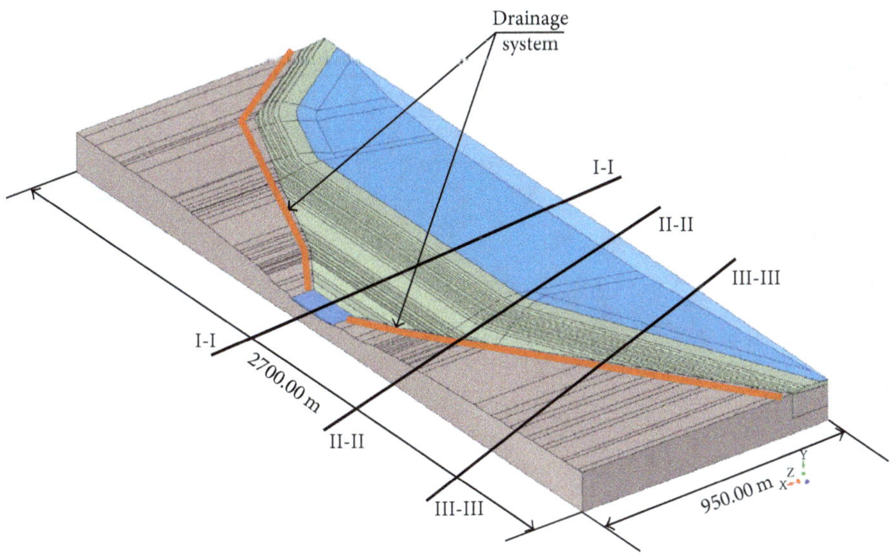

FIGURE 1: Solid approximant of the earth tailings dam.

where $(\vec{v}, \text{gradt}) = v_x(\partial t/\partial x) + v_y(\partial t/\partial y) + v_z(\partial t/\partial z)$ is a scalar product of seepage velocity vectors in x-, y-, and z-directions and temperature gradient; Ω_1, Ω_2 are the surfaces with the designated 2nd and 3rd category boundary conditions; t_{phase} is the temperature of phase transformations.

The proposed method for calculating the combined thermal-seepage equation based on heat-and-mass transfer and phase transition has been implemented by the authors in the FILTR_FAZ software system. This software allows solving both classic filtration tasks and heat conduction problems, as well as obtaining complex combined solutions. To date, there are also universal industrial computing systems (such as ANSYS [20], ANSYS CFD, MIDAS CFX, Feflow, and Plaxis Flow), but at the same time, their use is only possible to search for the simplest classical solutions. Without taking into account the complex physical processes taking place in permafrost, using FILTR_FAZ the authors have studied the thermal-seepage regime of hydraulic engineering structures in the permafrost (e.g., [5]). Some results of these studies are given below.

2.1. Problem Statement and Content. The facility is a thawed type hydraulic fill dam in terms of its design. The dam upstream fill is erected by aggregation, while the downstream one is filled with soil from open pits. The downstream face is fastened by means of mined rock with an effective thickness of 3.5 m. The medium fine gravel entrenchment is used as an anti-seepage cutoff.

The structure is erected in the severe climatic conditions. The average winter temperature is minus 24.8°C and the summer temperature is plus 15.6°C. The recorded absolute minimum is minus 44°C, the absolute anomalous maximum is plus 33°C, and the snow depth is over 1 m. The foundation permafrost thickness is over 100 m. Yearly average air temperature (according to the field studies) is within the range of minus 1.0 to minus 1.5°C.

The embank height is 72.00 m with an edge elevation of 360.00 m. The dam is expected to be topped by 2032 up to an edge elevation of 370.00 m. Therefore, the structure height will reach 82 m. This facility belongs to the 1st importance class.

The 3D simulation mathematic model has been developed to determine the combined thermal-seepage regime of the facility (see Figure 1). This model describes an earth dam, retaining prism, drainage system, lower manoeuvring basin, and pond beach.

The combined thermal-seepage regime was calculated by an improved procedure with the use of the FILTRT_FAZ and FILTR software system [5]. The phase transitions of interstitial water and heat-and-mass transfer in the direction of seepage flow were taken into account in the calculations. Multipurpose industrial software ANSYS APDL for mathematic modelling was used to analyze the seepage regime in process.

The purpose of the study was to determine the combined thermal-seepage regime in the 3D pattern based on the phenomena of heat-and-mass transfer and phase transitions within 10 years of the date of the facility erection up to a design elevation of 370.00 m.

To analyze the structure condition, the following problems were solved: seepage problem for a design area of 2.5 km^2; temperature problem in its classical pattern; combined thermal-seepage problem with due regard for heat-and-mass transfer by seepage flow; combined thermal-seepage problem based on heat-and-mass transfer by seepage flow and phase transitions of interstitial water.

2.2. Seepage Problem Solution. The calibration tests are carried out at the initial stage to determine the position of depression curve in the model similar to piezometers reading. The calculations were used to determine K_s seepage coefficient for the retaining prism and paving rock-fill blanket.

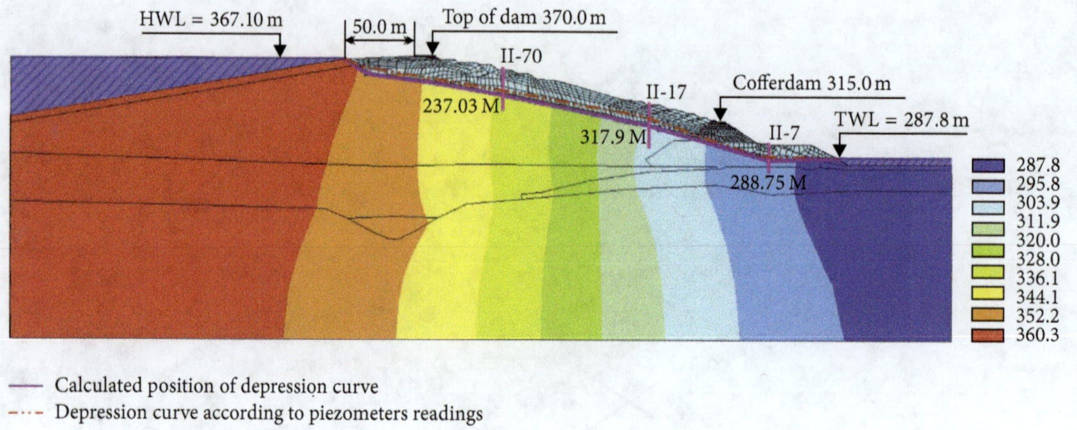

FIGURE 2: Calibration test results for the dam seepage regime at two K_s values of the rock-fill blanket: 50 m/day and 100 m/day.

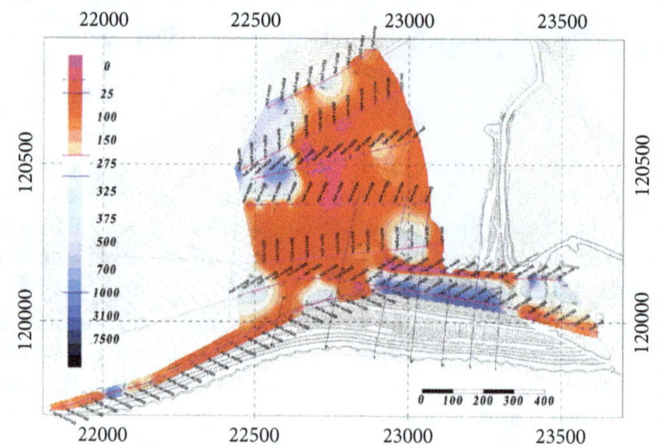

FIGURE 3: Field data, specific electric resistance isolines (3 meters from the ground surface).

The study results have shown that the retaining prism is most likely mudding at the upstream side, and, therefore, two K_s seepage coefficients were used in the calculations; that is, K_s for the mudding section is 50 m/day and K_s for other sections of the rock-fill blanket is 100 m/day. The seepage calculation results in dam section 1-1 are shown in Figure 2.

The seepage regime calculations based on the classical pattern have shown that the primary seepage flow is directed through the left-bank abutment (see Figure 3), which is confirmed by the field data.

In order to more accurately determine the depression surface position, it was identified iteratively with control of the finite-element grid within its expected location.

The seepage regime was corrected at every stage by time based on changes in the thermal regime of the previous stage. The lower values of K_s were obtained for the elements, whose temperature values of the thermal equation were within the range of minus 1.0°C and below and set by the absolute confining layer with temperatures of minus 0.5 to minus 1.0°C. Then the correction calculation was performed. The resulting seepage velocities at every stage were used to determine the structure thermal-seepage regime.

2.3. The Thermal Problem Solution Based on Heat-and-Mass Transfer and Phase Transitions. The initial and boundary conditions were set to solve the thermal problem. The initial condition (temperature distribution in the soil body at the calculation) was set according to the field data as of November of the current year. The condition of convective heat exchange (the 3rd category boundary condition) was set according to soil-air contact. The mean monthly air temperatures are given in Table 1. The 1st category boundary condition, $t_i = t_{pr}$ (where t_{pr} is minus 1.5°C and t_{pr} is the temperature of permafrost), was set in the permafrost location. The absolute thermal insulation was set according to the boundaries of design area (the special pattern of the 2nd category boundary condition). The 1st category boundary condition, $t_i = t_{water}$, where t_{water} is a water temperature in every design month, was specified according to the soil-air contact.

The thermal-seepage regime was successively calculated for every month.

The thermal problem was solved in several stages for 10 years' period of structure operation. The solution of absolute thermal problem was obtained at the first stage without

Table 1: Mean monthly temperatures of ambient air and water.

Parameter	Month of calendar year											
	I	II	III	IV	V	VI	VII	VIII	IX	X	XI	XII
Air temperature, °C	−32.3	−26.0	−17.1	−6.1	−3.7	13.5	16.8	12.3	4.9	−6.9	−22.9	−30.6
Wind velocity, m/sec	2.1	1.9	2.2	3.0	3.0	2.6	2.3	2.1	2.4	2.7	2.4	2.2
Upstream water temperature, °C	4.00	4.00	4.00	4.00	5.00	9.00	15.0	13.0	10.00	7.00	4.00	4.00
Downstream water temperature, °C	4.00	4.00	4.00	4.00	10.0	10.0	10.0	10.0	4.00	4.00	4.00	4.00

considering heat-and-mass transfer and phase transitions. The temperature patterns for dam I-I section in November for 1, 3, and 10 years of operation are shown in Figure 4.

In this case, a frozen core with temperatures of 0 to minus 0.5°C (higher than in the dam body) is formed in the centre of the thawed dam due to the boundary conditions being set according to the field data. The downstream face freezes to the permafrost rock top forming a negative temperatures field at the downstream side.

According to the temperature distribution in the soil body, it may be concluded that the results obtained from the classical pattern of the thermal problem do not comply with the field data. Thermometers located on the crest record positive soil temperatures ranging from 0 to plus 0.5°C. The seepage problem solution for this option showed the seepage flow occurring on the surface of the downstream face only, that is, in the area of seasonal freezing/defrosting. Also this fact does not comply with the true pattern of the seepage regime.

The solution of the combined thermal-seepage problem was considered at the second stage based on the heat-and-mass transfer phenomena. The temperature distribution pattern partly corresponding to the field data was obtained according to the calculation results (see Figure 5). The positive temperatures field ranging from 0 to plus 0.3°C is formed in the dam centre, but the temperature field between the thawed core and retaining prism is nearly unchanged. This is provided by severe exposure to winter negative temperatures that causes freezing to the soil body at the downstream face. Therefore, the seepage flow remains trapped in the dam centre, where "a pocket" is formed. In case of soil thawing in summer along the downstream face, high rate seepage takes place and it has no impact on the negative temperatures field smoothly flowing it around.

The defrosting effect due to heat-and-mass transfer occurs in the seasonal thawing area only. In comparison to the previous option, this area increases in size, but it completely froze in winter.

At the final stage of calculation, the combined thermal-seepage problem has been solved based on phase transitions of cohesive water in the soil (see Figure 6).

In the first years, while considering phase transitions, the soil body will be defrosted in the dam centre (as confirmed by instruments reading). A thawed area with passing seepage flow is formed beneath the layer of seasonally frozen soil. The thickness of the seasonally frozen area is 3.5–3.7 m that complies with climatological data of the region.

The frozen area from the downstream face formed at the initial stage gradually thaws out under the influence of heat-and-mass transfer and phase transitions of cohesive water. Since the foundation soils are characterized by the negative temperatures ranging from minus 1.0 to minus 1.5°C, the heat carried by the seepage flow is sufficient for soil thawing. Based on the fact that the structure was designed by the thawed type, this operation mode corresponds to the design one.

Retaining prism permafrost thawing is observed at the downstream face by the 10th year of operation causing additional setting. The most extensive temperature change is observed in the first 6-7 years of dam operation, and then the temperature field tends to be stabilized. However, some local areas of the negative temperatures in the thawed body remain by the 10th year of dam operation.

The key feature defined during the design studies is fast degradation of the permafrost in the way close to the soil phase transitions temperature; *it can be explained by the concept of incomplete water freezing in dispersive frozen soils [24, 25].* At any temperature rise (including the negative values field), frozen soils thaw out. This phenomenon caused accidents of many structures built in the permafrost conditions.

The results of problem solution as compared to the field data obtained in a dam temperature well for 2015 are given in Table 2.

Difference between obtained data and instrument readings is 0.5°C maximum that is less than 0.1% in terms of the annual variations amplitude. When considering phase transitions, the soil condition in the dam body may be exactly determined in contrast to other solutions.

The following conclusions are made based on the results of the calculation studies:

(1) The improved calculation procedure and software system FILTRT_FAZ created by the authors allow handling the combined thermal-seepage problems based on the phenomena of heat-and-mass transfer and phase transitions.

(2) The results of studying the temperature-filtration state of the structure based on the factors of heat-and-mass transfer and liquid phase transition have shown that the calculation results comply with the field data. Ignoring these factors or one of them distorts the real situation of the dam temperature-filtration conditions.

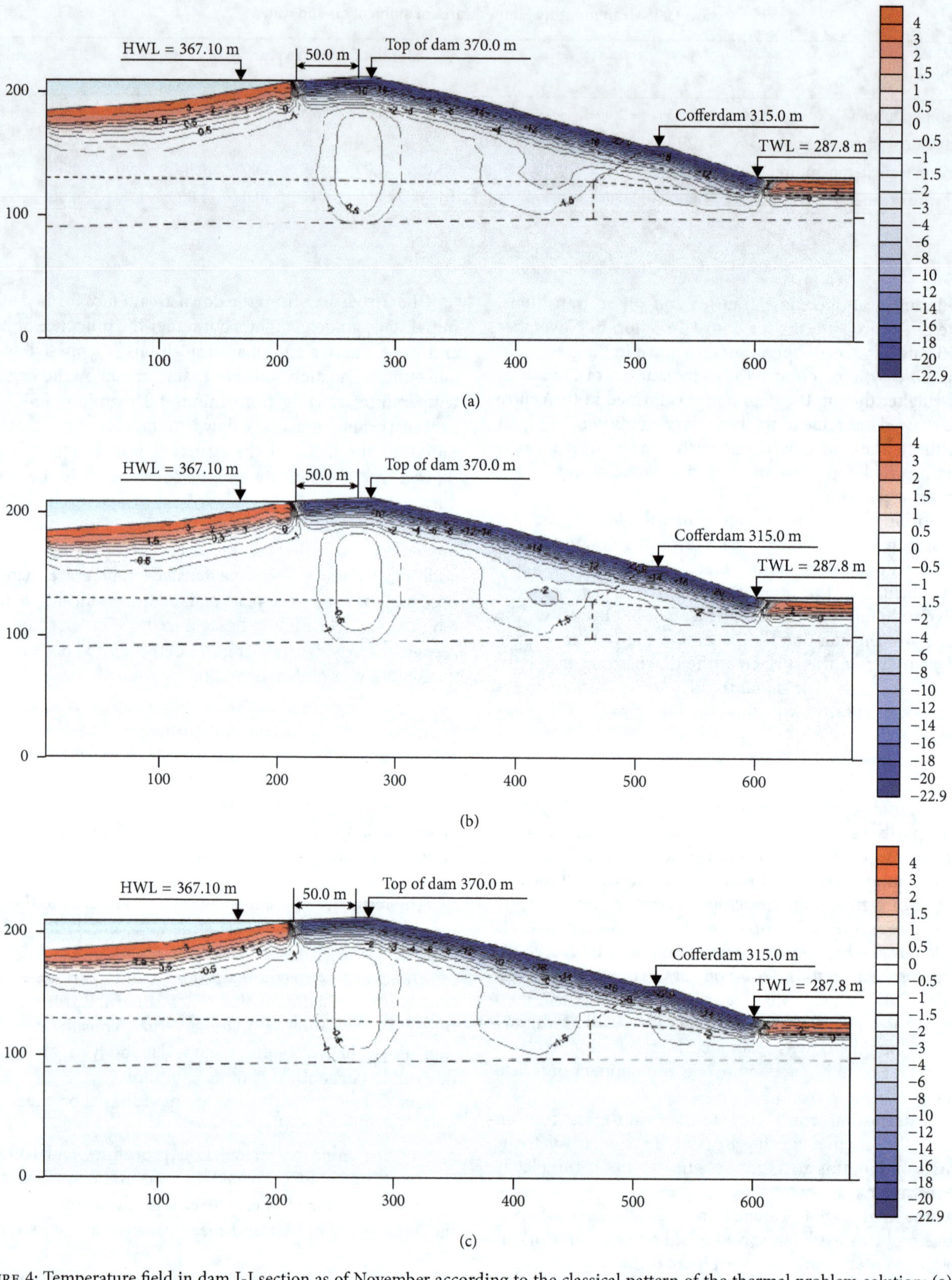

FIGURE 4: Temperature field in dam I-I section as of November according to the classical pattern of the thermal problem solution: (a) in 1 year of operation; (b) in 3 years; (c) in 10 years.

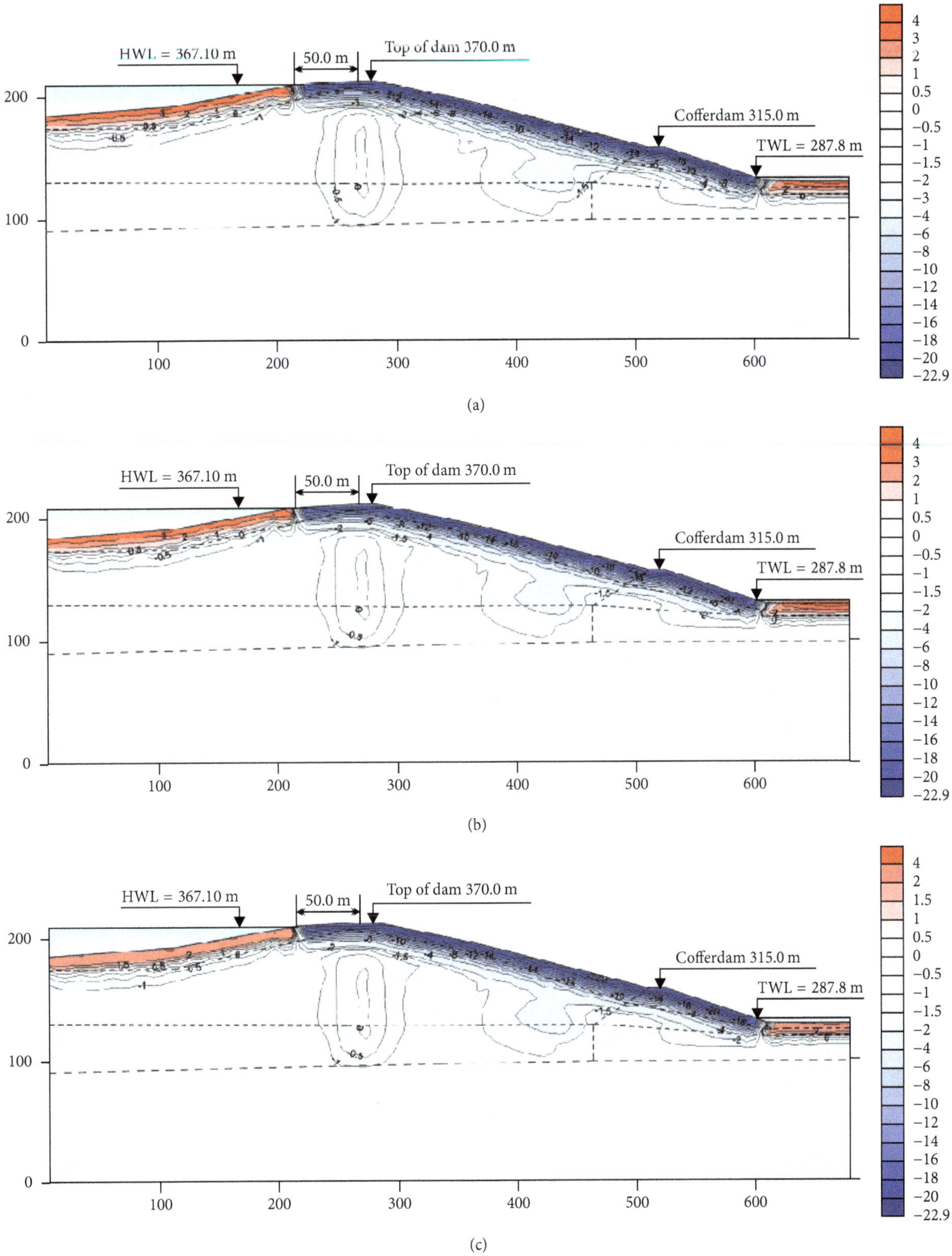

FIGURE 5: Temperature field in dam I-I section as of November according to the thermal problem solution considering heat-and-mass transfer: (a) in 1 year of operation; (b) in 3 years; (c) in 10 years.

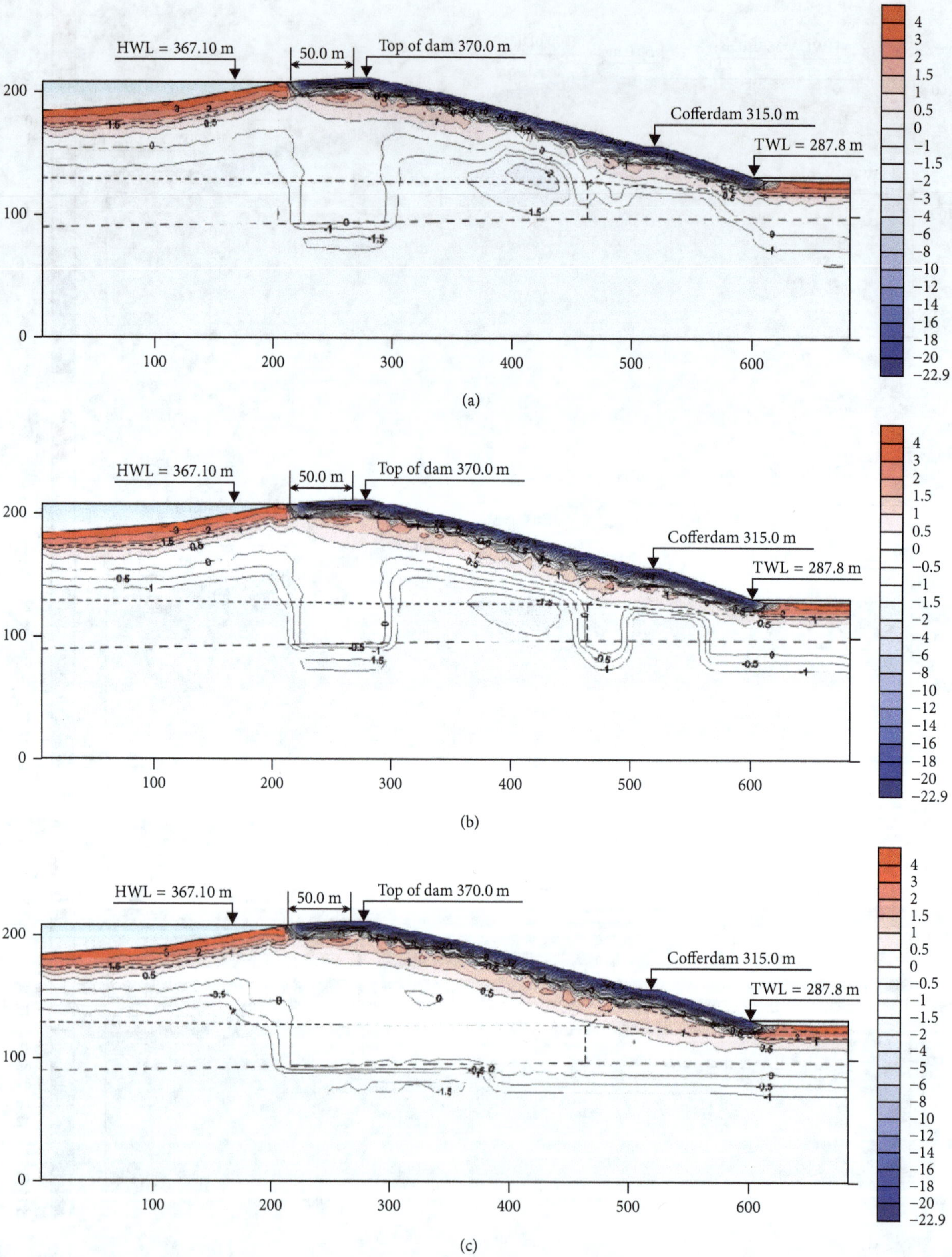

FIGURE 6: Temperature field in dam I-I section as of November according to the combined thermal problem solution considering heat-and-mass transfer and phase transitions: (a) in 1 year of operation; (b) in 3 years; (c) in 10 years.

TABLE 2: The obtained calculation results as compared to the field data.

Elev., m	Instrument readings, °C	Net temperature problem, °C	With heat-and-mass transfer, °C	With phase transitions, °C	Difference between obtained data and instrument readings, °C
207.5	−22.9	−22.9	−22.9	−22.9	0
168.86	0.7	−0.363	−0.175	0.18	0.52
134.25	0.3	−0.387	−0.2	0.08	0.22
128.27	0.3	−0.387	−0.2	0.10	0.20
112.69	0.3	−0.45	−0.325	0.09	0.21
101.06	0.3	−0.762	−0.7	0.10	0.21
97.38	0.3	−0.887	−0.825	0.07	0.23
94.23	−1.4	−1.4	−1.4	−1.09	−0.31
90.37	−1.4	−1.4	−1.4	−1.34	−0.06
50.19	−1.4	−1.4	−1.4	−1.40	0

Conflicts of Interest

The authors declare that there are no conflicts of interest regarding the publication of this paper.

References

[1] L. N. Rasskazov, N. A. Aniskin, V. V. Malakhanov, A. S. Bestuzheva, M. P. Sainov, and V. V. Tolstikov, Gidrotekhnicheskie sooruzheniya (rechnye). [Hydraulic structures(river)]. Moscow, ASV, 2011, p. 581.

[2] S. N. Guzenkov, D. V. Stefanishin, O. M. Finogenov, and S. G. Shul'man, *Nadezhnost' khvostovykh khozyaystv obogatitel'nykh fabric [Reliability tail farms processing plants]*, Vezelitsa, Belgorod, Russia, 2007.

[3] R. V. Zhang, "Monitoring of small and medium embankment dams on permafrost in changing climate," *Sciences in Cold and Arid Regions*, vol. 6, no. 4, pp. 348–355, 2014.

[4] R. V. Chzhan, "Geokriologicheskie printsipy raboty gruntovykh plotin v kriolitozone v usloviyakh menyayushchegosya klimata [Geocryological principles of operation of embankment dams on permafrost in a changing climate]," *Fundamental'nye Issledovaniya [Fundamental Research]*, vol. 9-2, pp. 288–296, 2015 (Russian).

[5] N. A. Aniskin, "Temperaturno-fil'tratsionnyy rezhim osnovaniya i plotiny Kureyskoy GES vo vtorom pravoberezhnom ponizhenii [Temperature-filtration mode of foundation and dam Kureiskaya HPP the second right bank lowering]," in *Proceedings of the Moscow State University of Civil Engineering [Vestnik MGSU]*, pp. 43–52, 2006.

[6] M. Foster, R. Fell, and M. Spannagle, "The statistics of embankment dam failures and accidents," *Canadian Geotechnical Journal*, vol. 37, no. 5, pp. 1000–1024, 2000.

[7] A. A. Korshunov, S. P. Doroshenko, and A. L. Nevzorov, "The Impact of Freezing-thawing Process on Slope Stability of Earth Structure in Cold Climate," in *Proceedings of the 3rd International Conference on Transportation Geotechnics, ICTG 2016*, pp. 682–688, September 2016.

[8] N. A. Aniskin and A. S. Antonov, "Development geo-seepage models for solving seepage problems of large dam's foundations on an example of ANSYS Mechanical APDL," *Advanced Materials Research - Trans Tech Publications - Switzerland*, vol. 1079-1080, pp. 198–201, 2015.

[9] Q. Chen and L. M. Zhang, "Three-dimensional analysis of water infiltration into the Gouhou rockfill dam using saturated-unsaturated seepage theory," *Canadian Geotechnical Journal*, vol. 43, no. 5, pp. 449–461, 2006.

[10] W. D. Liam Finn, "Finite-element analysis of seepage through dams," *Journal of the Soil Mechanics and Foundations Division*, vol. 93, no. 6, pp. 41–48, 1967.

[11] G. Rakhshandehroo and A. Bagherieh, "Three dimensional analysis of seepage in 15-Khordar dam after impoundment," *Iranian Journal of Science and Technology, Transaction B: Engineering*, vol. 30, no. B1, pp. 135–142, 2006.

[12] E. N. Bereslavskiy, "Matematicheskoe modelirovanie fil'tratsionnykh techeniy pod gidrotekhnicheskimi sooruzheniyami [Mathematical modeling of filtration currents under hydraulic engineering structures]," *Nauchnye Vedomosti BelGU [Bulletin of BSTU]*, vol. 5, pp. 32–46, 2009 (Russian).

[13] D. A. Krylov, "Matematicheskoe modelirovanie temperaturnykh poley s uchetom fazovykh perekhodov v kriolitozone [Mathematical simulation of temperature fields with phase transitions in the permafrost zone]," *Nauka i Obrazovanie [Science and Education]*, vol. 4, pp. 1–26, 2012 (Russian).

[14] O. K. Markhilevich, "Primenenie metodov modelirovaniya geofil'tratsii pri proektirovanii gidrotekhnicheskikh sooruzheniy [Application of methods of simulation of geofiltration in the design of hydraulic structures]," *Gidrotekhnicheskoe Stroitel'stvo [Hydrotechnical Construction]*, vol. 4, pp. 61–72, 2009 (Russian).

[15] B. Bussiere, R. P. Chapius, and M. Aubertin, "Unsaturated flow modelling for exposed and covered tailings dams / B.Bussiere, R.P.Chapius," in *Proceedings of the ICOLD Conference*, Montreal, Canada, 2003.

[16] S. Yousefi, A. Noorzad, M. Ghaemian, and S. Kharaghani, "Seepage investigation of embankment dams using numerical modelling of temperature field," *Indian Journal of Science and Technology*, vol. 6, no. 8, pp. 67–78, 2013.

[17] E. N. Gorokhov, "Virtual'nye 3D-modeli temperaturno-kriogennogo rezhima gruntovykh plotin v kriolitozone [A virtual 3D model of temperature-cryogenic regime of embankment dams on permafrost]," *Privolzhskiy Nauchnyy Zhurnal [Privolzhsky Scientific Journal]*, vol. 3, pp. 188–193, 2012 (Russian).

[18] S. A. Sazhenkov, "Issledovanie zadachi Darsi-Stefana o fazovykh perekhodakh v nasyshchennom poristom grunte [The problem of Darcy-Stefan on phase transitions in saturated

porous soil]," *Prikladnaya Mekhanika i Tekhnicheskaya Fizika [Journal of Applied Mechanics and Technical Physics]*, vol. 4, pp. 81–93, 2008 (Russian).

[19] K. A. Basov, *ANSYS: Spravochnik Pol'zovatelya*, DMK Press, Moscow, Russia, 2005.

[20] ANSYS Mechanical APDL, Documentation. Internet: http://www.ansys.com.

[21] A. Kamanbedast and A. Delvari, "Analysis of earth dam: Seepage and stability using ansys and geo-studio software," *World Applied Sciences Journal*, vol. 17, no. 9, pp. 1087–1094, 2012.

[22] L. Yuan and J. Lei, "The Analysis of the Seepage Characteristics of Tailing FLAC 3D Numerical Simulation," *The Open Civil Engineering Journal*, vol. 9, pp. 400–407, 2015.

[23] X. X. Zhao, B. L. Zhang, and Z. M. Wang, "Stability analysis of seepage flow through earth dam of huangbizhuang reservoir based on ANSYS/APDL," *Rock and Soil Mechanics*, 2005.

[24] N. A. Tsytovich, *Tsytovich Mekhanika merzlykh gruntov [Mechanics of frozen soils]*, High school, Moscow, Russia, 1973.

[25] V. I. Makarov, *Protivofil'tratsionnye mErzlotnye Zavesy v Gruntovykh Plotinakh Merzlogo Tipa. Stroitel'stvo i Ekspluatatsiya Gidrotekhnicheskikh Sooruzheniy v Zapadnoy Yakutii [Antifiltration Cryogenic Veils in Ground Dams of Frozen Type. Construction and Operation of Hydraulic Structures in Western Yakutia]*, Nauka, Novosibirsk, Russia, 1979.

20

System Identification Based Proxy Model of a Reservoir under Water Injection

Berihun M. Negash, Lemma D. Tufa, M. Ramasamy, and Mariyamni Bt. Awang

Universiti Teknologi Petronas, 32610 Seri Iskandar, Perak Darul Ridzuan, Malaysia

Correspondence should be addressed to Berihun M. Negash; bmamo.negash@utp.edu.my

Academic Editor: Feng Xiao

Simulation of numerical reservoir models with thousands and millions of grid blocks may consume a significant amount of time and effort, even when high performance processors are used. In cases where the simulation runs are required for sensitivity analysis, dynamic control, and optimization, the act needs to be repeated several times by continuously changing parameters. This makes it even more time-consuming. Currently, proxy models that are based on response surface are being used to lessen the time required for running simulations during sensitivity analysis and optimization. Proxy models are lighter mathematical models that run faster and perform in place of heavier models that require large computations. Nevertheless, to acquire data for modeling and validation and develop the proxy model itself, hundreds of simulation runs are required. In this paper, a system identification based proxy model that requires only a single simulation run and a properly designed excitation signal was proposed and evaluated using a benchmark case study. The results show that, with proper design of excitation signal and proper selection of model structure, system identification based proxy models are found to be practical and efficient alternatives for mimicking the performance of numerical reservoir models. The resulting proxy models have potential applications for dynamic well control and optimization.

1. Introduction

Reliable prediction is a natural requirement to the design of an optimal management strategy whereby important decisions on future investment, facility development, and expansion will be made [1]. Typically, in the petroleum industry, forecasting oil, gas, and water production rates is undertaken on a regular basis. Some of the factors that determine precision and accuracy of these forecasts include quality of geological and geophysical (G&G) data, quality and quantity of observed production rate and pressure data, and modeling procedures employed.

Decline curve analysis (DCA) and numerical reservoir simulation are classical methods that are frequently employed to forecast reservoir performance [2–6]. Each of these methods has strengths, limitations, and restrictions [7]. In practice, it is common to implement more than one method to reduce uncertainty and increase prediction accuracy. Decline curve analysis (DCA) is popular in the industry and is relatively a modest technique. It is one of the oldest methods, which was developed by Arps [2] and has been utilized even when there were no high-speed processors for handling large amounts of production and pressure data. The empirical Arps equations recognized for different curtailment trends are shown in (1), (2), and (3) [8]. The analysis is based on fitting one of these three empirical relations using measured production data.

For the exponential case ($b = 0$),

$$q = q_i e^{(-at)}. \tag{1}$$

For the hyperbolic case ($0 < b < 1$),

$$q = \frac{q_i}{(1 + ba_i t)^{1/b}}. \tag{2}$$

For the harmonic case ($b = 1$),

$$q = \frac{q_i}{(1 + a_i t)}, \tag{3}$$

where a is the nominal (instantaneous) decline rate, b is the decline exponent constant, q is the flow rate, and q_i is the initial flow rate

However, it is recognized that the application of DCA is limited to the early production stage of a well, where production is at steady state and the reservoir is under boundary-dominated flow. In addition, rock and fluid compressibility must be low and constant. In cases such as gas reservoir and solution gas derived oil production, where compressibility is higher, the decline is less severe and hence none of the three curtailment trends will fit. An improved DCA which considers reservoir parameters as part of the decline analysis was developed by Fetkovich [4] and is referred to as Fetkovich's type-curve analysis. Fetkovich showed that the value of b was an indication for efficiency of recovery, where zero was the lowest efficiency with small compressibility and 0.5 was the highest efficiency that is achievable from any known recovery mechanism with optimistic relative permeability. Analytical flow equations were applied to generate type curves for transient flow, and they were combined with the empirical decline curve equations documented by Arps [2]. However, Fetkovich's analysis has some notable limitations. It is applicable if the reservoir is homogeneous and has constant thickness, no-flow outer boundary, and constant flowing pressure at the wellbore. In addition, Fetkovich's type-curve analysis assumes a constant productivity index, which implies a straight-line relationship between pressure and flow rate. However, Evinger and Muskat [9] stated that for multiphase flow the relationship between pressure and flow rate is curved. Therefore, Fetkovich's type-curve analysis performs well if only the reservoir is saturated with a single-phase fluid of constant and small compressibility.

On the other hand, numerical reservoir simulation based on first principles provides a more accurate and robust solution to the task of forecasting reservoir performance. Reservoir simulation models are mathematical descriptions of an actual reservoir and are based on material balance, momentum balance, and constitutive equations such as Darcy's Law. Reservoir models are often used to analyze, optimize, and forecast both pressure and saturation dependent terms. Geological, geophysical, petrophysical, well log, core, and fluid data are required to construct reservoir simulation models. However, development of an accurate and representative numerical model is challenging and it involves utilization of a large number of uncertain data. This is because reservoirs are gigantic subsurface systems, which are both heterogeneous and anisotropic. Due to the relatively large size, samples of rock and fluid are collected from only a few locations within the boundary of the reservoir. Statistical techniques are then employed to populate and infer rock and fluid properties for the remaining portion. Moreover, contrary to direct measurement, most of the input data required to build a simulation model are inferred from physical and chemical measurements on rock and fluid samples taken at only few locations of the vicinity of the reservoir. For instance, fluid resistivity measurements are used to determine saturation of water within the reservoir and levels of natural gamma ray and spontaneous potential of a reservoir rock are used to determine the type of rock (lithology) and mineral content. Thus, reservoir property data used in modeling and forecasting are only best estimates and will not be exact representatives of the reservoir at large. Besides, the history-matching stage by which the inferred reservoir properties are adjusted until the numerical simulation model reasonably mimics the actual reservoir is a formidable task and has substantial limitations [3, 5].

History matching is the process of altering sensitive reservoir model parameters and developing a list of representative numerical models which are capable of reproducing observed production and pressure data [10, 11]. History matching is undertaken via manual alteration of sensitive parameters or automatically by using dedicated multiobjective optimization algorithms [12]. The former is time-consuming and requires expert knowledge and experience on the reservoir under study. In addition, manually adjusting reservoir parameters results in trial-and-error solutions; therefore, it is impossible to use all information; it is not suitable for complex reservoir models; it is time-consuming and it provides suboptimal results. A relatively new technique, which is referred to as assisted history matching (AHM), substantially reduces the time required. In AHM, global optimization algorithms help to reduce a misfit function between the observed and simulated production and pressure data. AHM techniques apply methods that allow for multiple realizations and are able to deal with the uncertainty in an optimal manner.

The Kalman Filter and its variants, such as the Ensemble Kalman Filter (EnKF) and Particle Filter, are currently popular among the research community [13–17]. Though AHM shortened the time taken during the history-matching phase, it is an ill-posed optimization problem, which is known to result in many realizations. Different sets of reservoir parameters may equally reproduce the historical data which results in a different prediction [6]. The so-called best realization could also produce a geologically inconsistent reservoir model [11]. The reader is advised to refer to [3] which has critically summarized the limitations of assisted history matching.

Once history matching is successfully completed, the models are used for production optimization and prediction of future reservoir performance under various development strategies. The process is repeated several times until a viable strategy is realized. For a medium sized reservoir model, each simulation run may require hours to complete. In spite of the demanding effort and time spent on model construction and history matching, the challenge encountered during optimization makes it difficult to exploit the full potential of numerical reservoir simulators. The use of proxy models that mimic the behavior of numerical reservoir models is currently on the rise and significantly eases the optimization stage. Once developed, these proxy models can be evaluated within seconds. Many commercial simulators that generate polynomial proxy models are being developed and used in both industry and academia. In all the simulators, response surface based proxy models are being employed. However, several runs are required to develop a representative proxy model that accurately mimics the response of numerical reservoir models. For instance, if there are ten factors (e.g., water injection wells) and three responses (oil, water, and gas production rates), then the Box-Behnken experiment design will require at least 170 runs. Again, if each run takes four hours to complete, then 680 computational hours are required to gather data for proxy modeling.

In this paper, we propose system identification based proxy models for optimization of a reservoir model. System identification is the art and science of model building that engages statistical methods to build mathematical models of dynamic systems from measured data. System identification involves proper design of excitation signal, model structure selection, parameter estimation, and validation of the identified model. By properly designing the excitation signal and by selection of an appropriate model structure, it is possible to obtain a representative proxy model using data from a single run of the numerical simulation model. System identification is extremely flexible in a sense that there are several model structures to capture linear as well as nonlinear relationships. In this study, a benchmark reservoir model originally developed to evaluate history-matching techniques is used to validate the efficacy of system identification based proxy models, which have the potential to be used for automatic control and production optimization. The proposed methodology assumes that the numbers of wells and operating conditions except for injection rates remain the same. In cases where infill drilling is required, it is necessary to redo the identification anew.

2. Materials and Methods

2.1. System Identification Based Proxy Models. System identification is a popular and steadily growing modeling approach. It is used to build analytical models of a dynamic system and forecast future behavior easily with less computational effort and time [18–20]. It is commonly used in engineering where prediction is part of an integrated technology such as model based control systems. System identification develops a uniquely structured mathematical model of a system through the application of certain model structure and conventional statistical methods. The number of researches and results reported related to the use of system identification in the oil and gas industry is increasingly growing. Aïfa [21] has presented applications of neural network based black-box model, to various problems that arise in the process of understanding the behavior of reservoirs. Sheremetov et al. [22] applied nonlinear autoregressive neural networks with exogenous input to forecast production rate of a naturally fractured reservoir. Olominu and Sulaimon [7] pointed out that autoregressive integrated moving average (ARIMA), which is a popular model structure in time series analysis, has a better efficacy than DCA. In addition, they showed that the model might be useful for short- to mid-long-term forecasting. However, ARIMA model structure does not consider external inputs, such as water injection. As such, the application is limited to primary recovery only. Elmabrouk et al. [23] studied the use of feedforward backpropagation artificial neural network in the prediction of oil production and employed real data from a Libyan oil field to test the hypothesis. The study proves the potential use of advanced empirical models for better prediction of oil production, which is essential for better reservoir management. However, the study does not incorporate the effect of external inputs.

Table 1: Polynomial system identification model structures.

	Name
$F(q) = D(q) = C(q) = A(q) = 1$	Frequency impulse response (FIR)
$F(q) = D(q) = C(q) = 1$	Autoregressive with external input (ARX)
$F(q) = D(q) = 1$	Autoregressive moving average with external input (ARMAX)
$D(q) = C(q) = A(q) = 1$	Output error (OE)
$A(q) = 1$	Box-Jenkins (BJ)

System identification is the process of developing mathematical models of a physical system from input and output data [18, 19]. In this study, the system is a reservoir under water injection, the inputs are either water injection rates or well bottomhole pressures, and outputs are production rates. During the lifetime of a reservoir, data such as wellhead flowing pressure, bottomhole flowing pressure, production rates, injection rate, gas/oil ratio, and water cut are recorded regularly to quantify production and evaluate field performance. Therefore, one can develop a dynamic model that assists in decision-making using system identification techniques and those readily available sets of data. In this study, linear polynomial model structures, which are among the many variations of system identification model structures, are used to build a proxy model, which can be used for production optimization. Equation (4) displays the general representation of a linear, dynamic time invariant system [24, 25]:

$$y(n) = q^{-k} G\left(q^{-k}, \theta\right) u(n) + H\left(q^{-k}, \theta\right) e(n), \tag{4}$$

where

$$G\left(q^{-1}, \theta\right) = \frac{B(q)}{F(q) A(q)}, \qquad H\left(q^{-1}, \theta\right) = \frac{B(q)}{F(q) A(q)}. \tag{5}$$

$A(q)$, $B(q)$, $C(q)$, $D(q)$, and $F(q)$ are polynomials with respect to the shift operator q^{-1}, where $q^{-k}u(k) = u(k-1)$. Depending on the nature of these polynomials, different structures can be formulated as shown in Table 1.

Noise integrated versions of the model structures listed in Table 1 are also available for system identification. These are special cases of general linear input-output models. They correspond to linear difference equations relating the input to the output under various noise assumptions. For ARX model structure, addition of noise integration leads to an "ARIX" model structure:

$$A(q) y(t) = B(q) u(t - nk) + \frac{C(q)}{1 - q^{-1}} e(t). \tag{6}$$

Similarly, the ARMAX and BJ structures can be enhanced to handle nonstationary disturbances and are referred to as ARIMAX and BJI, respectively.

TABLE 2: Groups and wells control data for water injection case.

Individual producer wells control					Individual injector wells control	
Minimum BHP	Maximum Q_L	Maximum WC	Maximum GOR	Minimum Q_O	Maximum BHP	Maximum Q_w
280	2000	0.9	200	20	350	7500

TABLE 3: Group wells control for water injection case.

Producers group control			Injectors group control
Maximum Q_L	Maximum Q_O	Maximum Q_w	Maximum Q_w
15500	15500	13950	65000

The development of the system identification based models comprises the following key procedures:

(a) Collection of input and output data from the physical model

(b) Choice of model structure that represents the physical system

(c) Estimating unknown model parameters in the selected model structure

(d) Validating the estimated model using cross validation and/or residual analysis

A detailed discussion on the steps of system identification and model development is available by Ljung [24] and Nelles [25].

2.2. Excitation Input Selection. Selection of a proper excitation input dictates efficacy in system identification. There are several input sequences, which can be used in modeling different complexities. Among them, a pseudorandom binary sequence (PRBS) is very popular and efficient in equally exciting all frequencies. PRBS is a sequence that switches between an upper and a lower bound at varying frequencies. PRBS can also have multiple amplitudes; that is, different upper and lower bounds may be specified at different times for a given input sequence. However, such PRBS inputs are often employed during nonlinear system identification. Since this article studies the application of linear polynomial model structures, the excitation signal used was PRBS that only switches between two amplitudes.

2.3. Model Validation. Model validation constitutes quantifying the deviation exhibited by the estimated model and making sure that all the information contained in the input and output data is exploited. In this article, percentage fit, whiteness test, and crossplot test were used to evaluate the efficacy of the identified model. The mathematical expression of percentage fit is shown in

$$\begin{aligned}\text{percentage fit} &= 1 - \text{NRMSE} \\ &= 1 - \frac{\|y - \hat{y}\|}{\|y - y_{\text{mean}}\|} * 100\%,\end{aligned} \tag{7}$$

where $\hat{y}$ is the predicted output, y is the observed output, y_{mean} is the mean of the observed output, and NRMSE is the normalized root mean square error.

Whiteness test of a residual indicates whether the model has extracted all the information or requires further refining. This is achieved by plotting a histogram of residuals. A good model is one that has a mean value of zero and some variance, indicating that the residuals are uncorrelated. On the other hand, a crossplot analysis is a graphical and visual method of comparing two sets of data whereby the plot of simulated output and validation data was compared graphically and visually, with a 45-degree straight line. The crossplot of a residual is an x-y plot obtained by plotting the simulated output on the x-axis and validation data on the y-axis. A good model will produce an output that gives a perfectly 45-degree straight line when plotted against the validation data set.

2.4. Description of the Case Study Reservoir Model. The reservoir model, which was used to generate the input and output data, is referred to as UNISIM-I-M and is based on a field located in Campos basin in Brazil [27]. The model was originally developed to serve as a benchmark case study on decision analysis regarding reservoir management. For a detailed description of the model, readers are referred to Avansi and Schiozer [26] and Gaspar et al. [27]. Figure 1 shows the porosity distribution while Figure 2 presents the reservoir geometry including the wells placement and vertical permeability distribution of the UNISIM-I-M reservoir model. All the red wells represent producer wells while blue wells represent injector wells. The reservoir volume is meshed into smaller 3D blocks, referred to as grid blocks. Each grid block has a thickness of 50 m, a width of 100 m, and a length of 100 m. There are a total of 93960 such active grid blocks with 81, 58, and 20 blocks arranged in the x, y, and z directions, respectively. The water-oil and gas-oil contacts are located at a depth of 3200 m and 3000 m, respectively. Initial oil originally in place is calculated to be 175000 m^3. The initial reservoir pressure was 300 bar. The total numbers of water injectors and producers are 11 and 14, respectively. The original well placement was slightly modified to optimize production; however, production constraints are kept the same as the original. Production and injection well constraints are described in Tables 2 and 3.

3. Result and Discussion

In developing a proxy model, the inputs consisted of water injection rate from individual wells and the output consisted

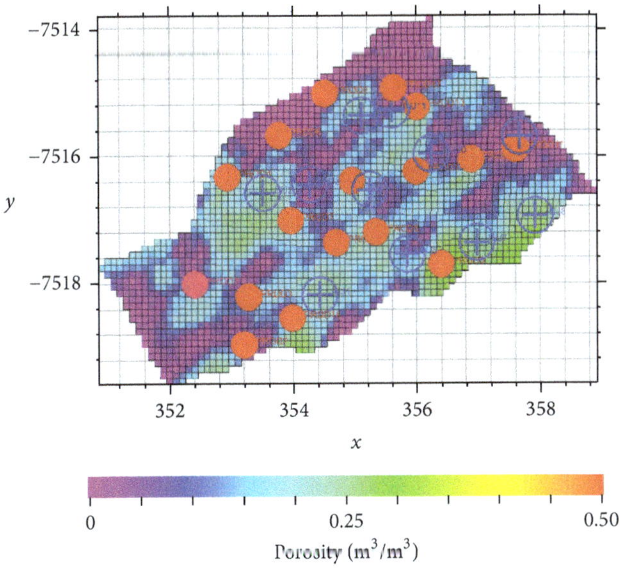

FIGURE 1: Porosity distribution of the first layer of the UNISIM-I-M reservoir model.

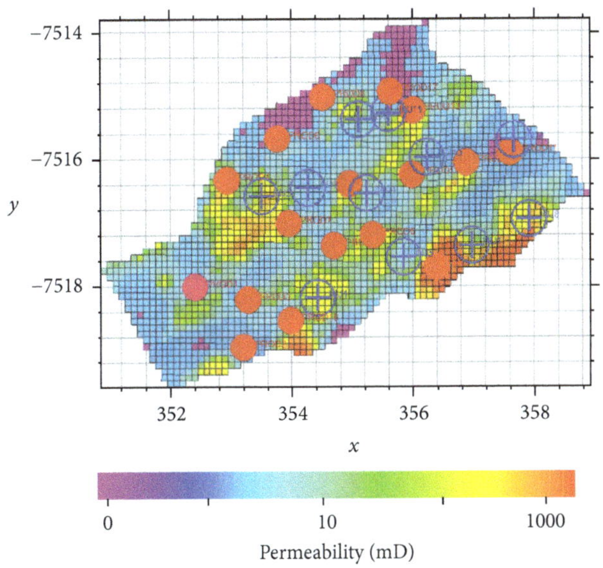

FIGURE 2: Vertical permeability distribution of the first layer of the UNISIM-I-M reservoir model.

of the total oil production rate. The design of a PRBS input was in such a way that each of the multiple inputs has different lower and upper bounds. The lower and upper bounds used to excite the numerical simulation model are shown in Table 4.

The response of the numerical reservoir model is presented in Figure 3. The figure shows that the total number of input and output data is 2807, of which 1500 were used for identification and the next 10 data points were removed to fulfill requirements set by the initial condition of prediction equations. The remaining 1297 data points were utilized for cross validation. Each of the 2807 data points presents the input and output calculated on a daily basis. The numerical simulation run time required to generate the outputs was 4 hours and 53 minutes on Intel® Core™ i5-3210 M CPU @2.5 GHz Processor with 6 GB RAM. This is considered very large compared to the few seconds required to run polynomial models, which are examined in this study. In addition, when carrying out multiple simulation runs to evaluate reservoir performance, the essence of utilizing reduced model or metamodel as a substitute for numerical reservoir simulator becomes extremely substantial.

TABLE 4: Lower and upper bounds of input design for modeling fluid production during water injection.

Well name	Lower bound (Sm^3/day)	Upper bound (Sm^3/day)
Well 1	1500	4500
Well 2	1750	4750
Well 3	2000	5000
Well 4	2250	5250
Well 5	2500	5500
Well 6	2750	5750
Well 7	3000	6000
Well 8	3250	6250
Well 9	3500	6500
Well 10	3750	6750
Well 11	4000	7000

3.1. Initial Model Estimation. A simple and yet useful system identification model structure, that is, ARX, and its noise integrated version were used to initiate the system identification process. This allows narrowing the model-order search region and determining time delay. Noise integrated ARX model (ARIX) with both poles and zeros equal to three and a time delay of zero was found to result in a percentage fit of 71.92%. The model was named ARIX (3-3-0). Model structure of ARIX model is

$$A(q)\,y(t) = B(q)\,u(t) + \frac{1}{(1 - q^{-1})} e(t), \qquad (8)$$

where $A(q)$ is a polynomial in the shift operator (q) of order na, $B(q)$ is a polynomial in the shift operator (q) of order nb, $y(t)$ is the output (oil production rate), $u(t)$ is the input (injection rates), na is the number of poles, and nb is the number of zeros.

Figure 4 presents the cross validation result of the ARIX (3-3-0) model. Akaike's final prediction error (FPE) and mean squared error (MSE) exhibited were 0.8157 and 0.8724. The MSE is a measure of the quality of an estimator; it is always nonnegative, and values closer to zero are better. Moreover, according to Akaike's theory, the most accurate model has the smallest FPE. The step response, which illustrates the time behavior of the outputs of ARX (3-3-0) when its inputs change from zero to one in a very short time, is shown in Figure 5. It can be seen that the model has responded without a time delay and has become stable.

3.2. Identification of a More Complex and More Accurate Model. Based on the analysis of the ARIX (3-3-0) model, more complex model structures were examined by

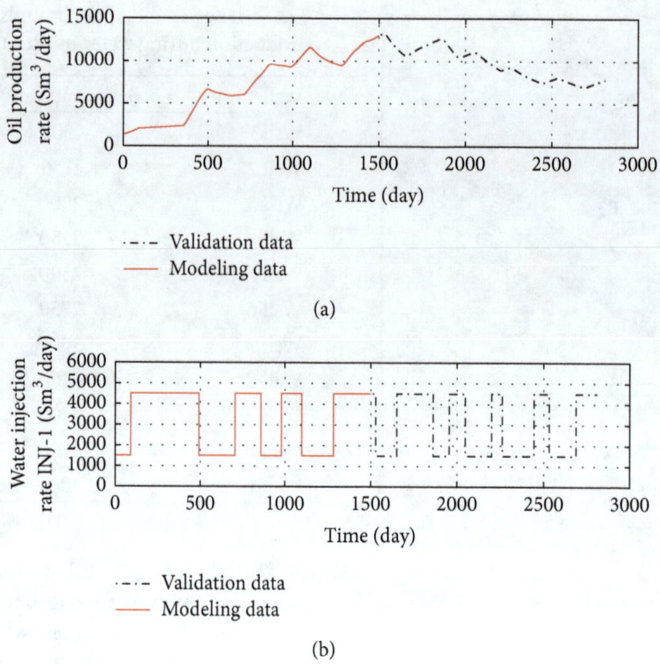

FIGURE 3: Oil production rate (a) and water injection rate of well INJ-1 (b), used during modeling and validation of linear time invariant model structures.

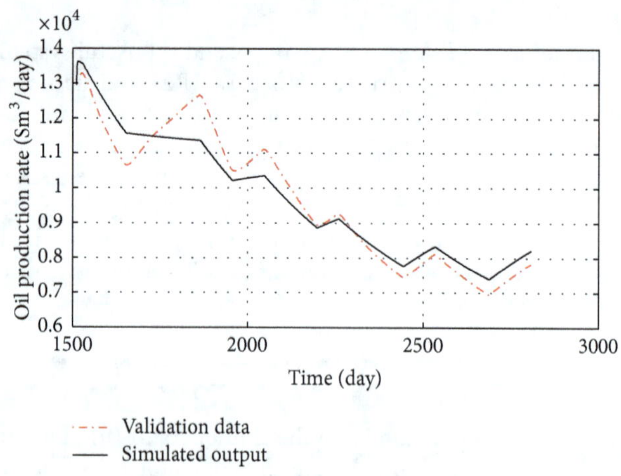

FIGURE 4: Validation of the ARIX (3-3-0) model for MISO system.

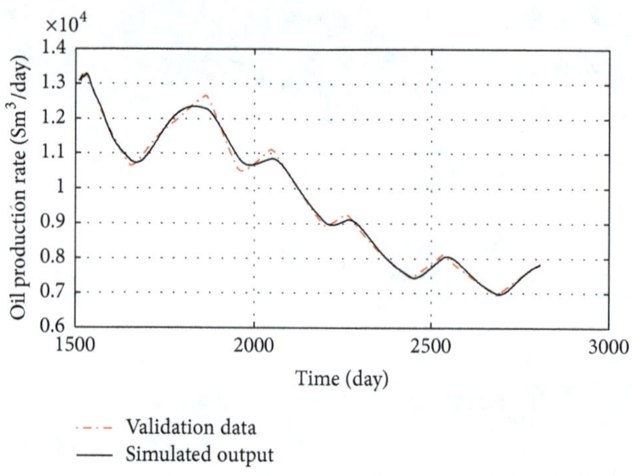

FIGURE 6: Comparison of validation data and BJI (4-4-4-4-2) model's simulated output.

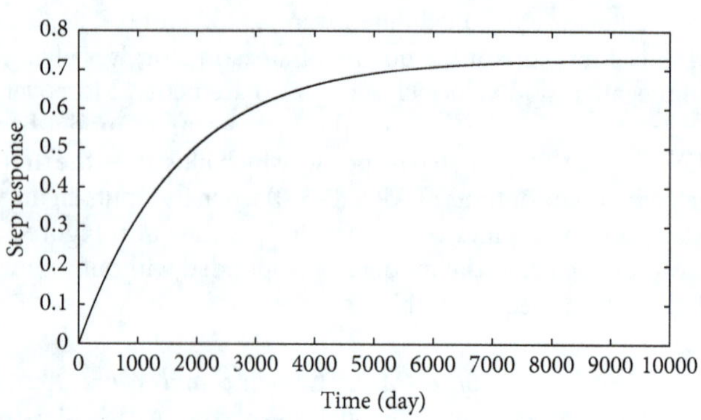

FIGURE 5: Step response of ARIX (3-3-0) model.

introducing a variation in model order and time delay close to the ARIX (3-3-0) model. The best model that resulted in the best percentage fit with the validation data set was found to be Box-Jenkins model with noise integration (BJI). The structure of the BGI model is

$$y(t) = \left[\frac{B(q)}{F(q)}\right] u(t) + \left[\frac{C(q)}{D(q)(1-q^{-1})}\right] e(t). \quad (9)$$

The number of poles and zeros of the deterministic part was found to be four for all the inputs. The stochastic part of the BJI model structure that gave the best performance was also of order four in both poles and zeros. The time delay is found to be two. The model can be referred to as BJI (4-4-4-4-2). Figure 6 presents a graphical comparison of simulated output and measured validation data. It can be seen that the simulated output has closely reproduced the validation data. The percentage fit, FPE, and MSE exhibited were 94.01%, 0.207, and 0.5632, respectively. The estimated polynomial functions of the BJI (4-4-4-4-2) model are

$$
\begin{aligned}
B1(q) &= 0.0003314q^{-2} - 0.0009479q^{-3} \\
&\quad - 0.001406q^{-4} - 0.001088q^{-5} \\
B2(q) &= -0.0003488q^{-2} - 1.278e-05q^{-3} \\
&\quad - 0.00049q^{-4} - 0.0001398q^{-5} \\
B3(q) &= -0.0003505q^{-2} - 2.031e-05q^{-3} \\
&\quad + 0.0001999q^{-4} + 0.000143q^{-5} \\
B4(q) &= 0.003034q^{-2} + 0.001483q^{-3} + 0.001071q^{-4} \\
&\quad + 0.0009769q^{-5} \\
B5(q) &= -0.0001085q^{-2} + 0.000201q^{-3} \\
&\quad + 0.0003909q^{-4} + 0.0003442q^{-5} \\
B6(q) &= -0.001266q^{-2} - 0.000845q^{-3} \\
&\quad - 0.0005904q^{-4} - 0.0008796q^{-5} \\
B7(q) &= -0.000173q^{-2} + 8.793e-05q^{-3} \\
&\quad + 0.0002124q^{-4} + 0.0001657q^{-5} \\
B8(q) &= -0.0001733q^{-2} + 8.761e-05q^{-3} \\
&\quad + 0.0002121q^{-4} + 0.0001654q^{-5} \\
B9(q) &= -0.0001734q^{-2} + 8.75e-05q^{-3} \\
&\quad + 0.000212q^{-4} + 0.0001653q^{-5} \\
B10(q) &= -0.0000.3q^{-2} + 1.986e-05q^{-3} \\
&\quad + 0.0001798q^{-4} + 0.0001357q^{-5} \\
B11(q) &= -0.0002676q^{-2} + 2.000e-05q^{-3} \\
&\quad + 0.0002q^{-4} + 0.0001157q^{-5} \\
C(q) &= 1 - 0.01847q^{-1} - 0.2401q^{-2} - 0.1495q^{-3} \\
&\quad - 0.03079q^{-4} \\
D(q) &= 1 - 1.85q^{-1} + 1.009q^{-2} - 0.2766q^{-3} \\
&\quad + 0.1215q^{-4} \\
F1(q) &= 1 - 1.937q^{-1} + 1.131q^{-2} - 0.3102q^{-3} \\
&\quad + 0.1171q^{-4} \\
F2(q) &= 1 - 1.936q^{-1} + 1.131q^{-2} - 0.3103q^{-3} \\
&\quad + 0.1166q^{-4} \\
F3(q) &= 1 - 1.936q^{-1} + 1.131q^{-2} - 0.3097q^{-3} \\
&\quad + 0.1175q^{-4} \\
F4(q) &= 1 - 1.937q^{-1} + 1.131q^{-2} - 0.3097q^{-3} \\
&\quad + 0.1177q^{-4} \\
F5(q) &= 1 - 1.936q^{-1} + 1.131q^{-2} - 0.3098q^{-3} \\
&\quad + 0.1174q^{-4} \\
F6(q) &= 1 - 1.936q^{-1} + 1.131q^{-2} - 0.3097q^{-3} \\
&\quad + 0.1176q^{-4} \\
F7(q) &= 1 - 1.937q^{-1} + 1.123q^{-2} - 0.3103q^{-3} \\
&\quad + 0.117q^{-4} \\
F8(q) &= 1 - 1.937q^{-1} + 1.123q^{-2} - 0.3103q^{-3} \\
&\quad + 0.117q^{-4} \\
F9(q) &= 1 - 1.937q^{-1} + 1.113q^{-2} - 0.3103q^{-3} \\
&\quad + 0.117q^{-4} \\
F10(q) &= 1 - 1.937q^{-1} + 1.113q^{-2} - 0.3101q^{-3} \\
&\quad + 0.1172q^{-4} \\
F11(q) &= 1 - 1.922q^{-1} + 1.123q^{-2} - 0.3101q^{-3} \\
&\quad + 0.1165q^{-4}.
\end{aligned}
\tag{10}
$$

In addition, the step response of the identified model is shown in Figure 7. It can be observed that the response of model BJI (4-4-4-4-2) is stable after an abrupt change in input. Knowing how the system responds to a sudden input is important because large and possibly fast deviations from the long-term steady state may have extreme effects on the prediction ability of the model. This will in turn influence decisions.

3.3. Residual Analysis. Two types of residual analysis were carried out: a whiteness test and crossplot analysis. The whiteness test was undertaken by plotting a histogram of the residuals, which were calculated by differencing simulated output and validation data. Figure 8 presents the histogram of the residuals. The mean and sample variance of the residual distribution are 0.0873 and 1272, respectively. According to the whiteness test, a good model that has exploited most of the information, which is inherently contained in the data, will have a white noise. Since the mean is almost zero, it can be considered that the model has captured most of the information in the input and output data.

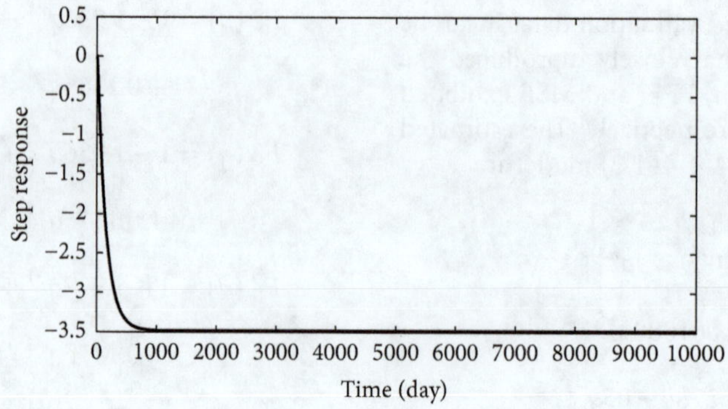

FIGURE 7: Step response of model BJI (4-4-4-4-2).

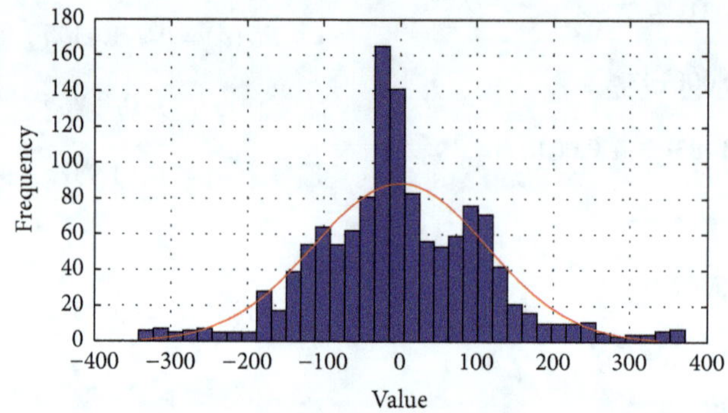

FIGURE 8: Histogram of residuals for model BJI (4-4-4-4-2).

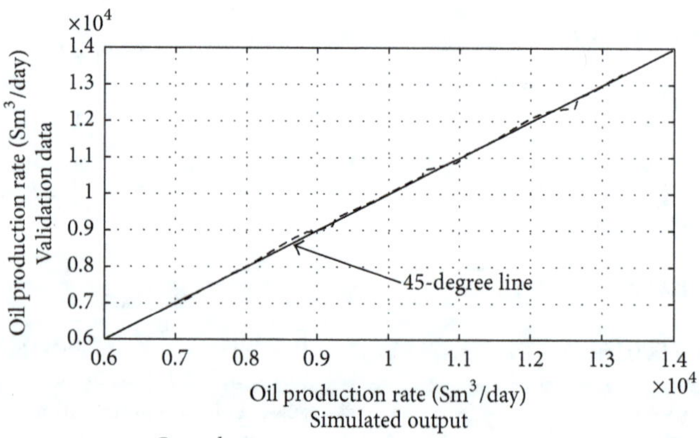

FIGURE 9: Crossplot of simulated output and validation data for model BJ (4-4-4-4-2).

In addition, a crossplot of simulated output versus validation data was plotted to allow further analysis on the identified model. The data set was concatenated and sorted in descending order based on the simulated output before it was plotted on x-y plane. A good model will have its crossplot aligned with the 45-degree line. A model that underestimates the output will have its crossplot below the 45-degree line and a model that overestimates the output will have its crossplot above the 45-degree line. Figure 9 presents a crossplot of model BJI (4-4-4-4-2). It can be seen that the model neither underestimates nor overestimates during the validation period. In other words, the prediction ability of the identified model is satisfactory.

4. Summary and Conclusion

Numerical reservoir modeling and simulation has many applications in the upstream oil and gas industry. However, due to the large computational time required, full exploitation of its potential use remains challenging. Currently, the use of response surface based proxy models is becoming a common trend. Nowadays, several simulation software packages incorporate a module for proxy modeling. While such proxy models are useful and significantly reduce optimization time, their development still requires several simulation runs. In this study, a system identification based proxy model that only requires a single run and properly designed excitation signal was proposed. A case study that involves water injection was used to investigate and establish a procedure for application of system identification techniques in reservoir simulation. Injection flow rates were designed in the form of a pseudorandom binary sequence that is known to excite all frequencies equally and allow successful identification. Input and output data were collected at a frequency of 1/day and were divided into modeling and validation sets. The modeling set was used to construct a proxy model. The validation set was used for cross validation and residual analysis. Residual analysis and crossplot test together with the calculation of percentage fit have shown that the identified model was able to accurately mimic the performance of the numerical reservoir simulator while predicting total oil production during oil recovery by water injection. In conclusion, with proper selection of input sequence and model structure, system identification can deliver a proxy model with just one simulation run.

Nomenclature

Q_L: Liquid flow rate
Q_O: Oil flow rate
DCA: Decline curve analysis
Q_W: Water flow rate
WC: Water cut
BHP: Bottomhole pressure
GOR: Gas-to-oil ratio
NRMSE: Normalized root mean square error
PRBS: Pseudorandom binary sequence
MISO: Multiple input-single output.

Competing Interests

The authors declare that there are no competing interests regarding the publication of this paper.

References

[1] P. Obidike, C. Emelle, and R. Vlaardingerbroek, "Integrated production forecasting for business planning," in *Proceedings of the 38th Nigeria Annual International Conference and Exhibition*, Nigeria, August 2014.

[2] J. Arps, "Analysis of decline curves," *Transactions of the AIME*, vol. 160, no. 1, pp. 228–247, 1945.

[3] M. Cancelliere, F. Verga, and D. Viberti, "Benefits and limitations of assisted history matching," Offshore Europe, 2011.

[4] M. Fetkovich, "Decline curve analysis using type curves," *Journal of Petroleum Technology*, vol. 32, pp. 1065–1077, 1980.

[5] Z. Tavassoli, J. N. Carter, and P. R. King, "Errors in history matching," *SPE Journal*, vol. 9, no. 3, pp. 352–361, 2004.

[6] Y. Tomomi, "Non-uniqueness of history matching," in *Proceedings of the SPE Asia Pacific Conference on Integrated Modelling for Asset Management*, Society of Petroleum Engineers, 2000.

[7] O. Olominu and A. A. Sulaimon, "Application of time series analysis to predict reservoir production performance," in *Proceedings of the SPE Nigeria Annual International Conference and Exhibition*, Society of Petroleum Engineers, Dallas, Tex, USA, 2014.

[8] H. C. Slider, *Worldwide Practical Petroleum Reservoir Engineering Methods*, PennWell Books, 1983.

[9] H. Evinger and M. Muskat, "Calculation of theoretical productivity factor," *Transactions of the AIME*, vol. 146, no. 1, pp. 126–139, 1942.

[10] H. B. Crichlow, *Modern Reservoir Engineering: A Simulation Approach*, Prentice Hall, 1977.

[11] D. S. Oliver, A. C. Reynolds, and N. Liu, *Inverse Theory for Petroleum Reservoir Characterization and History Matching*, Cambridge University Press, 2008.

[12] R. Rwechungura, M. Dadashpour, and J. Kleppe, "Advanced history matching techniques reviewed," in *Proceedings of the 17th Middle East Oil and Gas Show and Conference (MEOS '11)*, pp. 1729–1747, Manama, Bahrain, September 2011.

[13] A. A. Emerick and A. C. Reynolds, "Combining the ensemble Kalman filter with Markov Chain Monte Carlo for improved history matching and uncertainty characterization," *SPE Journal*, vol. 17, no. 2, pp. 418–440, 2012.

[14] J. Ping and D. Zhang, "History matching of channelized reservoirs with vector-based level-set parameterization," *SPE Journal*, vol. 19, no. 3, pp. 514–529, 2014.

[15] Y. Chen and D. S. Oliver, "History matching of the norne full-field model with an iterative ensemble smoother," *SPE Reservoir Evaluation & Engineering*, vol. 17, no. 2, pp. 244–256, 2014.

[16] A. A. Emerick and A. C. Reynolds, "History matching time-lapse seismic data using the ensemble Kalman filter with multiple data assimilations," *Computational Geosciences*, vol. 16, no. 3, pp. 639–659, 2012.

[17] M. Tarrahi, E. Gildin, J. Moreno, and S. Gonzales, "Dynamic integration of DTS data for hydraulically fractured reservoir characterization with the ensemble Kalman filter," in *Proceedings of the SPE Energy Resources Conference*, pp. 822–835, June 2014.

[18] B. George, *Time Series Analysis: Forecasting & Control*, Pearson Education, Uttar Pradesh, India, 3rd edition, 1994.

[19] M. Nerlove, D. M. Grether, and J. L. Carvalho, *Analysis of Economic Time Series: A Synthesis*, Academic Press, 2014.

[20] A. Zellner and F. Palm, "Time series analysis and simultaneous equation econometric models," *Journal of Econometrics*, vol. 2, no. 1, pp. 17–54, 1974.

[21] T. Aïfa, "Neural network applications to reservoirs: physics-based models and data models," *Journal of Petroleum Science and Engineering*, vol. 123, pp. 1–6, 2014.

[22] L. Sheremetov, A. Cosultchi, J. Martínez-Muñoz, A. Gonzalez-Sánchez, and M. A. Jiménez-Aquino, "Data-driven forecasting of naturally fractured reservoirs based on nonlinear autoregressive neural networks with exogenous input," *Journal of Petroleum Science and Engineering*, vol. 123, pp. 106–119, 2014.

[23] S. Elmabrouk, E. Shirif, and R. Mayorga, "Artificial neural network modeling for the prediction of oil production," *Petroleum Science and Technology*, vol. 32, no. 9, pp. 1123–1130, 2014.

[24] L. Ljung, *System Identification: Theory for the User*, PTR Prentice Hall Information and System Sciences Series, Prentice Hall, Englewood Cliffs, NJ, USA, 1987.

[25] O. Nelles, *Nonlinear System Identification: From Classical Approaches to Neural Networks and Fuzzy Models*, Springer, Berlin, Germany, 2001.

21

Numerical Investigation on Vortex Shedding from a Hydrofoil with a Beveled Trailing Edge

Seung-Jae Lee,[1] Jun-Hyeok Lee,[2] and Jung-Chun Suh[1,2]

[1] *Research Institute of Marine Systems Engineering, Seoul National University, Seoul 08826, Republic of Korea*
[2] *Department of Naval Architecture and Ocean Engineering, Seoul National University, Seoul 08826, Republic of Korea*

Correspondence should be addressed to Seung-Jae Lee; sj38.lee@gmail.com

Academic Editor: Jean-Michel Bergheau

To better understand the vortex shedding mechanism and to assess the capability of our numerical methodology, we conducted numerical investigations of vortex shedding from truncated and oblique trailing edges of a modified NACA 0009 hydrofoil. The hybrid particle-mesh method and the vorticity-based subgrid scale model were employed to simulate these turbulent wake flows. The hybrid particle-mesh method combines the vortex-in-cell and the penalization methods. We have implemented numerical schemes to more efficiently use available computational resources. In this study, we numerically investigated vortex shedding from various beveled trailing edges at a Reynolds number of 10^6. We then compared the numerical results with the experimental data, which show good agreement. We also conducted numerical simulations of wakes behind the hydrofoil at rest in periodically varying flows. Results reveal that vortex shedding is affected by the periodicity of a free-stream flow, as well as the trailing-edge shape.

1. Introduction

The vortex shedding phenomenon is encountered in many practical engineering applications and physical sciences, and it is an important characteristic of flows past a bluff body. A vortex sheet shed from a solid body consists of alternating vortices of strength Γ that produce their own velocity field, superimposed on the free-stream velocity. These shedding vortices cause an oscillating force component perpendicular to the direction of flow. This force can induce a vibration in the body. Vortex shedding behind circular and square cylinders is a benchmark problem that has been well investigated and is addressed in a vast amount of literature. Several scholarly reviews [1–4] are entirely devoted to the state of the art of this problem. In contrast, vortex shedding from a streamlined body such as a hydrofoil has been studied to a much lesser extent despite its being of direct relevance to practical engineering problems in hydraulic turbines, pumps, and marine propellers. Propeller-singing is well known as a critical vibration phenomenon generated by the interaction between a Kármán vortex-shedding mechanism from the trailing-edge of a blade and its natural frequencies [5].

Some laboratory experiments have been conducted on vortex shedding from the trailing edge of a hydrofoil. Bourgoyne et al. [6, 7] experimentally investigated the dominant features of flows over a two-dimensional (2D) hydrofoil with an antisinging trailing edge at Reynolds numbers ranging from 10^6 to 5×10^7. They concluded that turbulent fluctuations in the near wake are Reynolds-number dependent because of the varying strengths of the structured near-wake vortex shedding. Mulvany et al. [8] numerically simulated turbulent flows around a modified NACA 16 hydrofoil using four turbulent models available in the commercial computational fluid dynamics (CFD) code FLUENT. They compared their numerical results with those from the experimental data of Bourgoyne et al. [6]. The authors mainly discussed the capabilities of the four different models and did not report any vortex shedding phenomenon. Ausoni et al. [9] performed laboratory-scale experiments to investigate the von Kármán vortex street generated in the wake of a 2D hydrofoil with a truncated trailing edge. Their experiments were performed at a zero angle of attack for a Reynolds number ranging from 5.0×10^5 to 2.0×10^6, based on the hydrofoil chord length [9, 10]. Their experimental results showed that a lock-in phenomenon occurs, where the vortex shedding frequency is locked onto the natural frequency of a hydrofoil. Interestingly, it has been reported that cavitation has a minor effect on the wake dynamic. Recently, Zobeiri et al. [11] investigated two NACA 0009 hydrofoils with blunt and oblique trailing edges, respectively, for Reynolds numbers ranging from 5.0×10^5 to 2.9×10^6 and conducted high-speed visualization and flow-induced vibration measurements. They confirmed experimentally that flow-induced vibration was significantly reduced with an oblique trailing edge compared with a truncated edge. They also concluded that the collision between upper and lower vortices and the resulting vorticity redistribution were the main reasons for the obtained vibration reduction with the oblique trailing edge.

This study numerically investigated vortex shedding from truncated and oblique trailing edges of a hydrofoil to better understand the vortex shedding mechanism and to assess the capability of our numerical methodology. We employed the hybrid particle-mesh method and the vorticity-based subgrid scale model to undertake a numerical flow simulation. The hybrid particle-mesh method is a combination of the vortex-in-cell (VIC) method (see [12–15] and references therein) and the penalization technique [16], which has been developed in recent years (see [17–24] for a review). The hybrid particle-mesh method enables the use of fast and efficient techniques for computing differential operators, thereby enabling large scale simulations. However, techniques to save computational memory and CPU time consumption are still required. Thus, we have introduced the implementations of numerical schemes to more efficiently use available computational resources [25, 26]. With these numerical schemes, we numerically investigate vortex shedding from various beveled trailing edges of a NACA 0009 hydrofoil at a Reynolds number of 10^6. We then compare our numerical results with experimental data [10, 11] to assess the capability of our numerical methodology. In addition, we investigate the influence of periodically varying free-stream flows on vortex shedding to better understand the vortex shedding mechanism.

The organization of the remainder of this paper is as follows: we present a brief description of the numerical methods used for the vortex shedding simulation in Section 2. We describe our implementation of numerical schemes to efficiently use available computational resources in Section 3 and the computational procedure followed in Section 4. We present out vortex shedding results from a variety of beveled trailing edges and discuss the influence of periodic free-stream flows in Section 5. In Section 6, we provide a summary of this study and some perspectives for future work.

2. Governing Equations and Numerical Methods

The vorticity-velocity formulation of the Navier-Stokes equations allows a purely kinematical problem to be decoupled from the pressure term, which is eliminated by applying the curl operator. Pressure which can be evaluated in an explicit manner using the identified vorticity and velocity fields [27] is not part of the solution algorithm. For a 2D flow parallel to the xy-plane, the vorticity transport equation in terms of the vorticity can be expressed as

$$\frac{\partial \omega_z}{\partial t} + \mathbf{u} \cdot \nabla \omega_z = \nu \nabla^2 \omega_z, \tag{1}$$

where ω_z is the scalar plane component of the vorticity vector; that is, $\boldsymbol{\omega} \equiv \omega_z$ in two dimensions. The evolution of a flow is considered in discrete time steps. The viscous splitting algorithm can be expressed in a Lagrangian framework as

$$\frac{D\mathbf{x}}{Dt} = \mathbf{u} \tag{2a}$$

$$\frac{D\omega_z}{Dt} = \nu \nabla^2 \omega_z, \tag{2b}$$

where $D/Dt = \partial/\partial t + \mathbf{u} \cdot \nabla$ is the material derivative. N_p discrete Lagrangian fluid particles with a finite core size ϵ are linearly superposed to approximate the vorticity field as

$$\omega_z(\mathbf{x}, t) = \sum_{p=1}^{N_p} \Gamma_p \zeta_\epsilon (\mathbf{x} - \mathbf{x}_p), \tag{3}$$

where ζ_ϵ is a mollification of the Dirac-delta function [12]. Finite core size vortices are used instead of point vortices. Each particle is characterized by its position $\mathbf{x}_p$ and its strength Γ_p. For a given vortex particle, the circulation Γ_p is identical to the product of the vorticity and the area of the vortex particle ϵ^2 which also represents the contribution of the vortex particle to the vorticity field, $\Gamma_p = \int \omega_z dS \approx \omega_z \epsilon^2$. The vorticity-carrying fluid particle is advanced with the velocity $\mathbf{u}$ and is gradually diffused because of viscous effects.

2.1. Vortex-in-Cell (VIC) Method for Convection. Since the continuity equation and the definition of vorticity lead to $\nabla \cdot \mathbf{u} = 0$ and $\nabla \times \mathbf{u} = \boldsymbol{\omega}$, the velocity $\mathbf{u}$ of (2a) can be expressed as the gradients of the stream-function $\boldsymbol{\psi}$. The velocity vector is expressed as

$$\mathbf{u} = \mathbf{U}_\infty + \mathbf{u}_\omega = \mathbf{U}_\infty + \nabla \times \boldsymbol{\psi}, \tag{4}$$

where $\mathbf{U}_\infty$ is the free-stream velocity and $\mathbf{u}_\omega$ is the rotational velocity. The Poisson equation for the stream-function is $\nabla^2 \boldsymbol{\psi} = -\boldsymbol{\omega}$. In two dimensions, the stream function is scalar, ψ_z, and only the vorticity component perpendicular to the plane is nonzero $\boldsymbol{\omega} = \omega_z e_k$. The vector Poisson equation reduces to a scalar Poisson equation:

$$\nabla^2 \psi_z = -\omega_z. \tag{5}$$

Rotational velocity is defined by $u_x = \partial \psi_z / \partial y$ and $u_y = -\partial \psi_z / \partial x$. In the VIC method, stream functions to evaluate velocities are computed using a uniform Cartesian mesh. The fast Fourier transform- (FFT-) based Poisson solver reduces the computational cost of obtaining the velocity field to $\mathcal{O}(N \log N)$ where N is the number of grid points [12]. To compute the stream functions using the FFT-based Poisson solver, the vorticity ω_z is interpolated onto equally spaced Cartesian grid using

$$\omega_z(\mathbf{x}_g) = \sum_{p=1}^{N_p} \omega_z(\mathbf{x}_p) W\left(\frac{\mathbf{x}_p - \mathbf{x}_g}{h}\right), \tag{6}$$

where h is the mesh spacing, using the following third order interpolation kernel [28, 29]:

$$W(x) = M_4'(x) = \begin{cases} 0 & \text{for } |x| > 2 \\ \frac{1}{2}(2-|x|)^2(1-|x|) & \text{for } 1 < |x| \le 2 \\ 1 - \frac{5}{2}|x|^2 + \frac{3}{2}|x|^3 & \text{for } |x| \le 1 \end{cases} \tag{7}$$

in each coordinate direction. This kernel conserves the 0th, 1st, and 2nd order moments. The subscripts $_g$ and $_p$ denote the grid and particle quantities, respectively. A vortex particle contributes to the nearest 16 nodes through the M_4' scheme, and the total vorticity at each node is obtained by summing the vorticity contributions of all the vortex particles.

Boundary conditions for the stream function are required to solve the Poisson equation. If the computational domain boundaries are far enough from the particles, homogeneous Dirichlet boundary conditions ($\psi = 0$) may be used. However, using a larger domain is inefficient since it requires a regular grid with too many points. To compactly place a square FFT-domain Ω, we used nonhomogeneous Dirichlet boundary conditions in this study. Unknown stream functions at the domain boundaries ($\partial\Omega$) can be directly obtained from Green's function solution [27], whereby

$$\psi_z(\mathbf{x}_b) = -\frac{1}{2\pi}\sum_{p=1}^{N_p} \Gamma_p \ln\left(|\mathbf{x}_b - \mathbf{x}_p|\right) \quad b = 1, \ldots, M, \tag{8}$$

where $\mathbf{x}_b \in \partial\Omega$, $\mathbf{x}_p \in \Omega$, and $\mathbf{x}_g \neq \mathbf{x}_p$. M denotes the number of domain boundary points and the circulation strength of each vortex particle $\Gamma_p = \omega_z(\mathbf{x}_p)\epsilon^2$. We note here that all particle vorticity values should be considered. Fast evaluation of the stream functions at the domain boundaries is further discussed in Section 3.3.

2.2. Brinkman Penalization Method for Diffusion. The penalization method was initially designed to take into account solid obstacles in fluid flows [16]. The main point of the penalty term is to replace the usual vorticity creation algorithm for enforcing the no-slip condition for a solid body. By adding the penalization term to (1), the vorticity transport equation becomes

$$\frac{D\omega_z}{Dt} = \nu\nabla^2\omega_z + \nabla \times [\lambda\chi(\mathbf{u}_s - \mathbf{u})], \tag{9}$$

where $\mathbf{u}_s$ is the velocity of the solid body and χ denotes a mask function that yields 0 in a fluid and 1 in a solid [16, 18]. λ indicates the penalization parameter with the dimension $[s^{-1}]$. To avoid too small time step, the penalization term is evaluated by the implicit expression [16, 18, 30]:

$$\frac{\Delta\omega_z}{\Delta t} = \nabla \times \left[\lambda\chi\left(\mathbf{u}_s^n - \tilde{\mathbf{u}}^{n+1}\right)\right], \tag{10}$$

where

$$\tilde{\mathbf{u}}^{n+1} = \frac{\mathbf{u}^n + \lambda\Delta t\chi\mathbf{u}_s^n}{1 + \lambda\Delta t\chi}. \tag{11}$$

This scheme is unconditionally stable [19]. Equation (9) is rewritten by replacing its terms with those in (10) and (11) as follows:

$$\frac{\omega_z^{n+1} - \omega_z^n}{\Delta t} = \nu\nabla^2\omega_z^n + \nabla \times \left[\frac{\lambda\chi}{1 + \lambda\Delta t\chi}\left(\mathbf{u}_s^n - \mathbf{u}^n\right)\right]. \tag{12}$$

The penalization and diffusion terms of (12) are consecutively evaluated on a temporary grid. To reduce any error due to the penalization term, we used a mask function with second-order accuracy, for which the boundary is located at the midpoint of the grid, where the mask function jumps from 0 to 1 [18, 31].

2.3. Smagorinsky Turbulence Model. The filtered vorticity transport equation for a Lagrangian, vorticity-based large-eddy simulation (LES) in two-dimensions can be expressed as [32]

$$\frac{D\overline{\omega}_z}{Dt} = \nu\nabla^2\overline{\omega}_z + \nabla \cdot \Phi, \tag{13}$$

where the bar denotes spatial filtering at length scale ϵ. The subgrid-scale vorticity stress term Φ, which accounts for the effect of unresolved velocity and vorticity fluctuations, can be defined as $\Phi = \nu_T\nabla\overline{\omega}_z$. Smagorinsky eddy-viscosity model is one of most commonly used subgrid-scale (SGS) turbulence models. In spite of its simplicity, the Smagorinsky model is still extensively used in practice and has served as a basic tool for the developments of LES modeling [33]. In this model, eddy viscosity is defined by

$$\nu_T = (C_S\Delta)^2 |\overline{S}|, \tag{14}$$

where C_S is a nondimensional Smagorinsky constant and Δ is a subgrid characteristic length scale. $S_{ij} = (\partial u_i/\partial x_j + \partial u_j/\partial x_i)/2$ is the resolved strain-rate tensor and $|\overline{S}| = (2\overline{S}_{ij}\overline{S}_{ji})^{1/2}$ is its magnitude. Here $|\overline{S}|^2 = |\overline{\omega}_z|^2$ in incompressible flows ($\nabla \cdot \mathbf{u} = 0$), using the Caswell formula [34]. Finally, the vorticity transport equation for the LES in two dimensions is rewritten as

$$\frac{D\overline{\omega}_z}{Dt} = \left(\nu + (C_S\Delta)^2 |\overline{\omega}_z|\right)\nabla^2\overline{\omega}_z. \tag{15}$$

In 2D LES models, the filtering process is applied in only two directions, so $\Delta = (h_x h_y)^{1/2}$ is defined by the grid spacing. The Smagorinsky constant C_S may vary with the type of flow and the Reynolds number. Values ranging from 0.1 to 0.24 have been suggested in the literatures [33, 35, 36]. The constant C_S is prescribed and its standard value is typically 0.15 [37, 38].

3. Efficient Implementation of Numerical Methods

3.1. Particle-Based Domain Decomposition. Parallel machines and algorithms enable significant reduction in computing time and greater amount of available memory. For high-performance computing, we used the Message Passing Interface (MPI) for our distributed systems, in which each processor has its own local memory. The appropriate decomposition of particles and/or grids should be employed. In domain or spatial decomposition, a given domain Ω is geometrically decomposed by splitting it into several subdomains. Typically, the subdomains are uniform and particles are not evenly distributed among the subdomains. Differences in the number of particles assigned to each processor cause computational load imbalances. As some processors will sit idly while waiting for others, poor performance can result.

We introduce a simple idea to achieve a better load balance during the domain decomposition. In vortex particle methods, fluid particles must periodically be redistributed since their accumulation leads to inaccuracies in numerical solutions. This process is called particle remeshing or redistribution. The remeshing step creates a new set of particles on a uniform Cartesian grid, and then the randomly spaced old particles are removed. All the new particles are reindexed by $i = 1, 2, \ldots, N_p$ from upstream to downstream; that is, higher-indexed particles are located further downstream. Due to the position-dependent index of the particles, the entire spatial domain can be easily split to ensure an equal amount of particles in each subdomain (each processor). As a result, the horizontal extent of the domain is unequal, but the particles are evenly distributed in subdomains of different sizes. In every remeshing step, the entire domain is dynamically partitioned. In this way, each subdomain contains an equal number of particles and better load balance is maintained during simulation. This approach is very simple and effective. No additional algorithms are required to evenly distribute the particles or to identify the position of the particles. As shown in Figure 1, we obtained an approximately 30% reduction in computation time simply by load balancing during parallel computation. This technique is also valuable for the efficient use of available computational memory.

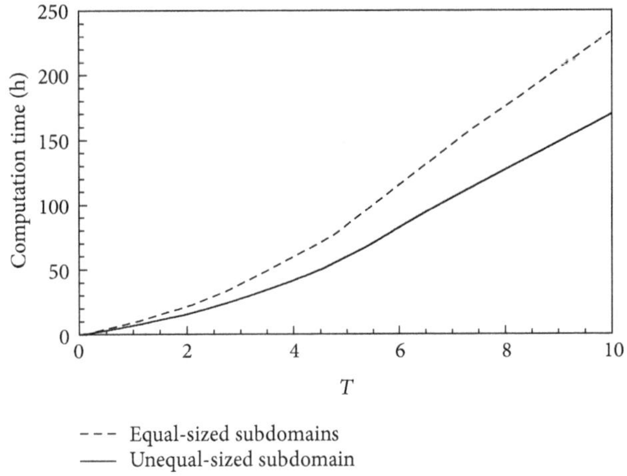

FIGURE 1: Computation times for equal- and unequal-sized subdomains.

3.2. Multidomain Approach. To avoid an excessive domain size, we considered multiple domains. In this study, the entire domain Ω is defined as the union of all the domains, covering all the fluid particles; that is, $\mathbf{x}_p \in \Omega$ where $\Omega = \Omega_1 \cup \Omega_2 \cup \cdots \cup \Omega_{N_d}$. The domain number N_d is not constant and is dependent on a spatially evolving flow. When the domain size exceeds a certain limit, a new domain is created, situated downstream from the domain. As a result, the first domain Ω_1 includes a solid body and the others are sequentially located downstream from the body. For the total circulation conservation, the grid spacing of the ith domain can be defined as $h_i = \epsilon$ (single-resolution) or $h_i = 2h_{i-1}$ (multi-resolution) where $h_1 = \epsilon$. Here ϵ denotes the particle size. To interpolate the vorticity values of the particles into the nodes on a grid with different grid spacing, (6) is rewritten as

$$\omega_z(\mathbf{x}_g) = \frac{\epsilon^2}{h_i^2}\sum_{p=1}^{N}\omega_z(\mathbf{x}_p)\,W\left(\frac{\mathbf{x}_p - \mathbf{x}_g}{h_i}\right), \quad (16)$$

where $\mathbf{x}_p \in \Omega_i$ and $\mathbf{x}_g \in \Omega_i$. The nodes are declared to be a 2D array so that the position of each node can be found straightforwardly from its index. The velocity and vorticity of each domain are sequentially and independently evaluated. Hence, relatively small amounts of working memory are required. This multidomain approach also allows more freedom in the choice of variable arrangements in the domains. The versatility of this approach was demonstrated in our previous study [26]. It can be achieved at the expense of a significant increase in computation time. This issue is further discussed in the next section.

3.3. Approximation of Far-Field Conditions. A substantial part of the overall computation time is spent on the calculation of the boundary stream functions to solve Poisson's equation for the stream function. This requires the order of $\mathcal{O}(MN_p/P)$ operations to obtain boundary values of domain Ω_i. Here, M denotes the number of nodes at the sides of domain Ω_i and N_p/P is the number of particles per processor. To achieve a nearly perfect load balance, particles are evenly distributed among the processors as discussed in Section 3.1 and each processor has all the boundary nodes M. It must also be very useful to avoid all unnecessary communications. N_p/P is reduced by increasing the number of processors involved in the computation whereas M is not. This is why the number of boundary nodes M is the most critical for determining the computation time.

In this study, we attempted to reduce the number of boundary nodes M using the cubic spline approximation. This is made possible by the smooth variation of the stream functions along the boundary. Only one-fourth of the boundary nodes were chosen as equidistant points for the direct calculations. That is, the stream functions at the selected boundary nodes were directly computed using (8), and the values at boundary nodes not selected were evaluated by interpolation, using cubic splines [39]. In this study, the error between the interpolated value and the true value is typically less than 0.01%. We have observed that the reduction in computational time is inversely proportional to the number of nodes selected for direct computation. We believe that this approach can be implemented with much lower computational effort because a complex data structure such as a quad-tree used in the fast multipole method [40] is not required. A rigorous comparison of the accuracy and computational costs is planned in the near future.

4. Computational Procedure

First, the entire simulation domain Ω, covering all the particles, is divided into small domains Ω_i ($i = 1, \ldots, N_d$) as discussed in Section 3.2. Each domain Ω_i is consecutively computed according to the following numerical procedure.

(1) Particle-based domain decomposition: as discussed in Section 3.1, domain Ω_i is decomposed into subdomains to ensure that the particles are evenly distributed among the processors. If redistribution has not been done prior to the time step, particles that have been removed from a subdomain are assigned to an adjacent processor without performing domain decomposition.

(2) Particle-to-grid interpolation: each processor interpolates the vorticity ω_z of its own particles into the grid nodes through the M_4' interpolation kernel.

(3) Calculation of the boundary condition: each processor computes the boundary stream functions by direct calculation and interpolation using cubic splines, as discussed in Section 3.3.

(4) Evaluation of the velocity field: the stream function on the grid is computed with the FFT-based Poisson solver, from an open-source library called FFTW (fastest Fourier transform in the west) [41, 42]. Thereafter, the rotational velocities on the grid nodes are computed from the definitions $u_x = \partial\psi/\partial y$ and $u_y = -\partial\psi/\partial x$ using the fourth order central finite difference scheme.

(5) Evaluation of the vorticity field: each processor evaluates the evolution of the vorticity field in time as follows:

  (a) The penalization term, that is, the second term on the right hand side of (12), is computed with the velocities on the grid. The spatial derivatives of the curl operators are evaluated using a centered difference approximation with a fourth-order error.

  (b) The diffusion term, that is, the first term on the right hand side of (12), is evaluated on the grid using a classical 9-point finite difference method.

  (c) The turbulent viscosity $(C_S\Delta)^2|\overline{\omega}_z|$ in (15) is approximated by the vorticity strength of the grid. Finally, the vorticity field ω_z^{n+1} is updated.

(6) Grid-to-particle interpolation: each processor interpolates the velocity and vorticity on the grid back to its own particle positions through the M_4' interpolation kernel.

Vortex particles are advanced in time using a midpoint predictor-corrector method as follows: particle positions are predicted by their velocities, $\mathbf{x}^* = \mathbf{x}^n + \Delta t\mathbf{u}(\mathbf{x}^n, \omega_z^n)$, then the particle positions are corrected by using combinations of the predicted and previous positions and velocity values, $\mathbf{x}^{n+1} = \mathbf{x}^n + 0.5\Delta t[\mathbf{u}(\mathbf{x}^n, \omega_z^n) + \mathbf{u}(\mathbf{x}^*, \omega_z^{n+1})]$. Particle redistribution is conducted every three time steps to ensure particle overlap. If a new particle has a small vorticity magnitude, that is, $|\Gamma_i| < 0.0001|\Gamma|_{\max}$, it is deleted to avoid having too many particles. Loss of circulation is equally distributed among the remaining particles to ensure the conservation of the total vorticity.

5. Numerical Results and Discussions

5.1. Model Description. We selected a NACA 0009 hydrofoil with a truncated trailing edge for the numerical simulations. This hydrofoil has the same cross-section as the experimental model used by Ausoni et al. [9], Ausoni [10], and Zobeiri et al. [11]. Ten percent of the original chord c_o was removed from the trailing-edge region of the NACA 0009 hydrofoil. The hydrofoil thickness distribution is written as

$$\begin{aligned}
&0.0 \le \frac{x}{c_o} \le 0.5 \\
&\frac{y}{c_o} = 0.1737\left(\frac{x}{c_o}\right)^{1/2} - 0.2422\left(\frac{x}{c_o}\right) + 0.3046\left(\frac{x}{c_o}\right)^2 \\
&\qquad - 0.2657\left(\frac{x}{c_o}\right)^3 \\
&0.5 < \frac{x}{c_o} \le 1.0 \\
&\frac{y}{c_o} = 0.0004 + 0.1737\left(1-\frac{x}{c_o}\right) - 0.1898\left(1-\frac{x}{c_o}\right)^2 \\
&\qquad + 0.0387\left(1-\frac{x}{c_o}\right)^3.
\end{aligned} \tag{17}$$

The hydrofoil geometry is further detailed in Ausoni [10]. The maximum thickness, as normalized by the chord length c, is $t_{\max}/c = 0.1$ and the thickness at the trailing edge t_{TE}/c is 0.0322. The bevel angle β is defined in Figure 2. The trailing edge with $\beta = 0^\circ$ denotes that the trailing edge is truncated in the vertical direction.

5.2. Numerical Setup. Numerical parameters can be determined using the stability condition $\Delta t = \epsilon^2/4\nu$ [18]. The grid convergence was achieved for $\epsilon = 0.00030$, and

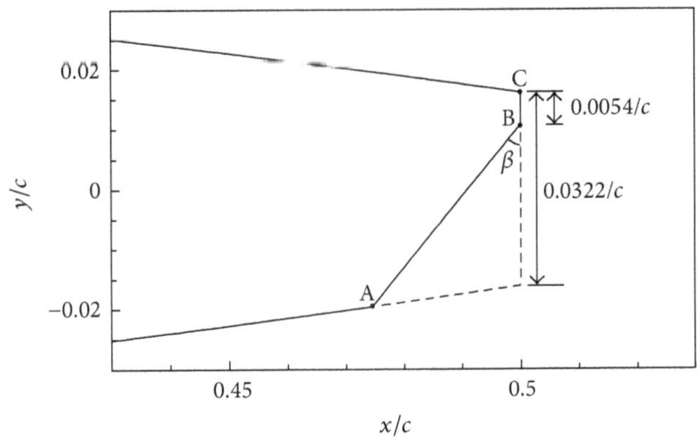

FIGURE 2: Trailing-edge shape. Note that the points A, B, and C indicate the knuckles of the chamfer and trailing-edge wedge, and the bevel angle is defined as β.

$\Delta t = 0.00015$ was chosen in order to guarantee numerical accuracy for a Reynolds number of 10^6. The grid spacing of the ith domain Ω_i is defined as $h_i = \epsilon 2^{i-1}$ to reduce computation time. The penalization parameter λ in (12) is set to be 10^8. The hydrofoil at a zero angle of attack is immersed in a Cartesian grid (the first domain) that does not conform to the surface of the body. The first domain, with 4096×1024 lattice nodes, is fixed and the other domains are dynamically determined depending on the spatially evolving flow. In this study, numerical simulations were conducted until $T = 10$. A new domain is created downstream of a body once the number of horizontal grid points exceeds 2048. Although there is no limit on the number of vertical grid points, it was typically less than or equal to 1024. During simulation, less than 0.8 million particles and four computational domains were created. The entire computation in each case, including start-up and printing times, took approximately 170 hours on 16 Intel Xeon64 3.3 GHz CPUs with 6 GB memory per processor.

5.3. Truncated Trailing-Edge. For an impulsively started flow past a hydrofoil at a chord-based Reynolds number of 10^6, the wake pattern can be classified into three regimes, based on the time evolutions shown in Figure 3. In the very early stages, the wake remained very symmetrical and consisted primarily of two large vortices in the near wake as shown in Figure 3(a). At that time, the drag force gradually decreased. As time went on, the wake became asymmetrical, as shown in Figure 3(b), and then shed a vortex, as shown in Figure 3(c). After a number of shedding periods, vortices were shed regularly from the symmetric trailing edge in the form of two trains of oppositezsign but equal-strength vortices as shown in Figure 3(d). Such regular vortex shedding induces periodic loading on the structure as shown in Figure 4. It is well known that drag-force oscillation during vortex shedding is much smaller than the lift force. Numerical results show that oscillations in the drag force occurred at twice the vortex-shedding frequency, due to the fact that two vortices were shed from alternate sides during one full period of wake oscillation.

From our simulation, the Strouhal number based on the trailing-edge thickness, $\mathrm{St}_t = f_s t_{\mathrm{TE}}/U_\infty$, is about 0.22 for $\mathrm{Re} = 10^6$. Ausoni [10] (also refer to Zobeiri et al. [11]) had a similar finding and experimentally gave $\mathrm{St}_t = 0.25$, which may be approximately read from their figure. They used a hydrofoil with a 100 mm chord length and 150 mm span length. The aspect ratio, defined as the ratio of the span length to the chord length, was 1.5. A lower aspect ratio leads to a three-dimensional (3D) effect, even though the hydrofoil is 2D. Ausoni [10] also observed experimentally that the shedding process was not induced in phase along the span. It was therefore speculated in [25] that the slight underestimation of the Strouhal number is due to 3D effects which were absent in our simulation.

5.4. Oblique Trailing-Edges. We conducted numerical simulations to observe vortex shedding from trailing edges with six different bevel angles of $\beta = 0$, 10, 20, 40, 60, and 70 degrees. Figure 5 shows the instantaneous vorticity distributions at $T = 10$ for $\mathrm{Re} = 10^6$. At $\beta = 0^\circ$, the vortex shedding is regular and periodic. As the bevel angle β increases, vortex shedding becomes increasingly disorganized. Such irregular shedding of vortices demonstrates the periodicity of oscillatory forces. Figure 6 shows the power spectral density function (PSD) of the drag and lift coefficients for different bevel angles. Compared with $\beta = 0^\circ$, the amplitude of the lift force oscillation for asymmetric trailing edges is significantly reduced, compared with the symmetric trailing edge at $\beta = 0^\circ$. The lift force amplitude is gradually decreased as the bevel angle increases, while the drag force amplitude increases and then decreases to between $\beta = 10^\circ$ and 60°. The periodicity of both forces is almost lost at $\beta = 60^\circ$, and there is no definite periodicity of the induced force components at $\beta = 70^\circ$.

At a bevel angle of 10° as shown in Figure 6(b), it is interesting that the dominant peak of the drag force is identical to that of the lift force. From a vibration standpoint, the forcing frequency is important for the prevention of resonance. The distance between the two vortices in Figure 5(b) does not stay the same, but the difference between the shedding vortices between $\beta = 0^\circ$ and 10° (Figures 5(a) and 5(b)) is not

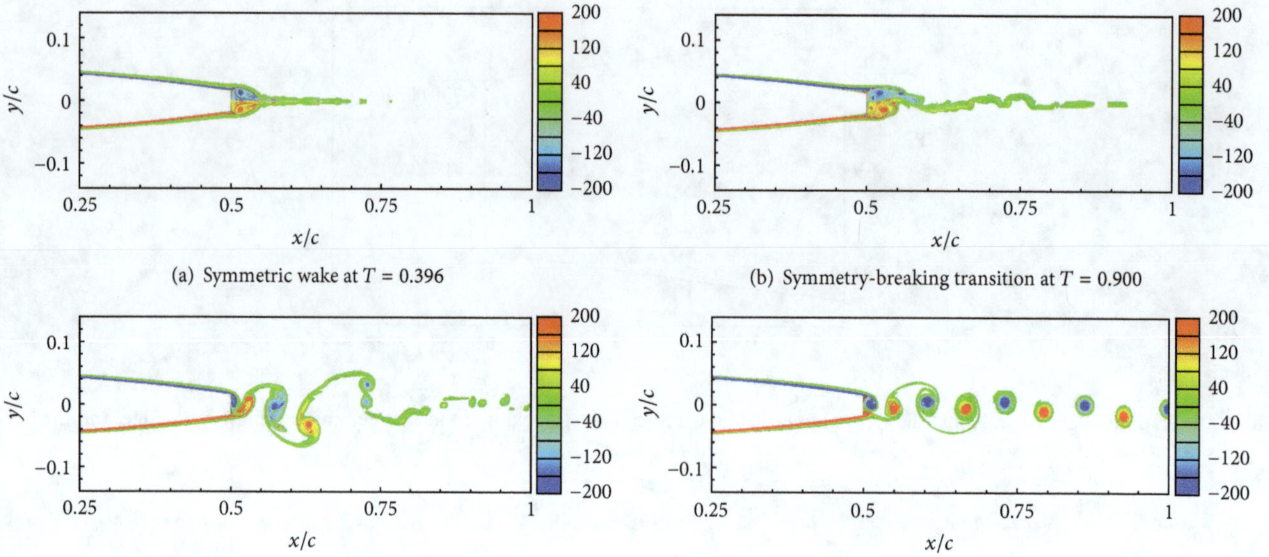

(a) Symmetric wake at $T = 0.396$

(b) Symmetry-breaking transition at $T = 0.900$

(c) Nonperiodic vortex shedding at $T = 1.098$

(d) Fully developed periodic shedding at $T = 3.500$

FIGURE 3: Wake patterns of the NACA 0009 hydrofoil with a truncated trailing-edge ($\beta = 0^\circ$) for Re = 10^6.

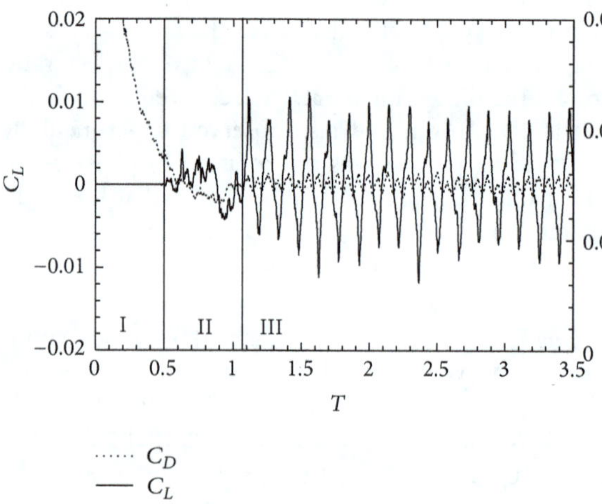

FIGURE 4: Time evolution of the lift and drag coefficients for the impulsively started NACA 0009 hydrofoil with a truncated trailing edge ($\beta = 0^\circ$) at Re = 10^6. Regime I denotes the developing regime of a symmetric wake, II a symmetry-breaking transition, and III periodic vortex shedding.

statistically significant. A possible reason for this is the phase difference between the vortices shed from the upper and lower sides. In the wake formation region, the two vortices shed from the symmetric trailing edge at $\beta = 0^\circ$ are almost exactly out of phase (180 degrees), whereas at $\beta = 10^\circ$ there is a phase difference of approximately 30 degrees. This means that before one vortex is fully shed from the body another vortex is created. At higher bevel angles, $\beta > 20^\circ$, the negative vortices from the upper side of the hydrofoil become stronger than the positive ones from the lower side. The oscillation force is induced mostly by the vortices shed from the acute (upper) side. As a result, the dominant frequency of the drag force oscillation appears to be the same as the vortex shedding frequency. We numerically observed that differences in strength and/or phase between two adjacent vortices cause nonperiodic and irregular vortex shedding.

Figure 7 shows the shedding frequency of vortices from different beveled trailing edges, compared with experimental results by Zobeiri et al. [11]. The numerically simulated shedding frequencies were approximately 6.8 and 6.4 for $\beta = 0^\circ$ and 60°, respectively. Numerical results show good qualitative agreements with the experimental results, but the numerical simulations underestimated the vortex-shedding frequency by about 10%. Figure 8 shows the mean value and standard deviation of the velocity components. We compared our numerical results with the experimental results measured by Zobeiri et al. [11]. The *truncated* and *oblique* trailing edges in Figure 8 indicate bevel angles of $\beta = 0^\circ$ and 60°, respectively. For the truncated trailing edge, the mean velocity profiles are more symmetric than those in the experimental data. The results from numerical simulation are in good agreement with the experimental results. We conclude that the vortex shedding phenomenon from the trailing edge of a 2D hydrofoil can be reasonably simulated using 2D LES which considerably reduces the computational cost.

5.5. Effect of a Periodically Varying Flow on Vortex Shedding. The propeller is generally located behind the ship's hull and is subjected to a nonaxisymmetric wake field. Considering a 2D flow, the propeller section effectively experiences periodic variation in flow incidence. We conducted numerical simulations to investigate vortex shedding with respect to the sinusoidal motions of the free-stream flow. The hydrofoil was at rest and the flow became the oscillator under these conditions. In the current study, the angle of attack varied

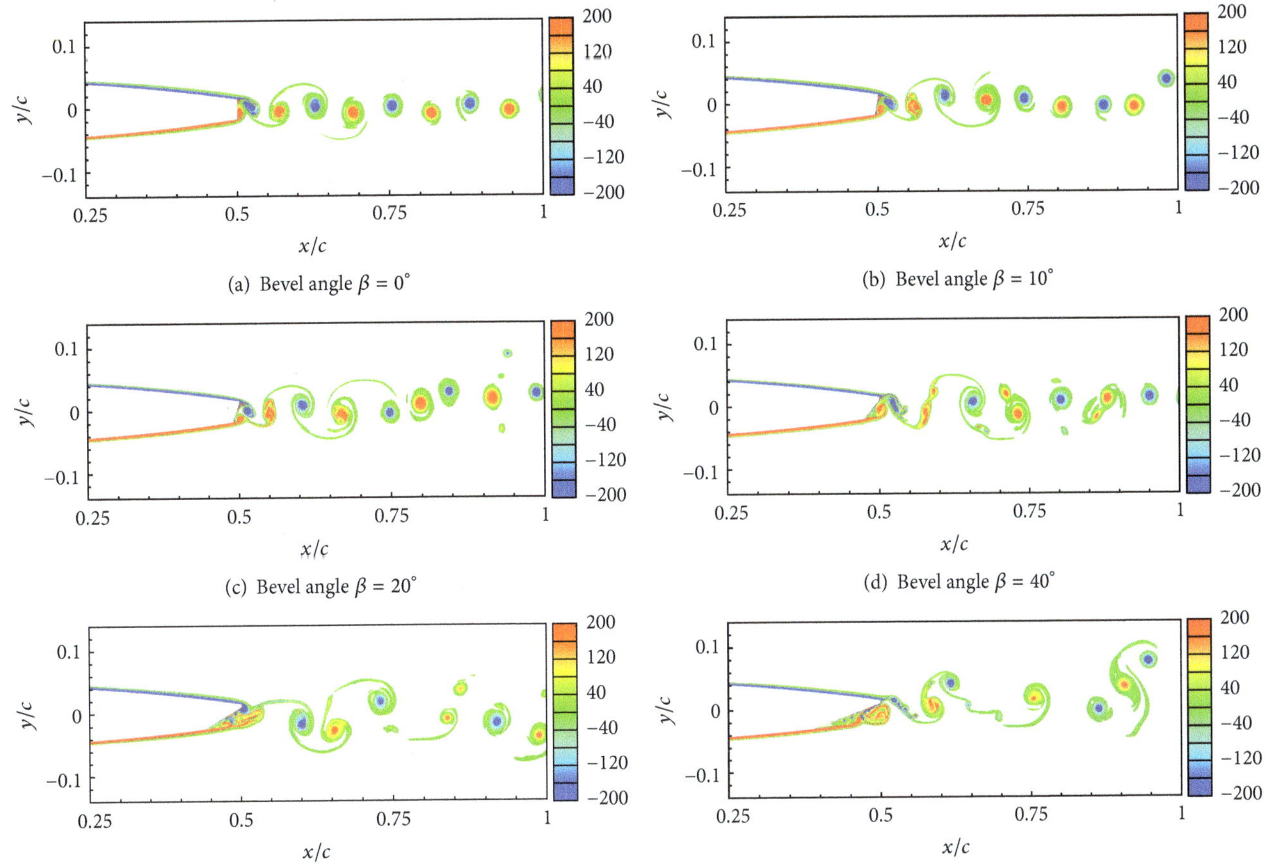

(a) Bevel angle $\beta = 0^\circ$

(b) Bevel angle $\beta = 10^\circ$

(c) Bevel angle $\beta = 20^\circ$

(d) Bevel angle $\beta = 40^\circ$

(e) Bevel angle $\beta = 60^\circ$

(f) Bevel angle $\beta = 70^\circ$

FIGURE 5: Instantaneous vorticity distributions at $T = 10$ for $\mathrm{Re} = 10^6$.

sinusoidally with time t; $\alpha = \alpha_o \sin(2\pi f_\infty t)$ where f_∞ was the frequency of the free-stream flow oscillation and its oscillation amplitude was restricted at $\alpha_o = 2^\circ$. The magnitude of the free-stream velocity was kept constant. The tested model is the NACA 0009 hydrofoil with a beveled trailing edge of $\beta = 60^\circ$.

Figure 9 shows the instantaneous vorticity distributions at $T = 10$ and the PSD of the drag coefficients at four different frequencies of the free-stream flow, that is, $f_\infty = 15$, 20, 25, and 30. There are two peaks in the PSD functions; the first peak appears around $f^* \approx 6$ and the second one is identical to the frequency of the free-stream flow oscillation. The obvious feature recognized in all the PSD functions, as shown in Figure 9, is that the first peak frequency is nearly identical to the vortex-shedding frequency from the hydrofoil in the uniform free-stream flow. Vortex formation in the near wake is thought to be less affected by small inflow oscillations. Most interestingly, at $f_\infty = 20$ vortices are regularly shed from the trailing edge, as shown in Figure 9(c). In the PSD functions, drag oscillation induced by vortex shedding is clearly observed. This means that a particular oscillation frequency in the free-stream velocity can cause regular and periodic vortex shedding. This is significant with respect to changing a marine propeller's angle of attack over time.

6. Concluding Remarks

We simulated impulsively started flows past a modified NACA 0009 hydrofoil at a Reynolds number of 10^6, using the hybrid particle-mesh method in conjunction with the vorticity-based subgrid scale model. We have achieved significant savings in both memory usage and computation time by making improvements in the numerical schemes: the particle-based domain decomposition, the use of multiple domains, and the approximation of domain boundary conditions. These techniques are simple and easy to implement but also effective in reducing the computational costs. The particle-based domain decomposition scheme is suitable for a distributed system in which each processor has its own private memory (distributed memory).

In this study, the numerical simulation results nearly match the experimental results. This demonstrates that vortex shedding from the trailing edge of a 2D hydrofoil can be reasonably simulated with a 2D LES. It is due to the fact that the flow is characterized by quasi-two dimensionality with vortex shedding from the trailing edge of the hydrofoil. We have considered that 2D vortex shedding simulations should not be dismissed so easily. With two-dimensionality, it is possible to solve the problem much less expensively, since

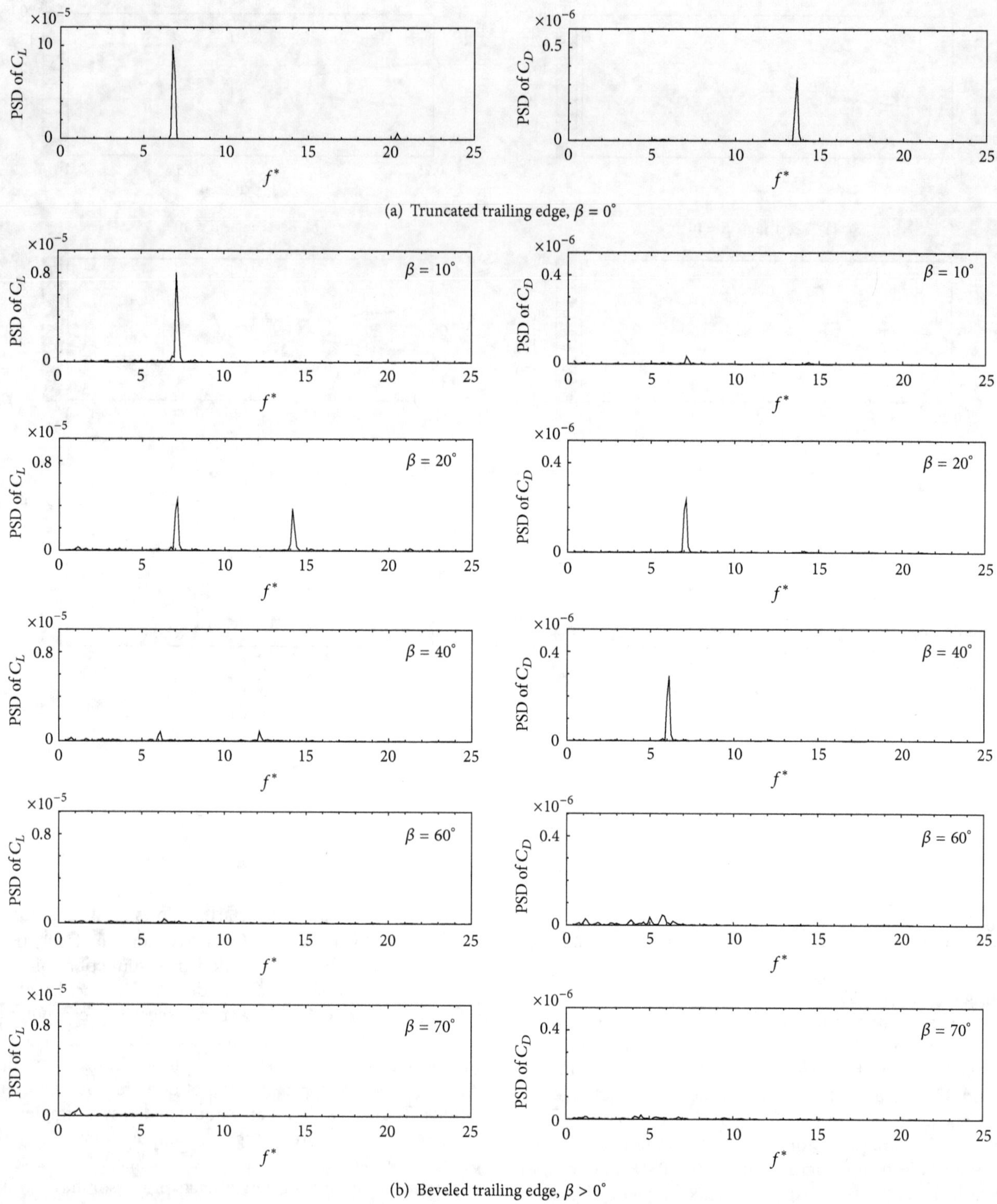

(a) Truncated trailing edge, $\beta = 0°$

(b) Beveled trailing edge, $\beta > 0°$

Figure 6: Power spectral density (PSD) of the lift and drag coefficients. f^* denotes the frequency normalized by the chord length and the undisturbed flow velocity.

3D problems require significant computational resources. However, the LES should be 3D since turbulence is inherently 3D. While we accept that more realistic 3D simulation is required physically, a 2D LES can be used as a basis for comparison prior to 3D implementation. Once the capability of a numerical algorithm is evaluated in two dimensions, it can be easily extended into three dimensions. This paper is intended as a guide for researchers and designers who are working on a similar problem: vortex shedding from a hydrofoil at high Reynolds numbers.

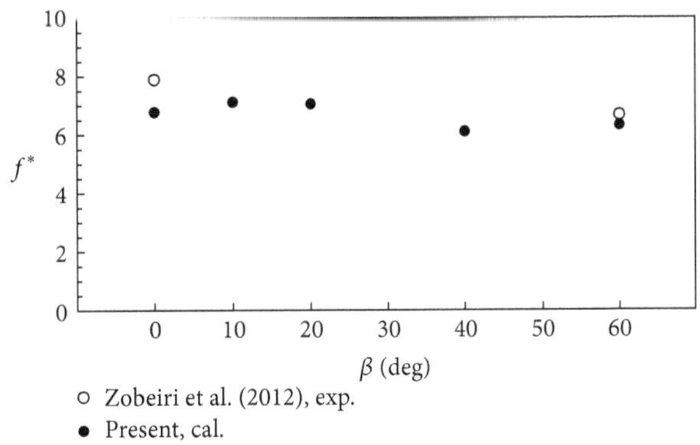

FIGURE 7: Shedding frequency of vortices from the different beveled trailing edges for Re = 10^6. The experimental data by Zobeiri et al. [11] were derived from LDV measurements, and the approximate shedding frequency, as read from their figure, is normalized by the chord length and the undisturbed flow velocity.

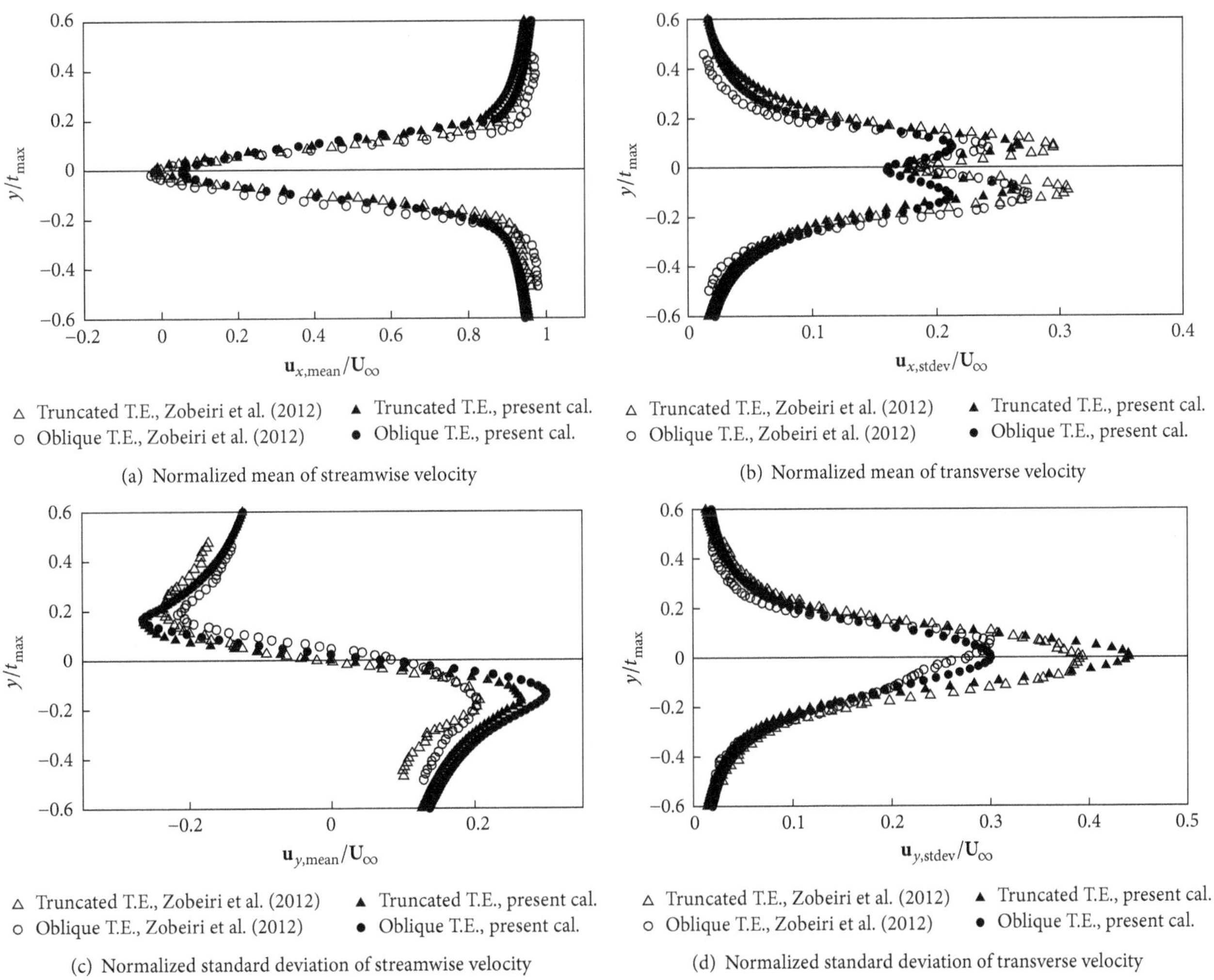

(a) Normalized mean of streamwise velocity

(b) Normalized mean of transverse velocity

(c) Normalized standard deviation of streamwise velocity

(d) Normalized standard deviation of transverse velocity

FIGURE 8: Mean and standard deviation of velocities at $x/c = 0.516$ for Re = 10^6. The experimental data by Zobeiri et al. [11] were measured at $x/c = 0.52$ and 0.528 for truncated ($\beta = 0°$) and oblique ($\beta = 60°$) trailing edges, respectively, and the reported Reynolds number of the flow is about 2×10^6.

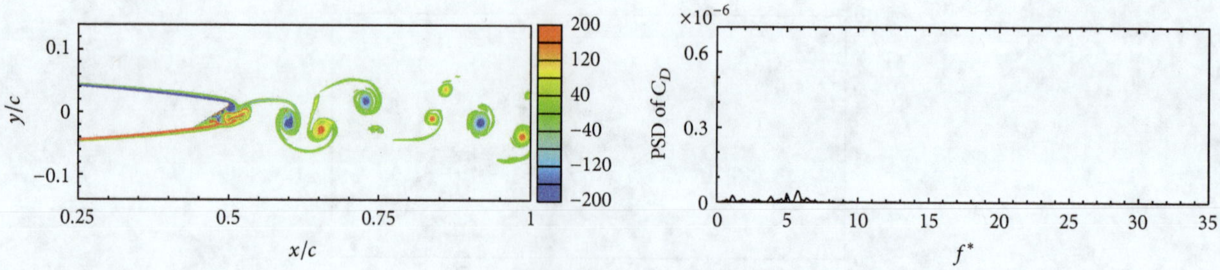

(a) Steady flow, $f_\infty = 0$

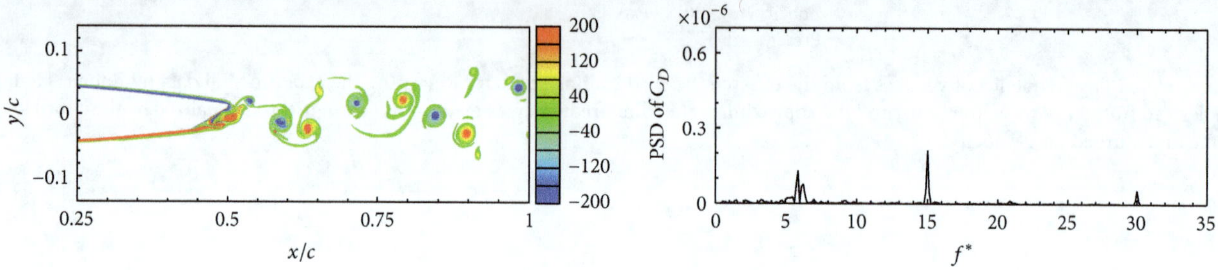

(b) Incoming flow frequency $f_\infty = 15$

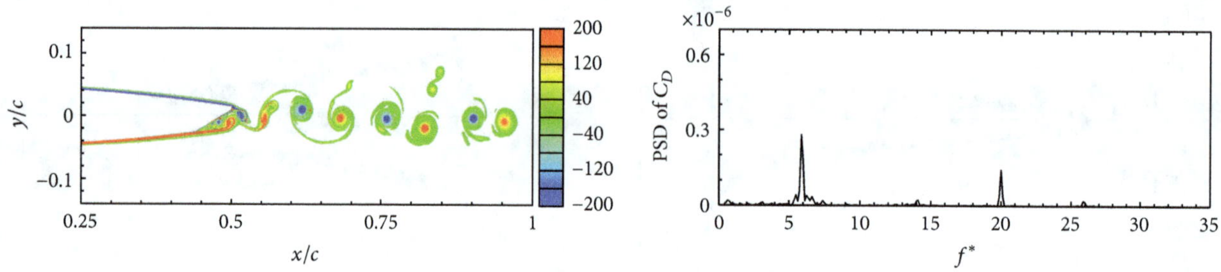

(c) Incoming flow frequency $f_\infty = 20$

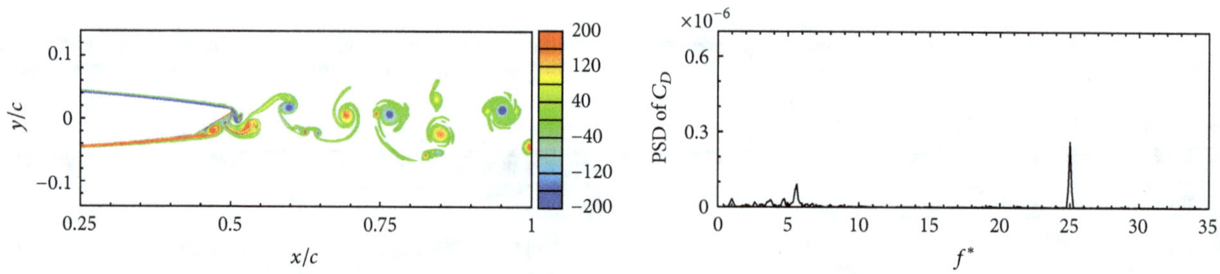

(d) Incoming flow frequency $f_\infty = 25$

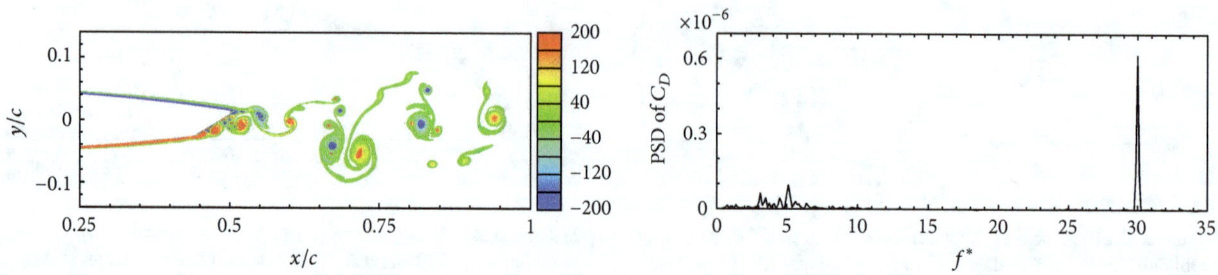

(e) Incoming flow frequency $f_\infty = 30$

FIGURE 9: Effect of the incoming flow frequency on vortex shedding from a beveled trailing edge of $\beta = 60^\circ$ at $f_\infty = 15, 20, 25$, and 30. The contour indicates the instantaneous vorticity distribution at $T = 10$, and the power spectral density (PSD) is calculated from the drag coefficients in the range $T = [3, 10]$.

We found in this study that the vortex formation characteristic in periodically varying inflows remains unchanged. A certain frequency of free-stream flow oscillation tends to make vortex shedding more regular again. This means that the periodicity of flow incidence can stimulate an inherent frequency of vortex shedding from a hydrofoil. However, the effect of inflow oscillation on vortex shedding is not yet fully understood. Future work will include experimental investigations, as well as a 3D flow simulation, to better understand the relationship between inflow oscillation and vortex shedding.

Conflict of Interests

The authors declare that there is no conflict of interests regarding the publication of this paper.

Acknowledgments

This research was supported by the Korean Institute of Ocean Science & Technology sponsored by the Ministry of Trade, Industry & Energy (no. 10033668) and the Industrial Convergence Strategic Technology Development Program (no. 10044499) funded by the Ministry of Trade, Industry & Energy (MI, Korea).

References

[1] E. Berger and R. Wille, "Periodic flow phenomena," *Annual Review of Fluid Mechanics*, vol. 4, no. 1, pp. 313–340, 1972.

[2] P. W. Bearman, "Vortex shedding from oscillating bluff bodies," *Annual Review of Fluid Mechanics*, vol. 16, pp. 195–222, 1984.

[3] C. H. K. Williamson, "Vortex dynamics in the cylinder wake," *Annual Review of Fluid Mechanics*, vol. 28, pp. 477–539, 1996.

[4] C. H. K. Williamson, "Vortex-induced vibrations," *Annual Review of Fluid Mechanics*, vol. 36, pp. 413–455, 2004.

[5] R. Fischer, "Singing propellers—solutions and case histories," *Marine Technology*, vol. 45, no. 4, pp. 221–227, 2008.

[6] D. A. Bourgoyne, J. M. Hamel, S. L. Ceccio, and D. R. Dowling, "Time-averaged flow over a hydrofoil at high reynolds number," *Journal of Fluid Mechanics*, vol. 496, pp. 365–404, 2003.

[7] D. A. Bourgoyne, S. L. Ceccio, and D. R. Dowling, "Vortex shedding from a hydrofoil at high Reynolds number," *Journal of Fluid Mechanics*, vol. 531, pp. 293–324, 2005.

[8] N. Mulvany, J. Tu, L. Chen, and B. Anderson, "Assessment of two-equation turbulence modelling for high Reynolds number hydrofoil flows," *International Journal for Numerical Methods in Fluids*, vol. 45, no. 3, pp. 275–299, 2004.

[9] P. Ausoni, M. Farhat, X. Escaler, E. Egusquiza, and F. Avellan, "Cavitation influence on von Kármán vortex shedding and induced hydrofoil vibrations," *Journal of Fluids Engineering, Transactions*, vol. 129, no. 8, pp. 966–973, 2007.

[10] P. Ausoni, *Turbulent vortex shedding from a blunt trailing edge hydrofoil [Ph.D. thesis]*, École Polytechnique Fédérale de Lausanne, Lausanne, Switzerland, 2009.

[11] A. Zobeiri, P. Ausoni, F. Avellan, and M. Farhat, "How oblique trailing edge of a hydrofoil reduces the vortex-induced vibration," *Journal of Fluids and Structures*, vol. 32, pp. 78–89, 2012.

[12] G.-H. Cottet and P. D. Koumoutsakos, *Vortex Methods: Theory and Practice*, Cambridge University Press, 2000.

[13] G. Cottet and P. Poncet, "Particle methods for direct numerical simulations of three-dimensional wakes," *Journal of Turbulence*, vol. 3, article N38, 2002.

[14] P. Chatelain, A. Curioni, M. Bergdorf, D. Rossinelli, W. Andreoni, and P. Koumoutsakos, "Billion vortex particle direct numerical simulations of aircraft wakes," *Computer Methods in Applied Mechanics and Engineering*, vol. 197, no. 13–16, pp. 1296–1304, 2008.

[15] Y.-C. Kim, J.-C. Suh, and K.-J. Lee, "Vortex-in-cell method combined with a boundary element method for incompressible viscous flow analysis," *International Journal for Numerical Methods in Fluids*, vol. 69, no. 10, pp. 1567–1583, 2012.

[16] P. Angot, C.-H. Bruneau, and P. Fabrie, "A penalization method to take into account obstacles in incompressible viscous flows," *Numerische Mathematik*, vol. 81, no. 4, pp. 497–520, 1999.

[17] M. El Ossmani and P. Poncet, "Efficiency of multiscale hybrid grid-particle vortex methods," *Multiscale Modeling & Simulation*, vol. 8, no. 5, pp. 1671–1690, 2010.

[18] J. T. Rasmussen, G.-H. Cottet, and J. H. Walther, "A multiresolution remeshed vortex-in-cell algorithm using patches," *Journal of Computational Physics*, vol. 230, no. 17, pp. 6742–6755, 2011.

[19] M. Coquerelle and G.-H. Cottet, "A vortex level set method for the two-way coupling of an incompressible fluid with colliding rigid bodies," *Journal of Computational Physics*, vol. 227, no. 21, pp. 9121–9137, 2008.

[20] F. Morency, H. Beaugendre, F. Gallizio, and S. Laurens, "Computation of ice shedding trajectories using cartesian grids, penalization, and level sets," *Modelling and Simulation in Engineering*, vol. 2011, Article ID 274947, 15 pages, 2011.

[21] M. M. Hejlesen, P. Koumoutsakos, A. Leonard, and J. H. Walther, "Iterative Brinkman penalization for remeshed vortex methods," *Journal of Computational Physics*, vol. 280, pp. 547–562, 2015.

[22] M. Gazzola, P. Chatelain, W. M. van Rees, and P. Koumoutsakos, "Simulations of single and multiple swimmers with non-divergence free deforming geometries," *Journal of Computational Physics*, vol. 230, no. 19, pp. 7093–7114, 2011.

[23] R. Chatelin and P. Poncet, "A hybrid grid-particle method for moving bodies in 3D stokes flow with variable viscosity," *SIAM Journal on Scientific Computing*, vol. 35, no. 4, pp. B925–B949, 2013.

[24] R. Chatelin and P. Poncet, "Hybrid grid-particle methods and penalization: a Sherman-Morrison-Woodbury approach to compute 3D viscous flows using FFT," *Journal of Computational Physics*, vol. 269, pp. 314–328, 2014.

[25] S. J. Lee, J. H. Lee, and J. C. Suh, "Vortex shedding frequency for a 2D hydrofoil with a truncated trailing edge," *Journal of the Society of Naval Architects of Korea*, vol. 51, no. 6, pp. 480–488, 2014 (Korean).

[26] S.-J. Lee and J. C. Suh, "A multi-domain approach for a hybrid particle-mesh method," *Journal of Computational Fluids Engineering*, vol. 19, no. 2, pp. 72–78, 2014.

[27] S.-J. Lee, J.-H. Lee, and J.-C. Suh, "Computation of pressure fields around a two-dimensional circular cylinder using the vortex-in-cell and penalization methods," *Modelling and Simulation in Engineering*, vol. 2014, Article ID 708372, 13 pages, 2014.

[28] J. J. Monaghan, "Extrapolating *B*-splines for interpolation," *Journal of Computational Physics*, vol. 60, no. 2, pp. 253–262, 1985.

[29] J. J. Monaghan, "Particle methods for hydrodynamics," *Computer Physics Reports*, vol. 3, no. 2, pp. 71–124, 1985.

[30] O. V. Vasilyev and N. K.-R. Kevlahan, "Hybrid wavelet collocation-brinkman penalization method for complex geometry flows," *International Journal for Numerical Methods in Fluids*, vol. 40, no. 3-4, pp. 531–538, 2002.

[31] W. Iwakami, Y. Yatagai, N. Hatakeyama, and Y. Hattori, "A new approach for error reduction in the volume penalization method," *Communications in Computational Physics*, vol. 16, no. 5, pp. 1181–1200, 2014.

[32] J. R. Mansfield, O. M. Knio, and C. Meneveau, "Towards lagrangian large vortex simulation," *ESAIM Proceedings*, vol. 1, pp. 49–64, 1996.

[33] G.-H. Cottet and P. Poncet, "Advances in direct numerical simulations of 3D wall-bounded flows by vortex-in-cell methods," *Journal of Computational Physics*, vol. 193, no. 1, pp. 136–158, 2004.

[34] B. Caswell, "Kinematics and stress on a surface of rest," *Archive for Rational Mechanics and Analysis*, vol. 26, no. 5, pp. 385–399, 1967.

[35] R. S. Rogallo and P. Moin, "Numerical simulation of turbulent flows," *Annual Review of Fluid Mechanics*, vol. 16, pp. 99–137, 1984.

[36] J. Meyers and P. Sagaut, "On the model coefficients for the standard and the variational multi-scale Smagorinsky model," *Journal of Fluid Mechanics*, vol. 569, pp. 287–319, 2006.

[37] J. R. Mansfield, O. M. Knio, and C. Meneveau, "A dynamic LES scheme for the vorticity transport equation: formulation and a priori tests," *Journal of Computational Physics*, vol. 145, no. 2, pp. 693–730, 1998.

[38] S. B. Pope, *Turbulent Flows*, Cambridge University Press, 2000.

[39] W. H. Press, S. A. Teukolsky, W. T. Vetterling, and B. P. Flannery, *Numerical Recipes in C++: The Art of Scientific Computing*, Cambridge University Press, 2nd edition, 2002.

[40] L. Greengard and W. D. Gropp, "A parallel verison of the fast multipoe method," *Computers & Mathematics with Applications*, vol. 20, no. 7, pp. 63–71, 1990.

[41] M. Frigo and S. G. Johnson, "FFTW: an adaptive software architecture for the FFT," in *Proceedings of the IEEE International Conference on Acoustics, Speech and Signal Processing (ICASSP '98)*, pp. 1381–1384, May 1998.

[42] M. Frigo and S. G. Johnson, "The design and implementation of FFTW3," *Proceedings of the IEEE*, vol. 93, no. 2, pp. 216–231, 2005.

22

Consistent Weighted Average Flux of Well-Balanced TVD-RK Discontinuous Galerkin Method for Shallow Water Flows

Thida Pongsanguansin,[1] Montri Maleewong,[2] and Khamron Mekchay[1]

[1] *Department of Mathematics and Computer Science, Chulalongkorn University, Bangkok 10330, Thailand*
[2] *Department of Mathematics, Kasetsart University, Bangkok 10900, Thailand*

Correspondence should be addressed to Montri Maleewong; montri.m@ku.ac.th

Academic Editor: Javier Murillo

A well-balanced scheme with total variation diminishing Runge-Kutta discontinuous Galerkin (TVD-RK DG) method for solving shallow water equations is presented. Generally, the flux function at cell interface in the TVD-RK DG scheme is approximated by using the Harten-Lax-van Leer (HLL) method. Here, we apply the weighted average flux (WAF) which is higher order approximation instead of using the HLL in the TVD-RK DG method. The consistency property is shown. The modified well-balanced technique for flux gradient and source terms under the WAF approximations is developed. The accuracy of numerical solutions is demonstrated by simulating dam-break flows with the flat bottom. The steady solutions with shock can be captured correctly without spurious oscillations near the shock front. This presents the other flux approximations in the TVD-RK DG method for shallow water simulations.

1. Introduction

Hyperbolic balanced law for one-dimensional problem is

$$U_t + F_x(U) = G(U), \tag{1}$$

where U, F, and G represent solution vector, flux function, and source terms, respectively.

In this work, the hyperbolic equation is the shallow water equations which can be expressed by

$$\begin{aligned} h_t + q_x &= 0, \\ q_t + \left(\frac{q^2}{h} + \frac{gh^2}{2}\right)_x &= -ghb_x, \end{aligned} \tag{2}$$

where h is the water depth, $q = uh$ is the discharge, u is the flow velocity in the x-direction, g is the acceleration due to gravity, and b is the bottom function. Equation (2) can be rewritten by setting

$$\begin{gathered} U = \begin{pmatrix} h \\ q \end{pmatrix}, \\ F(U) = \begin{pmatrix} q \\ \dfrac{q^2}{h} + \dfrac{gh^2}{2} \end{pmatrix}, \\ G(U) = \begin{pmatrix} 0 \\ -ghb_x \end{pmatrix}. \end{gathered} \tag{3}$$

The total variation diminishing Runge-Kutta discontinuous Galerkin (TVD-RK DG) method can be applied to solve the shallow water equations; see [1–5]. This scheme has several advantages. For instance, it can be used to handle complex geometries, and in the same time, adaptive strategies are easily applied. The accuracy of numerical solutions can be improved by increasing the polynomial degree of approximating polynomial and by the method for estimating the flux function at cell interface.

By the concept of discontinuous Galerkin method, numerical solutions need not to be continuous at cell interface. So, we require an efficient flux approximation. Generally, there are several types of approximations. The most famous one is the Harten-Lax-van Leer flux (HLL), while the higher order approximation is the weighted average flux (WAF); see [6–9]. This approach approximates the flux functions by averaging flux in each direction along the wave solutions at the half-time step. It originates from random flux scheme, which was shown to be second-order accuracy in space and time in statistical sense by Toro [6]. From previous work, the weighted average flux has been successfully applied to solve various types of problems, especially in the finite volume method [7, 10, 11]. It improves the accuracy of the finite volume method to be second-order without reconstruction process. But the weighted average flux is rarely applied in the discontinuous Galerkin method. So the main objective of this work is the application of the WAF method in the TVD-RK DG method. We have also shown that the present scheme has consistent property with the WAF approximation.

The steady solutions of (3) can be obtained by setting

$$\begin{aligned} hu &= \text{constant}, \\ \frac{1}{2}u^2 + g(h+b) &= \text{constant}. \end{aligned} \tag{4}$$

Flux gradients are nonzero, and it must be balanced with the bottom gradient. Usually, numerical schemes without balancing these two quantities produce oscillate steady solution. So a balancing numerical scheme is needed, and it is called the well-balanced scheme. This method is designed to preserve energy at steady state. Generally, a numerical scheme is said to be well-balanced if it satisfies the exact C-property which was first introduced by Bermudez and Vazquez [12]. The exact C-property means that the numerical solution must satisfy still water condition at steady state given by

$$\begin{aligned} u &= 0, \\ h+b &= \text{constant}. \end{aligned} \tag{5}$$

Thus, to obtain a well-balanced scheme, we must design the method that its steady solutions satisfy (5).

Recently, the second-order Runge-Kutta discontinuous Gelerkin method with well-balanced scheme is proposed by Kesserwani and Liang [13]. They presented the wetting and drying algorithms for solving one-dimensional problem. Xing and Shu [14] proposed a well-balanced finite volume method based on the weighted essentially nonoscillatory (WENO) scheme for solving one- and two-dimensional problems. They applied the decomposing source term technique into the sum of several terms to ensure the numerical balance between flux difference and source terms. Audusse et al. [15] presented a well-balanced finite volume with hydrostatic reconstruction to solve one-dimensional problem. Later, Noelle et al. [16] extended the well-balanced finite volume schemes proposed by [15] to any that desired the orders of accuracy.

In order to solve the shallow water equations with source terms, we modified the well-balanced discontinuous Galerkin scheme proposed by [15] and related work by Xing and Shu [14]. The weighted average flux (WAF) method [6, 7] has been applied in the modified DG scheme. We will show that the modified TVD-RK DG scheme with the WAF has consistent property. Various numerical experiments for both steady and unsteady flows are demonstrated to show the capability and accuracy of our numerical scheme.

Numerical scheme and its consistency property with WAF for the TVD-RK DG method have been shown in Section 2. The modified well-balanced TVD-RK DG scheme with source terms is presented in Section 3. Numerical examples are demonstrated in Section 4. Conclusions are finally made in Section 5.

2. Numerical Scheme for Shallow Water Equations without Source Term

The TVD-RK DG with weighted average flux method for the case of without source terms is presented in this section.

2.1. TVD-RK Discontinuous Galerkin Method. Consider the one-dimensional hyperbolic conservation law,

$$U_t + F_x(U) = 0. \tag{6}$$

The computational domain $(0, L)$ is divided into K cells. We denote the jth cell by $I_j = [x_{j-1/2}, x_{j+1/2}]$, for $j = 1, \dots, K$, with uniform cell size $\Delta_j = x_{j+1/2} - x_{j-1/2}$. The cell center is that $x_j = (x_{j+1/2} + x_{j-1/2})/2$, where $x_{j+1/2}$ and $x_{j-1/2}$ are the left and the right of cell boundaries, respectively.

Approximate solution is denoted by $U_h = \binom{h_h}{q_h}$.

Multiplying (6) by a test function, $v_h(x) \in P^N(I_j)$, where $P^N(I_j)$ is the polynomial space degree N on the interval I_j, and replacing U by U_h and then taking the integration by parts over I_j, we obtain a weak form of numerical scheme as

$$\begin{aligned} \int_{I_j} (\partial_t U_h) v_h dx - \int_{I_j} F(U_h) \partial_x v_h dx + \hat{F}_{j+1/2} v_h (x_{j+1/2}^-) \\ - \hat{F}_{j-1/2} v_h (x_{j-1/2}^+) = 0, \end{aligned} \tag{7}$$

where the flux function F at the cell interfaces is approximated by $\hat{F}$, which is the function of U_h^+ and U_h^- at $x_{j\pm1/2}$ as

$$\hat{F}_{j\pm1/2} = \hat{F}\left((U_h)_{j\pm1/2}^-, (U_h)_{j\pm1/2}^+\right). \tag{8}$$

Here $U_h|_{j\pm1/2}^-$ and $U_h|_{j\pm1/2}^+$ denote the approximate solutions at the left and the right of cell boundaries, respectively. If we apply the Legendre polynomials to be a local basis function, the approximate solution U_h can be written by

$$U_h(x,t) = \sum_{m=0}^{N} U_j^m(t) \varphi_m(x), \tag{9}$$

where $U_j^m(t)$ is called the temporal coefficient. Now, the basis function $\varphi_m(x)$ is defined by the Legendre polynomial $P_m(x)$ of degree m over $[-1, 1]$.

The test function $v_h(x)$ is typically chosen to be the basis function; that is, $v_h(x) = \{\varphi_l(x)\}_{l=0}^{N}$. So after applying Legendre's properties, (7) is simplified to

$$\frac{dU_j^l(t)}{dt} = \frac{2l+1}{\Delta_j}\int_{I_j} F(U_h)\,\partial_x\varphi_l(x)\,dx + \frac{2l+1}{\Delta_j}\left\{(-1)^l \widehat{F}_{j-1/2} - \widehat{F}_{j+1/2}\right\}, \quad (10)$$

for $j = 1, \ldots, K$, and $l = 0, \ldots, N$.

The time derivative term in (10) can be approximated by applying the high-order TVD Runge Kutta (TVD-RK) method; see [1, 3, 17]. Note that, when the polynomial degree N is applied, the TVD-RK method at least the order of $N+1$ must be applied to obtain the accuracy of order $O(\Delta x^{N+1})$ for smooth flows.

The TVD-RK DG method can be used to simulate shallow water flows with moving shocks. Unphysically oscillate behaviours are usually produced near the shock fronts. The slope limiter techniques can be applied to remove the oscillations. In this work, we apply the Monotonic Upstream-Centered Scheme for Conservation Laws (MUSCL) limiter (see [1, 3, 6, 18]) in the TVD-RK DG method. This approach limits the present solution slope by comparing the slope with neighbor cells.

2.2. Weighted Average Flux (WAF). The HLL numerical flux (Harten-Lax-van Leer) is usually used in the TVD-RK DG method. Another choice but higher order is the weighted average flux (WAF). It was first introduced by Toro; see [6, 7]. The WAF approximation is second-order accurate in both space and time in statistical sense [6]. This approximation has been extensively applied in the finite volume method, but it is rarely used in the TVD-RK DG method. So the main objective of this work is to try to modify the WAF in the TVD-RK DG method.

The weighted average flux, $\widehat{F}^{\text{WAF}}$ at the interface, $x_{j+1/2}$, is defined by the integral average of a flux function $F(U)$ at the half-time step,

$$\widehat{F}_{j+1/2}^{\text{WAF}} = \frac{1}{\Delta x}\int_{-\Delta x/2}^{\Delta x/2} F\left(U_{j+1/2}\left(x, \frac{\Delta t}{2}\right)\right) dx. \quad (11)$$

It can be written in the wave structure form as

$$\widehat{F}_{j+1/2}^{\text{WAF}} = \sum_{k=1}^{N_c+1} \omega_k F_{j+1/2}^{(k)}, \quad (12)$$

where N_c is the number of wave solutions in the Riemann problem and $F_{j+1/2}^{(k)}$ is the kth flux of the Riemann problem. For one-dimensional shallow water flows, we have $N_c = 2$, where $F_{j+1/2}^{(1)} = F(U_{j+1/2}^-)$ and $F_{j+1/2}^{(3)} = F(U_{j+1/2}^+)$. Flux component $F_{j+1/2}^{(2)}$ can be obtained from the HLL approach [6]. Weighted values, ω_k, are defined by $\omega_k = (1/2)(c_k - c_{k-1})$, where $c_k = S_k\Delta t/\Delta x$ is the Courant number of wave k, $c_0 = -1$, $c_{N_c+1} = 1$, and S_k is the speed of wave k.

To avoid spurious oscillations near a shock front, the WAF method will be modified by enforcing a total variation diminishing (TVD) scheme [6, 7, 9, 11, 19]. The TVD-WAF version is that

$$\widehat{F}_{j+1/2}^{\text{WAF-TVD}} = \frac{1}{2}\left(F\left(U_{j+1/2}^-\right) + F\left(U_{j+1/2}^+\right)\right) - \frac{1}{2}\sum_{k=1}^{N_c} \operatorname{sgn}(c_k)\,\phi_{j+1/2}^{(k)}\Delta F_{j+1/2}^{(k)}, \quad (13)$$

where

$$\Delta F_{j+1/2}^{(k)} = F_{j+1/2}^{(k+1)} - F_{j+1/2}^{(k)}. \quad (14)$$

Here $\phi_{j+1/2}^{(k)}$ is a WAF limiter function. There are various choices; see more details in [6, 7, 9, 11, 19]. In this work, we choose the basic one of minmod type.

2.3. Consistency of WAF with TVD-RK DG Method. In this section, we show the consistency of the weighted average flux in the TVD-RK discontinuous Galerkin method.

Lemma 1. *The TVD Runge-Kutta discontinuous Galerkin method is consistent with the weighted average flux for smooth flows.*

Proof. Considering the weak form of RKDG method for the conservation law,

$$\int_{I_j}\left(\frac{\partial}{\partial t}U\right) v(x)\,dx - \int_{I_j} F(U)\,v'(x)\,dx - F\left(U_{j+1/2}\right) v\left(x_{j+1/2}^-\right) + F\left(U_{j-1/2}\right) v\left(x_{j-1/2}^+\right) = 0. \quad (15)$$

Solution U is approximated by U_h^*. The test function v is estimated by v_h. Flux function at cell interfaces is approximated in terms of numerical flux, $\widehat{F}_{j\pm1/2}$, so we obtain a numerical scheme of the following form:

$$\int_{I_j}\left(\frac{\partial}{\partial t}U_h^*\right) v_h(x)\,dx - \int_{I_j} F(U_h^*)\,v_h'(x)\,dx - \widehat{F}_{j+1/2}\left(\left(U_h^*\right)_{j+1/2}^-, \left(U_h^*\right)_{j+1/2}^+\right) v_h\left(x_{j+1/2}^-\right) + \widehat{F}_{j-1/2}\left(\left(U_h^*\right)_{j-1/2}^-, \left(U_h^*\right)_{j-1/2}^+\right) v_h\left(x_{j-1/2}^+\right) = 0. \quad (16)$$

Substituting U_h^* by U in (16) and subtracting with (15), we obtain a truncation error term T_1 as

$$T_1 = \left|\int_{I_j}\frac{\partial}{\partial t}[U - U]\,v_h(x)\,dx - \int_{I_j}[F(U) - F(U)]\,v_h'(x)\,dx\right.$$

$$- \left[F\left(U_{j+1/2}\right) - \widehat{F}_{j+1/2}\left(U_{j+1/2}^{-}, U_{j+1/2}^{+}\right)\right] v_h\left(x_{j+1/2}^{-}\right) + \left[F\left(U_{j-1/2}\right) - \widehat{F}_{j-1/2}\left(U_{j-1/2}^{-}, U_{j-1/2}^{+}\right)\right] \cdot v_h\left(x_{j-1/2}^{+}\right)\Big| . \quad (17)$$

When the solution is smooth, it must be continuous at $x_{j+1/2}$ or $U_{j+1/2}^{+} = U_{j+1/2}^{-} = U_{j+1/2}$. □

The TVD weighted average flux is defined by

$$\widehat{F}_{j+1/2}\left(U_{j+1/2}^{-}, U_{j+1/2}^{+}\right) = \frac{1}{2}\left(F\left(U_{j+1/2}^{-}\right) + F\left(U_{j+1/2}^{+}\right)\right) - \frac{1}{2}\sum_{k=1}^{2} \operatorname{sgn}\left(c_k\right)\phi^k \Delta F^{(k)}, \quad (18)$$

where $F^{(1)} = F(U_{j+1/2}^{-}) = F(U_{j+1/2})$, $F^{(3)} = F(U_{j+1/2}^{+}) = F(U_{j+1/2})$, and the component $F^{(2)} = F(U_{j+1/2})$ which can be obtained from the HLL method [4, 6, 7]. Thus,

$$\widehat{F}_{j+1/2}\left(U_{j+1/2}^{-}, U_{j+1/2}^{+}\right) = \frac{1}{2}\left(F\left(U_{j+1/2}\right) + F\left(U_{j+1/2}\right)\right) - \frac{1}{2}\sum_{k=1}^{2} \operatorname{sgn}\left(c_k\right)\phi^k \Delta F^{(k)} = F\left(U_{j+1/2}\right). \quad (19)$$

Similarly, we have $\widehat{F}_{j-1/2}(U_{j-1/2}^{-}, U_{j-1/2}^{+}) = F(U_{j-1/2})$. Then, the weighted average flux is consistent.

For the TVD-RK DG method, the approximate solution U_h^* in (16) is defined by

$$U_h^*(x,t) = \sum_{k=0}^{N} U_k(t)\varphi_k(x), \quad (20)$$

where $\varphi_k(x)$ is the Legendre polynomial degree k. Since the considering solution is smooth, we know from Theorem (3.1) in [1] that

$$\left\|U - U_h^*\right\|_{L^2(0,1)} \le O\left(\Delta x^{N+1/2}\right), \quad \text{for } N \ge 0. \quad (21)$$

This norm is defined as a truncation error term due to approximating polynomial, denoted by T_2.

After substituting the approximate solution into the numerical scheme (16), we obtain an ODE system,

$$\frac{d}{dt}U_h^* = L\left(U_h^*\right). \quad (22)$$

Next, we apply the TVD Runge-Kutta scheme for integrating in time. We approximate U_h^* at each time step by $\widehat{U}_h$. The order of accuracy for time integration depends on the TVD-RK order [1, 3, 17]. If the TVD-RK order $N+1$ is applied, a truncation error term T_3 is given by

$$T_3 = \left|U_h^* - \widehat{U}_h\right| \le O\left(\Delta t^{N+1}\right). \quad (23)$$

Combining all of the truncation error terms together, the total truncation error term T is estimated by

$$T = T_1 + T_2 + T_3 \le O\left(\Delta x^{N+1/2}, \Delta t^{N+1}\right). \quad (24)$$

The total truncation error term is approaching zero as $\Delta x \to 0$ and $\Delta t \to 0$. Hence, the TVD-RK DG with the WAF method is consistent.

3. Well-Balanced TVD-RK DG with WAF Scheme

In this section, we develop a well-balanced scheme for the TVD-RK DG with the WAF method. The main purpose is to present a modified scheme for solving the shallow water equations with source term. Also, this scheme must preserve exactly stationary solution when bottom slope exists. Let us start by considering the standard TVD-RK DG method,

$$\int_{I_j} \left(\partial_t U_h\right) v_h dx - \int_{I_j} F\left(U_h\right)\partial_x v_h dx + \widehat{F}_{j+1/2} v_h\left(x_{j+1/2}^{-}\right) - \widehat{F}_{j-1/2} v_h\left(x_{j-1/2}^{+}\right) = \int_{I_j} G\left(U_h\right) v_h dx, \quad (25)$$

with initial condition

$$\int_{I_j} U_h(x,0)\, v_h(x)\, dx = \int_{I_j} U_0(x)\, v_h(x)\, dx, \quad (26)$$

where $\widehat{F}_{j\pm1/2}$ are the weighted average flux in (13). Source term is given by

$$G(U) = \begin{pmatrix} G_1(U,x) \\ G_2(U,x) \end{pmatrix} = \begin{pmatrix} 0 \\ -ghb_x \end{pmatrix}. \quad (27)$$

We will derive a well-balanced scheme based on [14], but the weighted average flux is applied in the TVD-RK DG instead of using the Lax-Friedrichs (LF) flux. The main modification is the source term treatments in the WAF method. Assume that the source terms can be written by the summation of some functions as

$$G_p(U,x) = \sum_i s_i\left(a(U,x)\right) t_i'(x), \quad p = 1,2, \quad (28)$$

where s_i and t_i are some functions which will be determined later.

If the solution at steady state is stationary, then $a(U, x)$ in (28) can be decomposed by $a_1(U, x)$ and $a_2(U, x)$, such that

$$a_1 \equiv h + b = \text{constant},\qquad a_2 \equiv u = 0. \tag{29}$$

Since G_1 is zero, we consider only G_2,

$$G_2(U, x) = -ghb_x = -g(h+b)b_x + \frac{1}{2}g\left(b^2\right)_x. \tag{30}$$

From (28), we have that

$$s_1 = s_1(a_1) = -g(h+b),\qquad s_2 = \frac{1}{2}g,\qquad t_1(x) = b,\qquad t_2(x) = b^2. \tag{31}$$

To balance the flux gradient and the source term approximation at steady state, it is required that

$$\frac{d}{dx}\left(F(U(x)) - \sum_i s_i(a(U, x))t_i(x)\right) = 0, \tag{32}$$

or

$$F(U(x)) - \sum_i s_i(a(U, x))t_i(x) = \alpha, \tag{33}$$

where α is arbitrary constant.

The integration of source term in (25) can be approximated by

$$\begin{aligned}\int_{I_j} G_2(U_h, x)v_h dx &= \int_{I_j}\sum_i s_i(a(U, x))t_i'(x)v_h(x)dx\\ &= \sum_i\left[\frac{1}{2}\left(s_i\left(a(U_h, x)_{j-1/2}^+\right) + s_i\left(a(U_h, x)_{j+1/2}^-\right)\right)\right]\left[t_i(x_{i+1/2})v_h\left(x_{i+1/2}^-\right) - t_i(x_{i-1/2})v_h\left(x_{i-1/2}^+\right) - \int_{I_j} t_i(x)v_h'(x)dx\right]\\ &+ \sum_i\left[\int_{I_j}\left\{s_i(a(U_h, x)) - \frac{1}{2}\left(s_i\left(a(U_h, x)_{j-1/2}^+\right) + s_i\left(a(U_h, x)_{j+1/2}^-\right)\right)\right\}t_i'(x)v_h(x)dx\right].\end{aligned} \tag{34}$$

Functions $a(u_h, x)$ and t_i on the RHS of (34) can be approximated by $a_h(U_h, x)$ and $(t_i)_h$, where $a_h(U_h, x)$ and $(t_i)_h$ are the L^2 projection of $a(U_h, x)$ and t_i into the space of V_h. Then, we obtain $(t_1)_h = b_h$ and $(t_2)_h = b_h^2$. Thus,

$$F(U_h(x)) - \sum_i s_i(a(U_h, x))(t_i(x)_h) = \alpha. \tag{35}$$

Then,

$$\begin{aligned}\int_{I_j} G_2(U_h, x)v_h dx &= \sum_i\left[\frac{1}{2}\left(s_i\left(a_h(U_h, x)_{j-1/2}^+\right) + s_i\left(a_h(U_h, x)_{j+1/2}^-\right)\right)\right]\\ &\cdot\left[\left(\hat{t}_i\right)_{h,j+1/2}v_h\left(x_{i+1/2}^-\right) - \left(\hat{t}_i\right)_{h,j-1/2}(x_{i-1/2})v_h\left(x_{i-1/2}^+\right) - \int_{I_j}(t_i)_h(x)v_h'(x)dx\right]\\ &+ \sum_i\left[\int_{I_j}\left\{s_i(a_h(U_h, x)) - \frac{1}{2}\left(s_i\left(a_h(U_h, x)_{j-1/2}^+\right) + s_i\left(a_h(U_h, x)_{j+1/2}^-\right)\right)\right\}(t_i)_h'(x)v_h dx\right].\end{aligned} \tag{36}$$

To satisfy the weighted average flux in (12), $\left(\hat{t}_i\right)_{h,j+1/2}$ must be modified. Here, we assume that

$$\left(\hat{t}_i\right)_{h,j+1/2} = \sum_{k=1}^{N_c+1}\omega_k t_{h,j+1/2}^{(k)}, \tag{37}$$

or if the TVD version in (13) is applied, this term can be written by

$$\left(\hat{t}_i\right)_{h,j+1/2} = \sum_{k=1}^{N_c+1}\operatorname{sgn}(c_k)\phi^{(k)}t_{h,j+1/2}^{(k)}. \tag{38}$$

Note that ω_k are weighted values in the WAF approximations and $\phi^{(k)}$ is the WAF limiter function. So $t^{(k)}_{h,j+1/2}$ can be defined similar to $F^{(k)}_{j+1/2}$.

For one-dimensional shallow water equations, we have $N_c = 2$, where $t^{(1)}_{h,j+1/2} = t_h(x^-_{j+1/2})$ and $t^{(3)}_{h,j+1/2} = t_h(x^+_{j+1/2})$. The approximation of $t^{(2)}_{h,j+1/2}$ in the intermediate region can be obtained by the HLL approach,

$$t^{(2)}_{h,j+1/2} = \begin{cases} t_h\left(x^-_{j+1/2}\right), & \text{if } S_L \geq 0, \\ \dfrac{S_R t_h\left(x^-_{j+1/2}\right) - S_L t_h\left(x^+_{j+1/2}\right) + S_R S_L\left(t_h\left(x^+_{j+1/2}\right) - t_h\left(x^-_{j+1/2}\right)\right)}{S_R - S_L}, & \text{if } S_L < 0 < S_R, \\ t_h\left(x^+_{j+1/2}\right), & \text{if } S_R \leq 0. \end{cases} \tag{39}$$

Here S_L and S_R are signal velocities in the Riemann problem.

At steady state, the solution is assumed to be stationary. It implies that $h + b = c$, where c is constant and $u = 0$. From (35), we have that

$$\begin{aligned} F(U_h) &- \sum_i s_i\left(a_h(U,x)\right) t_i(x) \\ &= \frac{1}{2}g(h_h)^2 - \left[\frac{1}{2}gb_h^2 - g(h_h + b_h)b_h\right] \\ &= \frac{1}{2}g(h_h + b_h)^2 = \frac{1}{2}gc^2 = \alpha. \end{aligned} \tag{40}$$

This shows the suitable choices of $a(U_h, x)$ and t_i in the TVD-RK DG with the WAF approximation.

Next, we will show that the TVD-RK DG with the WAF method preserves the well-balanced property.

Proposition 2. *The TVD-RK DG with the WAF scheme has preserved exactly stationary solutions at steady state.*

Proof. Since $a_h(U_h, x)$ and $a_h(U_h, x)^{\pm}_{i\mp1/2}$ are equal to the same constant c at each Gauss-Lobatto point over cell I_j, thus

$$\sum_i\left[\int_{I_j}\left\{s_i\left(a_h\left(U_h,x\right)\right) - \frac{1}{2}\left(s_i\left(a_h\left(U_h,x\right)^+_{j-1/2}\right) + s_i\left(a_h\left(U_h,x\right)^-_{j+1/2}\right)\right)\right\}(t_i)'_h(x)\,v_h dx\right] = 0. \tag{41}$$

Hence, we obtain a truncation error term T as

$$\begin{aligned} T = &\int_{I_j} F\left(U_h(x,t)\right) v'_h(x)\,dx - \widehat{F}_{j+1/2} v_h\left(x^-_{j+1/2}\right) \\ &+ \widehat{F}_{j-1/2} v_h\left(x^+_{j-1/2}\right) + \sum_i\left[\frac{1}{2}\left(s_i\left(a_h\left(U_h,x\right)^+_{j-1/2}\right)\right.\right. \\ &\left.\left. + s_i\left(a_h\left(U_h,x\right)^-_{j+1/2}\right)\right)\right]\left[\left(\widehat{t}_i\right)_{h,j+1/2} v_h\left(x^-_{i+1/2}\right)\right. \\ &- \left(\widehat{t}_i\right)_{h,j-1/2}\left(x_{i-1/2}\right) v_h\left(x^+_{i-1/2}\right) \\ &\left. - \int_{I_j}(t_i)_h(x)\,v'_h(x)\,dx\right]. \end{aligned} \tag{42}$$

Using $\widehat{F}$ in (12), $(\widehat{t}_i)$ in (37), and $a_h(U_h, x)^{\pm}_{j\mp1/2} = c$, we have that

$$\begin{aligned} T = &\int_{I_j}\left[F\left(U_h(x,t)\right) - \sum_i s_i(c)(t_i)_h\right] v'_h(x)\,dx \\ &- \left[\sum_{k=1}^{3}\omega_k F^{(k)}_{j+1/2} - \sum_i s_i(c)\left(\sum_{k=1}^{3}\omega_k t^{(k)}_{h,j+1/2}\right)\right] \\ &\cdot v_h\left(x^-_{j+1/2}\right) \\ &+ \left[\sum_{k=1}^{3}\omega_k F^{(k)}_{j-1/2} - \sum_i s_i(c)\left(\sum_{k=1}^{3}\omega_k t^{(k)}_{h,j-1/2}\right)\right] \\ &\cdot v_h\left(x^+_{j-1/2}\right) \\ = &\int_{I_j}\left[F\left(U_h(x,t)\right) - \sum_i s_i(c)(t_i)_h\right] v'_h(x)\,dx \\ &- \left[\sum_{k=1}^{3}\omega_k\left(F^{(k)}_{j+1/2} - \sum_i s_i(c)\,t^{(k)}_{h,j+1/2}\right)\right] v_h\left(x^-_{j+1/2}\right) \\ &+ \left[\sum_{k=1}^{3}\omega_k\left(F^{(k)}_{j-1/2} - \sum_i s_i(c)\,t^{(k)}_{h,j-1/2}\right)\right] v_h\left(x^+_{j-1/2}\right). \end{aligned} \tag{43}$$

After applying condition (35) and rearranging terms, yields

$$\begin{aligned} T = &\,\alpha\left[v_h\left(x^-_{j+1/2}\right) - v_h\left(x^+_{j-1/2}\right)\right] - \alpha v_h\left(x^-_{j+1/2}\right) \\ &+ \alpha v_h\left(x^+_{j-1/2}\right) = 0, \end{aligned} \tag{44}$$

where α is arbitrary constant. Hence, the TVD-RK DG with the WAF scheme preserves exactly the stationary solution at steady state. □

TABLE 1: RMS errors when $N = 0$, $N = 1$, and $N = 2$.

N	Numerical flux	$K = 50$	$K = 100$	$K = 200$
$N = 0$	HLL	0.0280	0.0215	0.0165
	WAF	0.0182	0.0127	0.0094
$N = 1$	HLL	0.0130	0.0102	0.0072
	WAF	0.0107	0.0093	0.0067
$N = 2$	HLL	0.0103	0.0079	0.0053
	WAF	0.0100	0.0076	0.0052

Remark 3. One can show that the developed well-balanced TVD-RK DG with WAF scheme in the case of existing source term is consistent, by showing that the approximation of the source term is also consistent.

4. Numerical Results

In this section, various test cases have been investigated to demonstrate the accuracy of the present scheme, not only steady, but also unsteady flows.

4.1. Dam Break Flow. It has been shown in the previous section that the weighted average flux is consistent with the TVD-RK DG method. We apply this modified scheme to solve the shallow water equations without source terms in this subsection. The accuracy of numerical solutions is shown and compared with the standard TVD-RK DG when using the HLL method.

The computational domain is that $-5 \le x \le 5$. The initial water depth is given by

$$h(x,t) = \begin{cases} 1, & \text{if } x \le 0, \\ 0.6, & \text{if } x > 0. \end{cases} \tag{45}$$

The initial velocity is assumed to be zero. The boundary conditions are transmissive boundaries. We perform 50, 100, and 200 cells in the numerical experiments. Polynomial degree zero, one, and two are applied as a local basis in the TVD-RK DG method. Simulation time is $t = 2$, with $\Delta t = 0.005$. The root mean squared errors (RMS) are shown in Table 1.

When N and K are fixed, the accuracy of numerical solution obtained from the WAF is higher than those obtained from the HLL method. The RMS errors decrease as K increases.

If we fix K and vary N, the RMS error decreases as the polynomial degree increases.

The water depth profiles using the HLL and the WAF methods at $t = 2$ when $N = 1$ and $K = 100$ are shown in Figures 1 and 2, respectively. The front of moving shock can be captured correctly by the HLL and the WAF methods. But the scheme using the WAF can capture shock and rarefaction wave more precisely than those using the HLL method. This investigation shows the accuracy of the TVD-RK DG with the WAF method. It has been generalized from the finite volume method.

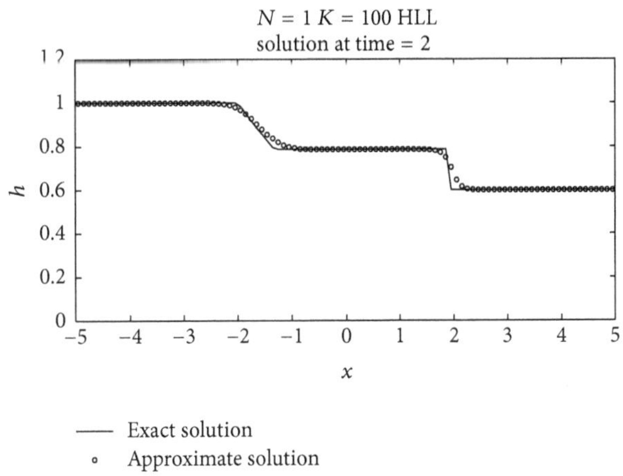

FIGURE 1: Exact solution and water depth profile obtained by the TVD-RK DG HLL method.

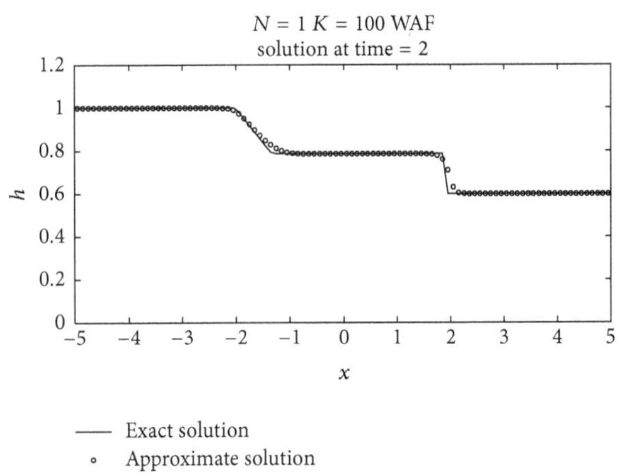

FIGURE 2: Exact solution and water depth profile obtained by the TVD-RK DG WAF method.

4.2. Flow over Irregular Bed. The uniform channel is length of 1500 m. The bottom elevation is irregular that is shown in Figure 3. This problem is proposed by [5] for testing the accuracy of numerical scheme at stationary state. The boundary conditions are transmissive. The initial water depth is that $h + b = 16$, with zero initial velocity. We set $\Delta t = 0.01$ and run simulation until $t = 100$.

It can be seen from Figure 3 that the well-balanced scheme (dot) gives exactly the stationary solution while the non-well-balanced scheme (dash line) gives solution error especially in the high gradient area of bottom elevation.

4.3. Steady Flow over a Bump. We consider the shallow water flows over a bump in a rectangular channel with length of 25 m. The bump elevation is given by

$$b(x) = \begin{cases} 0.2 - 0.05(x - 10)^2, & \text{if } 8 < x < 12, \\ 0, & \text{otherwise.} \end{cases} \tag{46}$$

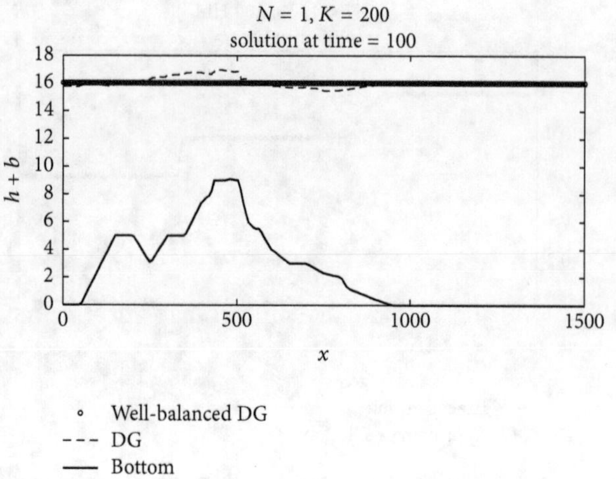

FIGURE 3: Stationary flow over irregular bed by well-balanced and non well-balanced TVD-RK DG with WAF method.

TABLE 2: RMS errors when $N = 1$ for subcritical flow over a bump.

	$K = 50$	$K = 100$	$K = 200$	$K = 400$
HLL	$1.6954e-03$	$4.5563e-04$	$1.2546e-04$	$3.5900e-05$
WAF	$1.5816e-03$	$4.1648e-04$	$1.1474e-04$	$3.3019e-05$

At steady state, classical flows can be characterised by subcritical flow, transcritical flow with shock, and transcritical flow without shock. The TVD-RK DG with the WAF method is performed to solve these problems. The accuracy of numerical solutions can be checked by comparing with the existing analytical solutions; see [20].

Subcritical Flow over a Bump. The upstream boundary is imposed by $q = 4.42\,\mathrm{m}^2/\mathrm{s}$, while the downstream boundary is set by $h = 2$ m. The initial water depth is $h+b = 2$ with zero initial velocity. Time step is that $\Delta t = 0.01$. The RMS errors obtained from the HLL and the WAF methods are shown in Table 2. This shows that the RMS errors obtained by the TVD-RK DG WAF are less than those obtained by the HLL for all K.

The water depth and the bump profiles are shown in Figure 4. The numerical solution is very close to the analytical solution. We also found in this test case that the well-balanced scheme converges to the steady solution faster than the scheme that is not well-balanced.

Transcritical Flow with Shock over a Bump. The upstream boundary is given by $q = 0.18\,\mathrm{m}^2/\mathrm{s}$, while the downstream boundary is set by $h = 0.33$ m. The initial condition is that $h + b = 0.33$ m. The comparison of water surfaces is shown in Figure 5. The numerical result is in good agreement with the analytical result. This shows the accuracy of the well-balanced scheme that can capture the shock front without any oscillations. It is also found that the well-balanced scheme converges to the steady solution faster than the non-well-balanced scheme.

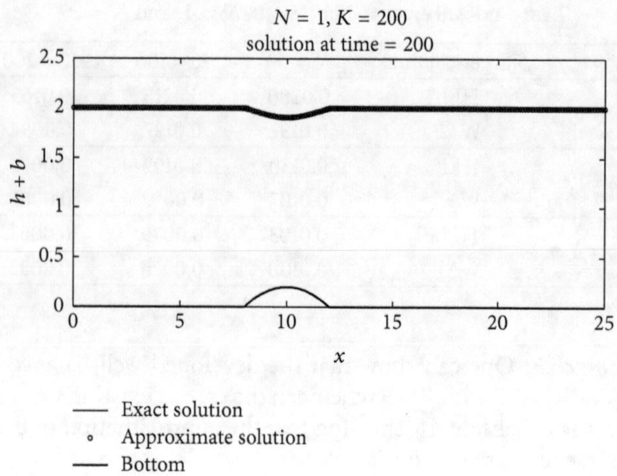

FIGURE 4: Water depth and bump profiles for subcritical flow.

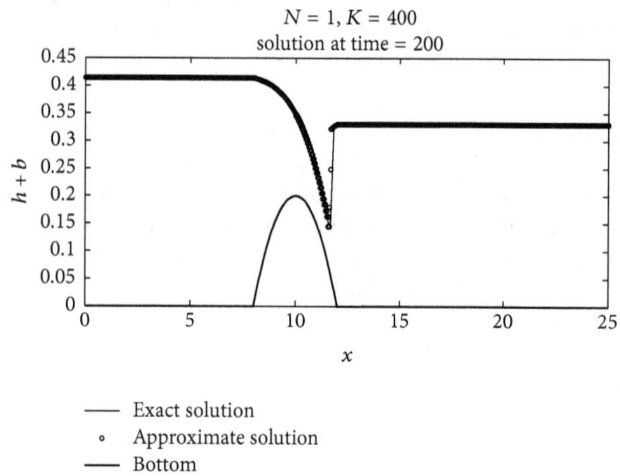

FIGURE 5: Transcritical flow with shock over a bump.

Transcritical Flow without Shock over a Bump. The upstream boundary is prescribed by $q = 1.53\,\mathrm{m}^2/\mathrm{s}$, while the downstream boundary is not specified. The initial conditions is that $h + b = 0.4$ m with zero initial velocity.

The water depth profiles are shown in Figure 6. The numerical result agrees well with the analytical solution. These results show the accuracy of the well-balanced scheme for solving transcritical flow problem.

4.4. Small Perturbation of Steady State Water. This problem is first proposed by [18, 21, 22]. It can be used to study the capability of numerical scheme for solving small perturbation in shallow water flow. The bottom topography is defined by

$$b(x) = \begin{cases} 0.25\left(\cos\left(10\pi\left(x-1.5\right)\right)+1\right), & \text{if } 1.4 < x < 1.6, \\ 0. & \text{otherwise.} \end{cases} \quad (47)$$

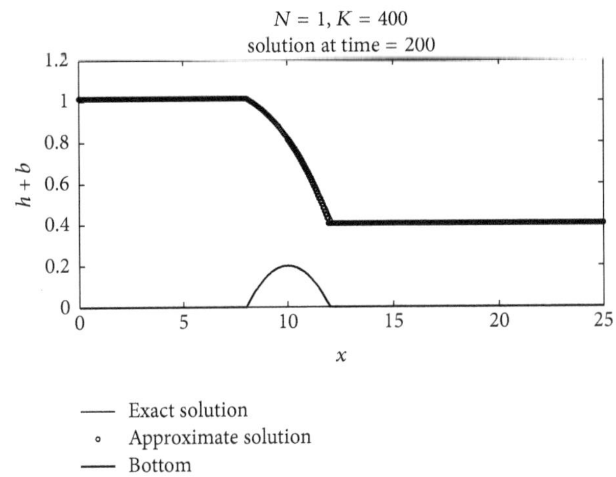

FIGURE 6: Transcritical flow without shock over a bump.

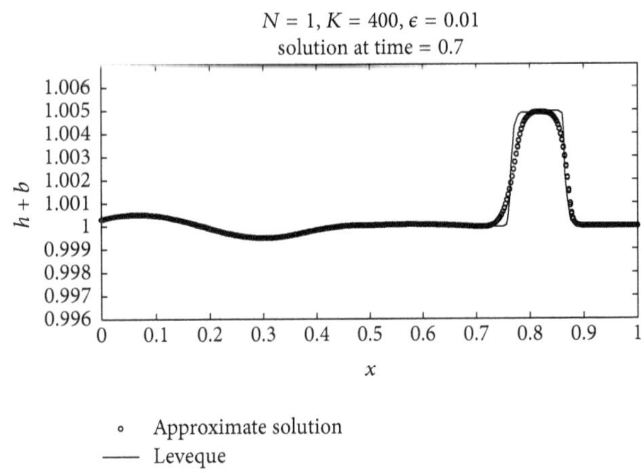

FIGURE 7: Quasi-stationary flow when $\epsilon = 0.01$.

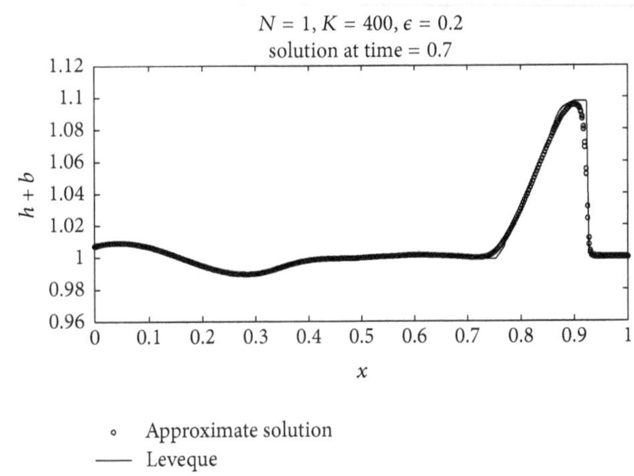

FIGURE 8: Quasi-stationary flow when $\epsilon = 0.2$.

The initial conditions are specified by

$$q(x,0) = 0,$$
$$h(x,0) = \begin{cases} 1 - b(x) + \epsilon, & \text{if } 1.1 < x < 1.2, \\ 1 - b(x), & \text{otherwise,} \end{cases} \qquad (48)$$

where ϵ is a nonzero perturbation constant. The boundary conditions are transmissive boundaries. In this work, we consider the cases of $\epsilon = 0.2$ and 0.01. The disturbance of initial water depth from small ϵ should split the initial wave into two waves. They propagate to the left and the right with characteristic speed $\pm\sqrt{gh}$ at the early stage. A standard scheme which is not well-balanced usually faces with some difficulties to capture correctly the wave speed.

In our simulation, we use 400 uniform grid cells and polynomial degree one in the TVD-RK DG with the WAF method. The simulation time is that $t = 0.7$.

The comparison of water depths between our results and the Leveque's solutions is shown in Figures 7 and 8 for $\epsilon = 0.01$ and 0.2, respectively. They are in good agreement for both amplitude and wave speed. These test cases show the ability of our numerical scheme for solving the quasi-stationary flow with initial disturbance.

4.5. Flow over Nonhorizontal Bed. This test case is presented by [8]. The main propose is to study the ability of numerical scheme for solving unsteady flow over topography. The uniform channel is length of 30 m. The bed elevation is defined by

$$b(x) = \begin{cases} 0, & \text{if } 0 \le x \le 10, \\ 0.1(x-10), & \text{if } 10 \le x \le 20, \\ 1, & \text{if } 20 \le x \le 30. \end{cases} \qquad (49)$$

The initial conditions are given by

$$h(x) + b(x) = \begin{cases} 4, & \text{if } 0 \le x < 5, \\ 2, & \text{if } 5 \le x \le 30. \end{cases} \qquad (50)$$

Here, we set $\Delta t = 0.01$ and use 200 uniform grid cells with polynomial degree one to simulate the unsteady flow. The boundary conditions are transmissive boundaries.

Our numerical results when comparing with Toro's solutions [8] at $t = 1$ s and 4 s are shown in Figure 9. Wave speed and shock profiles are very close. These results show the accuracy of the present scheme for solving unsteady flow with source term.

5. Conclusions

In this work, we present the TVD-RK discontinuous Galerkin method (TVD-RK DG) for solving nonlinear shallow water equations. Most of the TVD-RK DG methods in the literatures usually approximate intercell flux by applying the

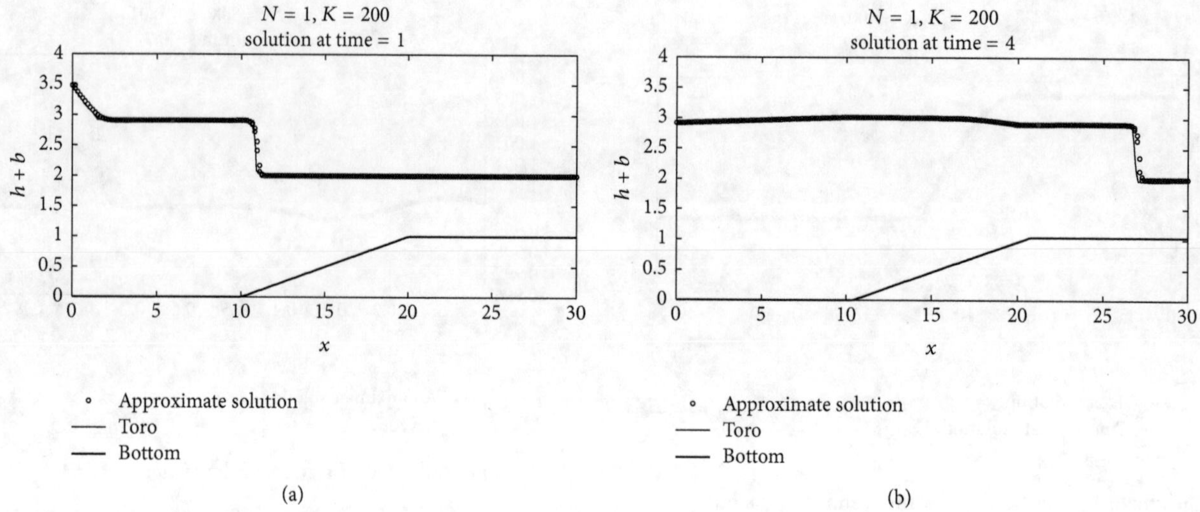

FIGURE 9: Flow over non-horizontal bed at time 1 s (a) and 4 s (b).

HLL method. But here we apply another approach called the weighted average flux (WAF) in the TVD-RK DG. We have also shown the consistent property of the TVD-RK DG with the WAF approximation. Then the well-balanced TVD-RK DG scheme with the WAF approximation is developed. The present method can be used to simulate not only steady flows, but also unsteady flows. The accuracy of modified numerical scheme is demonstrated by various test cases, flow over irregular bed, steady flow over a bump, quasi-stationary, and flow over nonhorizontal bed. The well-balanced TVD-RK DG with the WAF method can be used to solve all the kinds of these problems. Moreover, if we restrict at steady state, the scheme using the WAF method converges to the steady solution faster than the scheme using the HLL method. In addition, the well-balanced scheme converges to the steady solution faster than the scheme that is not well-balanced. Due to its advantages of numerical accuracy, simplicity, and well-balanced property, the present scheme can be modified and extended to simulate two-dimensional problems. However, depending on the types of elements, for example, triangles or rectangles, it is not trivial to extend for two-dimensional problems due to the polynomial basis functions and the WAF fluxes at element interfaces and is not considered in this paper.

Conflict of Interests

The authors declare that there is no conflict of interests regarding the publication of this paper.

Acknowledgments

This research is partially supported by the Science Achivement Scholarship Thailand (SAST) to the first author and financially supported by the Thailand Research Fund (TRF) under the Grant no. RSA5680038.

References

[1] B. Cockburn, C. W. Shu, C. Johnson, and E. Tadmor, *Advanced Numerical Approximation of Nonlinear Hyperbolic Equations: Lectures Given at the 2nd Session of the Centro Internazionale Matematico Estivo (C.I.M.E.) held in Cetraro, Italy, June 23–28, 1997*, Springer, 1998.

[2] D. Schwanenberg, R. Kiem, and J. Kongeter, "A discontinuous Galerkin method for the shallow-water equations with source terms," in *Discontinuous Galerkin Methods: Theory, Computations and Applications*, B. Cockburn, G. E. Karniadaki, and C.-W. Chu, Eds., Lecture Notes in Computational Science and Engineering, pp. 1419–1424, Springer, 2000.

[3] J. S. Hesthaven and T. Warburton, *Nodal Discontinuous Galerkin Methods: Algorithms, Analysis, and Applications*, Springer, 2008.

[4] P. A. Tassi and C.-A. Vionnet, "Discontinuous Galerkin method for the one dimensional simulation of shallow water flows," *Mecànica Computational*, vol. 22, pp. 2403–2428, 2003.

[5] W. Lai and A. A. Bouchut, "Discontinuous Galerkin method for 1D shallow water flow with water surface slope limiter," *International Journal on Civil and Environmental Engineering*, vol. 3, pp. 167–176, 2011.

[6] E. F. Toro, *Riemann Solvers and Numerical Methods for Fluid Dynamics: A Practical Introduction*, Springer, Berlin, Germany, 1999.

[7] E. F. Toro, *Shock-Capturing Methods for Free-Surface Shallow Flows*, John Wiley & Sons, New York, NY, USA, 2001.

[8] E. F. Toro, "Riemann problems and the WAF method for solving the two-dimensional shallow water equations," *Philosophical Transactions of the Royal Society of London A: Mathematical, Physical and Engineering*, vol. 338, no. 1649, pp. 43–68, 1992.

[9] M. Maleewong, "Modified predictor-corrector WAF method for the shallow water equations with source terms," *Mathematical Problems in Engineering*, vol. 2011, Article ID 178491, 17 pages, 2011.

[10] E. F. Toro and P. L. Roe, "A hybridised high-order random choice method for quasi-linear hyperbolic systems," in *Proceedings of the 16th International Symposium on Shock Tubes and Waves*, p. 701, Gronig, Aachen, Germany, July 1987.

[11] S. J. Billett and E. F. Toro, "On WAF-type schemes for multidimensional hyperbolic conservation laws," *Journal of Computational Physics*, vol. 130, no. 1, pp. 1–24, 1997.

[12] A. Bermudez and M. E. Vazquez, "Upwind methods for hyperbolic conservation laws with source terms," *Computers & Fluids*, vol. 23, no. 8, pp. 1049–1071, 1994.

[13] G. Kesserwani and Q. Liang, "Well-balanced RKDG2 solutions to the shallow water equations over irregular domains with wetting and drying," *Computers & Fluids*, vol. 39, no. 10, pp. 2040–2050, 2010.

[14] Y. Xing and C.-W. Shu, "High order well-balanced finite volume WENO schemes and discontinuous Galerkin methods for a class of hyperbolic systems with source terms," *Journal of Computational Physics*, vol. 214, no. 2, pp. 567–598, 2006.

[15] E. Audusse, F. Bouchut, M.-O. Bristeau, R. Klein, and B. Perthame, "A fast and stable well-balanced scheme with hydrostatic reconstruction for shallow water flows," *SIAM Journal on Scientific Computing*, vol. 25, no. 6, pp. 2050–2065, 2004.

[16] S. Noelle, N. Pankratz, G. Puppo, and J. R. Natvig, "Well-balanced finite volume schemes of arbitrary order of accuracy for shallow water flows," *Journal of Computational Physics*, vol. 213, no. 2, pp. 474–499, 2006.

[17] C.-W. Shu, "Total-variation-diminishing time discretizations," *SIAM Journal on Scientific and Statistical Computing*, vol. 9, no. 6, pp. 1073–1084, 1988.

[18] R. J. Leveque, *Finite Volume Method for Hyperbolic Problem*, Cambridge University Press, 2005.

[19] E. D. Fernández-Nieto and G. Narbona-Reina, "Extension of WAF type methods to non-homogeneous shallow water equations with pollutant," *Journal of Scientific Computing*, vol. 36, no. 2, pp. 193–217, 2008.

[20] O. Delestre, C. Lucas, P.-A. Ksinant et al., "SWASHES: a compilation of shallow water analytic solutions for hydraulic and environmental studies," *International Journal for Numerical Methods in Fluids*, pp. 346–365, 2012.

[21] R. J. Leveque, *Numerical Methods for Conservation Laws*, Birkhäuser Press, 1992.

[22] R. J. LeVeque, "Balancing source terms and flux gradients in high-resolution Godunov methods: the quasi-steady wave-propagation algorithm," *Journal of Computational Physics*, vol. 146, no. 1, pp. 346–365, 1998.

Permissions

All chapters in this book were first published in MSE, by Hindawi Publishing Corporation; hereby published with permission under the Creative Commons Attribution License or equivalent. Every chapter published in this book has been scrutinized by our experts. Their significance has been extensively debated. The topics covered herein carry significant findings which will fuel the growth of the discipline. They may even be implemented as practical applications or may be referred to as a beginning point for another development.

The contributors of this book come from diverse backgrounds, making this book a truly international effort. This book will bring forth new frontiers with its revolutionizing research information and detailed analysis of the nascent developments around the world.

We would like to thank all the contributing authors for lending their expertise to make the book truly unique. They have played a crucial role in the development of this book. Without their invaluable contributions this book wouldn't have been possible. They have made vital efforts to compile up to date information on the varied aspects of this subject to make this book a valuable addition to the collection of many professionals and students.

This book was conceptualized with the vision of imparting up-to-date information and advanced data in this field. To ensure the same, a matchless editorial board was set up. Every individual on the board went through rigorous rounds of assessment to prove their worth. After which they invested a large part of their time researching and compiling the most relevant data for our readers.

The editorial board has been involved in producing this book since its inception. They have spent rigorous hours researching and exploring the diverse topics which have resulted in the successful publishing of this book. They have passed on their knowledge of decades through this book. To expedite this challenging task, the publisher supported the team at every step. A small team of assistant editors was also appointed to further simplify the editing procedure and attain best results for the readers.

Apart from the editorial board, the designing team has also invested a significant amount of their time in understanding the subject and creating the most relevant covers. They scrutinized every image to scout for the most suitable representation of the subject and create an appropriate cover for the book.

The publishing team has been an ardent support to the editorial, designing and production team. Their endless efforts to recruit the best for this project, has resulted in the accomplishment of this book. They are a veteran in the field of academics and their pool of knowledge is as vast as their experience in printing. Their expertise and guidance has proved useful at every step. Their uncompromising quality standards have made this book an exceptional effort. Their encouragement from time to time has been an inspiration for everyone.

The publisher and the editorial board hope that this book will prove to be a valuable piece of knowledge for researchers, students, practitioners and scholars across the globe.

List of Contributors

Behrouz Gordan, Azlan Adnan and Mariyana A. K. Aida
Faculty of Civil Engineering, Universiti Teknologi Malaysia, 81310 Skudai, Johor, Malaysia

Jiale Huang
School of Civil Engineering and Mechanics, Huazhong University of Science and Technology, Wuhan 430074, China

Fei Xiao and Yang Zhang
Jiangsu Research Institute of Building Science Co. Ltd, Nanjing 210008, China

Xi Tu, Jin Di, Cunjun Pang and Xiaoqing Xu
Key Laboratory of New Technology for Construction of Cities in Mountain Area (Chongqing University), Ministry of Education, Chongqing, China
College of Civil Engineering, Chongqing University, Chongqing, China

Erlei Yao and Yu Miao
School of Civil Engineering and Mechanics, Huazhong University of Science and Technology, Wuhan 430074, China

Guobo Wang
Hubei Key Laboratory of Roadway Bridge & Structure Engineer, Wuhan University of Technology, Wuhan 430074, China

Andrea Sorzia
Dipartimento di Scienze e Metodi dell'Ingegneria (DISMI), Universitá di Modena e Reggio Emilia, 42122 Reggio Emilia, Italy

Michalina Markousi, Dimitrios K. Fytanidis and Johannes V. Soulis
Fluid Mechanics/Hydraulics Division, Department of Civil Engineering, Democritus University of Thrace, 67100 Xanthi, Greece

Elmiloud Chaabelasri
National School of Applied Sciences, First Mohammed University, Al-Hociema, Morocco
LME, Faculty of Sciences, First Mohammed University, Oujda, Morocco

Liyan Zhang and Jian Ma
School of Civil Engineering, Suzhou University of Science and Technology, 1701 Binhe Road, New District, Suzhou, Jiangsu 215011, China

Department of Civil and Environmental Engineering, University of Wisconsin, 2312 Engineering Hall, 1415 Engineering Drive, Madison, WI 53706, USA

Bin Ran
Department of Civil and Environmental Engineering, University of Wisconsin, 2312 Engineering Hall, 1415 Engineering Drive, Madison, WI 53706, USA

Lixin Yan
College of Transportation and Logistics, East China Jiaotong University, Changbei Open and Developing District, Nanchang, Jiangxi 330013, China

Luciano Feo and Rosa Penna
Department of Civil Engineering, University of Salerno, Via Ponte don Melillo, 84084 Fisciano, Italy

Xiaohong Long, Wei Wang and Jian Fan
School of Civil Engineering and Mechanics, Huazhong University of Science and Technology, Wuhan 430074, China
Hubei Key Laboratory of Control Structure, Huazhong University of Science and Technology, Wuhan 430074, China

Afshin Mansouri
Department of Civil Engineering, The Islamic Azad University of Roudehen Branch, Tehran, Iran

Babak Aminnejad
Department of Civil Engineering, Faculty of Civil Engineering, The Islamic Azad University of Roudehen Branch, Tehran, Iran

Luciano Feo and Rosa Penna
Department of Civil Engineering, University of Salerno, Via Giovanni Paolo II, 132, 84084 Fisciano, Italy

V. P. Berardi
Department of Civil Engineering, University of Salerno, Via Ponte don Melillo, 84084 Fisciano, Italy

Yong Zhang, Tingsheng Zhao, Aijiao Zhou, Zhengzhu Zhang and Wen Liu
School of Civil Engineering & Mechanics, Huazhong University of Science and Technology, No. 1037, Luoyu Road, Hongshan District, Wuhan 430074, China

Beatrice Pomaro
Department of Civil, Environmental and Architectural Engineering, University of Padova, via F. Marzolo 9, 35131 Padova, Italy

Mahmoudreza Keymanesh
Faculty of Transport and Traffic Engineering, Payam Noor University of North Tehran, Tehran, Iran

Hasan Ziari
Faculty of Transport and Traffic Engineering, Iran University of Science & Technology, Tehran, Iran

Samira Roudini and Ali Nasrollahtabar Ahangar
Payam Noor University of North Tehran, Tehran, Iran

Monalisa Mallick and Abinash Mohanta
PhD Scholar, Department of Civil Engineering, National Institute of Technology Rourkela, Odisha, India

Awadhesh Kumar
Professor, Department of Civil Engineering, National Institute of Technology Rourkela, Odisha, India

Vivek Raj
M Tech Scholar, Department of Civil Engineering, National Institute of Technology Rourkela, Odisha, India

M. Hafezolghorani Esfahani, F. Hejazi, R. Vaghei and D. C. J. Tze
Department of Civil Engineering, University of Putra Malaysia, Serdang, Malaysia

E. Nikbakht
Department of Civil Engineering, Universiti Teknologi Petronas, Malaysia

Aniskin Nikolay Alekseevich
National Research Moscow State University of Civil Engineering (NRU MGSU), 26 Yaroslavskoye Shosse, Moscow 129337, Russia
Hydrotechnical and Energy Construction Department, NRU MGSU, Moscow, Russia

Antonov Anton Sergeevich
National Research Moscow State University of Civil Engineering (NRU MGSU), 26 Yaroslavskoye Shosse, Moscow 129337, Russia
JSC "Institute Hydroproject", Moscow, Russia

Berihun M. Negash, Lemma D. Tufa, M. Ramasamy and Mariyamni Bt. Awang
Universiti Teknologi Petronas, 32610 Seri Iskandar, Perak Darul Ridzuan, Malaysia

Index

A
Adobe Masonry, 135, 137, 139
Alkali-silica Reactions, 157
Analytical Hierarchy Process, 164, 172
Approach Azimuthal Angle, 53, 55, 57-58
Artificial Intelligence, 141
Artificial Neural Network, 209, 215
Asphalt Pavement Structure, 9-11, 13, 15
Assisted History Matching, 208, 215
Asynchronous Integration Strategy, 77, 93
Autoregressive Integrated Moving Average, 209
Autoregressive with External Input, 209

B
Black Spots, 164-169, 171-172

C
C-shaped Model, 174, 176-177, 184
Carbon Nanotubes, 100, 133, 135
Clay Brick Masonry, 135, 139
Collapse Seismic Fragility Analysis, 101, 115, 117-118
Computational Fluid Dynamics, 52, 76, 128, 173-174, 176, 184
Crack Tip, 46
Cross-sectional Plane, 96-97

D
Dam-foundation System, 197-198
Decline Curve Analysis, 207, 215
Disaggregation Method, 80
Discrete Element Method, 102, 120

E
Elastic Equilibrium, 96-97, 130-132
Elastic Nanobeams, 130
Elastic Seismic Response, 101-102, 113
Ensemble Kalman Filter, 208, 215
Environmental Kuznets Curve, 141

F
Fibre Reinforced Plastics, 135
Fibre-reinforced Concrete, 45
Finite Element Method, 6, 66, 75-76, 101-102, 107, 111, 113, 120, 135, 158, 176
Finite Particle Method, 101-102, 107-109, 111, 113, 117, 120
Fractional Derivative Model, 46

G
Gamma-rays, 154-157, 159
Global Stiffness Matrix, 102, 111
Grid Blocks, 207, 210

H
Heat-and-mass Transfer, 197-201, 203, 205
Hilbert Transform, 36-37, 39, 42
Hydraulic Engineering Structures, 197, 199, 205

I
Interfacial Transition Layer, 17-18, 33
Interfacial Transition Zone, 16, 34

L
Large Eddy Simulation, 53, 64
Lattice Discrete Particle Model, 160
Longitudinal Bar, 189-190, 193, 195
Low-head Dams, 197
Low-relaxation Prestressing Strand, 188, 194

M
Mesoscale Extended Finite Element, 160

N
Nanocantilever, 97-98, 130-132
Nonlocal Continuum Mechanics, 100, 130, 133
Normalized Difference Vegetation Index, 141
Nuclear Power Reactors, 154, 158
Numerical Reservoir Models, 207-208

P
Peak Ground Accelerations, 38
Peak Ground Velocity, 38
Photovoltaic Structures, 52-53
Plan-shaped Building, 173
Power Spectral Matrix, 36
Precast Prestressed Concrete Segmental, 186-187
Precast Seismic Structural Systems, 186
Pressure Coefficient, 174, 176
Prestressing Tendon, 187, 189, 191
Probability Density Function, 8-9, 11, 13
Proxy Models, 207-209, 214-215

R
Radiation Shielding, 154, 161-162
Rebar Diameter, 16, 23, 26, 30-31, 33-34

Reference Volume Element, 135
Reinforced Concrete Structures, 16-17, 35, 161
Renormalization Group, 175-176
Reynolds Averaged Navier Stokes, 53

S
Seepage Flow, 198-201, 206
Sobczyk Model, 37-38, 40
Spatially Correlated Earthquake, 36, 39, 43
Stock-flow Diagram, 143

T
Thermal-seepage Regime, 197-200
Torsional Rotations, 96, 98
Traffic Multiresolution Modeling, 77-78

V
Viscoelastic Operator, 45-46

W
Wastewater Treatment Capacity, 143, 146-147, 149, 151
Wind Load Reduction, 52-53, 57, 61, 64-65
Wind Pressures, 173-174, 184

www.ingramcontent.com/pod-product-compliance
Lightning Source LLC
LaVergne TN
LVHW081013210926
848485LV00004B/857